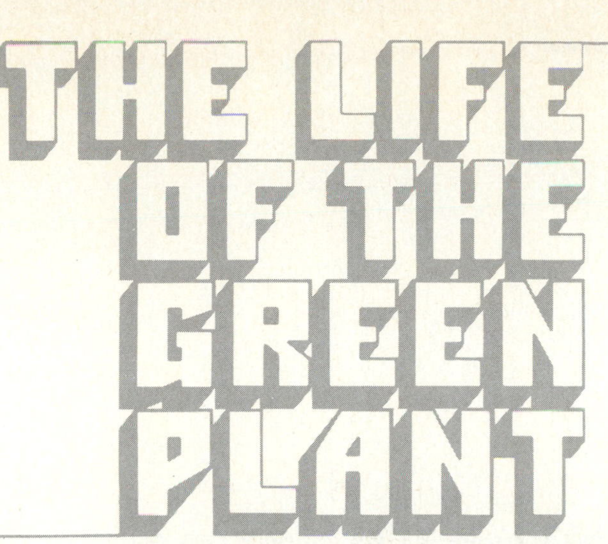

THE LIFE OF THE GREEN PLANT

THIRD EDITION

Arthur W. Galston, Ph.D. YALE UNIVERSITY
Peter J. Davies, Ph.D. CORNELL UNIVERSITY
Ruth L. Satter, Ph.D. YALE UNIVERSITY

PRENTICE-HALL, INC., ENGLEWOOD CLIFFS, NEW JERSEY 07632

Library of Congress Cataloging in Publication Data

GALSTON, ARTHUR WILLIAM (date)
The life of the green plant.

(Foundations of modern biology)
Bibliography: p.
Includes index.
1. Plant physiology. I. Davies, Peter J., 1940– joint author. II. Satter, Ruth L., joint author. III. Title.
QK711.2.G34 1979 581.1 79-16227
ISBN 0-13-536326-8
ISBN 0-13-536318-7 pbk.

Printed in the United States of America

10 9 8 7 6 5 4

Editorial/Production Supervision by
Ted Pastrick and Guy Lento
Chapter Opening Design by Janet Schmid
Cover Design by © A Good Thing, Inc.
Cover Photograph by John Hamel, Rodale Press, Inc.
Manufacturing Buyer: Ed Leone

Prentice-Hall International, Inc., *London*
Prentice-Hall of Australia Pty. Limited, *Sydney*
Prentice-Hall of Canada, Ltd., *Toronto*
Prentice-Hall of India Private Limited, *New Delhi*
Prentice-Hall of Japan, Inc., *Tokyo*
Prentice-Hall of Southeast Asia Pte. Ltd., *Singapore*
Whitehall Books Limited, *Wellington, New Zealand*

CONTENTS

2 The Green Plant Cell

5 Respiration and Metabolism: Supplying Energy and Building Blocks 113

6 Water Economy 137

7 Mineral Nutrition

8 Translocation: the Redistribution of Nutrients

9 Hormonal Control of Rate and Direction of Growth

10 Hormonal Control of Dormancy, Senescence, and Stress 257

11 Control of Growth by Light 281

PREFACE

This is a completely rewritten, greatly enlarged and freshly illustrated version of a successful textbook last revised in 1964. The original author has joined forces with two colleagues having different backgrounds, research specialties, expertise and outlooks. The result is a wide-ranging presentation of the facts of life about green plants, incorporating many topics not usually covered in books on this subject. Examples include the physiology of stress in plants, plant protection against pests, the relation of plant physiology to horticultural and agricultural practice and a discussion of potentially important new economic plants.

The first edition, written about twenty years ago, appeared at a time when the great ferment in biology demanded maximum flexibility. Prentice-Hall's Foundation of Modern Biology Series achieved this through the preparation of a series of short specialized volumes from which the teacher could select to fit almost any style of course. Now, in a different climate of opinion, teachers demand larger individual volumes that can present an integrated treatment of a single subject. This new book, now more than double its previous length, can meet this need.

As our knowledge of biology increases, our ability to generalize improves, and much of what was considered advanced material becomes easily incorporated into an elementary presentation. Accordingly, we have been able to include in this book much material, properly updated, from *Control Mechanisms in Plant Development* by A. W. Galston and P. J. Davies, published in 1970 for advanced courses in botany and plant physiology. We have then attempted to equalize the level of the various parts of the book by adding substantially to those chapters not dealt with in the Galston-Davies text. The result is, we feel, a comprehensive, modern, yet basic description of the functioning of the green plant, easily understood by persons without an advanced background in biology or chemistry.

This book can be appropriately used in one of four ways: (1) As the sole textbook for a short course in plant physiology or functional botany. Such courses, on the increase, should appeal to those who want to understand the higher green plant without at the same time studying its evolution or relation to other plant types. This text will prove particularly valuable to students of agriculture and horticulture who need to understand the functioning of the plants they grow, without becoming professional biochemists in the process. (2) As the functional botany text in a comprehensive course in botany, being combined with a survey book such as Harold Bold's *The Plant Kingdom*. (3) As the botanical portion of a year course in biology. (4) As a book to be read by non-specialists who want to gain an appreciation of the functioning of the green organisms around them. In such a context, the reader should be able to retain understanding of the general explanations given for various plant life processes without necessarily remembering the chemical or even the ultrastructural details. For any of these purposes, the questions at the end of the chapters should aid recall of major facts and help evaluate the importance of the various points discussed in the chapters. The suggested readings at the end of each chapter will guide the interested student or reader to further source materials, frequently at a more advanced level.

The green plant is a fascinating creature. Studying its intricate life processes has given each of us many hours of pleasure. We hope this book communicates to our readers at least some of our enthusiasm for the subject.

New Haven, Connecticut
Ithaca, New York

Arthur W. Galston
Peter J. Davies
Ruth L. Satter

PEANUTS

WHY DON'T TREES HAVE LEAVES IN THE WINTER?

BOY, YOU SURE ASK SOME STUPID QUESTIONS!

2-21
Tm. Reg. U. S. Pat. Off.—All rights reserved
© 1968 by United Feature Syndicate, Inc.

EVEN STUPID QUESTIONS HAVE ANSWERS!
SCHULZ

1

The Green Plant in the Economy of Nature

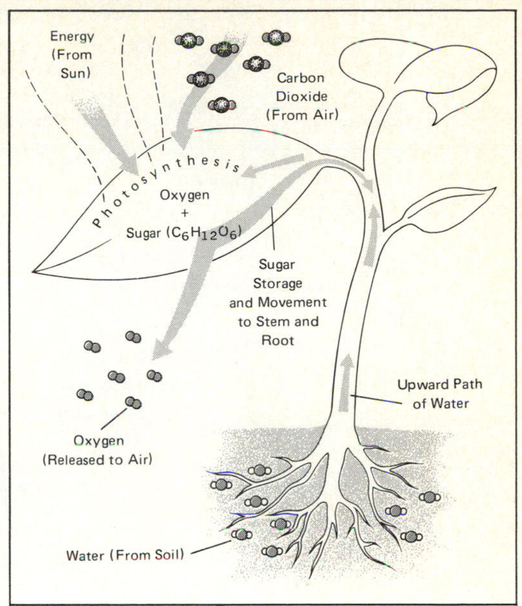

The story of life on earth, like the story of the earth itself, begins with the sun. Except for man's recent harnessing of the power generated by interactions of atomic nuclei, the sun is the sole source of energy for almost all forms of life. Every machine requires some source of energy to make it go: a watch uses the energy of a coiled spring; a hydroelectric plant employs the kinetic energy of falling water; an automobile runs on gasoline, releasing the chemical energy of the fuel molecules through the process of oxidation. Similarly, all living cells obtain their energy from oxidizable fuels called *foods*.

Food molecules are very diverse chemically, but we can get an idea of their characteristics by examining a common one—the simple sugar, glucose ($C_6H_{12}O_6$), which contains 6 carbon atoms, 12 hydrogen atoms, and 6 oxygen atoms. All sugars belong to the class of chemicals known as *carbohydrates*, so called because their chemical formulae all include carbon, as well as the elements hydrogen and oxygen in the same 2 to 1 ratio found in water. Glucose is formed and gaseous oxygen liberated when the simple substances carbon dioxide and water are united through the operation of the sun-powered photosynthetic machinery of a living green cell (Fig. 1-1). The energy thus stored in the glucose molecule can be liberated in the process of *respiration*, in which glucose is oxidized to yield carbon dioxide and water. This reaction is chemically the reverse of photosynthesis.

When a green plant grows, it is in fact tapping solar energy. Since man consumes either green plants or creatures that eat green plants, he too is drawing on solar energy, indirectly. Even the gasoline-powered automobile

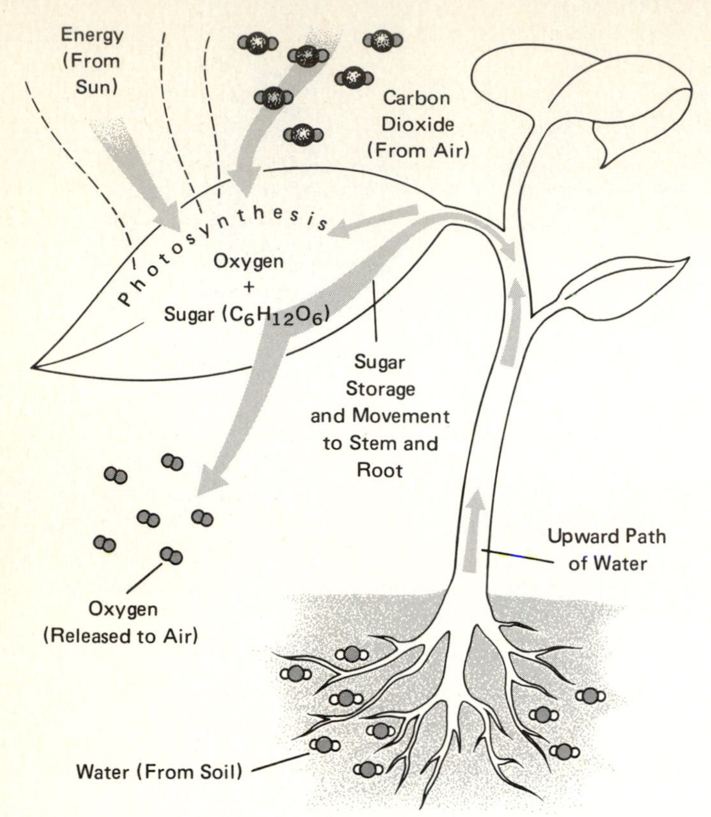

Figure 1-1 The green plant is man's major link with the energy of the sun.

and the coal-driven power plant use "fossilized" solar energy captured through photosynthesis by organisms that died millions of years ago. Were there no green plants to function as solar energy converters, practically all life on earth would cease. The only exceptions might be certain bacteria that derive their energy from the oxidation of unusual substrates such as ferrous iron. Even these organisms, however, are dependent on solar energy for the fixation of carbon used in building their bodies and, in any event, they constitute only a small fraction of the earth's living creatures.

The Sun as a Thermonuclear Device

Since the end of World War II, when the first atomic bombs were exploded, man has become increasingly familiar with the tremendous quantities of energy that can be released through the interactions of atomic nuclei. Although the Atomic Age originated with the creation of *fission* reactions, in which large atoms such as uranium are degraded to smaller atoms and subatomic particles, much of the modern emphasis in nuclear energy research involves *fusion* reactions, in which small units such as protons (hydrogen nuclei) are fused into larger units such as alpha particles (helium atoms). This reaction, the basis of the hydrogen bomb, is also being studied in an attempt to achieve controlled thermonuclear fusion as a source of commercial energy.

The sun, in fact, is a kind of hydrogen bomb. It is a thermonuclear "reactor" in which four hydrogen atoms of approximately mass 1 are fused into helium of approximately mass 4 through a complicated series of reactions involving other nuclei as intermediates (Fig. 1-2). The overall equation may be symbolized as:

$$4H \longrightarrow He$$

Actually, each of the four hydrogen atoms participating in the thermonuclear fusion has a mass of 1.008, whereas the helium atom produced by their fusion has a mass of 4.003. Since more mass enters into the reaction ($4 \times 1.008 = 4.032$) than emerges from it ($1 \times 4.003 = 4.003$), the equation is unbalanced. The difference in mass (0.029 units) is converted into energy, according to the Einstein equation ($E = Mc^2$), in which E is the energy produced (in ergs),

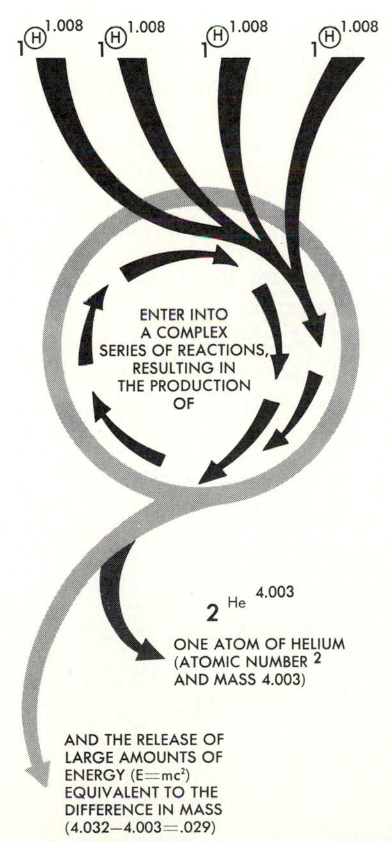

Figure 1-2 Fusion of hydrogen atoms to form helium.

M is the mass of matter transformed (in grams), and c is the velocity of light (3×10^{10} cm/sec). Although the erg is a very small unit (it takes more than 40 million to make one calorie), the equation shows that large quantities of energy are released by the conversion of very small amounts of mass. Researchers estimate that deep within the sun matter vanishes at the rate of 120 million tons per minute and is converted into tremendous quantities of energy that are radiated into space.

Of this solar radiation, the earth's surface receives annually approximately 5.5×10^{23} calories, or about 100,000 cal/cm^2/yr. About one-third of this total is expended in evaporating water, leaving roughly 67,000 cal/cm^2/yr for photosynthesis and other purposes. Every year, photosynthesis by green plants converts 200 billion tons of carbon from atmospheric carbon dioxide to sugar—about 100 times the combined mass of all the goods that man produces in a year. Even though photosynthesis is the most extensive chemical process on earth, green plants are relatively inefficient in utilizing the sun's radiant energy. Annual photosynthesis for the entire earth averages only about 33 cal/cm^2, which means that photosynthesis converts only about $\frac{1}{2000}$ of the available energy. This figure presents a somewhat inaccurate picture of photosynthetic efficiency, however, for much solar radiation reaches the earth at locations devoid of vegetation. If we include in our calculations only the radiation actually absorbed by green plants, the overall efficiency of photosynthesis (radiant energy stored/radiant energy absorbed) rises to several percent.

Radiant Energy

As hydrogen is converted to helium in the solar thermonuclear furnace, energy is liberated as many different kinds of radiation. Forming a continuous spectrum of energy, the kinds of radiation can be distinguished by their different wavelengths (Fig. 1-3). To express these wavelengths numerically, it is convenient to use the *nanometer* as the fundamental unit (one nanometer

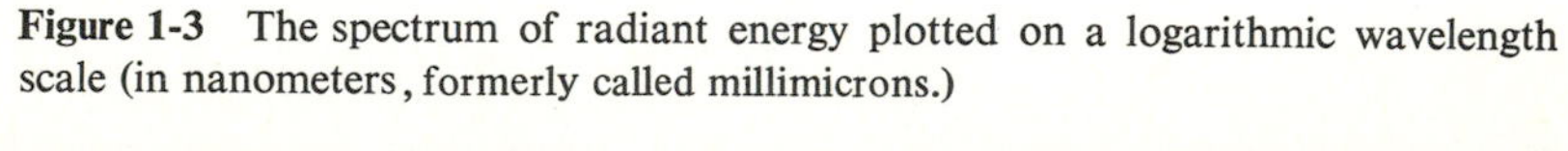

Figure 1-3 The spectrum of radiant energy plotted on a logarithmic wavelength scale (in nanometers, formerly called millimicrons.)

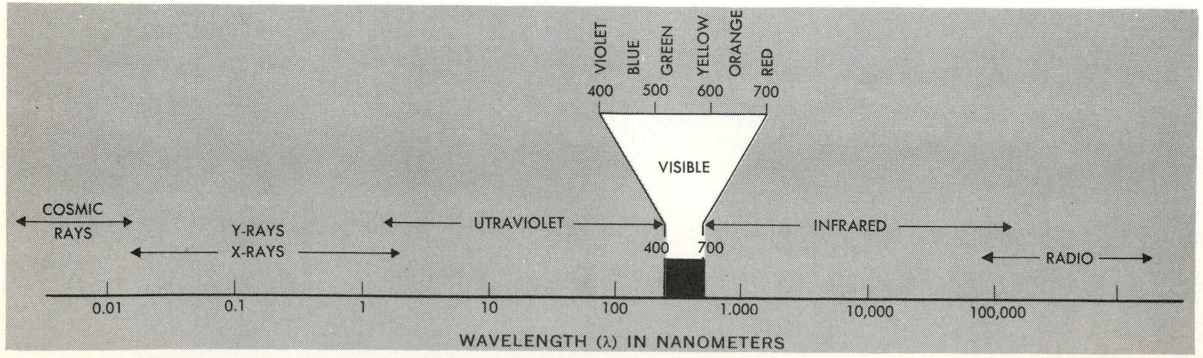

is one billionth of a meter, or 10^{-9} meters). The visible spectrum includes wavelengths of about 400 to about 700 nanometers. (Some people can see farther into the ultraviolet and the infrared, but these limits can be taken as a reasonable average). The blue-violet end of the spectrum is at 400 nanometers and the red end at 700—the colors of the spectrum ranging in order from violet, blue, green, yellow, and orange, to red. Plants are sensitive to almost exactly the same range of radiation as is the human eye, although certain kinds of bacteria can utilize infrared radiations we cannot see.

Early in the twentieth century, German physicist Max Planck established that radiant energy occurs in "packets" called *quanta*, or *photons*, and that the energy content of these quanta is directly proportional to the frequency of the radiation. In other words:

$$E = h \cdot \nu \text{ (Greek } nu\text{)}$$

E	$=$	h	$\cdot$	ν (Greek *nu*)
Energy of the quantum		Planck's constant		Frequency of the radiation

Since all radiation travels at the same speed (3×10^{10} cm/sec) and since frequency times wavelength equals velocity of light,

$$\lambda \text{ (Greek } lambda\text{)} \cdot \nu = c$$

λ (Greek *lambda*)	$\cdot$	ν	$=$	c
Wavelength of light		Frequency		Velocity of light

the frequency can be deduced from a known wavelength and vice versa. Clearly, the longer the wavelength of light the lower the frequency and the less energetic the quantum. Thus a quantum of ultraviolet light has more energy than a quantum of blue light, which in turn is more energetic than a quantum of red light.

When a quantum of radiant energy strikes a molecule, it may be absorbed by that molecule. The energy of the absorbed quantum produces an "excited state" in the absorbing molecule, which may then be able to enter into reactions which would have been virtually impossible for the same molecule at its lower energy level. In order for a quantum to drive a particular chemical reaction, its energy content must exceed a certain critical value, specific for that reaction. For example, quanta of the frequency of X-rays and short-wavelength ultraviolet are able to remove electrons from atoms, converting the atoms to ions. Quanta in the visible range are not sufficiently energetic to produce such ionizations, but when absorbed by pigments of the chloroplast, can effect the conversion of CO_2 into glucose. Quanta in the infrared (heat) range cannot perform either of these reactions, but can cause other, less energetic molecular excitations.

The complex of radiations from the sun is greatly altered by the time it reaches the earth's surface. For example, the ozone (O_3) of the atmosphere absorbs strongly in the ultraviolet region. This is fortunate, since the undiminished ultraviolet radiation of the sun would severely damage living terrestrial systems. Recently, some scientists have warned that certain products of man's activities, such as the exhaust from supersonic aircraft and the fluorocarbons that drive spray cans, may diminish the ozone layer; such an effect would have serious consequences for life on earth. Solar infrared radiation is absorbed largely by water vapor and partially by the small amount of carbon dioxide in the atmosphere; this absorption helps keep the earth's temperature within the proper range for living systems. The energy that penetrates the atmosphere and finally hits the earth is mostly in the visible and infrared range, but also extends somewhat into the ultraviolet. It is this radiation that constitutes the basis for the energetics of all living systems on earth, and it is the green plant that is able to capture and store a portion of this radiant energy through the process of photosynthesis.

Human Populations and Food Supply

We live in unusual times. The earth's human population, now more than 4 billion people (it was only about 3 billion when this book was first written, in 1960!) is increasing at an unparalleled annual rate (births minus deaths) of about 1.8% (Fig. 1-4). At this rate, the earth must support about 72 million additional consumers of food every year. The *daily* increase approximates 200,000 people, or *more than two additional mouths to feed every passing second*! Furthermore, as the total number of people on earth increases, so does the annual population increment. As public health measures are extended and improved, thus lowering the death rate, the *rate* of increase in population *itself* rises. We can therefore expect the human population to just about double every 35 years. Should this continue for another millenium, the weight of the people on the earth would approximately equal the weight of the planet itself. Obviously something must change before the "population bomb" engulfs us all.

Since all animals, including man, are dependent for their metabolic fuel on the solar energy trapped by green plants, any calculation of how many people the earth can comfortably support ultimately depends on the amount of energy that can be trapped by photosynthesis. To what extent can we increase the 200 billion tons of carbon stored per year through photosynthesis? Even herculean efforts to expand the acreage of agriculturally productive land can do no more than double the area under cultivation, but since the best land is already in use, this increased effort will not double productivity. Most estimates of photosynthetic productivity reveal that the waters of the earth contribute markedly to the total; at least half, and possibly up to 80% of the photosynthesis on earth occurs in marine and freshwater environments. Can we, then, "farm" the seas, or perhaps cultivate various algae for food in great "tank cultures" of highly fertilized aqueous media? Although now economically unfeasible, such a system may one day be necessary, when the

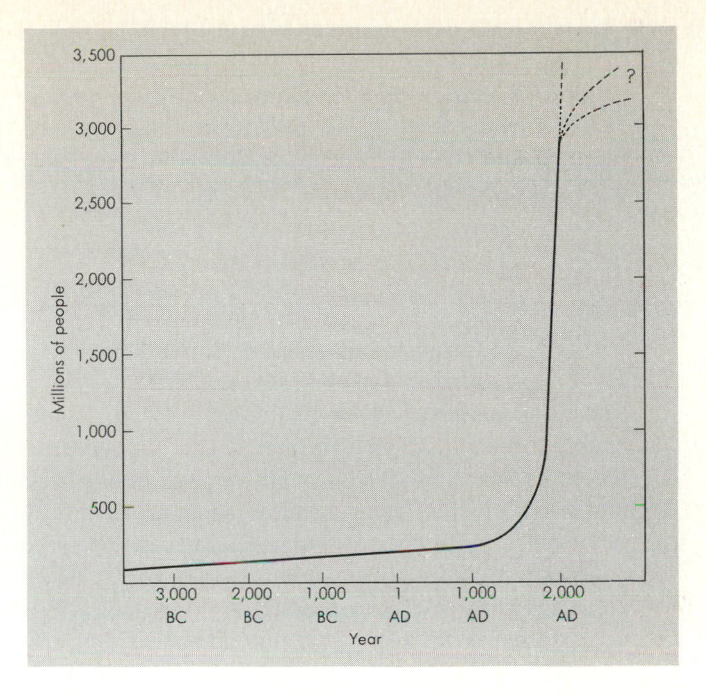

Figure 1-4 The growth of the human population on earth.

world is bulging with hungry people. In this new kind of agriculture, botanical "knowhow" will obviously play a vital role.

Another way to increase food production is to improve the plant itself. A good deal has been accomplished in this direction already by the development of scientific agriculture and the "Green Revolution," to which botanists have contributed greatly. The geneticist has given us increasingly better kinds of plants to work with; the plant physiologist has taught us to care for the nutritional needs of the plant, and to alter its growth habit and kill weeds by specific chemical treatments; the plant pathologist has shown us how to ward off insect and fungus pests; the soil scientist has demonstrated how to enrich and preserve the complex soil environment of pulverized rock, organic matter, and organisms, and the agronomist has taught us how to operate farms efficiently. Some day we may understand the mechanism of photosynthesis well enough to control and improve its efficiency within the plant, or even duplicate it efficiently outside the living cell. New prospects for improving plants further are discussed in Chapters 14 and 16.

Even if the earth produces more food, however, the gain will be more than negated by an unrestricted increase in the human population. Why struggle to double agricultural productivity when in 35 years that achievement will be completely erased by a doubling of the number of mouths to feed? Clearly, man must some day decide how many people can comfortably be accommodated by the earth's resources; then means must be found to restrict the population to that number. Although such a proposal raises a tremendous number of religious, political, and sociological questions, human populations will eventually have to be limited in some way, either through peaceful, planned, and voluntary means or through the violence and chaos imposed by starvation, disease, or war.

SUMMARY

Virtually all forms of life on earth derive their energy directly or indirectly from thermonuclear fusion reactions in the sun in which hydrogen is converted to helium. During this solar reaction, some mass is converted to energy, which is radiated through space. A small fraction of this energy hits the earth and is absorbed by chlorophyll and associated pigments in green plants. Through intricate chemical mechanisms in the chloroplast of the green plant cell, this energy is stored in the form of carbohydrates [$C_x(H_2O)_y$] made from carbon dioxide (CO_2) and water (H_2O) in the process of *photosynthesis*; oxygen is released as a by-product of this reaction. When these carbohydrates are oxidized in the process of *respiration*, yielding CO_2 and H_2O, the stored energy is released, and can be used by the organism.

Since man eats either green plants or animal products derived from them, the ultimate limit on the earth's human population is set by the net amount of photosynthetic product stored. The human population now exceeds 4 billion, and at its present rate of increase of more than 70 million per year, will double about every 35 years. Even with more efficient plant growth and photosynthesis, and with the conversion of more land to agricultural use, it is unlikely that plant productivity can forever keep pace with human reproduction. For this and other reasons, it will eventually be necessary to limit the size of the human population on earth.

SELECTED READINGS

Textbooks in Plant Physiology and Biochemistry

BIDWELL, R. G. S. *Plant Physiology*, 2nd ed. (New York: Macmillan, 1979). A competent, thorough treatment, stressing biochemical interpretations.

BONNER, J., and J. VARNER, eds. *Plant Biochemistry*, 3rd ed. (New York: Academic Press, 1976). A collection of articles by experts on important aspects of the chemistry of plants.

GOLDSBY, R. A. *Cells and Energy* (New York: Macmillan, 1967) Basic chemistry and biochemistry, especially helpful to beginners.

GOODWIN, T. W., and E. I. MERCER. *Introduction to Plant Biochemistry* (Oxford: Pergamon Press, 1972). The plant as seen by a biochemist.

KROGMANN, D. W. *The Biochemistry of Green Plants* (Englewood Cliffs, N.J.: Prentice-Hall, 1973). A simplified, integrated treatment of plant biochemistry.

SALISBURY, F. B., and C. ROSS. *Plant Physiology*, 2nd ed. (Belmont, Calif.: Wadsworth, 1978). A chatty, discursive survey with many personal interpretations and comments.

Reviews of Advances in Plant Physiology

Annual Review of Plant Physiology (Stanford, Calif., Annual Reviews, Inc.). Approximately 15 topics are covered each year, starting in 1950. Comprehensive references are provided, and collective indices help guide the uninitiated into the advanced literature.

Encyclopedia of Plant Physiology, New Series (Berlin–Heidelberg–New York: Springer, 1976 and later). Each volume covers a different area of plant physiology, updating earlier volumes published in the 1950's. When completed, this series will provide a thorough analysis of all topics covered in *The Life of the Green Plant*.

Plants, Food, and Agriculture

"FOOD and AGRICULTURE," *Scientific American* 235 (3): 30–196 (1976). An entire issue with 12 articles surveying the problems and prospects of agriculture.

ABELSON, P. H., ed. "Food: Politics, Economics, Nutrition and Research" (Washington, D.C.: American Association for the Advancement of Science, 1975). A collection of articles by experts on the basic scientific and social scientific background for coping with the world's food problems.

QUESTIONS

1-1. Green plants have long been successful at using solar energy. Give an overall description of the process by which they store the sun's energy and comment on their efficiency in that process.

1-2. What is the meaning of the biblical quotation "All flesh is as grass"?

1-3. The sun is a thermonuclear furnace 93 million miles from the earth. Describe briefly the source of the sun's energy, the radiation it produces and the nature of its radiation reaching the earth.

1-4. What is meant by a quantum? What is the relation between the frequency of a radiation and its energy per quantum?

1-5. Given the frequency of a radiation, one can calculate the wavelength, and vice versa. What is the equation that permits this calculation?

1-6. What is the wavelength range in metric units of the portion of the spectrum (a) visible to the human eye? (b) of use to the green plant?

1-7. What is the approximate size of the current human population? At what rate is it increasing? At that rate, how long will it take to double?

1-8. Thomas Robert Malthus' "Essay on Population," published in 1798, pointed out that the attainment of a happy society will always be limited by the fact that while population tends to increase geometrically, the means of subsistence can only increase arithmetically. Is Malthus' generalization still valid? What factors may operate to delay the onset of famine occasioned by the increase of population beyond our means to feed it?

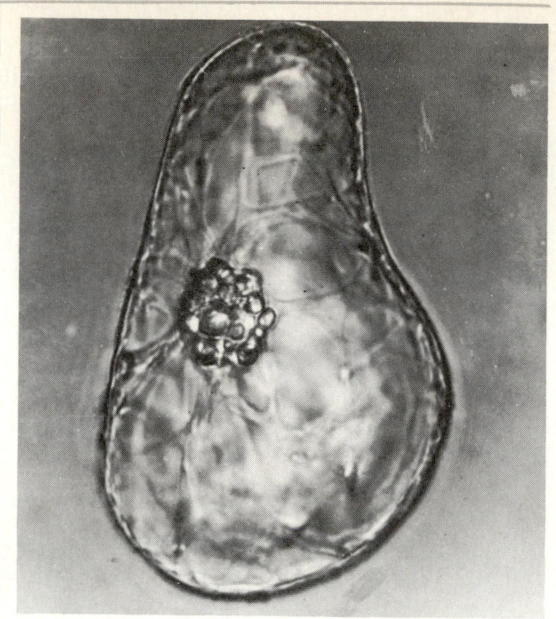

The Green Plant Cell

Scientists generally try to reduce the problems in their field to the simplest possible terms. For more than a century, biologists have agreed that the cell is the basic structural and functional unit in all organisms. For unicellular forms, of course, the cell *is* the organism; the multicellular state introduces further problems of organization and differentiation, and of cooperation and competition among cells. Although no responsible biologist will claim that an understanding of the cell furnishes complete knowledge of the organism, almost all biologists agree that the cell is a logical starting place for a meaningful study of an organism.

Finding Out About the Cell

To gain information about the cell, biologists have used a variety of techniques. Of these the most obvious is direct visual observation, but this is possible with only the largest cells. Smaller cells require magnification with either a simple lens (about 10 X) or a compound microscope bearing lenses in tandem (up to about 1000 X). Under a compound light microscope, a living, three-dimensional cell can look very complex, turbulent, and confusing. To simplify and stabilize the picture, biologists have learned to kill and preserve cells by immersion in a *fixative* such as formaldehyde. The killed and fixed cell may then be washed, dehydrated by passage through alcohol, embedded in a solid supporting medium such as paraffin or plastic, then cut into thin slices with a very sharp knife or razor in a device called a *microtome*. Slices with thicknesses of 1–10 μm can easily be cut on such a device under favorable conditions. To define and highlight cellular structures, these thin slices are placed successively in various dyes, which, by virtue of their different solubilities and molecular charges, will attach to different cellular structures. With care and skill, one may produce a polychromatic display of the different cellular structures, with, for example, nuclear material stained pink, cytoplasmic structures stained various shades of green or violet, and the cell wall unstained or bearing a third color. These techniques of fixation, embedding, slicing, and staining have enabled scientists to discover most of what we currently know about the cell.

The electron microscope provides still higher magnifications (up to 1 million X). By using an electron beam instead of a light beam, one obtains greater resolution, since resolving power is inversely related to the wavelength of the radiation used in the viewing process, and the electron beam uses a shorter wavelength than visible light. Theoretically, the highest potential resolution of the electron microscope has not yet been approached, since the magnetic lenses used to focus the beam suffer from electrical "noise" that makes the image unstable. New developments involving hypercooling of these lenses to reduce the noise should eventually yield still greater resolution.

For electron microscopy, thinner slices are cut from plastic-embedded cells with diamond or glass knives. These sections are then exposed to agents such as osmium tetroxide, which attach differentially to cell structures, producing a differential opacity to the electron beam. The details are then visualized on either a photographic plate or a fluorescent screen. No color is yet possible on electron micrographs; only black, white, and shades of gray appear.

Another technique for gaining information about cells involves chemical experimentation with isolated cell organelles. If one gently grinds a mass of cells in an appropriate medium, at least some of the organelles will be released intact. Organelles of different densities can then be harvested by deposition at successively greater forces in a centrifugal field (Fig. 2-1). For example, heavy particles such as nuclei and chloroplasts sediment at relatively low speeds which give rise to forces of 1000–3000 × gravity (g), mitochondria at about 10,000 × *g*, ribosomes at about 30,000 × *g*, while still smaller particles and large molecules may require hundredfold greater rates of centrifugation. Together with stepwise filtration, physical absorption and elution, and separation on the basis of electrical charge, differential centrifugation has permitted the harvesting of large masses of identical cell organelles or organelle fragments. These masses of chloroplasts, mitochondria, ribosomes, and membrane and other fragments are then used in experiments to define the chemical nature and biochemical activities of each of the isolated fractions.

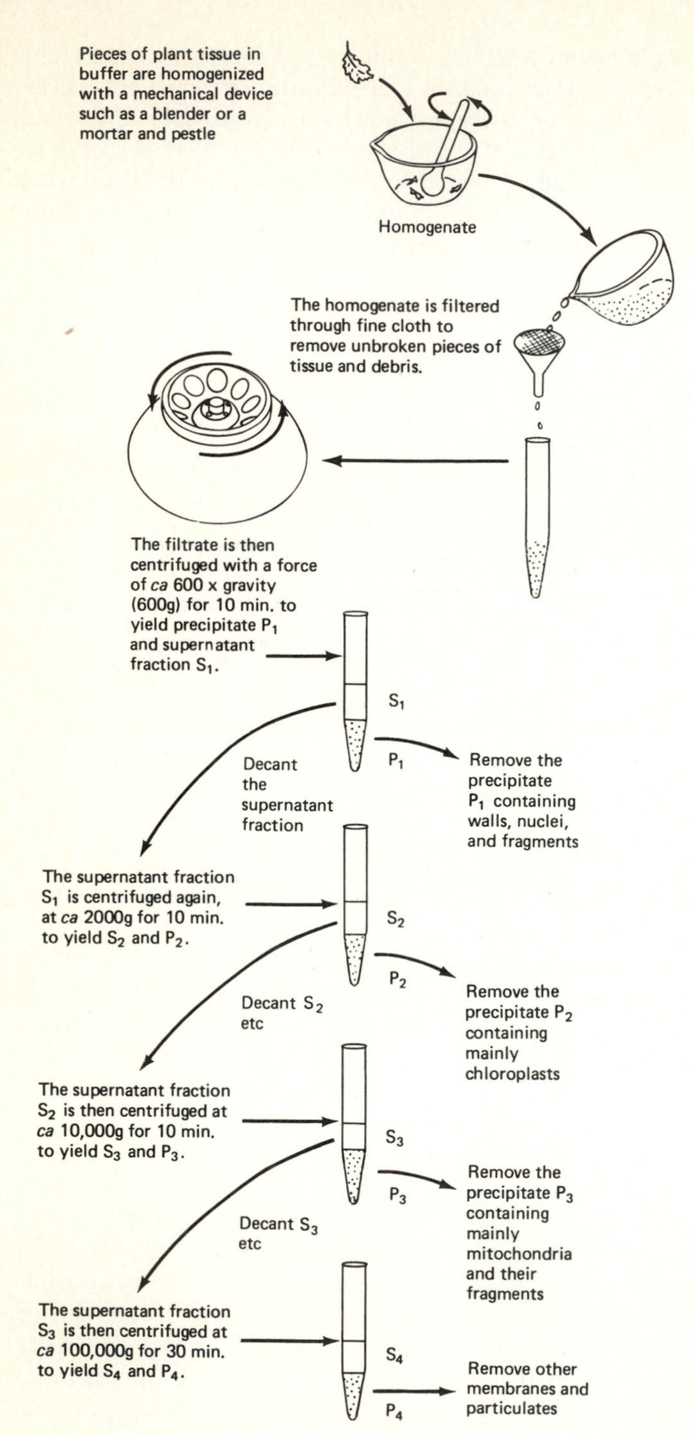

Figure 2-1 Separation of cellular fractions by differential centrifugation.

Cell Size and Shape

Cells vary tremendously in size, from bacterial cells less than a micrometer (μm) in diameter to elongated cells several millimeters (mm) in length. Even a relatively small bacterial cell contains on the order of 10^{12} molecules. The tremendous complexity of this basic biological unit should make it clear why the methods used and precision achieved in biological research are usually quite different from those in physics or chemistry, where the units may be as elementary as a single proton or an individual quantum.

A plant cell grown in isolation generally assumes a roughly spherical shape (Fig. 2-2), but when surrounded by other cells, it is pressed into a polyhedron. A cell from the elongating zone of a stem or root assumes approximately the shape of a box, about 50 μm long by 20 μm wide by 10 μm deep, with a volume of about 10,000 μm^3. One hundred million such cells, if tightly packed, would fit in a volume of 1 cubic centimeter (cm^3). The complex and highly differentiated structure of a plant cell can be divided into three main regions: (1) the *cell wall*, a relatively rigid, presumably nonliving, highly structured and chemically complicated mixture of substances secreted by (2) the *protoplast*, or living portion of the cell, which contains all the cell organelles suspended in a complex solution, and (3) *vacuoles*, the nonliving, membrane-limited storage tanks containing an aqueous solution of absorbed inorganic salts and organic molecules produced by the metabolic activities of the cell. The wall of the cell serves as the plant's skeleton, by contributing to the rigidity and form of the plant body. The vacuoles aid in this function by

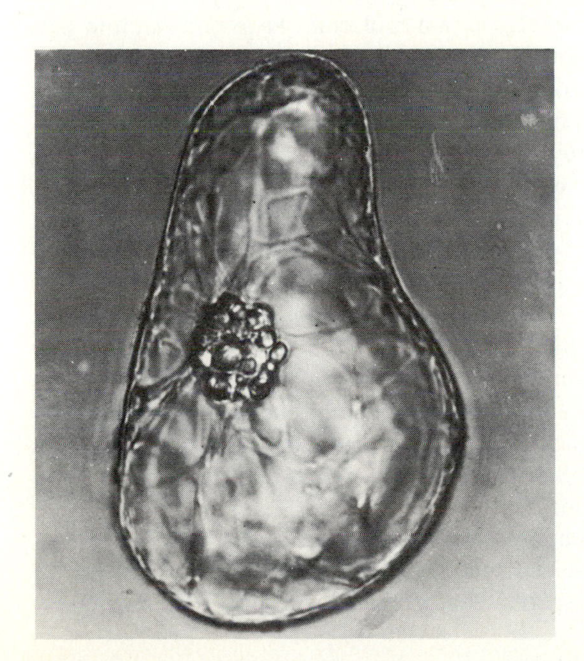

Figure 2-2 A large single isolated cell of pea grown in hanging drop culture. Note the cell wall, the large central nucleus surrounded by starch grains, and the cytoplasmic threads suspending it in a large vacuole. (From J. G. Torrey, in *Cell, Organism and Milieu*, D. Rudnick. The Ronald Press 1959.)

exerting pressure on the cytoplasm and wall. Vacuoles also function as a sort of excretory system, since material deposited within them is effectively removed from the scene of active chemical transformations in the cell. It is to the protoplast, then, that we must look for the locale of the ceaseless activities that characterize the highly organized and dynamic state we call *life*.

Membranes

The protoplast is surrounded by boundaries called *membranes*; the *plasmalemma* is the membrane bounding the protoplast at its interface with the cell wall, while the *tonoplast* separates the protoplast from the vacuole. When properly stained and magnified a millionfold or more with the electron microscope, these plant membranes appear as two dark parallel lines 6–10 nm apart (Fig. 2-3).

The protoplast also contains several kinds of bodies called *organelles*, the most prominent being the single large nucleus and the numerous smaller mitochondria and chloroplasts (Figs. 2-3 and 2-4). Each kind of organelle has its own set of functions, performed in a unique internal environment generated by the *differential permeability* and other special properties of the membranes that surround the organelle and separate it from the remainder of the protoplasm. *Differential permeability* means that different kinds of molecules cross a given membrane at different rates, primarily because of their different solubilities in the components of the membrane. Membranes also contain "pumps", i.e., systems that actively use energy to transport specific molecules across the membrane. As a result, certain molecules have non-random patterns of distribution; for example, potassium is usually much more concentrated within the protoplast than outside it, while the related element sodium is virtually excluded from the protoplast of most plants.

Membranes isolated in bulk from plant cells by differential centrifugation have been analyzed chemically. Since they are large, thin, fragile surfaces, they fracture during isolation, but the fragment ends often seal to form spherical vesicles. Analysis of the vesicles derived from many different membranes reveals two main constituents, *protein* and *phospholipid*. While the lipid components tend to be rather similar, each type of membrane contains unique proteins consistent with the physiological functions of that membrane in the cell. For example, active proteins (enzymes) that regulate mineral transport into and out of the protoplast tend to be located on the plasmalemma or tonoplast, while enzymes required for photosynthesis are found in the green chloroplast membranes, and enzymes that regulate the oxidative reactions of respiration are localized in mitochondrial membranes.

When appropriate phospholipids and proteins are mixed and spread on a water surface, membrane-like structures of a thickness similar to that of biological membranes form spontaneously. Such synthetic membranes, pre-

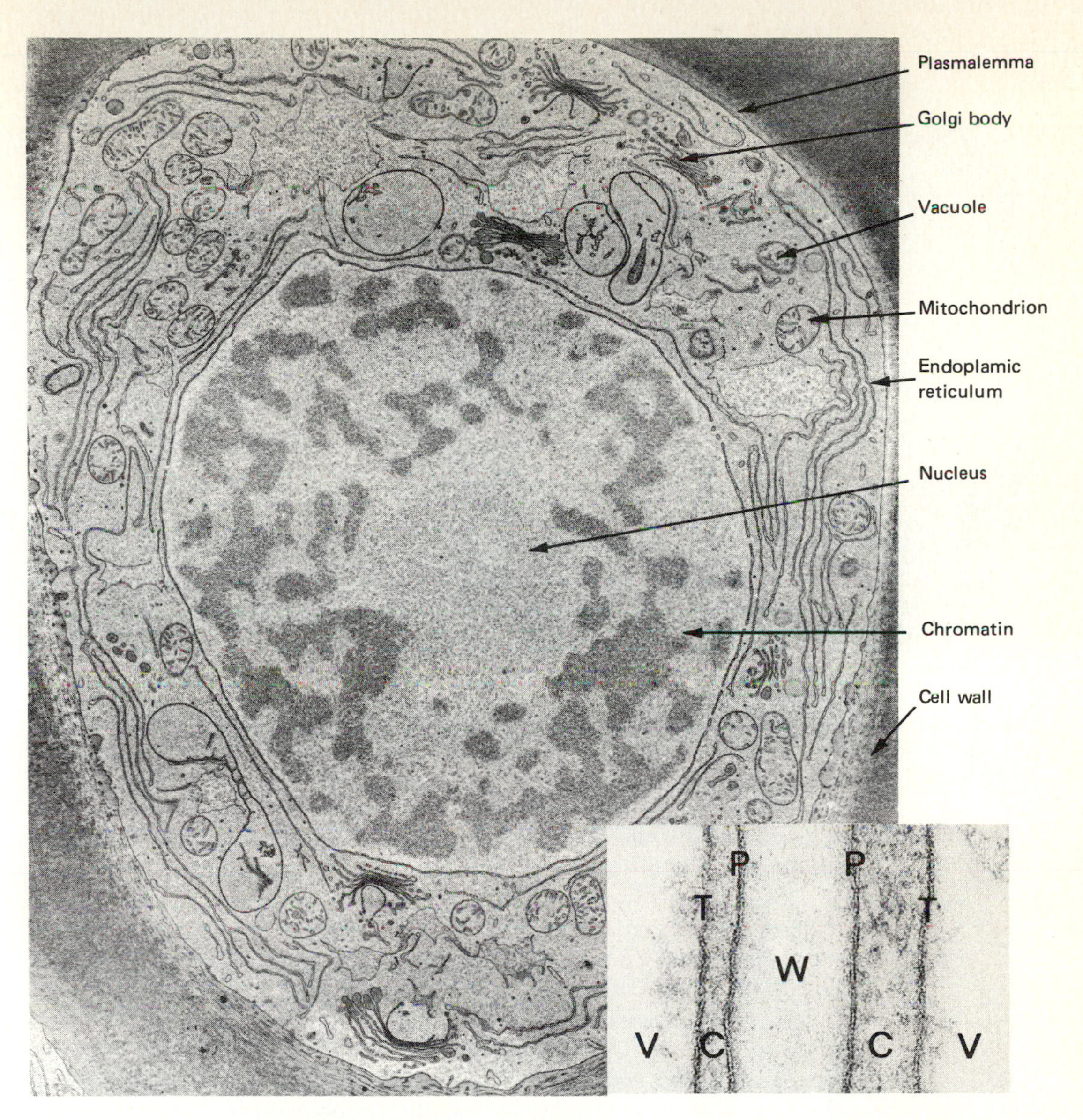

Figure 2-3 An electron microscopic view of a cell of a maize root tip. This is a transverse section, taken 75μm from the apex. The large central body is the nucleus; note the darker chromatin areas and the many pores in the double nuclear membrane. The elongated canals running through the cytoplasm are portions of the endoplasmic reticulum, which represents projections of the nuclear double membrane. The numerous club-shaped or rounded bodies with internal projections are the mitochondria with their cristae. The close groups of short canals with vesicular ends are Golgi bodies; the lighter stippled-appearing areas in the cytoplasm are the vacuoles. (Courtesy W.G. Whaley and the University of Texas Electron Microscope Laboratory.) The insert at the right shows the plasmalemma (P) and tonoplast (T) membranes of two adjacent cells at higher magnification. V = vacuole, C = cytoplasm, W = wall common to both cells. Note that each membrane appears as two electron dense parallel lines (the hydrophilic exterior) separated by an electron transparent space (the hydrophobic interior). (Courtesy of H.W. Israel, Cornell University.)

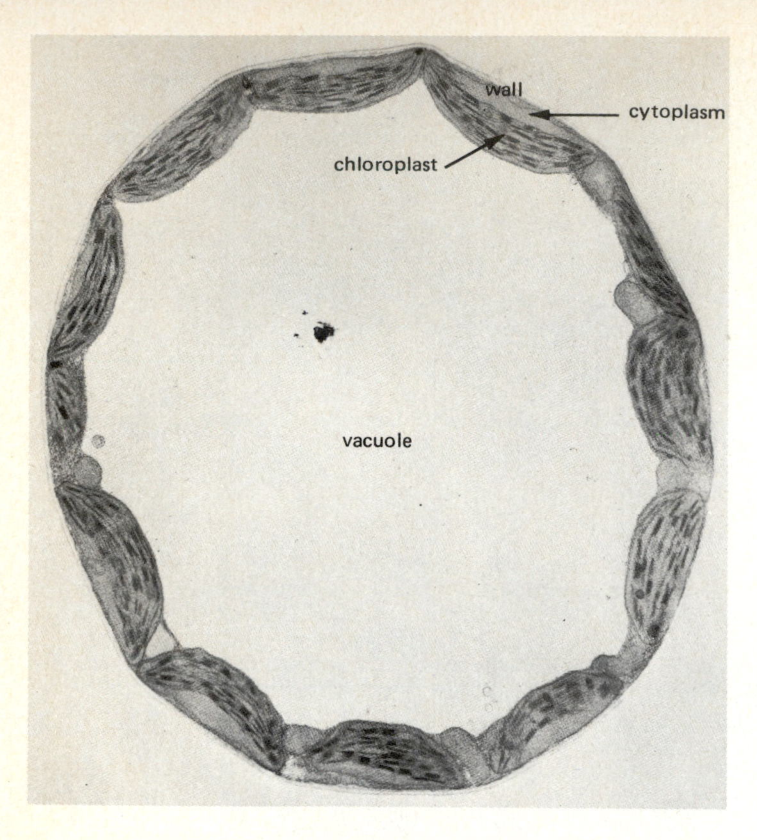

Figure 2-4 Electron micrograph of a cell in a leaf, the principal photosynthetic tissue. Numerous chloroplasts surround the large central vacuole in the cytoplasm. (Courtesy of H.W. Israel, Cornell University.)

pared from proteins and lipids found in natural membranes, have been useful in increasing our understanding of the structure and function of biological membranes. Synthetic membranes of this type can have different permeabilities for different ions, depending on their constituent proteins and lipids. Striking effects have been obtained by adding certain antibiotics to synthetic membranes. For example, valinomycin has a structure with appropriate dimensions and electric charge to attract and hold K^+ ions but not Na^+ ions (Fig. 2-5). When valinomycin is added to a synthetic membrane that separates solutions of K^+ and Na^+ from pure water, it increases the movement of K^+ ions across the membrane severalfold, but has little effect on the movement of Na^+ ions. By contrast, gramicidin, another antibiotic with different dimensions and structure, increases movement of both Na^+ and K^+ through membranes. Artificial membranes have also been used to reveal mechanisms whereby light and hormones control plant growth, as described in Chapter 11.

Lying adjacent to the plasmalemma are the cell wall and the cytoplasm, both aqueous regions presumed to be in contact with hydrophilic, charged regions of the membranes. Since proteins have more charged residues than do lipids, early models of membrane structure proposed that plasmalemma membranes are composed of two outer protein layers (the two dark lines in Fig. 2-3) with lipid molecules sandwiched between them. This description of membrane structure was generally accepted until the early 1970's, when several types of new evidence suggested that the model required modification. For

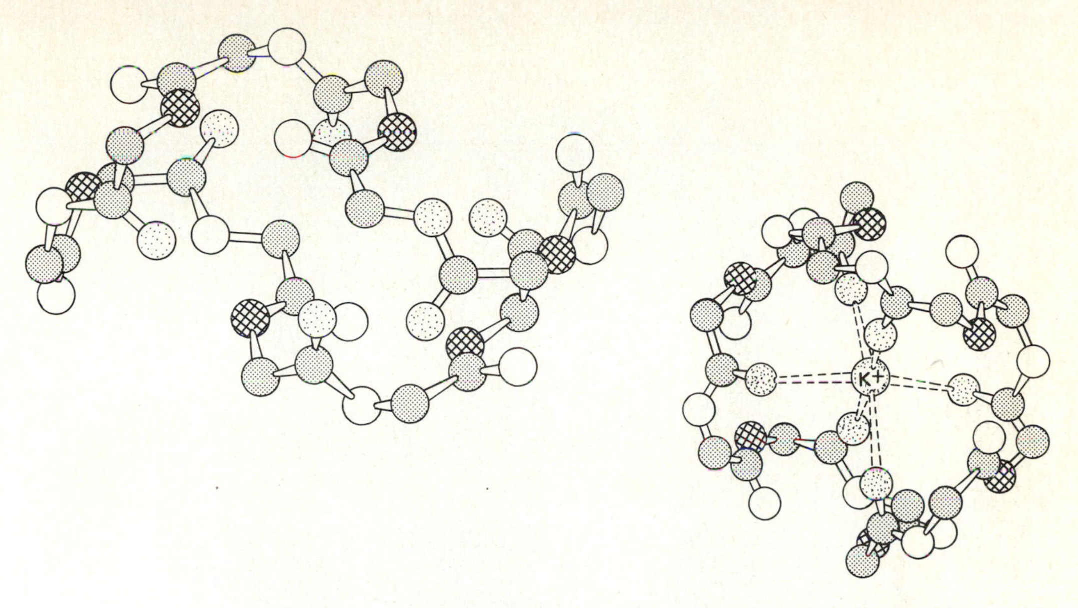

Figure 2-5 Model of valinomycin prior to binding (left) and after complexing with K^+ (right). Note the change in the conformation of the antibiotic. (Adapted from L. Stryer, *Biochemistry*, San Francisco: W.H. Freeman and Co., 1975.)

example, an electron microscopy technique called *freeze-etching* revealed internal features of membranes inconsistent with the "sandwich" model. Tissue for freeze-etch analysis is frozen in liquid nitrogen, then fractured with a blunt microtome knife that cleaves the membrane in a plane parallel to its surface (Fig. 2-6). The frozen tissue water is then evaporated away (sublimed) and the dry residue shadowcast with carbon or metal to reveal its surface details (Fig. 2-7). In such preparations large particles, shown to be globular

Figure 2-6 To prepare membrane fragments for freeze fracture electron microscopy, the fragments are rapidly frozen in liquid nitrogen, and then cleaved with a microtome knife. Cleavage usually occurs parallel to the outer surfaces, and the membrane splits apart, revealing the bilayer interior. The ice is then removed by sublimation, and the exposed regions are shadowed with carbon and platinum. Both particles and depressions are seen.

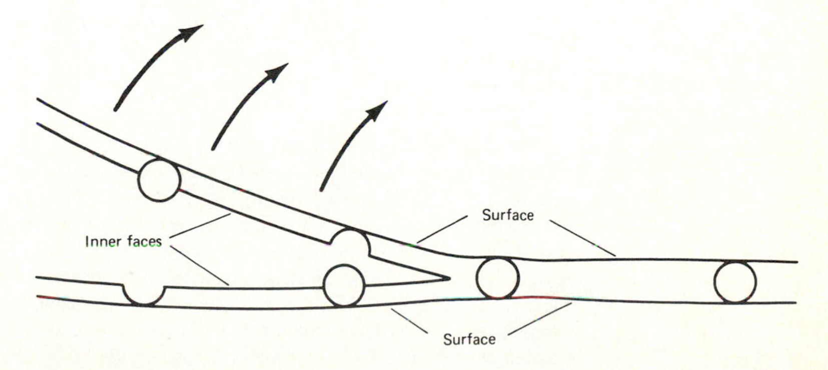

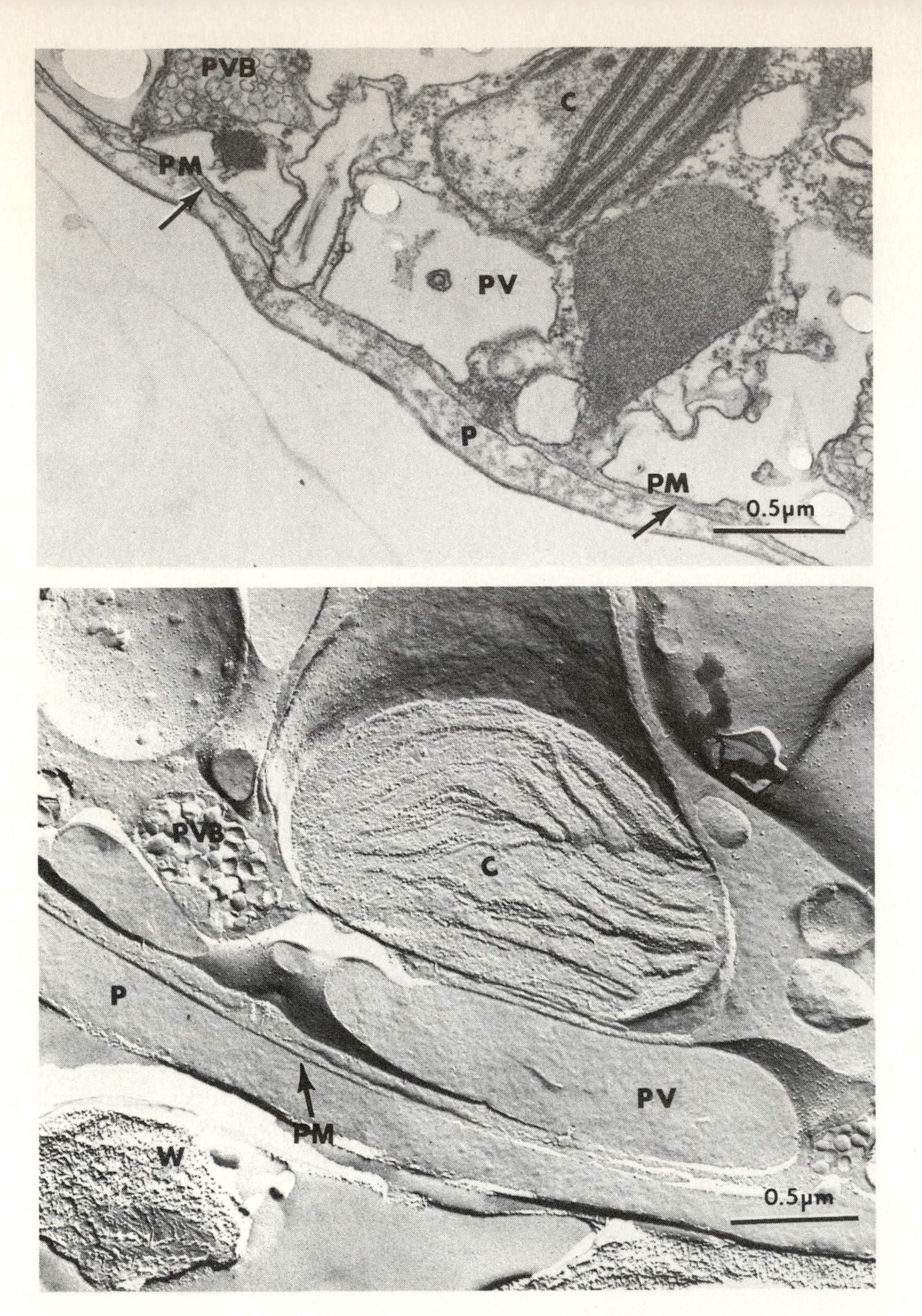

Figure 2-7 An electron micrograph of fixed tissue (upper) and a freeze fracture replica of unfixed tissue (lower) of *Gonyaulax polyedra*, a dinoflagellate (unicellular alga with two flagella, each located in a furrow). The freeze fracture process has cleaved several of the cell's membranes, exposing their interiors. The small round protrusions in the freeze fracture micrograph are considered to be membrane-localized proteins. Some of these are evident in the cytoplasmic membrane while many others are in the chloroplast, where they are probably localized on thylakoid membranes. P = pellicle, the principal protective coat of the cell; PM = cytoplasmic membrane; PVB = a polyvesicular body; C = chloroplast; W = part of a cellulose plate; PV = part of the peripheral vesicle. Both micrographs ×40,000. (From B.M. Sweeney. 1976. J. Cell Biol. *68*: 451–461.)

proteins, interrupt the otherwise smooth surface of membranes. In another procedure that indicates the patterned nature of the membrane surface, specific proteins called *lectins*, obtained from seeds of plants, attach to and reveal discrete and specific glycoprotein receptors on the membrane surface.

Evidence obtained by these and other techniques now suggests that membranes are *mosaics* composed of a lipid matrix in which proteins are interspersed (Fig. 2-8). This model takes into account that not all regions of protein molecules are hydrophilic, nor are lipids exclusively hydrophobic. It proposes that the charged (polar) residues of both protein and lipid components are on the outer membrane surface, in contact with cell water, while the uncharged (nonpolar) portions form the hydrophobic interior. It further suggests that some proteins are loosely bound to the outer membrane surface, while others (called *integral proteins*) traverse the entire membrane. This view is supported by biochemical experiments indicating that some proteins can be separated from membranes quite readily, while others can be isolated only by drastic fragmentation of membrane structure.

Although early studies suggested that membranes have a stable and rigid structure, evidence is rapidly accumulating that at least the lipid portions of membranes are fluid and mobile. In addition, studies with radioactively

Figure 2-8 The fluid mosaic membrane model: schematic three-dimensional and cross-sectional views. The large bodies represent proteins, some of which traverse the entire membrane, while others are more loosely attached. Each small sphere with its two attached vertical lines represents a phospholipid. The sphere represents the hydrophilic portion, located at the exterior surface, while the vertical lines represent the hydrophobic long chain fatty acids. (From S.J. Singer and G.L. Nicolson. 1972. Science *175*, 720–731.)

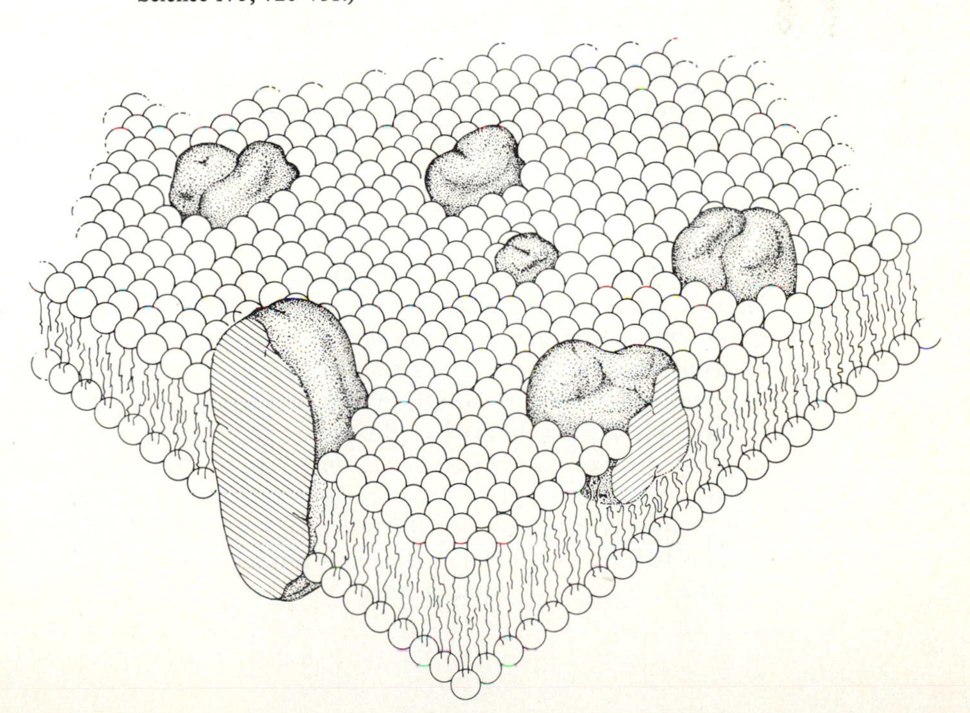

labeled precursors of membrane components have shown that some parts of the membrane have a high rate of metabolic turnover, being continuously degraded and resynthesized. When these labeled precursors are "fed" to mature tissue, they are incorporated into the membranes, remain there for some hours, then disappear from the membranes and show up in some other part of the cell.

The Nucleus, Ribosomes, and Protein Synthesis

Within the last several decades, biologists have become reasonably certain of the structure, chemistry, and functional significance of the major organelles of the protoplast. The largest organelle of the cell is the *nucleus* (Fig. 2-3). This globular body, generally about 5–10 μm in diameter, contains most of the basic genetic information of the cell, encoded in the form of long strands of a complex chemical material called *deoxyribose nucleic acid*, or DNA. The DNA in the nondividing cell appears as a netlike mass of *chromatin*, a complex substance containing mainly negatively charged DNA and positively charged proteins called *histones*. By the time the cell is ready to divide, the chromatin "network" reveals its true nature by condensing into distinct, identifiable rodlike bodies called *chromosomes*, of which each cell has a basic and fixed number. For example, each cell of the pea plant has 14 chromosomes—7 chromosomes derived from each of the two parents. In man, the basic chromosome number is 46, of which 23 come from the mother and 23 from the father. The nuclear division that replicates the chromosomes and preserves their double complement in all daughter cells is called *mitosis* (see box p. 21).

The full double complement of chromosomes is referred to as the *diploid* (2n) number, and the basic number obtained from each parent through the sex cell the *haploid* (n) number. All the cells of the higher plant body, with the exception of the haploid sex cells, are at least diploid. The haploid sex cells are found mainly in mature pollen grains and in the embryo sac of the ovules of the flower. In the life cycle of the plant, the haploid number is attained from the diploid by reduction division, or *meiosis* (see box p. 21), carried out in male and female *spore mother cells* in the anther and ovule of the flower. The haploid cells so produced divide to form male and female haploid, or *gametophyte*, generations, which ultimately produce sex cells, or *gametes*, called *sperms* and *eggs*. When, in sexual reproduction, the gametes of opposite sex fuse to produce the *zygote*, the diploid number of the *sporophyte* generation is restored. Thus, in terms of chromosome content and DNA, the flowering plant goes through a cycle in which diploidy alternates with haploidy, and in which the fusion of haploid cells of different genetic origins to form new diploid individuals produces organisms with new combinations of genetic characteristics.

Mitosis

Prior to visible mitotic activity, the cell replicates its chromosomal DNA. Mitosis starts when the chromosomes become visible as already longitudinally doubled threads, each half being called a *chromatid.* The nucleolus and nuclear membrane disappear, the chromosomes line up on the cell's equator, and spindle fibers, running from equator to poles, pull apart the sister chromatids, which become separate but identical chromosomes. Two typical interphase nuclei, with membrane, nucleolus, and chromatin network are reconstituted. One nucleus has thus divided into two, and the final phase of cell division, the formation of a cell wall across the equatorial plate (*cytokinesis*), follows shortly thereafter. Actual mitosis takes about 1–2 hr, while the new DNA synthesis and other reactions preparatory to the next division may take another 6 hr. (See illustration on the following page.)

Meiosis

Meiosis differs from mitosis in that the homologous, already doubled chromosomes approach each other closely along their entire length (*synapsis*). At this time, the four closely appressed chromatids may entangle and exchange parts (*crossing-over*); only two chromatids may cross over at any one point. Now the chromosome pairs line up on the equator and, as in mitosis, are pulled apart poleward, and separated into individual cells. In each daughter cell, a new mitosislike meiotic division occurs in which the sister chromatids are separated. The product is four cells, each with a reduced (*haploid*) chromosome number, and possibly showing new chromosome types, unlike the parentals, because of crossing-over. (See illustration on the following page.)

Until recently, all the somatic cells of the plant body were thought to be diploid, but new evidence has revealed the existence of pockets of cells with multiples of the diploid condition, such as 4n, 6n, 8n, etc. These *polyploids* probably arise as a result of nuclear divisions that are not followed by separation of the daughter nuclei into two cells. The ploidy of cells can sometimes be controlled by the use of natural and synthetic chemicals of various types. For example, colchicine, an alkaloid extracted from corms (short, fleshy stems) of the meadow saffron, causes polyploidy by preventing the mitotic spindle from moving the two sets of chromosomes to opposite ends of the dividing cell. Sometimes *aneuploids* arise by the accidental loss or gain of one

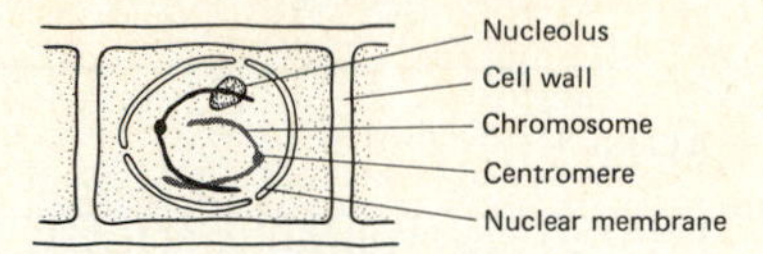

1. In *interphase*, the period between nuclear divisions, the porous nuclear membrane surrounds the elongate *chromosomes*. Each chromosome has a *centromere*, a region at which spindle fibers will later attach, and by contractile movements, pull the chromosomes to the poles of the cell. A *nucleolus* is also visible.

2. Just before *prophase*, each chromosome replicates longitudinally into two identical sister *chromatids*, which remain attached at the centromere. The doubled chromatids contract along their long axis, and the nuclear membrane and nucleolus disappear.

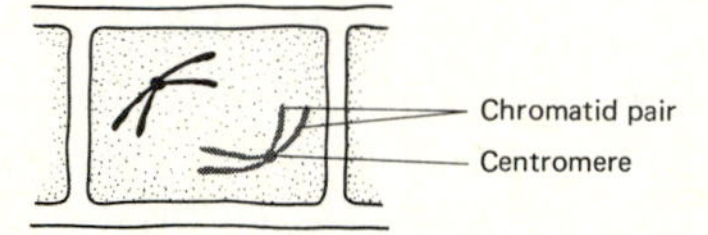

In these diagrams, only one pair of chromosomes is being depicted per *diploid* cell, i.e. 2n = 2, with the black chromosome coming from one parent and the grey from another. Each chromosome pair would behave identically. In pea, 2n = 14; in humans, 2n = 46.

3. In *mitosis*, each pair of doubled chromatids lines up on the equator (*metaphase*) and the centromere divides.

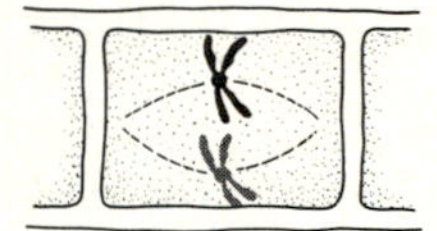

4. One chromatid of each pair is then pulled to an opposite pole by contraction of attached spindle fibers (*anaphase*).

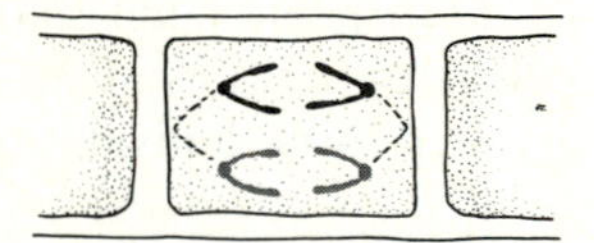

5. After all chromatids have become separated in this way, two identical daughter nuclei are reconstituted (*telophase*) and a new cell plate (wall) forms, giving rise to two cells, each with 2n = 2.

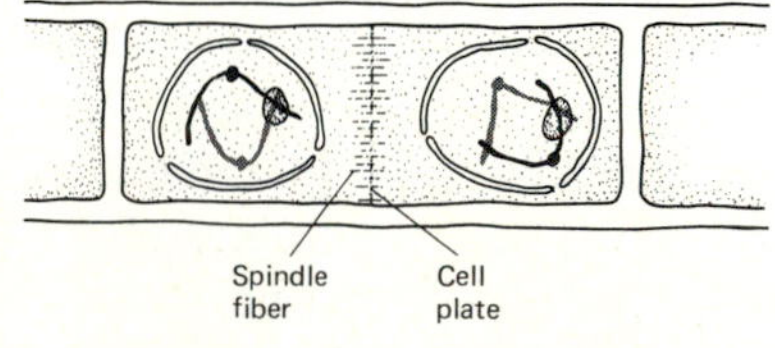

Two identical diploid (2n = 2) products of mitosis.

3′. In *meiosis*, the homologous pairs of double chromatids approach each other (*synapsis*) and frequently exchange parts (*crossing-over*).

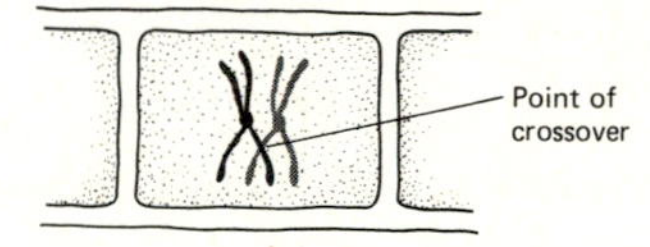

4′. At the *first* meiotic metaphase, there is a pair-by-pair separation.

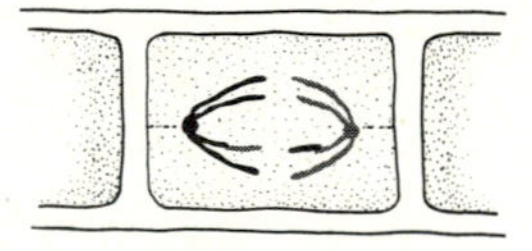

5′. At the *second* meiotic metaphase, the individual chromatids of each pair separate, giving rise to 4 *haploid* cells, i.e., n = 1.

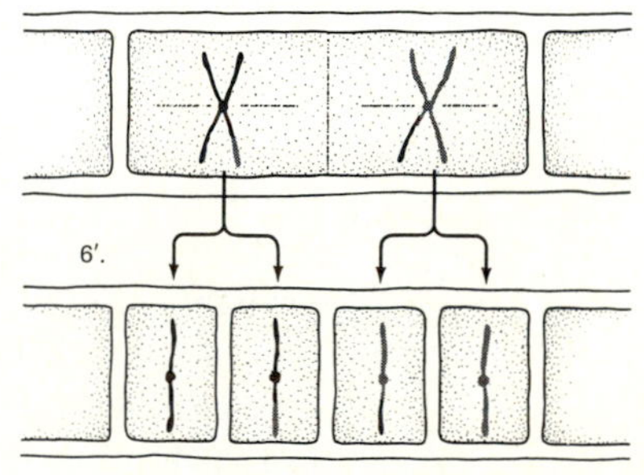

Because a crossover has occurred, each of the four haploid cells is now different. This provides variety in subsequent sexual fusion of haploid cells.

or more chromosomes. Such 2n + 1 or 2n − 1 individuals are usually abnormal, but may survive.

The molecules of DNA found in the chromosomes are linear aggregates of four types of units called *nucleotides,* and the order in which the nucleotides occur in the DNA chain determines the genetic information carried by the chain. Thus, the plant's inheritance consists essentially of a pattern of nucleotides in the DNA molecules of nuclear chromosomes and of other organelles such as chloroplasts and mitochondria. The discovery of the way in which information carried by the DNA molecules is transcribed into the related material *RNA* (*ribose nucleic acid*), whose information is in turn translated into the characteristics of newly synthesized *proteins* constitutes one of the most exciting chapters of modern biochemistry. Because of its basic importance in the specification of cell structure and function, we shall describe this process in some detail, although much of this information has become common knowledge in the intelligent lay public.

For most of the life of a cell, the nucleus is separated from the rest of the cytoplasm by a nuclear membrane. This membrane is double and contains numerous large pores as well as long projections into the cytoplasm (Fig. 2-3). These projections are often continuous with the *endoplasmic reticulum* (ER), a multiply-branched membrane network that penetrates throughout the cytoplasm. *Ribosomes,* spheroidal particles about 0.2 μm in diameter and composed largely of high-molecular-weight RNA and protein, may be located on the endoplasmic reticulum, or may be free in the cytoplasm. When attached to the ER, the ribosomes are said to be part of a "rough ER" system and are probably involved in synthesizing protein that is secreted into the ER *lumen,* the enclosed space between ER membranes. When free in the cytoplasm, clusters of ribosomes are sometimes seen attached along strandlike *messenger RNA*; such *polyribosomes,* as well as ER-bound ribosomes, are active in protein synthesis as long as they are in contact with messenger RNA. Proteins, large molecules composed of *amino acids* in a specific order, are synthesized on the surface of ribosomes. This intricate mechanism involves *transcription* of the information of DNA into RNA, followed by a *translation* of ordered RNA units into ordered amino acids that make up the protein. Since the active protein molecules called *enzymes* control the basic biochemistry of the cell, the control of protein specificity by DNA is the key to cellular regulation.

Nucleic Acids and Protein Synthesis

DNA occurs in the form of long double strands of intertwined helices (Fig. 2-9). Each strand is a linear aggregate of four different kinds of *nucleotides,* each composed of one nitrogen-containing base (adenine, guanine, cytosine, or thymine) attached to a *sugar* (deoxyribose), and *phosphate.* That portion of the DNA that specifies one complete protein is referred to as a *gene.* The order of the nucleotides in the chain deter-

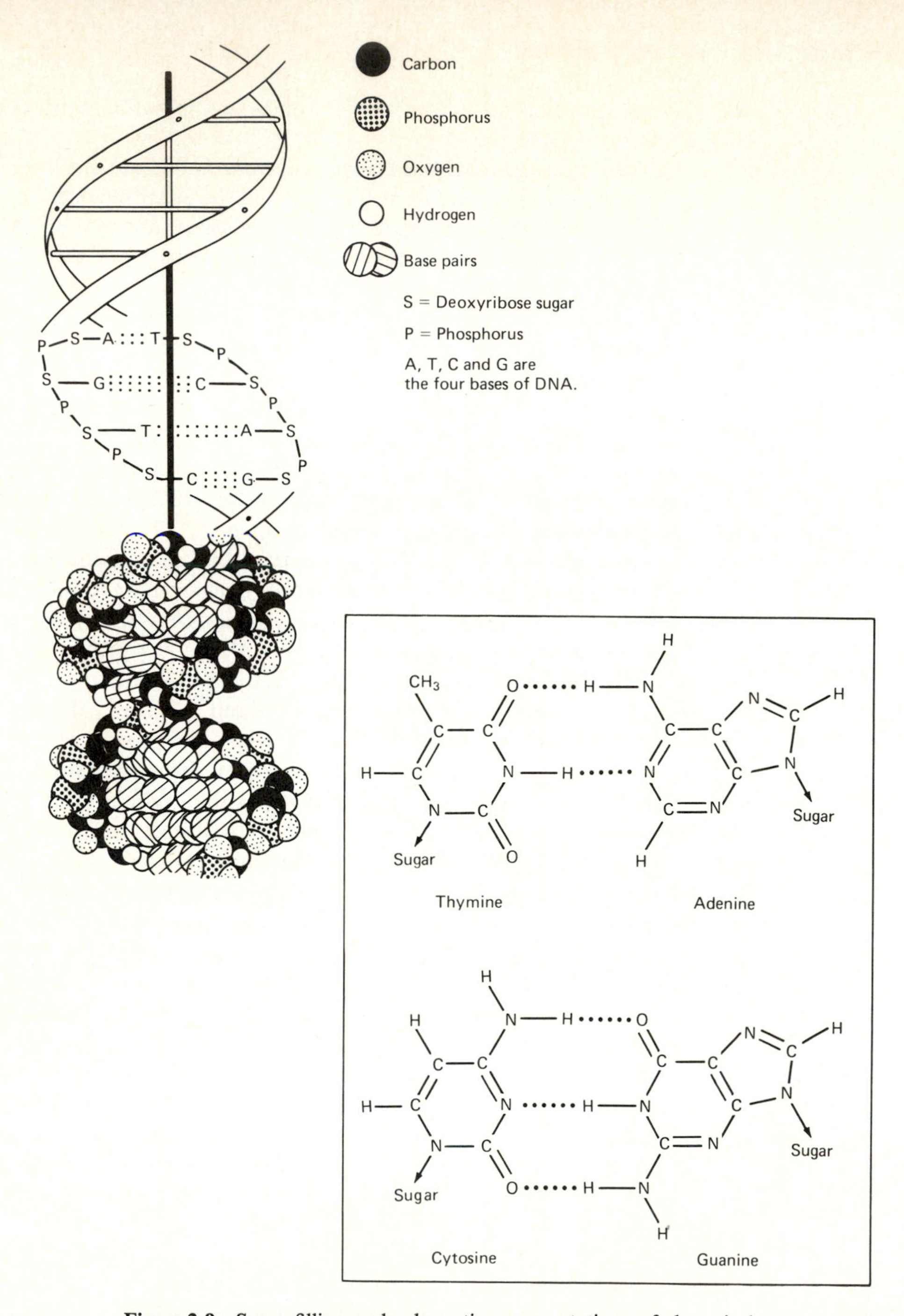

Figure 2-9 Space-filling and schematic representations of the spiral structure of the DNA molecule. The twin spirals are long, linear, polynucleotide chains. The letters P and S denote the phosphate-sugar backbone of each spiral; the letters A, T, G, and C represent the four bases adenine, thymine, guanine and cytosine; the dotted lines represent hydrogen bonds that hold the spirals together. The insert, lower right, shows the two hydrogen bonds between adenine and thymine and three hydrogen bonds between guanine and cytosine. (After C.P. Swanson, *The Cell*, 2nd ed., Englewood Cliffs, N.J.: Prentice-Hall, 1964.)

mines the genetic information carried by the molecule, much as the order of letters in a word determines its meaning. Thus, if the four nucleotides are represented by A, G, C, and T, the initials of the four bases, a linear aggregate in the order -ACGT- will constitute different genetic information from -AGCT- or -ATCG- chains.

The information in DNA serves two different and important roles in the cell. First, it is the basis for the continuity of DNA from cell to cell, serving as a template for its own replication. Second, it provides the information for the synthesis of specific cellular proteins through the intermediation of various types of ribosenucleic acid (RNA). RNA differs from DNA only in the fact that its sugar (ribose) contains one more oxygen atom than does the deoxyribose of DNA, an apparently small difference that leads to large differences in the configuration and geometry of the molecule.

The interrelated coiling of the two helical strands of DNA arises from the specific weak hydrogen-bonding between pairs of bases. Thus adenine and thymine are drawn together because two hydrogen atoms oscillate between nitrogen and oxygen atoms found in $-\overset{\overset{\displaystyle H}{|}}{N}-$ and $-\overset{\overset{\displaystyle O}{\|}}{C}-$ groups of adjacent A and T rings (Fig. 2-9, insert). Thus, the A—T pairing arises naturally from the structure of the molecules themselves. In the same way, guanine and cytosine form a specific G—C pairing through three hydrogen bonds. Given this specificity, we can see how information encoded in a template strand leads inevitably to complementary information on an adjacent, newly synthesized strand. If the new strand is in turn used as a template, the original template will have been duplicated.

When the two strands of the DNA molecule are tightly coiled about one another they are inactive synthetically, but when the spiral uncoils or opens up, various synthetic activities can occur. In the presence of the enzyme *DNA polymerase* and a mixture of the four deoxyribonucleotides (present as their more energetic triphosphates), a new DNA strand, complementary to the preexisting strand, is made. At another point in the cell cycle, in the presence of the enzyme *RNA polymerase* and a mixture of the four ribonucleotides (also as triphosphates), RNA, instead of DNA, is synthesized on the same basic DNA template. The only difference in basic coding is that in RNA the base uracil (U) is substituted for the thymine (T) of DNA. Thus, A—U hydrogen bonds in RNA replace the A—T hydrogen bonds of DNA; the G—C pairing is common to both types of nucleic acid (Fig. 2-10).

Three different kinds of RNA are made on the DNA template. *Ribosomal RNA* (r-RNA) and protein are components of the *ribosomes*, bipartite bodies formed of two unequal flattened spheres pressed

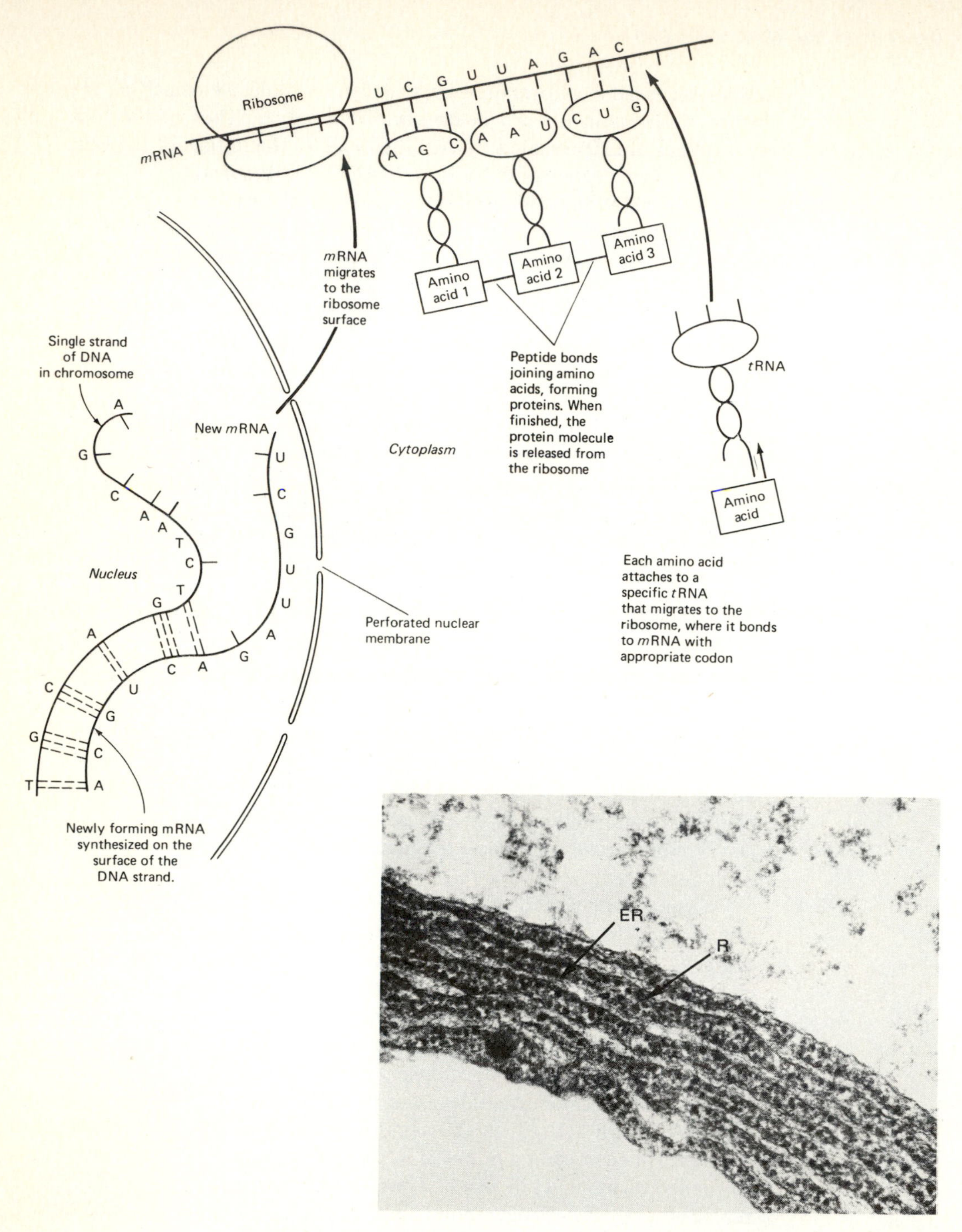

Figure 2-10 The control of protein synthesis by RNA. The insert (lower right) shows an electron micrograph of ribosomes (R) lined up along the endoplasmic reticulum (ER) in cotyledons of *Phaseolus vulgaris.* In such a situation, the newly formed protein is secreted directly into the ER lumen. X 53,000. (Courtesy of M. Boylan, Yale University.)

against one another. Ribosomes attach to *messenger RNA* (m-RNA) by the smaller of their two units; a group of such ribosomes attached to a strand of m-RNA constitute a *polyribosome* (or polysome), a protein synthesis aggregate ready for action. The correct ordering of amino acids in the protein is guaranteed by the fact that each of the 20 amino acids is first attached to a molecule of amino-acid-specific transfer RNA (t-RNA). The order of three nucleotides at a critical point in the t-RNA molecule "recognizes" (via hydrogen bonds) a complementary nucleotide triplet on the m-RNA strand attached to the ribosome and becomes attached at that point. In this way each loaded t-RNA delivers its amino acid in a specified order to the growing linear chain of attached amino acids that will become the newly synthesized protein; when its amino acid is unloaded, the specific t-RNA detaches and is ready for another round of amino acid transport. At some point on the m-RNA chain a triplet code signal terminates synthetic activity and detaches the protein from the ribosome. Each unit of the assembly team (ribosome, t-RNA, and m-RNA) may complete this same cycle many times, although all are susceptible to breakdown and must be resynthesized periodically.

Amino Acids and The Genetic Code

Proteins are linear chains of α-amino acids joined by peptide bonds. The general structure of an amino acid is:

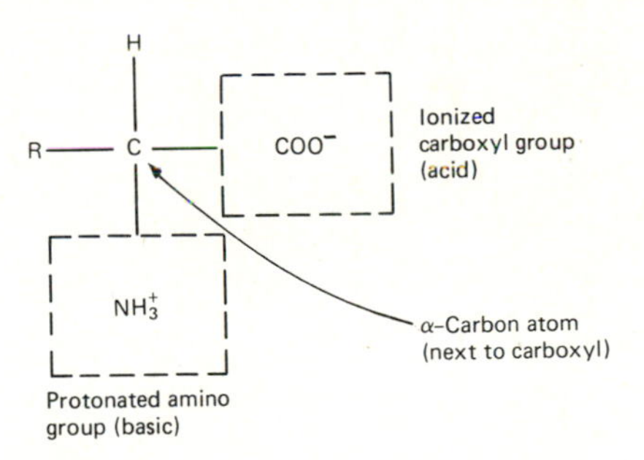

R represents any one of 20 different chemical groupings giving the amino acid its specific properties. The major amino acids of proteins are as follows:

R	Amino acid
H—	glycine
CH_3—	alanine
HO—CH_2—	serine
HS—CH_2—	cysteine

R	Amino acid
$CH_3-CH(OH)-$	threonine
$CH_3-CH(CH_3)-$	valine
$CH_3-CH(CH_3)-CH_2-$	leucine
$CH_3-CH_2-CH(CH_3)-$	isoleucine
$CH_3-S-CH_2-CH_2-$	methionine
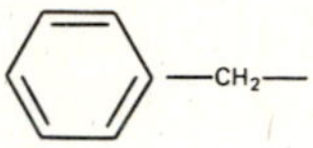 $C_6H_5-CH_2-$	phenylalanine
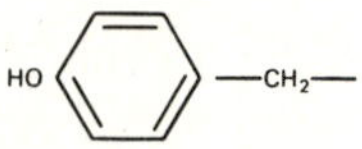 $HO-C_6H_4-CH_2-$	tyrosine
$HOOC-CH_2-$	aspartic acid
$HOOC-CH_2-CH_2-$	glutamic acid
$H_2N-CO-CH_2-$	asparagine
$H_2N-CO-CH_2-CH_2-$	glutamine
$H_2N-CH_2-CH_2-CH_2-CH_2-$	lysine
$H_2N-C(=NH)-NH-CH_2-CH_2-CH_2-$	arginine
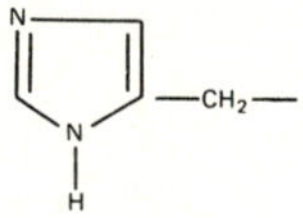 (imidazole ring: N, N—H) $-CH_2-$	histidine
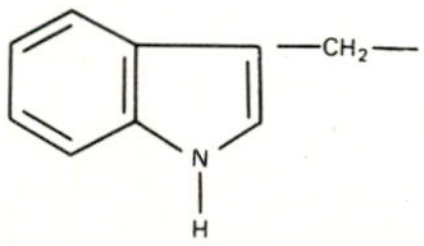 (indole ring: N—H) $-CH_2-$	tryptophan

The final amino acid, proline, is cyclized so that its amino group is substituted by its own R chain. Its structure is:

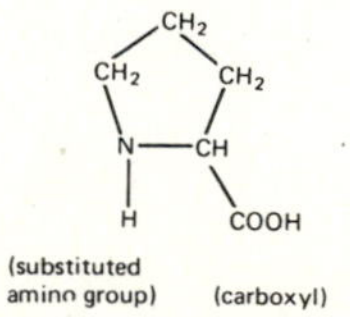

CH_2 CH_2 CH_2

N—CH

H COOH

(substituted amino group) (carboxyl)

When peptide bonds form, the carboxyl group of one amino acid joins with the amino group of another amino acid to form the $-\overset{\overset{\displaystyle O}{\|}}{C}-\overset{\overset{\displaystyle H}{|}}{N}-$ linkage common to all proteins. The order in which the amino acids are inserted into the growing polypeptide chain being synthesized on the ribosome is dictated by the order of nucleotides of the messenger RNA attached to the ribosome, which in turn is stipulated by the order of nucleotides in the DNA. Groups of three consecutive nucleotides, called triplets or *codons*, determine the insertion of a single amino acid in the chain. The codons for the 20 amino acids are sometimes unique, sometimes not. The 64 possible combinations of 3 nucleotides and their "meaning" in amino acid language are shown in the accompanying table, where U, C, A, and G designate uracil, cytosine, adenine and guanine, respectively. This code is believed to be universal in all forms of life.

First Letter	Second Letter				Third Letter
	U	C	A	G	
U	phenylalanine	serine	tyrosine	cysteine	U
	phenylalanine	serine	tyrosine	cysteine	C
	leucine	serine	——*	——*	A
	leucine	serine	——*	tryptophan	G
C	leucine	proline	histidine	arginine	U
	leucine	proline	histidine	arginine	C
	leucine	proline	glutamine	arginine	A
	leucine	proline	glutamine	arginine	G
A	isoleucine	threonine	asparagine	serine	U
	isoleucine	threonine	asparagine	serine	C
	isoleucine	threonine	lysine	arginine	A
	methionine	threonine	lysine	arginine	G
G	valine	alanine	aspartic acid	glycine	U
	valine	alanine	aspartic acid	glycine	C
	valine	alanine	glutamic acid	glycine	A
	valine	alanine	glutamic acid	glycine	G

**At these codons, the chain is terminated, and the finished polypeptides are split off the ribosome.*

Enzymes

Many of the proteins synthesized on ribosomes regulate the rate of specific chemical reactions occurring in the cell. Such proteins, possessing specific catalytic properties, are called *enzymes*. Their role is a crucial one, for they

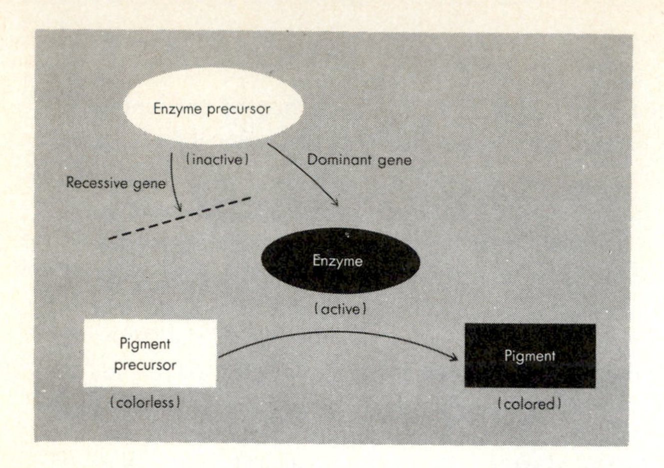

Figure 2-11 Genes may control the synthesis of specific enzymes which in turn control cell chemistry.

determine the ultimate nature of the cell by controlling the chemical reactions that synthesize the cell's constituents. For example, in certain species, the genetic difference between red and white flowers arises from differences in a single gene pair. The petal cells of the red variety, but not the white, have an enzyme that can convert a colorless precursor substance to a red pigment. The DNA of the nuclear genetic material determines the color of the petals by regulating the synthesis of the cytoplasmic enzyme that produces a colored substance from a colorless one (Fig. 2-11). This nuclear control is transferred to the cytoplasm through messenger RNA made in the nucleus, but active on cytoplasmic ribosomes.

There are thousands of enzymes in any cell, and each controls one chemical reaction or a group of related reactions. Many of these enzymes have been extracted from the cell and then purified, isolated, and finally crystallized. All are composed largely or exclusively of protein. Certain enzymes contain a small, nonprotein part called a *prosthetic group*, while others can be dissociated into two parts, a protein (the *apoenzyme*) and a smaller molecule (the *coenzyme*) (Fig. 2-12). In such instances, neither the apoenzyme nor the coenzyme alone can function catalytically; if the two are dissociated, catalytic activity ceases, and if they are recombined, the activity can be completely restored. Minute quantities of certain essential mineral elements and vitamins

Figure 2-12 Some enzymes consist of a large protein apoenzyme and small coenzyme. In these cases, only the apoenzyme-plus-coenzyme complex is active.

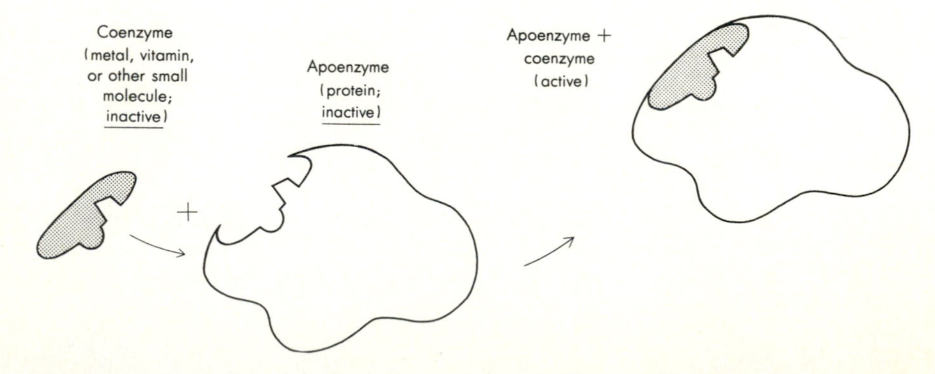

are crucially important in physiological processes because they serve as coenzymes for specific enzymatic molecules; if they are lacking, the enzyme cannot function, and the cell's biochemistry becomes abnormal, sometimes diseased. Some enzymes contain carbohydrate, lipid, or other components in addition to protein; these are referred to as glycoprotein, lipoprotein, etc.

Among the kinds of molecules that function as coenzymes are (1) such *metals* as iron, manganese, copper, zinc, molybdenum, and magnesium; and (2) such *vitamins* as thiamin, riboflavin, nicotinic acid, and pyridoxine. In both metallic and vitamin-type coenzymes, the active coenzyme may be somewhat more complicated than the metal or vitamin alone; for example, iron may be present as *heme*, a complex organic molecule with an iron center found in hemoglobin and several important oxidation enzymes. Similarly, thiamine, riboflavin, and nicotinic acid may be present as phosphorylated derivatives needed for the operation of several respiratory enzymes. Certain enzymes, called *metalloflavoproteins*, have more than one type of coenzyme; in aldehyde oxidase, for example, the active enzymatic molecule contains (in addition to the basic structural protein) free iron, iron in heme, and riboflavin in a complex form called *flavinadenine dinucleotide*. All these coenzymes are required for activity, and each must be attached at the correct position on the protein if it is to be effective.

Enzymes catalyze a wide variety of chemical reactions, including syntheses, degradations, hydrolyses, oxidations, reductions, and transfers of chemical groups such as amino, methyl, and phosphate. In general, one enzyme catalyzes only one reaction or type of reaction. All enzymes seem to function by first forming a chemical complex with the substances upon which they act (*substrates*). The enzyme–substrate complex then undergoes some internal rearrangements that cause alterations of the substrate and the eventual release of the products of the reaction (Fig. 2-13a). For example, suppose that two small molecules, A and B, slowly unite to form a larger molecule AB, and that this reaction is speeded (*catalyzed*) by the enzyme, E. The overall reaction,

$$A + B \xrightarrow{E} AB$$

can be shown to consist of the following steps:

$$E + A \longrightarrow E \cdot A$$
$$E \cdot A + B \longrightarrow E \cdot A \cdot B$$
$$E \cdot A \cdot B \longrightarrow E + AB$$

If these equations are added together, the net reaction is $A + B \longrightarrow AB$, and the enzyme does not appear as a constituent of the overall reaction. The regeneration of the enzyme in the final step accounts for its catalytic role in the overall reaction, and only small amounts of the enzyme are required to produce large total changes in substrate and product level (Fig. 2-13b). In many cases, the velocity, v, of an enzyme-controlled reaction varies with the

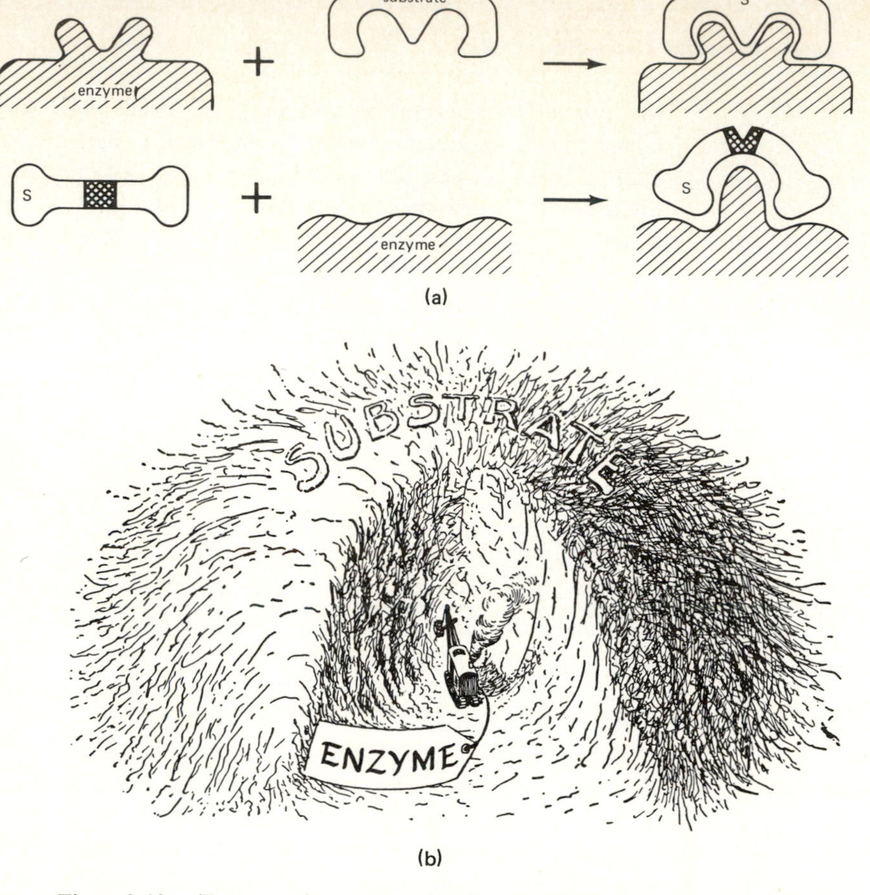

Figure 2-13a Enzyme-substrate complex. Some enzymes and substrates fit together as adjacent pieces of a jigsaw puzzle (upper). In other cases the enzyme changes shape upon binding substrate (lower). The new conformation also induces changes in the shape of the substrate. (Adapted from D.E. Koshland and K.E. Neet. 1968 Ann. Rev. Biochem. *37*: 359–410.) **2-13b** Enzyme molecules perform feats out of all proportion to their size. From J. Bonner and A.W. Galston, *Principles of Plant Physiology*, San Francisco: W.H. Freeman and Co., 1952.

concentration of substrate, [S], as shown in Fig. 2-14. When the concentration of enzyme is held constant, v varies almost linearly with [S] when [S] is small and each added unit of S makes more ES, but v is almost independent of [S] when [S] is large and almost all E is already present as ES. In a double-reciprocal plot of $1/v$ against 1/S, the curve becomes a straight line; the advantage of such a plot is that it permits the analysis of biochemical reactions in terms of interference with enzyme function. For example, certain inhibitors which compete with substrate for attachment to the active site of the enzyme cause a change in the slope, but not the y intercept. Such analysis can be useful in devising corrective procedures for blocked enzymes.

The enzymes of the cell, located in the various cell organelles and in the nonparticulate area of the cytoplasm, are thus the direct superintendents of

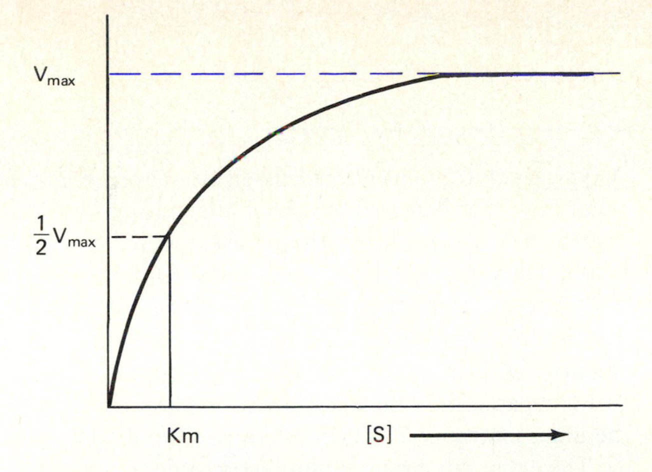

Figure 2-14a A graph showing the relationship between the velocity of an enzyme controlled reaction and the substrate concentration, when the enzyme concentration remains constant. V_{max} is the maximum velocity attained when all the sites on the enzyme are associated with substrate. Km is the substrate concentration at $\frac{1}{2}$ V_{max}.

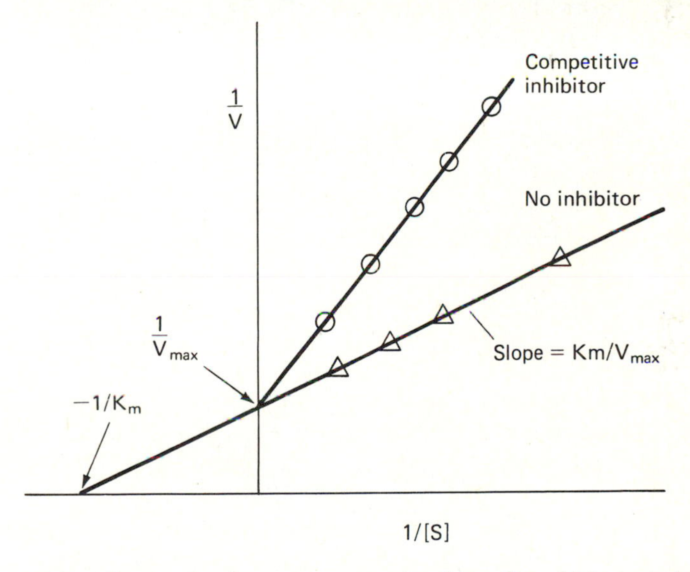

2-14b The graph of an enzyme-catalyzed reaction yields a straight line if the reciprocal of the velocity is plotted against the reciprocal of the substrate concentration (double reciprocal plot). The line intercepts the Y-axis at $1/V_{max}$, while its extrapolation intercepts the x-axis at $-1/K_m$. A competitive inhibitor (one that competes with the substrate for a particular site on the enzyme) alters the slope of the line, but does not alter V_{max}.

the cell's chemical machinery. All cells are what they are by virtue of their chemistry; their chemistry is determined by their enzymes; the nature of the enzymes is determined by cytoplasmic RNA; and the specificity of the RNA is in turn determined by the DNA of the nucleus and other organelles.

Mitochondria

A typical cell contains numerous pickle-shaped *mitochondria*, the respiratory organelles, distributed throughout the cytoplasm. They range from one to several micrometers in length and are approximately one-half micrometer in width, but especially active cells may contain larger and more numerous mitochondria than the average. Each mitochondrion is surrounded by a double membrane, with the inner one convoluted into numerous platelike *cristae* (Fig. 2-15). The inner membrane thus divides the mitochondrion into two compartments: the space between the inner and outer membranes, and the inner matrix bounded by the inner membrane.

The outer membrane is quite permeable to most small molecules and ions. Various organic molecules, including the 3-carbon compound pyruvate, generated by the breakdown of 6-carbon sugars in the cytoplasm (see Chapter 5), pass through this membrane and are then oxidized to CO_2 and H_2O by a series of enzymatic reactions. During the mitochondrial oxidations, the energy released is stored in the form of special energy-rich phosphate bonds of adenosine triphosphate (ATP) (see Chapter 4). Since the cell can use the energy in these bonds to perform work of various kinds, the continuing mitochondrial oxidations supply the cell with its required "energy currency" of ATP.

Figure 2-15 Schematic drawing of a mitochondrion. The drawing lower right is at higher magnification.

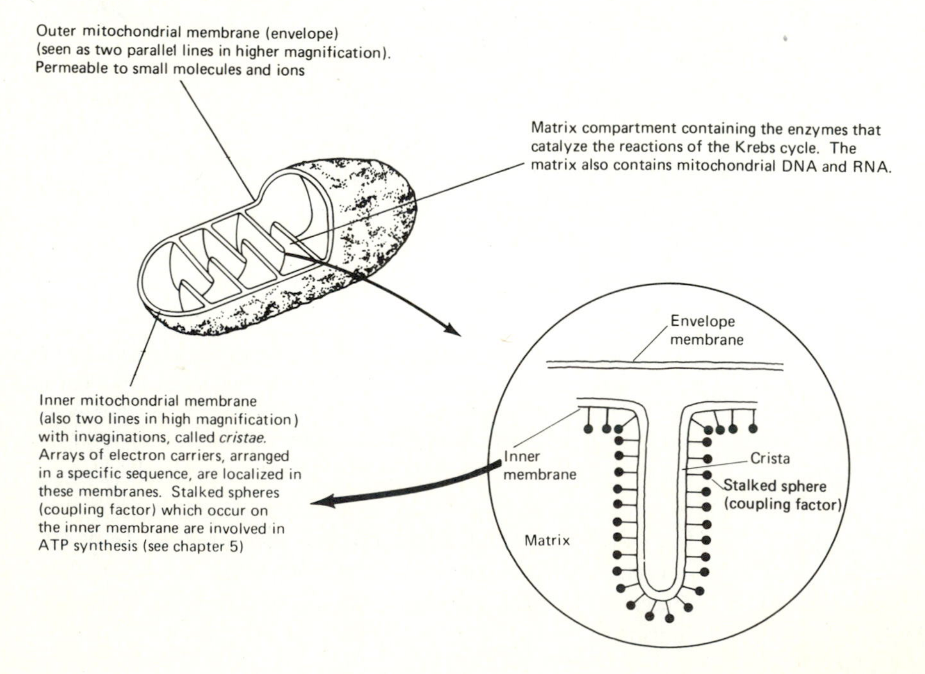

Pyruvate that enters the mitochondria is first dismembered to yield CO_2 and "activated acetate," a 2-C fragment which then enters a cyclical series of oxidative reactions called the *Krebs cycle*, discussed in Chapter 5. Each reaction in this cycle is catalyzed by a specific enzyme located in the inner matrix of mitochondria. During this cyclic process, electrons and protons are transferred from Krebs-cycle intermediates to respiratory enzymes containing such vitamin derivatives as NAD^+ (a nicotinic acid derivative) and FAD (a riboflavin derivative). These carriers, which are thereby converted to their reduced forms NADH and $FADH_2$, subsequently pass their electrons and protons to still other electron carriers in the respiratory assembly, and finally to oxygen, which is thus converted to H_2O. The electron carriers, arranged in a specific order, are located in the inner membranes of the mitochondria. As electrons and protons are transferred from one carrier to the next, the energy released is used to join ADP to inorganic phosphate (P_i), thereby generating ATP. When ATP breaks down to ADP and P_i, the released energy can be used to drive other reactions or processes.

Chloroplasts and Other Plastids

Plant cells are unique in that they contain a variety of cytoplasmic bodies called *plastids*. Of foremost importance in the green plant cell is the *chloroplast* (Figs. 2-16 and 2-17), site of photosynthetic activity and of all the chlorophyll and accessory pigments associated with this process.

In higher plant cells, the chloroplast is a lens-shaped body roughly 5–8 μm in diameter by about 1 μm in width. Each chloroplast is bounded by a double membrane with differentially permeable properties, and contains as well an extensive internal membrane system. The basic structural unit is a thin, flat disc called a *thylakoid* that contains chlorophyll, other pigments, and enzymes that participate in the photochemical reactions. The discs are arranged in stacks called *grana* between which are the nongreen areas called *stroma*, containing many of the enzymes involved in CO_2 fixation. Approximately 50 chloroplasts are present in an average cell, and, as far as we know, each chloroplast is derived from a *proplastid*. Proplastids are presumably able to replicate by some sort of division process, thus increasing their number in the cell; mature chloroplasts, while sometimes capable of such replication, do so less often. Proplastids in dark-grown plants are called *etioplasts*.

Proplastids are smaller than chloroplasts and do not have their characteristic layered structure. They possess instead a *prolamellar body*, an ordered, "paracrystalline" center of canals that, under proper stimulation by light, reorient into parallel layers (Fig. 2-18). In the flowering plants, or angiosperms, the mature chloroplast will not develop from the proplastid unless the plant is illuminated, while in certain gymnosperms, the transformation can be accomplished in total darkness. Another difference between most gymno-

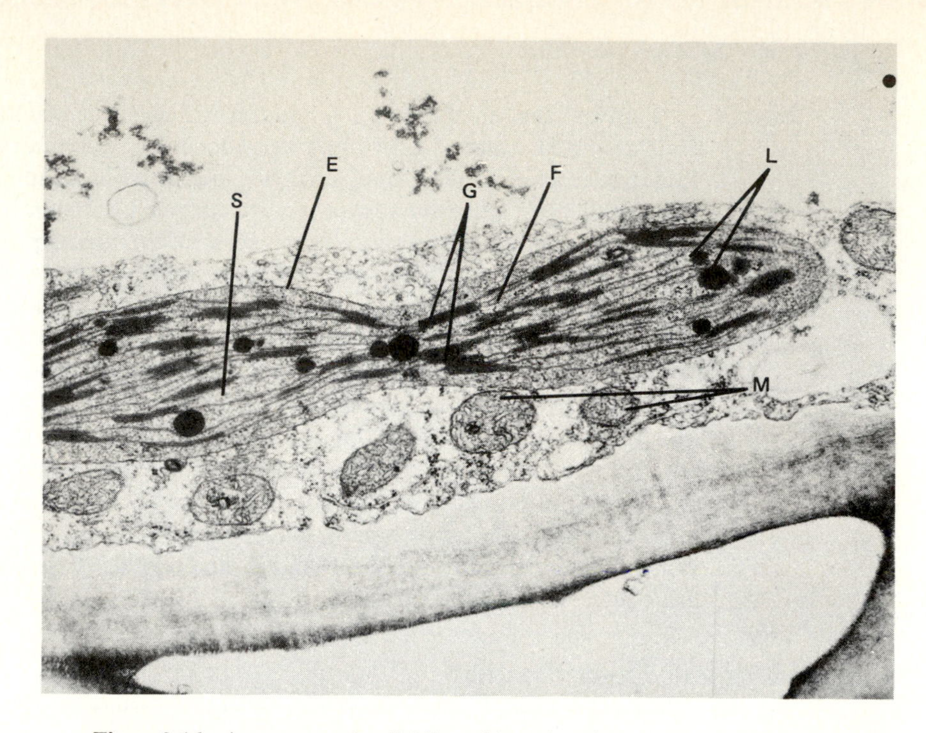

Figure 2-16 An apparently dividing chloroplast in *Samanea saman*. Note the two membranes of the chloroplast envelope (E), the grana stacks (G) composed of thylakoid units, the fret membranes (F) that interconnect the grana, the stroma (S) and the numerous lipid globules (L). Several mitochondria (M) can be seen in the cytoplasm that surrounds the chloroplast. (X16,000). (Courtesy of M. J. Morse, Yale University.)

Figure 2-17 Chloroplast in *Samanea* containing several large starch granules (S). (X 16,000). (Courtesy of M. J. Morse, Yale University.)

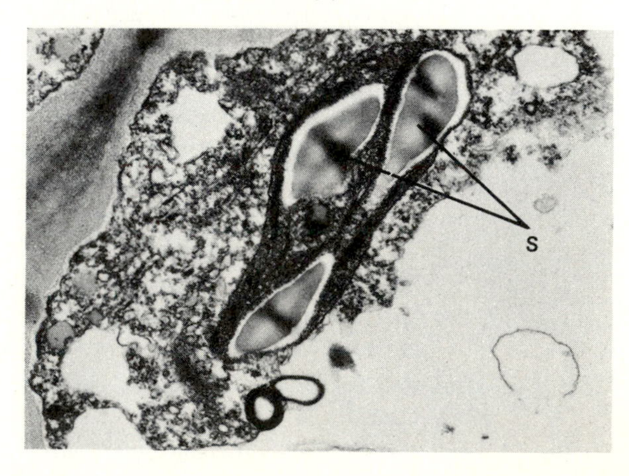

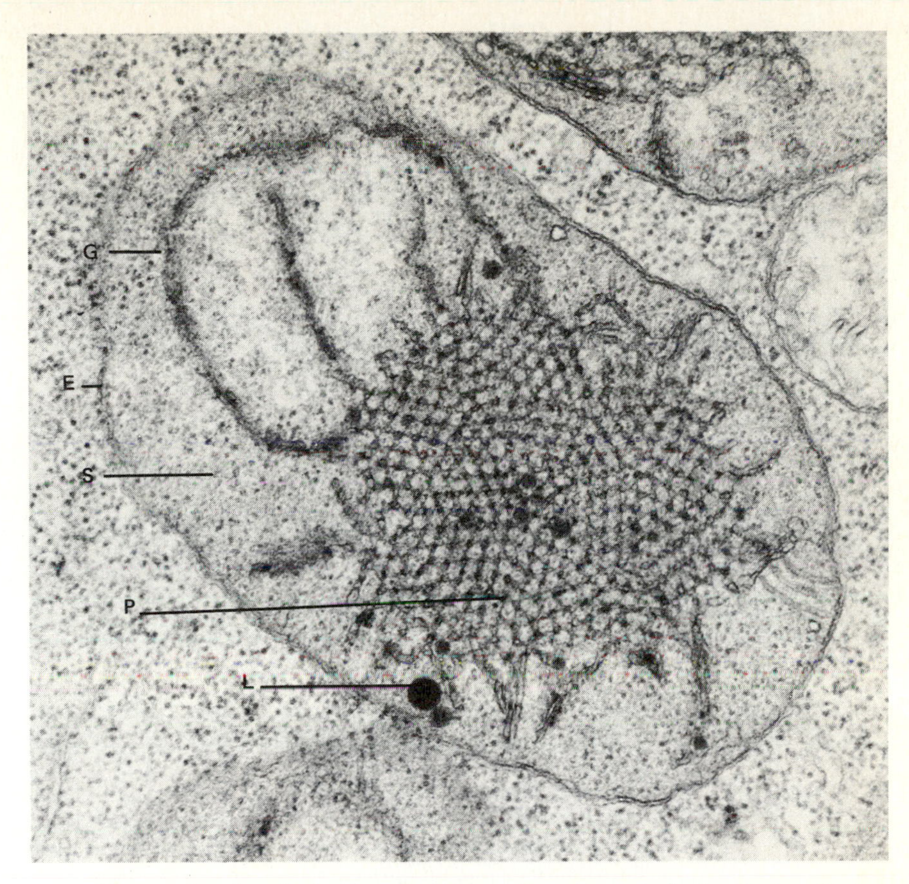

Figure 2-18 An etioplast (proplastid) from the primary leaf of a 10-day-old dark grown pea plant. Note the highly ordered structure of the prolamellar body (P) which will develop into stacks of grana when the plant is illuminated. One such stack (G) is beginning to develop at the upper left. Note the two membranes of the plastid envelope (E), the stroma (S) and the lipid globule (L), X44,000. (Courtesy of W. J. Hurkman, Purdue University.)

sperms and angiosperms lies in their comparative ability to transform the pigment protochlorophyll to chlorophyll, a change that involves the addition of two hydrogen atoms and a phytol chain to the protochlorophyll molecule (Fig. 2-19). In all angiosperms, this step is dependent on light but in some gymnosperms it can proceed in the dark. Chloroplasts do not usually develop in root tissues, even those exposed to light, although in certain species, such as carrot and morning glory, root cells may "green-up" upon illumination. Why the plastids do not develop to maturity in the cytoplasm of some cells is unknown.

Chloroplasts contain their own specific DNA, which differs from nuclear DNA and is transmitted genetically through proplastids in the maternal (egg) cytoplasm; no chloroplast DNA is inherited from the male (pollen) parent. Enzymes present in the chloroplast can be coded by either nuclear or plastid

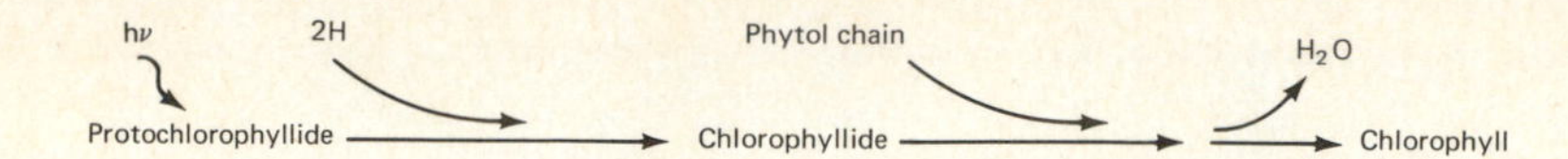

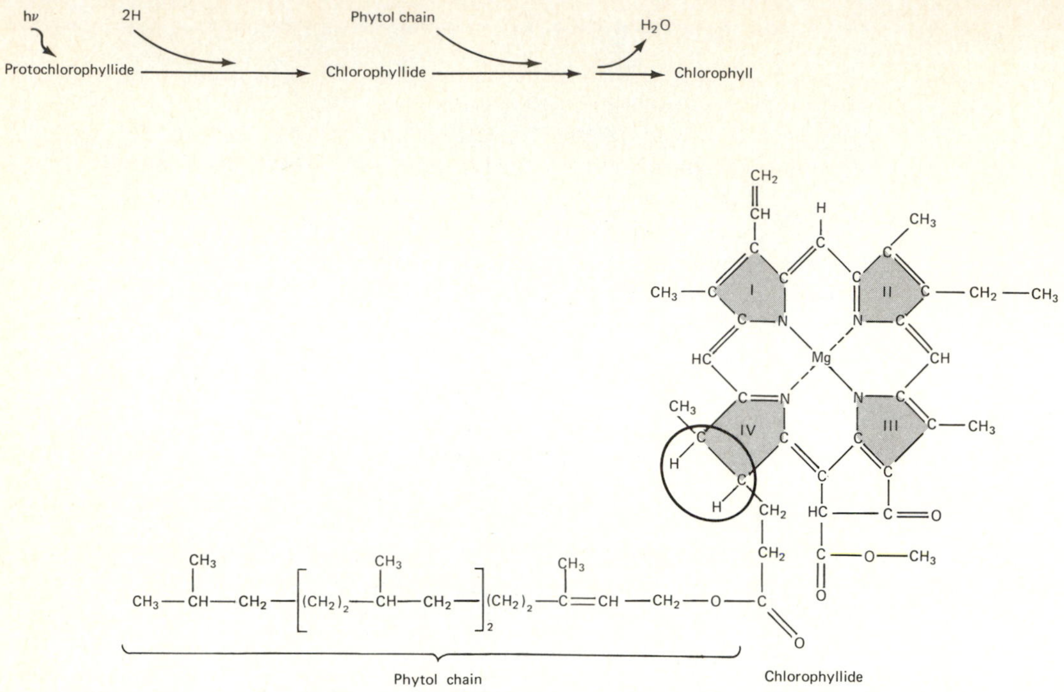

Figure 2-19 In angiosperms, light causes two hydrogen atoms to add to protochlorophyllide, thereby converting it to chlorophyllide. Subsequent attachment of the phytol chain converts chlorophyllide to chlorophyll. The chemical structure of chlorophyll *a* is shown below. Note the magnesium atom in the center and the four pyrrole rings (I–IV) attached to it. This structure resembles heme, which has iron at the center bound to four pyrrole rings. The circled group in ring IV is replaced by C = C in protochlorophyll.

DNA, and some enzymes, such as ribulose bisphosphate (RuBP) carboxylase, contain two protein parts, one coded by nuclear and the other by plastid DNA. Chloroplasts also contain the ribosomes, RNA, amino acids, and enzymes required for protein synthesis. Some biologists therefore view the semi-independent chloroplast as a sort of invading organism whose ancestors by chance found their way into previously nongreen cells and thereby made them autotrophic (self-feeding via photosynthesis). According to this theory, single-celled photosynthetic organisms lacking membrane-bounded organelles (*prokaryotes*) became associated by chance with heterotrophic organisms (nonphotosynthetic and thus forced to rely upon preformed food molecules for their nutrition). As it was evolutionarily advantageous, this symbiotic (mutually beneficial) association was perpetuated as the modern *eukaryotic* plant cell, containing membrane-bounded nuclei and organelles. This viewpoint is supported by ultrastructural studies which reveal striking similarities between chloroplasts of higher plants and cells of blue-green algae, which lack a nucleus and membrane-bound chloroplasts, but have lamellae that extend through the cytoplasm (Fig. 2-20). Since mitochondria also contain unique DNA and protein-synthesizing machinery, they too may be descendants of independent organisms. Mitochondria are in some ways structurally similar to certain primitive bacteria, and their DNA shows some chemical affinities with that of prokaryotic organisms.

Proponents of this theory point out that if certain cells, such as the unicellular flagellate *Euglena*, are grown at fairly high temperatures for several successive generations, the main body of the cell replicates faster than the chloroplast, producing successively paler and paler green cells. Totally nongreen cells may eventually be produced by the "dilution out" of the last chloroplast or proplastid. Such cells are permanently nongreen and cannot regain the autotrophic habit. The disappearance of chloroplasts may also be induced by treatment with streptomycin or other chemicals. Thus, the cell can be "cured" of the "invading" chloroplasts by high-temperature therapy or by chemical treatment.

Biologists can isolate intact chloroplasts from fragmented cells by techniques of differential centrifugation, and can demonstrate that they retain for some time all the attributes of the photosynthetic apparatus of the cell.

Figure 2-20 Electron micrograph of the blue-green alga *Synechococcus lividus*. Note that photosynthetic lamellae (L) and stroma (S) are distributed throughout the cell. Disc-shaped phycobilisomes (arrow) containing several different pigments are localized on the lamellae (×54,000). (By M. R. Edwards, published in E. Gantt. 1975. BioScience *25*, 781–788). Note similarities between this cell and the chloroplast of a higher plant (Fig. 2-16).

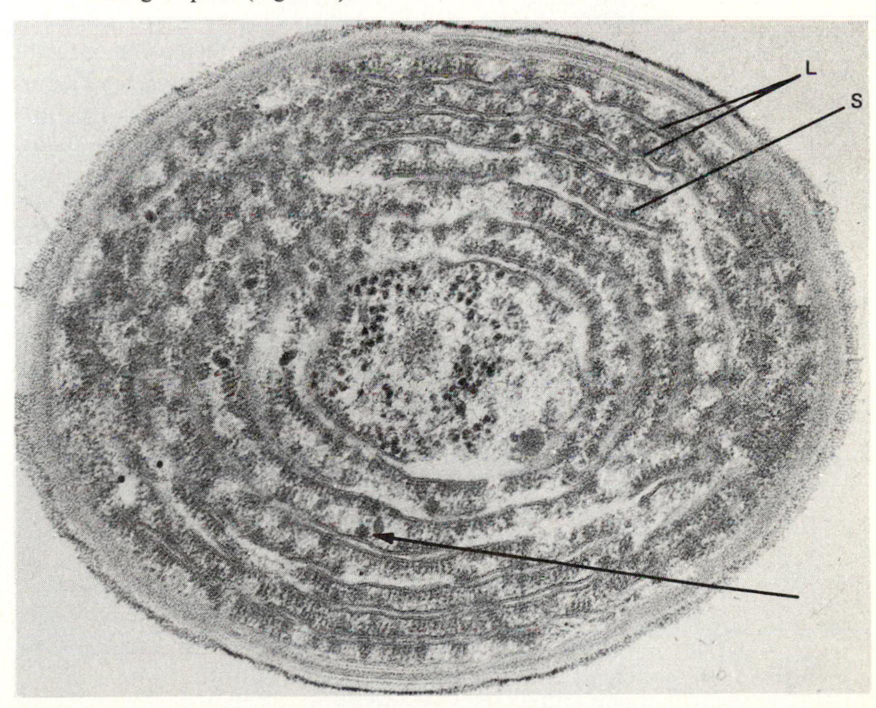

Isolated intact chloroplasts have been shown to fix CO_2, liberate O_2, and generate energy-rich phosphate bonds in the presence of light. However, the chloroplast cannot maintain itself or reproduce outside the cell. If the chloroplast is really an "invader from without," it has become markedly dependent on the remainder of the cell for many aspects of its existence.

The chloroplast contains several different pigments. *Chlorophyll*, the most important because it is the only one that participates directly in photosynthesis, exists in several forms, each with a slightly different absorption spectrum. Some algal chloroplasts are especially rich in blue or red *phycobilins*, while the chloroplasts of most higher plants contain yellow, orange, and red *carotenoids*. The carotenoids appear to protect chlorophyll from degradation by molecular oxygen when exposed to light. They also increase photosynthetic efficiency by absorbing energy of wavelengths not absorbed by chlorophyll, then passing this energy to chlorophyll, where it is used for photosynthesis. The carotenoids are not noticeable during most of the growing season, since their pigmentation is masked by higher concentrations of chlorophyll; however, in autumn, when chlorophyll concentration decreases as leaves senesce, the brightly colored carotenoids become evident and are responsible for much of the pigmentation of fall foliage. Other pigments from outside the chloroplast, such as the bright red anthocyanins in cell vacuoles, also contribute to autumn leaf colors.

Absorption Spectra and Pigments

Pigments are molecules whose electronic structure permits them to absorb radiations at particular wavelengths in the visible region. A plot of the relation between absorbancy and wavelength is called an *absorption spectrum*. It can be so unique that it absolutely characterizes a compound; at the very least it can tell something about the structure of the absorbing molecule.

Absorption spectra are constructed by producing monochromatic light by a prism, a diffraction grating or a light filter. This beam of light is then passed through a solution of the pigment, and the emergent light energy compared with the energy of the original light beam. Energy absorption depends upon the *absorption coefficient* (α) of the pigment, its concentration (c) and the length of the light path (l). Frequently called the *Beer-Lambert law*, this relationship can be expressed as follows:

$$\ln \frac{I_0}{I} = \alpha c l$$

where I_0 is the energy of the incident light beam and I the energy of the emergent light beam. For solutions it is usual to employ logarithms to the base 10 and to transform α (the absorption coefficient) to ϵ (the extinction coefficient).

Now $$\log_{10} \frac{I_0}{I} = \epsilon c l$$

where c = concentration in moles/liter
l = optical pathway in centimeters, and
ϵ = molar extinction coefficient, in units of liter/mol/cm

$\log_{10} \frac{I_0}{I}$ is also called *absorbance* (A) or *optical density* (O. D.)

A *spectrophotometer* is a machine that produces monochromatic light, passes it alternately through a solution of a substance and its pure solvent, measures I_0 and I, and records the logarithm of their ratio. A typical absorption spectrum for chlorophyll b in organic solvent is shown below:

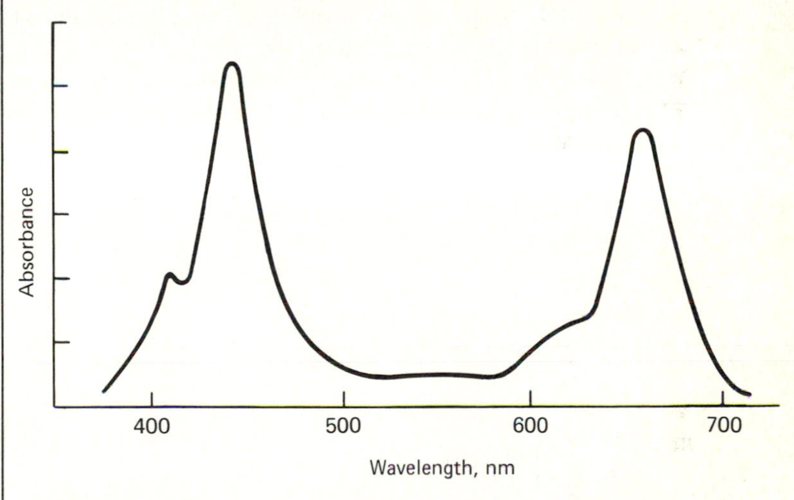

The pigment seems green because it absorbs in the blue (400–470 nm) and red (600–700 nm) regions, but transmits green and yellow light.

In addition to chloroplasts, cells of higher plants contain other types of plastids that lack the layered structure and photosynthetic apparatus of chloroplasts. These bodies include *leucoplasts*, which are colorless, and *chromoplasts*, which usually contain a high concentration of carotenoid pigments. Like chloroplasts, these bodies seem to be transmitted from one cell generation to the next by means of proplastidlike structures in the maternal cytoplasm. The leucoplast serves as a storage center for the cell's reserve food materials, such as starch grains, and therefore probably has the enzymatic machinery necessary to synthesize such materials from smaller precursor molecules. Chromoplasts probably facilitate reproduction and seed dissemination by producing in flowers and fruits the bright colors that attract animals. They generally do not develop in the cell if chloroplasts are present. In the ripening process of fruits such as tomato, the green–yellow–red transition reflects three successive stages of development: the dominance of chloroplasts, the decline of chloroplasts, and the rise of the carotenoid-laden chromoplasts. The cause of these transitions is not well understood.

The Vacuole

The vacuole is a large central compartment in the cytoplasm, surrounded by a differentially permeable membrane (the *tonoplast*) and containing a watery solution of salts, organic molecules, and waste products of the cell's metabolism. When a cell is young, it contains numerous small vacuoles which together constitute only a small percentage of its total volume (Fig. 2-21). As the cell grows, the vacuoles expand through the inward diffusion of water (see Chapter 6), and ultimately coalesce into one large central vacuole occupying 90% or more of the cell's total volume. This vacuole exerts pressure on the cytoplasm and wall, thus contributing to plant form and rigidity.

Figure 2-21 Young parenchyma cells in culture. These cells have begun to enlarge by the development of one or more vacuoles (V) in the cytoplasm. There are many small vacuoles in the upper cell, while vacuoles in the lower cell are coalescing to form a single large vacuole. Note the fusion of two vacuoles at X, and breakdown of the membranes and cytoplasm between them (×11,400). (Courtesy of H. T. Wilson, University of Alabama at Huntsville).

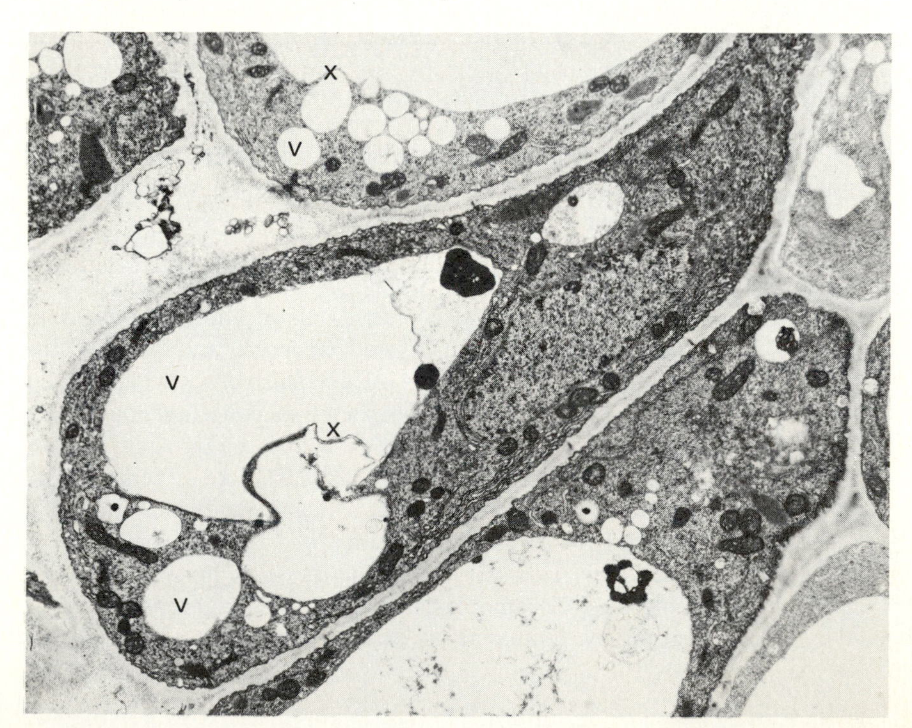

The tonoplast can form in several different ways. It often arises from preexisting membranes of the endoplasmic reticulum or the *Golgi apparatus*, described later in this chapter. Under some circumstances it appears to form by localized hydration of a region of the cytoplasm, followed by synthesis of a new membrane. Electron micrographs indicate that the tonoplast is thicker and stains more densely than the plasmalemma. Since the composition of the solute within the vacuole is very different from that of the cytoplasm, the tonoplast and the plasmalemma obviously must have different permeability characteristics. They probably also contain different *ion pumps*, specialized membrane-localized proteins that use the energy of ATP to move solutes across the membrane barrier. The pH (see box, p. 44) of the vacuolar sap of most plants lies between 3.5 and 5.5, but in some plants is as low as 1.0, whereas that of the cytoplasm is close to 7.0. This large difference in H^+ ion concentration suggests that the tonoplast contains pumps that transport H^+ ions from the cytoplasm to the vacuole. H^+ ion pumps probably help maintain pH control in the cytoplasm; such control is crucially important, since the cytoplasm contains most of the enzymes that regulate metabolism, and enzyme activity is very dependent upon pH. Thus the vacuole may store or accumulate ions and other materials, such as organic acids or their salts (calcium oxalate is frequently found), pigments such as anthocyanins, and phenolic substances such as tannins, that would otherwise interfere with metabolism.

Our present knowledge of the properties of the tonoplast is based primarily on ultrastructural studies, and on evidence of differences in the composition of the vacuole and the cytoplasm. Attempts to separate the tonoplast from other membrane fractions were unsuccessful until the recent development of new techniques for isolating intact vacuoles from the remainder of the cell (Fig. 2-22). The first step in this method is the formation of spherical

Figure 2-22 A photomicrograph of vacuoles isolated from two colored *Tulipa* species. To prepare the vacuoles, flower petals were chopped into small fragments with a razor blade, and the fragments incubated in an enzyme solution that dissolves the cell walls, releasing the protoplasts. The protoplasts were then incubated in a solution that causes swelling and ultimate rupture of the plasmalemma. Gentle stirring then released the vacuoles and other organelles from the cytoplasm. (From G. J. Wagner and H. W. Siegelman. 1975. Science *190*: 1200.)

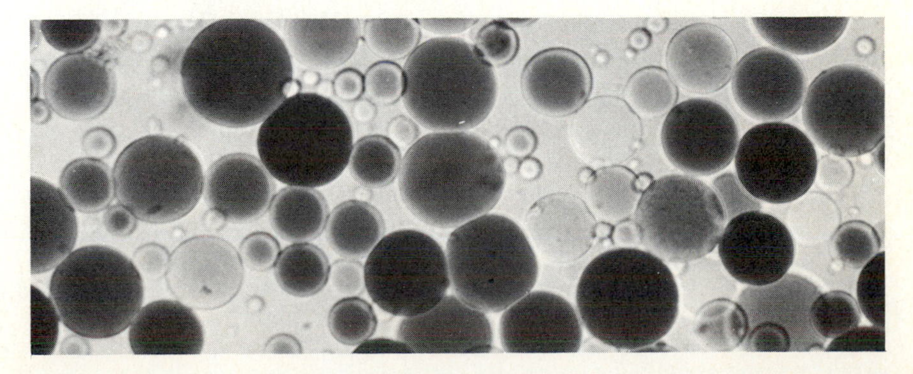

protoplasts by enzymatic degradation of the cell walls in the presence of high concentrations of an osmotically active solute. When the protoplasts are transferred to a less concentrated (hypotonic) medium, they absorb water, rupture, and release the vacuoles. Vacuoles and organelles are then separated from each other and from the incubation medium by differential centrifugation. The first analyses of such vacuoles indicate that enzymes regulating salt transport are highly concentrated in the tonoplast. Researchers in several laboratories are now conducting additional experiments to determine the permeability properties and enzymatic composition of the tonoplast; these should significantly increase our understanding of the role of the tonoplast in regulating cell metabolism.

Acidity, pH and Buffers

Acidity and Alkalinity An acid is a substance capable of releasing a proton (H^+), whereas a base is a substance capable of combining with a proton. If an acid is added to an aqueous system, it will increase the concentration of protons; if a base is added, it will combine with protons, lowering their concentration. This will in turn increase the OH^- concentration, because H^+, OH^-, and H_2O are in equilibrium as water dissociates.

$$H_2O \rightleftharpoons H^+ + OH^-$$

Thus in aqueous systems: acidity $= [H^+]$ and basicity (alkalinity) $= [OH^-]$, where [ ] symbolizes concentration in gram atoms or molecules per liter, i.e., molarity. In aqueous solutions $[H^+] \times [OH^-] = 10^{-14}$. As $[H^+]$ and $[OH^-]$ must vary inversely, one need know only the value of $[H^+]$ to determine $[OH^-]$ as well. In a solution that is neutral (equal acidity and alkalinity), $[H^+] = [OH^-] = 10^{-7}$ M.

pH The range of acidity of the cell sap in plants is 10^{-1} to 10^{-7} molar, that is, a millionfold difference in $[H^+]$. In order to express differences conveniently over such a range, a logarithmic expression, pH, is used. pH is the negative logarithm to the base 10 of the H^+ ion concentration in moles per liter.

$$\text{pH} = -\log [H^+]$$

Thus, if $[H^+] = 10^{-x}$ molar, then pH $= x$. Accordingly, decreasing the pH by a whole number represents a tenfold increase in acidity, and raising pH one unit represents a tenfold decrease in acidity.

Acidity is very important to plants and their growth. Enzymes function optimally only over a very narrow pH range, usually near neutrality. The pH of the cytoplasm is kept near 7 by buffers (see below), and by pumps that transport H^+ into and out of the vacuole. The vacuole of a plant is a reservoir for organic acids, so that its pH is usually in the region of 3.5–5.5, although in certain fruits it may go as low as 1.0. The ATP-generating systems in chloroplasts and mitochondria rely on pH differences across their membranes for their functioning (see chapters 4 and 5). When biochemists extract organelles or molecules from plant cells, they are careful to add a proper buffer to the extraction medium (usually about pH 6–8) so that the acids in the cell vacuoles are counteracted, and not allowed to denature the enzymes in the cells. The growth of plants in soil is very dependent on pH—best growth generally occurs near neutrality, chiefly because pH controls the availability (and, at pH extremes, the toxicity) of the nutrient elements in the soil (see Chapters 7 and 14).

Titratable Acidity The actual acidity of a plant sap (above) is expressed by its H^+ ion concentration, while *titratable acidity* represents the amount of base required to titrate the sap to neutrality (pH 7). This cannot be predicted from the H^+ ion concentration, because plant saps contain weak acids that are poorly ionized. Thus actual acidity equals titratable acidity only if the acid is completely dissociated, that is, if it is a *strong acid.* By definition,

1.0 N HCl = 1 gm atom/l titratable acidity.

$[H^+]$ would equal 1, or 10^0 molar

If fully dissociated, its pH would be $-\log 1$, or 0.

(Actually, the dissociation is not quite complete and 1.0 N HCl gives a pH of 0.10.)

A *weak acid* is one in which only a small fraction of the molecules are dissociated. An example is acetic acid, in which the 1N acid gives a pH of 2.37, because only 0.42% of the molecules are dissociated.

$$[H^+] \text{ in 1N acetic acid} = \frac{0.42}{100} \times 1 = .0042, \text{ or } 10^{-2.37}$$

Thus, pH = 2.37, because -2.37 is the log of 0.0042.

Although 1 N HCl is more than 200 times more acid than 1N acetic acid, the same volumes of alkali would be required to neutralize equal volumes of these solutions. This is true because as each small amount of alkali combines with the H^+ in solutions of acetic acid, some of the previously undissociated acetic acid dissociates, to maintain 0.42% of the total in the dissociated form. Thus, there is a continuous supply of H^+ ions until all the potential H^+ ions in the acetic acid have combined with the alkali.

Buffers *Buffers,* combinations of substances that tend to maintain a stable pH, exist in plant cells. A weak acid has nearly all its titratable hydrogen ions in the form of *potential* H^+ ions. Thus, when mixed with its salt, it maintains a nearly constant pH even though large amounts of H^+ or OH^- ions are added. Such a mixture is called a buffer because of its ability to prevent pH changes. For a weak acid HA (where A is any anion, such as acetate)

$$\frac{[H^+] \times [A^-]}{[HA]} = K \text{ (dissociation constant)}$$

At equilibrium there is a constant ratio of undissociated molecules and ions; for a weak acid, it might be $9HA : 1H^+ : 1A^-$. Addition of a foreign base disturbs the equilibrium by adding OH^-, which combines with H^+. The equilibrium ratio is immediately restored by dissociation of some of the large reserve of HA, as a result of which the $[H^+]$ is nearly unaltered.

In order to prevent pH changes upon the addition of H^+, the buffer must also have a reserve of A^-. This is achieved by the addition of the salt of a weak acid, which is highly ionized. For example, for the salt NaA, the ratio might be $1NaA : 9Na^+ : 9A^-$. If the acid and salt are mixed, a new equilibrium will be produced, but the content of HA (which provides H^+ to combine with OH^-) and A^- (which combines with H^+) will remain high. Equal quantities of acid and salt will hold the pH constant. For example, if 1 drop of concentrated HCl is added to 1 liter of water the pH is lowered by 3.7 units, but if 1 drop of concentrated HCl is added to 1 liter of buffer of 0.1N weak acid and its salt, the added HCl has a negligible effect on the pH. Cell vacuoles are buffered in this way by organic acids, for example, citric acid, malic acid, and their salts.

Buffers in Protoplasm In protoplasm, proteins and amino acids are the major buffers, because amino acids can have either a net positive or negative charge, depending on the pH. If H^+ is added, it will combine with carboxyl groups of the molecule, and if OH^- is added, it will combine with H^+ released from amino groups of the molecule. We can represent this with a general formula for all amino acids, where R is the chemical grouping that specifies any particular amino acid (see box, p. 27).

Dictyosomes (The Golgi apparatus)

The *dictyosome*, or *Golgi apparatus*, is a large disc-shaped organelle composed of closely stacked membrane-bounded *cisternae* involved in secretory processes (Fig. 2-23). Associated with the cisternae are dense *vesicles* and larger vacuoles which are thought to arise by a process of "budding-off" from the cisternae. Since the dictyosome can be seen only in electron micrographs, and

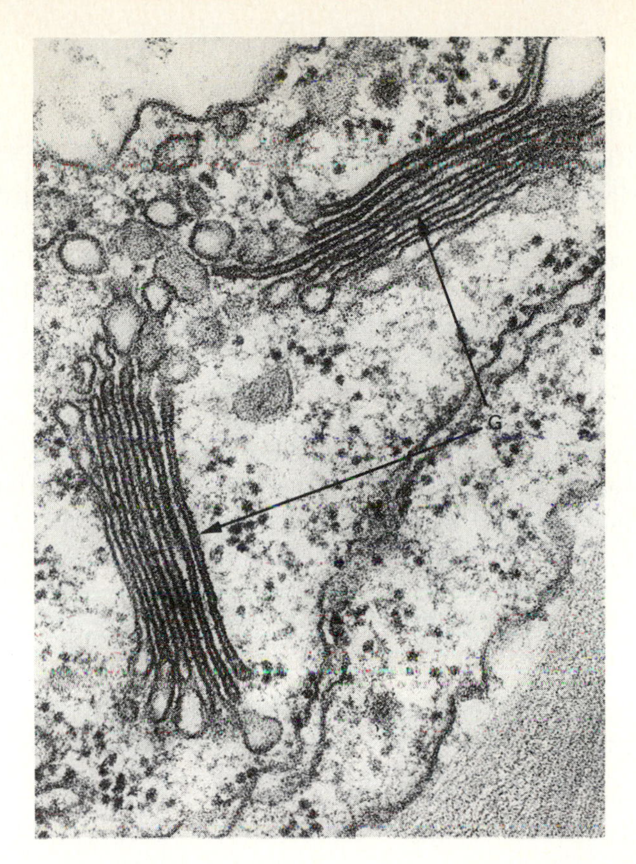

Figure 2-23 Golgi bodies (G) in *Chamaedora oblogonta*, seen with aid of the electron microscope. (Courtesy of M. V. Parthasarathy, Cornell University).

since cells must be killed prior to examination by electron microscopy, relative movement of dictyosomal parts cannot be seen directly. However, sequences of electron micrographs seem to indicate that chemical synthesis of wall polysaccharides occurs in the cisternae, and that vesicles transport the synthesized material to the periphery of the cell. Here the vesicle fuses with the plasma membrane, and the synthesized material is deposited outside the protoplast. New cell wall material is believed to follow this path from synthetic center to place of deposition. As discussed below, the cell wall also contains some protein whose production and secretion involves cooperation between rough ER and dictyosomes.

Each flattened, membrane-bounded cisterna is about 15 nm and each intracisternal space about 20–30 nm thick. The thickness of the total assembly is variable, depending on the number of cisternal plates in the complex. Some cell biologists believe that there is a continuous, ever-changing network of membranes all the way from the nuclear membrane, through the endoplasmic reticulum and dictyosomes, to the plasmalemma. The "membrane network" proposed by this still-controversial theory would provide a mechanism for rapid communication between nucleus and exterior of the cell.

Lysosomes, Peroxisomes and Glyoxysomes

Many cells, including some of higher plants, include small membrane-bounded spherical bodies containing concentrations of particular enzymes or groups of enzymes. The number, distribution, and enzymatic activity of such particulates change greatly, depending on the conditions to which the cell has been exposed. *Lysosomes* generally contain enzymes engaged in *lytic* (breakdown) activity; examples are *proteases*, which break down proteins, *nucleases*, which break down DNA and RNA, and *lipases*, which digest fats. As long as the lysosome is intact, these enzymes cannot reach the cell's contents, but following injury, lysosomes may rupture and release their enzymes. The subsequent biochemical reactions may be deleterious or may be used to combat invading pathogens such as viruses, bacteria, or fungi.

Peroxisomes and *glyoxysomes* contain concentrations of enzymes involved in biochemical pathways not generally considered as central to the cell's metabolism, but important under certain conditions. Peroxisomes contain both peroxide-generating and peroxide-utilizing systems, as well as enzymes that oxidize 2-carbon acids produced during photosynthesis (*photorespiration*) and by other reactions to be discussed later. The physiological role of these bodies is not well understood. Glyoxysomes contain the enzymes of the glyoxylate cycle (see Chapter 5), which is involved in the breakdown of lipids.

The Cell Wall

Plant cells differ from all other cells in having a fairly rigid outer case that resembles a box. Indeed, it is no great exaggeration to say that "the plant cell lives in a wooden box," for the chemical components of the cell wall include those that give wood its rigidity and high tensile strength. The wall of a mature cell is probably a nonliving secretion product of the protoplast, formed in successive layers during the cell's developmental stages (Fig. 2-24). Nonetheless, there is some protein, even enzymatically active protein, in the cell wall. The first layer formed after cell division is the *middle lamella*, at first composed largely of jellylike pectin compounds, but later infiltrated with tougher cellulose, polysaccharides of various types, and in woody tissue, even with lignin.

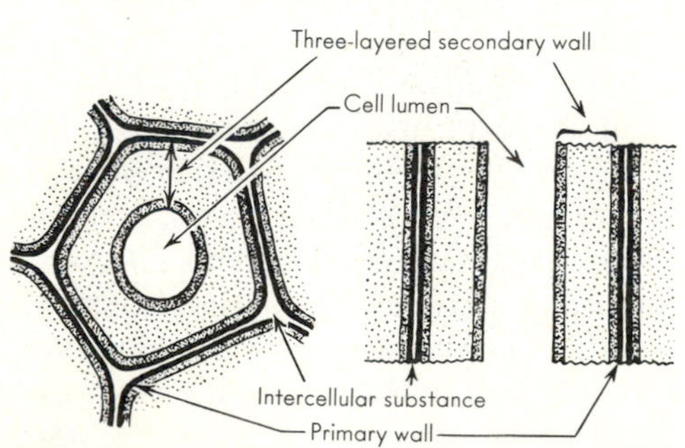

Figure 2-24 The various layers in the plant cell wall. (From K. Esau, *Plant Anatomy*, 2nd edition, New York: John Wiley and Sons, 1965.)

The next layer, the *primary wall*, is laid down as the cell grows in volume. During growth of the primary wall, the cell contents exert great pressure against the wall, which both stretches and incorporates new materials to accommodate the growth. Some of this expansion is reversible (elastic) and some is irreversible (plastic). The primary wall thickens as well as elongates, by incorporating new components into the framework of the existing wall (*intussusception*) or by adding new layers alongside the older layers (*apposition*). In some cells, such as mesophyll, wall development ceases after the cell has attained its maximum size. In other tissues, a rigid *secondary wall* is deposited on the inner side of primary cell walls that have ceased enlarging. Since the cell is no longer growing, the thickening of the wall decreases the volume of the protoplast, and in fibers and tracheids, the wall may become so thick that it occupies almost the entire volume of the cell. If this happens, the protoplast dies, leaving the now-hollow wall cylinders to serve in support or conduction.

The secondary wall is perforated by numerous *pits*, thin areas in which no additional wall deposition occurs on top of the middle lamella and primary wall (Fig. 2-25). In mature, living cells, such as parenchyma cells, these pits

Figure 2-25a An electron micrograph of a simple pit in the primary wall of a parenchymatous cell. Note the randomly oriented cellulose microfibrils making up the wall and their changed orientation near the periphery of the pit. (Courtesy of James Cronshaw, U. of California, Santa Barbara.) The photo lower right shows microfibrils at higher magnification. (Courtesy of H. W. Israel, Cornell University.)

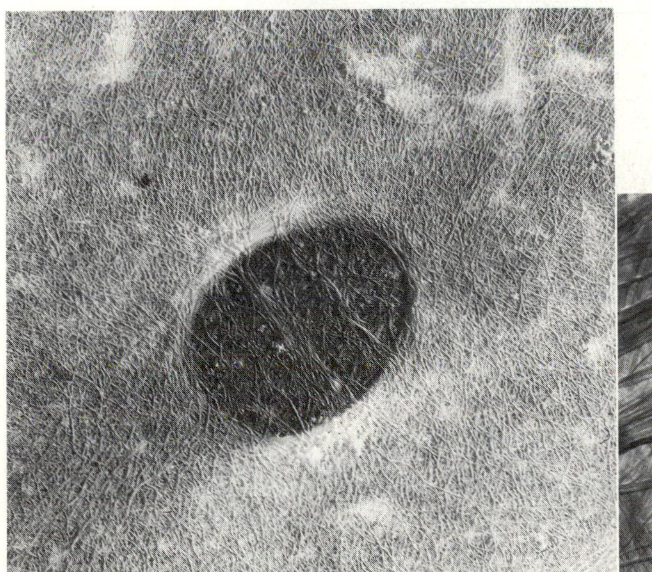

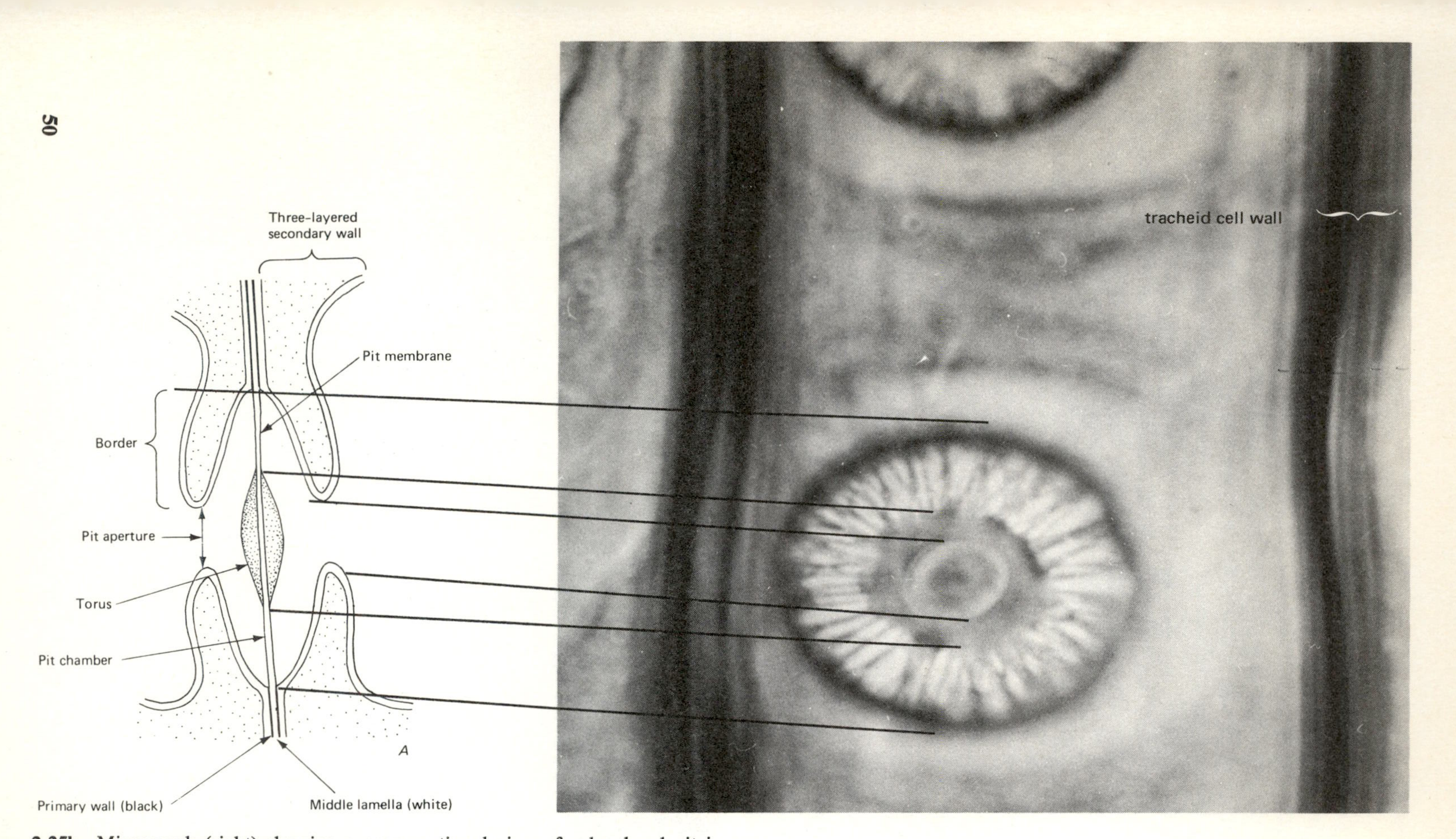

2-25b Micrograph (right) showing a cross-sectional view of a bordered pit in a tracheid of *Picea sitchensis* (Sitka spruce). (Courtesy of G. Berlyn, Yale University). The diagram at the left shows the pit in longitudinal view. The pit membrane consists of two primary walls and the intercellular lamella but is thinner than the same triple structure in the unpitted part of the wall. The torus is formed by thickenings of the primary wall. Note that the outer and inner layers of the secondary wall are connected into a rim about the pit aperture. (Adapted from K. Esau, *Plant Anatomy*, first edition, New York: John Wiley and Sons, 1953.)

are simple cylinders of open space that run from the innermost portion of the secondary wall to the outermost portion of the primary wall. They are frequently found grouped in thin parts of the primary wall, called *pit fields*. In conifer cells that are dead at maturity, such as tracheids, the cylindrical area is covered by a flange of secondary wall material called a *border*, and the intervening primary wall is thickened to form a lens-shaped *torus*. Such *bordered pits* may constitute pressure valves within the conducting system of trees, since the torus can be displaced by a pressure gradient, fitting snugly against the border and sealing off that cell.

The cell wall, both secondary and primary, is also perforated by holes through which pass strands of cytoplasm, the *plasmodesmata*. These strands bind the protoplasts of neighboring cells together into one large community, the *symplast*. As discussed below, the symplast is an important pathway of transport.

Cellulose fibers form the framework of both primary and secondary walls. Cellulose is a giant polymer composed of bundles of *glucan* chains, each of which is in turn a polymer of the 6-carbon sugar glucose (Fig. 5-3). Cellulose microfibrils in the primary wall are about 4 nm in diameter, while those in the secondary wall are almost six times as thick. Compounds surrounding the cellulose fibrils cement them together (Fig. 2-26); these include *hemicelluloses* (composed of long chains of the 5-carbon sugars xylose and arabinose with side chains containing other sugars), *glycoproteins*, and *pectins*. Pectins, polymers of sugarlike units, form either gels or viscous solutions with water; since these changes are reversible with changes in temperature and other environmental conditions, they can have an important effect on the texture of the wall. The major rigid wall component is *lignin*, the typical component of wood. Made by polymerization of oxidized subunits of typical aromatic plant alcohols (coniferyl, sinapyl, coumaroyl), lignin is chemically resistant and adds considerably to the wall's rigidity and strength. Typically, the protoplasts of lignin-containing cells die, leaving the hollow wall tubes as residues. Corky cell walls are impregnated with waterproof *suberin*.

Materials for the construction of cell walls, including polysaccharides composed of thousands of sugar molecules, as well as proteins, are synthesized in the cytoplasm and transported to the wall. The question of how such giant molecules could move through the plasmalemma to the wall puzzled plant physiologists for many years. Recent studies indicate that the rough endoplasmic reticulum and the dictyosome form a functionally integrated membrane system (Fig. 2-27) involved in wall synthesis. Polypeptides of cell wall proteins are synthesized on the rough ER, as are other proteins; however, wall polypeptides are then segregated by release into the lumen of the ER. They then move through the lumen to the dictyosome, undergoing structural modifications in the process. Both the proteins and the cell wall polysaccharides synthesized in the dictyosome are packaged into its secretory vesicles. These vesicles migrate to the plasmalemma, fuse with it, then open toward

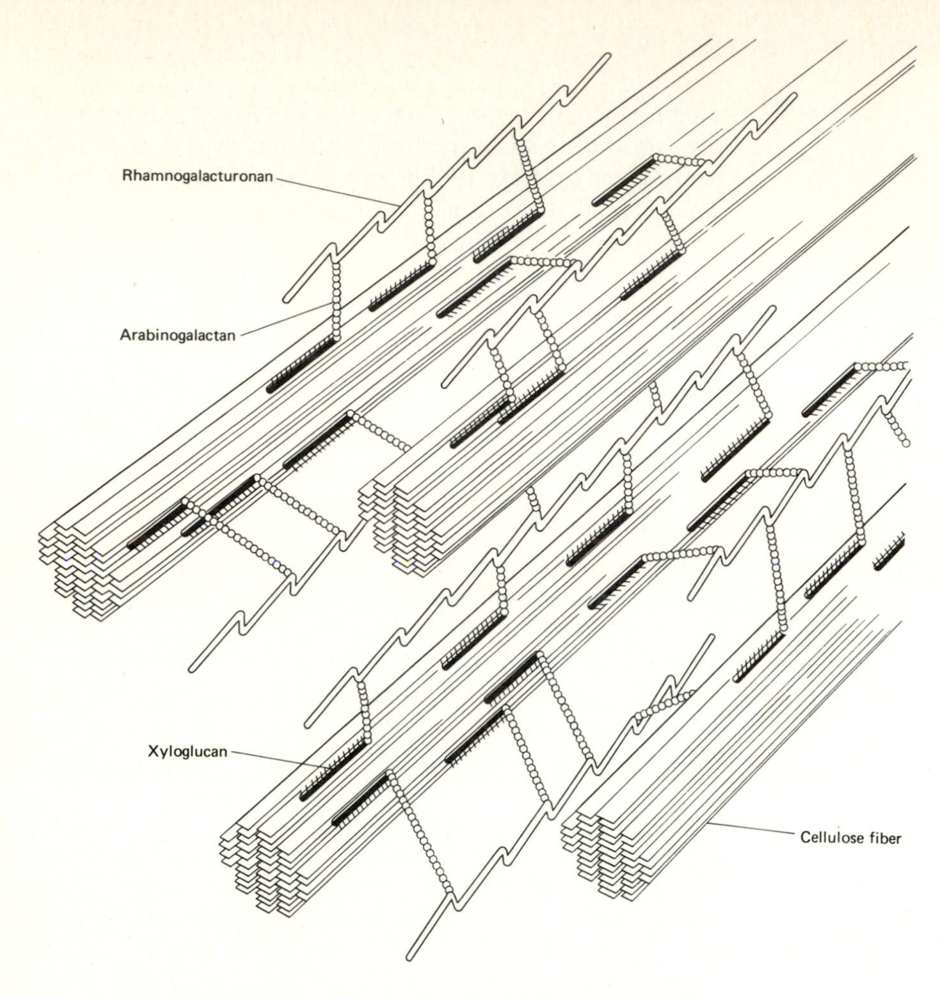

Figure 2-26 This model of the cell wall depicts cellulose fibers held together by three other polysaccharides: xyloglucan, arabinogalactan and rhamnogalacturonan. As a result of extensive cross-linking by these polysaccharides, the cellulose fibers are held together in a relatively rigid matrix. Glycoproteins and pectins attached to the polysaccharides aid in binding these structural units. (Adapted from P. Albersheim. 1975. *Sci. Amer. 232*: 80–95.)

the exterior, releasing their contents to the cell wall. This process, called *reverse pinocytosis*, can be visualized with the aid of Fig. 2-27. (Ordinary pinocytosis involves the *uptake* of particles *into* cells by an engulfing action of the plasma membrane.)

Hormonal control of cell enlargement and wall synthesis

The growth hormone called *auxin*, known chemically as indole-3-acetic acid (IAA), plays an important role in promoting cell enlargement. The growth of many immature tissues, whether on the plant or surgically excised,

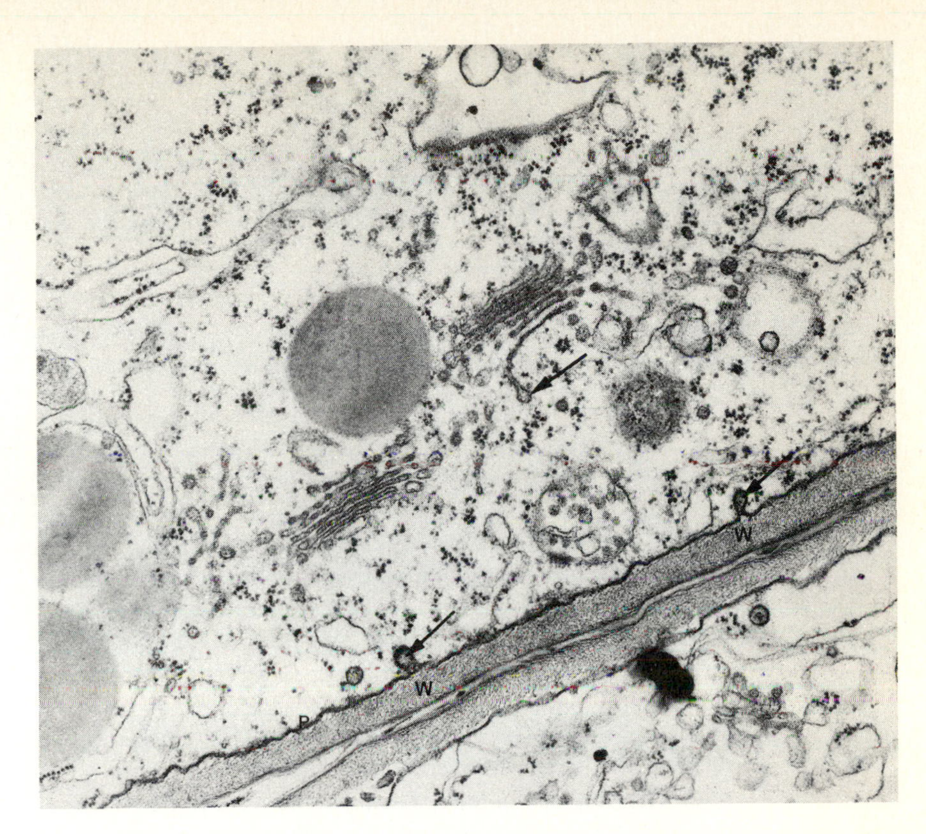

Figure 2-27 Participation of the Golgi apparatus in formation of the cell wall in cotton fibers. The arrows show a vesicle budding from the Golgi apparatus and a vesicle recently merged with the plasmalemma (P) emptying its contents into the wall (W). (From Ramsey and Berlin. 1976. Amer. J. Bot. *63*: 872–876.)

can frequently be increased six- to eight-fold by the addition of IAA. The mechanism by which IAA promotes cell elongation has long been a mystery, but recent studies have focused on the fact that part of the effect of IAA can be duplicated by incubating the tissue at low pH. IAA may therefore activate some acidification mechanism, such as a H^+ ion pump in the plasmalemma. By moving H^+ ions from the cellular interior to the wall region, such a pump could produce a locally lowered pH, which in turn might stimulate enzymes that promote cell wall loosening by breaking bonds between molecules in the wall. For osmotically driven cell extension to occur, an increase in water uptake must follow, accompanied by an increase in cytoplasmic synthesis of wall material, its transport through the plasmalemma, and incorporation into the wall. Thus, increase in cell turgor and biochemical changes in the cell wall are both necessary for permanent cell enlargement and growth (see Chapter 9).

The wall, because of its rigidity, gives a certain form and minimum size to the plant cell and thus serves as a kind of skeleton for the plant. In cells with heavily lignified secondary walls, such as those in wood, this skeleton can maintain the size and shape of the cell in the absence of any other supporting forces. In thin-walled cells, such as those in the leaf, the walls are too pliable to hold their shape without the support of the contents of the cell. This support comes mainly from pressure generated by the vacuole, as is further discussed in Chapter 6.

Plasmodesmata

We have seen that the plasmalemma effectively separates the protoplast of a cell from the wall which surrounds it. Yet each protoplast is not an island, completely separated by membrane boundaries from other protoplasts in the tissue; the protoplasts of neighboring cells must communicate rapidly and effectively with each other, since growth is a highly integrated process. Rapid communication is facilitated by numerous *plasmodesmata* (singular, *plasmodesma*), cytoplasmic bridges that penetrate the plasmalemma and cell wall and connect the cytoplasm of a cell with that of adjacent cells (Fig. 2-28). Plasmodesmata are narrow tubelike structures which range from 20 to 100 nm in width. Since structures smaller than about 500 nm cannot be resolved by the light microscope, our knowledge of the structure of plasmodesmata has been obtained only recently, with the development of appropriate techniques of electron microscopy. Each plasmodesma appears to contain a channel, called a *desmotubule*, through which molecules can move from one cell to another. Some biologists believe that the desmotubule is a modified form of the endoplasmic reticulum, since it appears to be continuous with endoplasmic reticulum in the cells it connects, but others believe that it more closely resembles structures called *microtubules*, described later in this chapter.

Plasmodesmata connect the protoplasts of most cells in higher plants. Their frequency is highly variable, even in different walls of the same cell, ranging from 100,000 to 50 million per mm^2 of cell surface. It is not surprising to find that they are largest in diameter as well as most numerous between cells that interchange large quantities of metabolites. Where the concentration of plasmodesmata is extremely high, the surrounding wall is often very thin, a formation called a *pit field*.

Cytoplasmic Movements

With so many discrete organelles, each bounded by a differentially permeable membrane and separated from neighboring organelles by relatively large distances, how does the cell accomplish the necessary interchange of materials between the organelles? Part of the answer lies in diffusion, and part in *cyclosis*, a fairly rapid movement of the contents of many plant cells. In this process, the entire cytoplasm rotates around the inner surface of the cell wall, either clockwise or counterclockwise, carrying the various organelles along

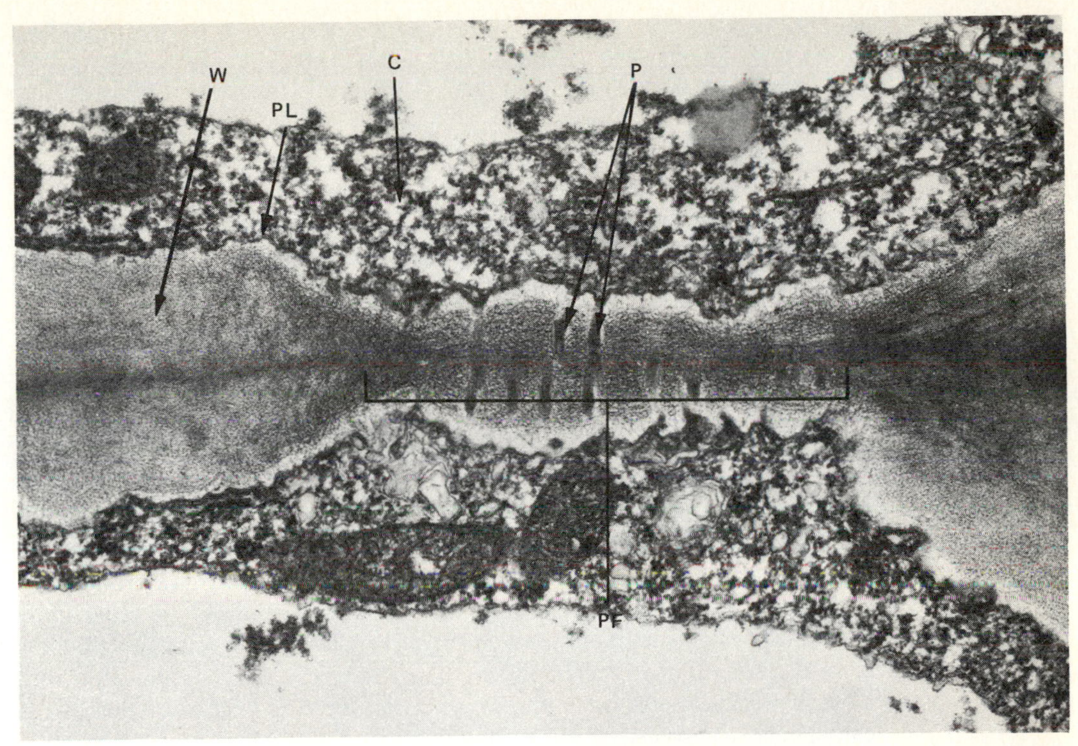

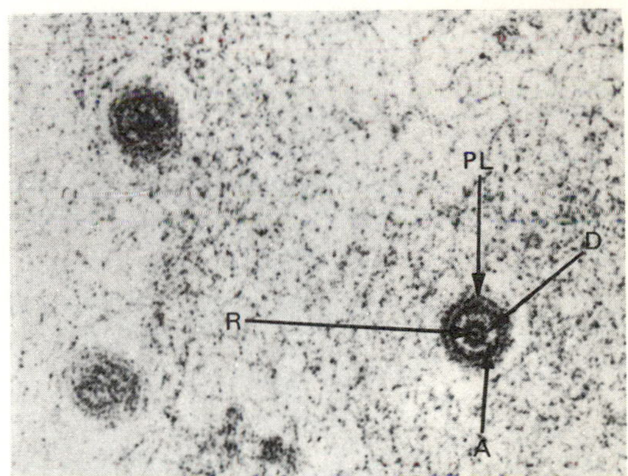

Figure 2-28 Electron micrographs of two cells in the *Samanea* pulvinus and their common wall (W). The thin region of the wall, called a pit field (PF) contains several plasmodesmata (P), seen here in longitudinal view C = cytoplasm; PL = plasmalemma (X42,000).

Details of the internal structure of a plasmodesma can be seen in the figure, lower right, which shows a transverse view of a plasmodesma at higher magnification. It is bounded by the plasmalemma (PL) limiting the cytoplasm, and contains an internal channel called a desmotubule (D), and a central rod (R). Note that there are two possible pathways through which solutes might move: the interior of the desmotubule, and the cytoplasmic annulus (A) that lies between the desmotubule and the plasmalemma (X160,000). (Courtesy of M. J. Morse, Yale University.)

with it. Sidecurrents and countercurrents also exist, and in certain instances, as in the stamen hairs of *Tradescantia*, actively streaming cytoplasmic strands extend through the vacuole.

Although the mechanism which produces cytoplasmic streaming is not completely understood, organelles called *microfilaments* are somehow involved. These filaments appear to contain both *actin* and *myosin*, two proteins whose movement relative to one another under the influence of the energy-rich molecule ATP causes muscular contraction in animals. Thus, cyclosis is sensitive to ATP levels in the cell, and will proceed vigorously only under conditions which produce ATP. Cyclosis can be inhibited by agents that change the structure of microfilaments; for example, the drug cytochalasin B, which causes microfilaments to aggregate, inhibits streaming in many plant cells as well as the movement of giant chloroplasts in various algae. (Some chloroplasts can move within the cytoplasm and orient their broad surfaces either parallel or perpendicular to the surface of the leaf, usually in response to altered light intensity; see Chapter 11). Movements inhibited when cells are incubated in cytochalasin B are restored when the drug is washed out of the tissue.

The movement of chromosomes during mitosis and meiosis is similarly controlled by contractile elements called *microtubules*. These elongated hollow elements, several micrometers in length, are only 15–25 nm in diameter, with a "wall" thickness of about 6 nm. They contain a protein, *tubulin*, whose shape can change in response to chemical stimuli such as Ca^{++} ions. Microtubules attach to the *kinetochore* region of chromosomes and help pull them to the opposite poles during cell division. Ciliated cells of algae and motile sex cells (*gametes*) of various plant groups move by virtue of the contractions of microtubules attached to them. In cross section, cilia typically contain an outer ring of 9 pairs of microtubules surrounding an inner core of two.

The plasmalemma and possibly also the tonoplast are constantly moving. The plasmalemma can form "blebs" that surround and engulf external particles or large molecules and in the process of *pinocytosis* transport them into the cytoplasm as small membrane-bounded vesicles. Similarly, materials from inside the cell can be transported to the exterior and excreted by the reverse process.

In the living cell, as throughout nature, form and function are closely correlated. Each detail of the special architecture of every cell organelle is uniquely adapted to the performance of specific functions to be considered in the following chapters.

Overview of Plant Cell Structure

In common with other eukaryotes, the cells of higher plants contain membrane-bounded nuclei, endoplasmic reticulum, dictyosomes, and mitochondria. They have ribosomes free in the cytoplasm, attached to the ER, and in certain organelles. DNA replication, DNA-directed RNA synthesis, and RNA-directed protein synthesis occur as in other cells. Many of the proteins are catalytically active enzymes, while others are important structural components of the cell.

The unique features of the plant cell can be summarized as follows:

1. The numerous *chloroplasts* enable the cell to convert radiant energy into chemical energy, thus rendering the cell autotrophic.

2. The *cell wall* surrounding each protoplast serves as the skeleton of the plant body, encases each protoplast in a rigid framework, and allows high local turgor pressures to build up.

3. The *central vacuole*, which occupies about 90% of the plant cell, not only collects waste products of plant metabolism, but also by virtue of its membrane-limited content of solutes, facilitates the osmotic uptake of water. This water is absorbed without the expenditure of energy by the plant, since the diffusion of the water molecules themselves causes their net movement into the vacuole.

4. *Plasmodesmata*, narrow channels through which small molecules can diffuse from cell to cell without crossing cell membranes, provide protoplasmic continuity throughout most of the plant body. The width and frequency of these structures varies from one cell wall to another, and these variations may play an important role in regulating chemical communication among cells of the plant body.

SUMMARY

Almost all large organisms are subdivided into microscopic units called *cells*, containing still smaller units called *organelles*. To study the complex structure of cells, biologists have learned to immobilize or fix them in a chemical preservative, embed them in a supporting matrix of paraffin or plastic, cut them into fine slices on a microtome, stain them with various dyes to highlight the different features, and then observe them under compound light microscopes yielding about a thousandfold magnification, or electron microscopes yielding about a millionfold magnification. To gain some insight into the chemical role and biological activities of each of the observed structural units, each type must be obtained in pure form in large quantities. This is generally accomplished by disrupting large numbers of cells and then depositing each type of organelle from the *homogenate* as a *pellet* in a tube through progressive, stepwise increase of centrifugal force in a *centrifuge*. The deposited organelles can then be collected and their chemical nature and biochemical activities studied.

Plant cells, generally about 50 μm in diameter, contain a *nucleus* in which most of the cell's hereditary information is stored in the form of *deoxyribose nucleic acid* (DNA) contained in the rodlike *chromosomes*. Chromosomes divide longitudinally at each cell division (*mitosis*), providing a stable chromosome number and equal qualitative distribution of DNA to each daughter cell. Before sexual reproduction, a special reduction division (*meiosis*) occurs, yielding *haploid* cells with one half the usual *diploid* chromosome number. When the sex cells (*gametes*) fuse in fertilization to yield the *zygote*, the diploid number is restored.

Segments of the DNA, called *genes*, determine the nature of cellular *proteins* by virtue of a specific ordering of their four component *nucleotides*, containing either *adenine* (A), *thymine* (T), *guanine* (G), or *cytosine* (C). Three consecutive nucleotides specify which of the 20 *amino acids* should be introduced into a growing *protein* chain. Proteins are synthesized on the surface of bipartite *ribosomes*, made mainly of *ribonucleic acid* (RNA) and protein, to which particular *messenger RNA* (mRNA) chains are attached. These mRNA units are made from the DNA template in the process of *transcription*, and contain the basic information of the DNA in a new nucleotide language, that of RNA. A third type of RNA, *transfer RNA* (tRNA), picks up individual amino acids and carries them to the ribosome-mRNA combination, where, in the process of *translation*, the amino acid is inserted into the growing polypeptide chain, consisting of amino acids joined by peptide bonds.

Many of the synthesized proteins are *enzymes*, specific catalysts of cellular activities. Enzymes may be entirely protein, or protein attached to smaller *coenzymes*, of which certain *metals* and *vitamins* can form a part. Enzymes can be located in organelles or may be free in the cellular *cytoplasm*.

Plant cells are bounded by a semirigid *wall* made mainly of *cellulose*, but containing also hemicelluloses, jellylike *pectins* gluing cells together, as well as *lignin*, a component of wood, and *suberin*, characteristic of cork.

Inside the wall is the *differentially-permeable plasma membrane*, made of proteins and phospholipids, which encloses all of the cytoplasm. Individual organelles such as *chloroplasts* (sites of photosynthesis) and *mitochondria* (sites of respiration) are also membrane-bounded, and interconnected secretory membranes, the *endoplasmic reticulum*, permeate most parts of the cell. Stacks of membrane-bounded discs, the *Golgi apparatus*, or *dictyosomes*, probably contribute to the origin of *vacuoles*, membrane-bounded bodies containing organic and inorganic solutes. The internal structure of membranes can be seen by the technique of *freeze fracture*, in which frozen cells, struck by a blunt knife, fracture along natural cleavage planes, usually membranes.

Chloroplasts and mitochondria contain their own DNA, which together with nuclear DNA specifies the precise arrays of enzymes equipped to transport electrons between more reducing and more oxidizing *carriers*. In this process, energy is conserved in the form of special bonds in *adenosine triphosphate* (ATP). ATP is made from *adenosine diphosphate* (ADP) and inorganic phosphate (P_i), the extra bond containing the stored energy from electron transport. When ATP breaks down to these two precursor substances, it releases the energy stored in the third phosphate bond. ATP is thus equipped to supply the energy when chemical work must be done in the cell.

SELECTED READINGS

Plant Cell Structure and Function

ALBERSHEIM, P. "The Wall of Growing Plant Cells," *Scientific American* 232 (4): 80–95 (1975).

GUNNING, B. E. S., and A. W. ROBARDS. *Intercellular Communication in Plants: Studies on Plasmodesmata* (Berlin–Heidelberg–New York: Springer-Verlag, 1976).

HALL, J. L., T. J. FLOWERS, and R. M. ROBERTS. *Plant Cell Structure and Metabolism* (New York: Longmans, 1974).

SMITH, H., ed. *The Molecular Biology of Plant Cells* (Berkeley: University of Calif. Press, 1977).

SWANSON, C. P., and P. L. WEBSTER. *The Cell*, 4th ed. (Englewood Cliffs, N. J.: Prentice-Hall, 1977).

See also selected readings for Chapter 3.

Biochemistry Background

BAKER, J. J., and G. E. ALLEN. *Matter, Energy and Life* (Palo Alto, Calif.: Addison–Wesley, 1965).

GOLDSBY, R. A. *Cells and Energy* (New York: Macmillan, 1967).

LEHNINGER, A. L. *Biochemistry* (New York: Worth Publications, 1970).

STRYER, L. *Biochemistry* (San Francisco: W. H. Freeman, 1975).

QUESTIONS

2-1. Describe the structural similarities between animal and plant cells. In what ways do they differ? What differences in function are related to these differences in cell structure?

2-2. What experimental techniques developed during the twentieth century have had an important impact upon our understanding of cell structure and function? What kinds of information does each of these methods reveal?

2-3. The protoplast and each of the organelles within it are surrounded by membranes. (a) Describe features common to all cell membranes. (b)Describe differences in the membranes of different organelles. How are these differences related to organelle function?

2-4. How do prokaryotes and eukaryotes differ? What is the adaptive value of eukaryotic organization?

2-5. Chloroplasts and mitochondria have internal membrane systems as well as outer envelope membranes. What are the functions of the internal membranes?

2-6. Certain antibiotics such as valinomycin and gramicidin alter membrane function. How might this alteration be responsible for their effectiveness as antibiotics?

2-7. Is the presence of an enzyme in an extract of a plant conclusive proof that such an enzyme is active in the plant? Is the absence of a given enzymatic activity in a homogenate of a plant tissue conclusive proof this enzyme is absent from the plant? Why?

2-8. In what way do enzymes speed chemical reactions? How is it possible for an enzyme to effect the decomposition of many molecules of its substrate without itself being substantially altered?

2-9. Certain enzymes are much less labile to heat and pH alterations when in the presence of their substrate. How can one explain this apparent protective action of substrate?

2-10. The economically important phloem fibers of the flax plant are best prepared free of impurities by permitting the stems to "ret," that is, to lie submerged in water until most of the nonfibrous material has disappeared. Could enzymes be involved in this retting? How could one test this hypothesis?

2-11. Mutant pea plants deficient in carotenoid pigments have low rates of photosynthesis. What are two possible reasons for this effect?

2-12. Fungi heterotrophic on green plants are usually a rich source of extracellular cellulase (an enzyme that decomposes cellulose). Would you expect this? Why?

2-13. Plasmodesmata have been described as "structures which elevate a plant from a mere collection of individual cells to an interconnected commune of living protoplasts." What is the basis for this statement?

2-14. The drug cytochalasin B causes microfilaments to aggregrate. How would you expect application of this drug to plant tissue to affect plant growth? Explain.

2-15. What property does pH measure, and what, numerically, is it a measure of? Why is it important that the pH of the cytoplasm be closely regulated? How is this accomplished?

2-16. What components might a pH 4.0 buffer solution contain? How might each component act to constrain a change in pH?

2-17. Describe the major processes which occur in each of the organelles of a cell of a leaf.

2-18. Describe the structure and function of the cell wall. How is the cell wall synthesized?

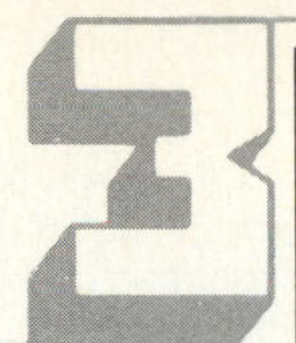

Plant Growth and Form: an Overview

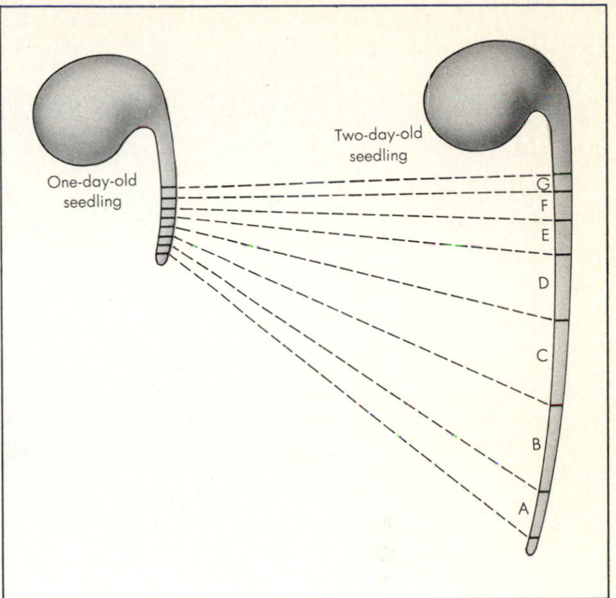

The development of a seed into a mature plant is a remarkable process, involving cell division; growth by cell extension; differentiation of organs such as roots, stems, leaves, and flowers; and a long series of complicated yet well integrated chemical changes. The final form of the plant is a product of both the inherent genetic patterns present in the fertilized egg and the modifying effects of the forces of the environment. While the genes set the ultimate limits between which the plant may vary, environment determines where within these limits the developmental pattern shall be set (Fig. 3-1).

A seed contains an embryo plant, surrounded and protected by a seed coat and supplied with a source of stored food (Fig. 3-2). The plant embryo is a bipolar axis, containing terminal growing points for root and shoot and laterally positioned cotyledon(s). In some instances, the cotyledons are elongate, thin, and leaflike, serving first to digest the stored food of the *endosperm* tissue, and later expanding into green photosynthetic organs. In other instances, they are fleshy storage organs, above or below ground, that absorb the endosperm before maturation of the seed; such cotyledons rarely become leaflike or photosynthetic. Fleshy cotyledons usually fall off after their food reserves have been transported to the growing regions of the seedling.

At the onset of germination, the seed absorbs large quantities of water, and the ensuing chemical changes stimulate the growing points into mitotic activity. For reasons we do not understand, the root almost invariably begins development first; growth activity at the shoot growing point may be delayed by hours, days, or even weeks. At both apices, growth depends on the forma-

Figure 3-1 Influences of heredity and environment of the growth of pea plants. The varieties Alaska, a tall growing type (left and center) and Midget, a dwarf variety (right) are shown here. All of the seedlings are 14 days old. The Alaska seedlings in the center photo and the Midget seedlings were both grown under optimal conditions and are achieving their maximum genetic potentials. The Alaska peas on the left were grown with inadequate light, water and minerals, and are dwarfed for these reasons.

tion of new cells by the *meristematic* (dividing) areas of the growing point, followed by elongation and differentiation of these cells. In the root, the processes of cell division, elongation, and differentiation occur in fairly well defined regions that overlap considerably. Since the root must grow downward through a firm and resistant soil medium, its tender growing point requires protection against abrasion. This is furnished by the *root cap*, a group of several thousand cells wrapped around the meristem. This cap is produced by divisions of an adjacent cap meristem; its cells are continuously flaked off and replaced. In the root of corn, one cap meristem can produce about 10,000 cells daily, rapidly enough to completely replace the cap each day. The disintegrating root cap cells liberate a mucilage that lubricates the root, easing its downward passage through the soil.

One marked difference between plants and animals is that plants grow in restricted and localized areas near the meristems, whereas animals tend to

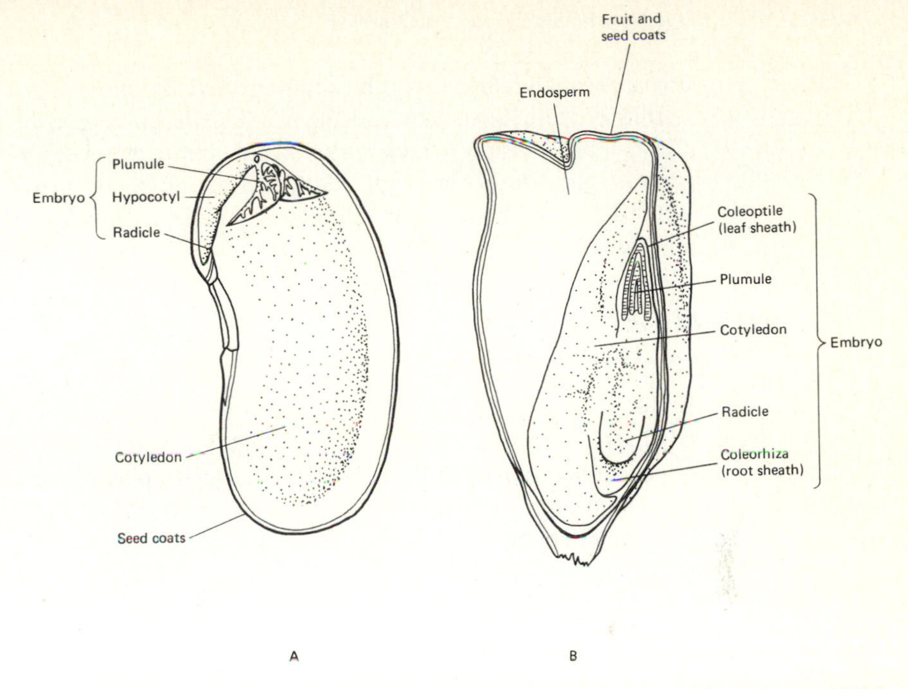

Figure 3-2 Longitudinal sections of seeds of common bean (left) and corn (right). Both seeds contain a miniature plant (embryo), sufficient stored food to get the plant started (endosperm or cotyledon) and a protective layer (seed coat). Bean is dicotyledonous (the seed contains two cotyledons) while corn is monocotyledonous (the seed contains a single cotyledon).

have growth zones distributed all over the body. The restricted growth in plants can be visualized by the simple technique of marking the surface of the root or stem with equidistant lines of some nontoxic substance, such as charcoal in lanolin paste (Fig. 3-3). After some days, the lines will have moved

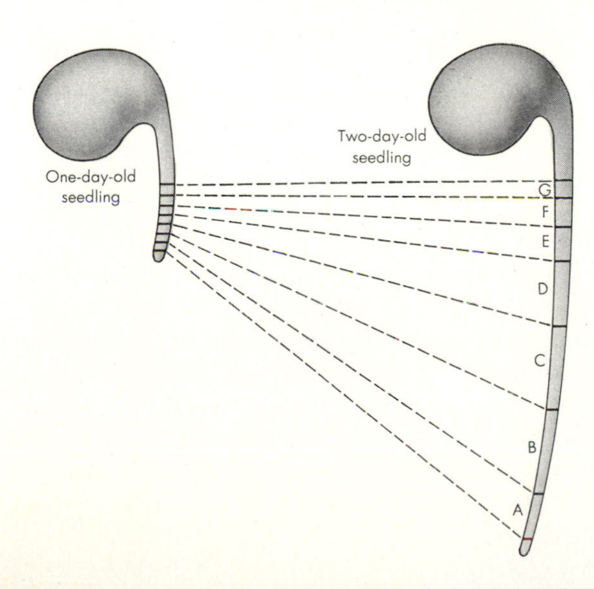

Figure 3-3 Detection of zones of elongation by the parallel-line marking technique. Zones B, C, and D immediately behind the apex have elongated the most.

apart unequally, and it will be apparent that the most rapid growth has occurred in the area just behind the tip. This is the region of cell elongation. Clearly, cell division, which occurs at the apical meristem, does not by itself increase the size of the plant body; it does, however, provide new units of potential elongation, which enlarge at some later time when repeated apical divisions have moved the tip some distance forward.

The Kinetics of Growth

If the height or the weight of the plant is plotted as a function of time in days after germination, the curve of Fig. 3-4 results. This S-shaped, or *sigmoid*, curve is typical of the growth of all organs, plants, populations of plants or animals, and even of civilizations of men. It consists of at least four distinct components: (a) an initial lag period, during which internal changes occur that are preparatory to growth; (b) a log phase, during which the logarithm of the growth rate, when plotted against time, yields a straight line; (c) a phase in which growth rate gradually diminishes; and (d) a phase in which the organism reaches maturity, growth ceases, and size is stabilized.

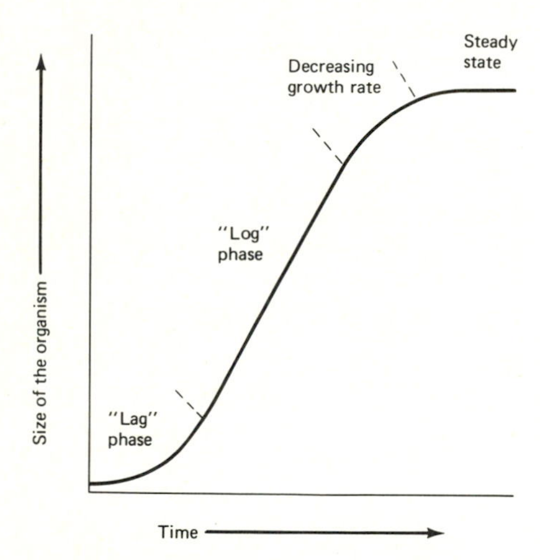

Figure 3-4 The sigmoid growth curve. This curve is characteristic of single cells, tissues, organs, organisms, and populations.

If the curve is prolonged further, a time may arrive when senescence and death of the organism will add several additional components to the overall growth curve.

In certain plants, senescence and death are not a necessary part of the developmental cycle. Some bristlecone pines and *Sequoias* of the western United States can attain ages of well over 3000 years. That they die eventually is probably due to accidents such as lightning damage, the onset of infection, or the weakening of the mechanical base from which they spring.

If such adverse conditions are somehow prevented, growth can continue potentially indefinitely. The techniques of tissue culture, in which parts of plants are excised and grown upon artificial media, have been employed to demonstrate the potential immortality of plant cells. For example, in 1937 several research workers in France showed that portions of the interior of carrot roots, if removed aseptically and transferred to appropriate sterile chemical media, resume cell division and produce a rapidly growing undifferentiated mass of tissue called *callus*. If such a callus is subdivided and transferred to new flasks at frequent intervals, its growth continues at a constant rate and shows no sign of diminishing, even after 40 years. A carrot is normally a biennial plant, so the plant from which the original carrot tissue was taken certainly must have died many years previously. The normal cessation of growth in a plant, usually followed by senescence and death, must then be due to the onset of some inhibition whose removal or circumvention could result in a constantly growing, potentially immortal plant. Research in this area continues to yield interesting and important results.

Growth curves are useful because they provide evidence for the existence of physiological growth controls of various kinds. The length of the lag period, for instance, can be used to analyze the nature of the changes that must occur before the inception of growth. With many seeds the lag period is only a few hours long, indicating that all systems needed for growth are present and need only water for activation; with other seeds, the lag period may be days, weeks, or even months. In such dormant seeds, inhibitory substances are usually present in the tissue, and growth cannot begin until these substances have either been destroyed metabolically or removed by leaching or other mechanical means. The rate of growth during the log phase is often determined by hormonal substances to be discussed in a later chapter. An analysis of the slope of this curve can frequently tell us much about the genetic background of the plant's growth potential, as well as the adequacy of the environment in which the plant is growing. The total height of the plant and the time of the onset of the stationary phase are also frequently genetically controlled, but these characteristics are also susceptible to alteration by the environment. Finally, the onset of senescence and the death of the organism may not be necessary consequences of its genotype, but can be regulated by the environment.

Cell Division, Elongation, and Differentiation

The plant is a community of different cell types formed by cell division and subsequent differentiation. The actively dividing or meristematic cells, localized in regions called *meristems* (Fig. 3-5), retain their ability to divide repeatedly and give rise to new cells throughout the life of the plant. They are found at the apex of the roots and shoots of all plants, and at the base of

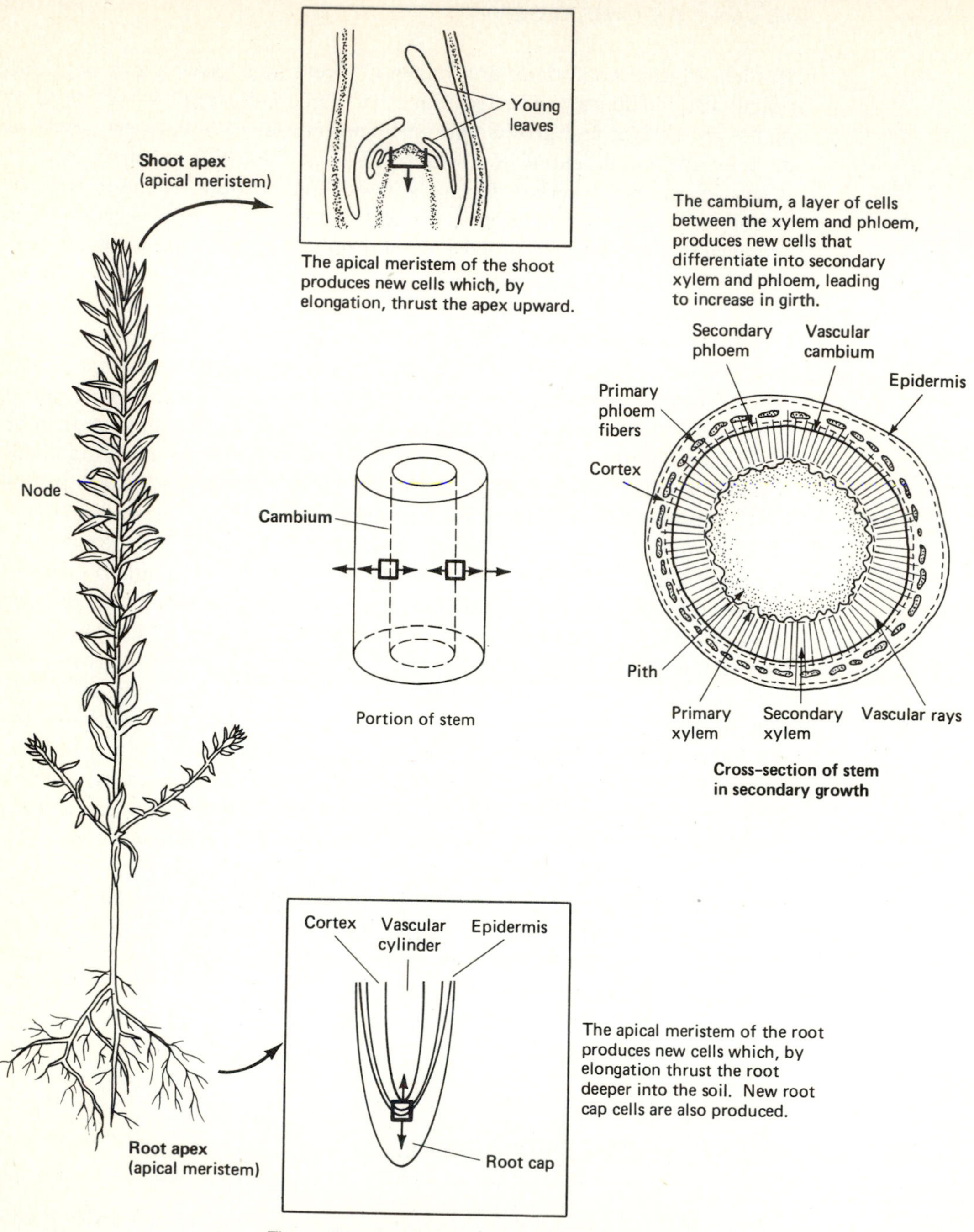

Figure 3-5 The organization of a vascular plant. (Adapted from K. Esau, *Plant Anatomy*, first edition, New York: John Wiley and Sons, 1953.)

leaves or stem internodes of some monocotyledonous plants. Thus, dicot stems grow from the tip, while monocots may grow either from the tip or base of the stem or internode. *Nodes*, regions of the stem at which leaves are attached, may also contain meristematic cells. Axillary buds are found in the angle (*axil*) between leaf and stem, and under appropriate conditions these buds can be activated to produce branches. This development of branches at nodes is regulated hormonally, as discussed in chapters 9 and 10. Increase in girth of dicotyledonous stems and roots is produced by tangential division of meristematic cells in the *cambium*, a one-cell-thick cylinder of tissue that lies between the *xylem* of the wood and the *phloem* of the bark. The cells produced by cambial divisions differentiate into either secondary xylem (inward) or secondary phloem (outward).

In the highly organized context of a plant root or shoot, each cell goes through an orderly series of developmental phases. Meristematic cells are small, approximately cubical in shape, thin-walled, and multivacuolate, and their nucleus is large relative to the rest of the cell (Fig. 3-6). Increase in cell size, especially in length, is mainly a consequence of water uptake into the vacuoles. As the many separate vacuoles expand, they ultimately fuse into one large, central, membrane-limited vacuolar tank. The rest of the cell keeps pace with the enlarging vacuole by stretching of the cell wall, synthesis of wall and cytoplasmic material, and by division and growth of the various types of cell organelles. The major part of extension growth and increase in fresh weight, under hormonal control, is thus accomplished in the area of cell elongation.

Usually together with elongation, but sometimes after it has ceased, *differentiation* occurs. Differentiation is the process by which cells that appear to be similar assume different morphological patterns and physiological roles (Fig. 3-6). Differentiation is one of the most difficult puzzles in biology. If, as we believe, each cell of the multicellular organism has arisen by division of the original 2n cell, or zygote, then each cell should have an exactly equal complement of genes. As we have seen, this is not strictly true, since different cells in the same plant may have different numbers of chromosomes as a result of certain aberrations in the mitotic process. However, there seems to be no qualitative difference in the gene complements of cells that ultimately assume vastly different forms, and entire plants have been regenerated in culture from single cells derived from leaf, stem, and root. Thus, differentiation appears not to change the basic genetic information in the cell. A newly formed cell has certain broad potentials and may develop along any of several morphological and physiological lines, depending on physical, chemical, and spatial influences around it. Once a cell has differentiated along a specific pathway, it ordinarily does not revert to the undifferentiated state or assume a new form, but under conditions of tissue culture, dedifferentiation and redifferentiation may occur. Current thinking centers about the theory that in all cells only a portion of the total genetic potential is expressed, some genes

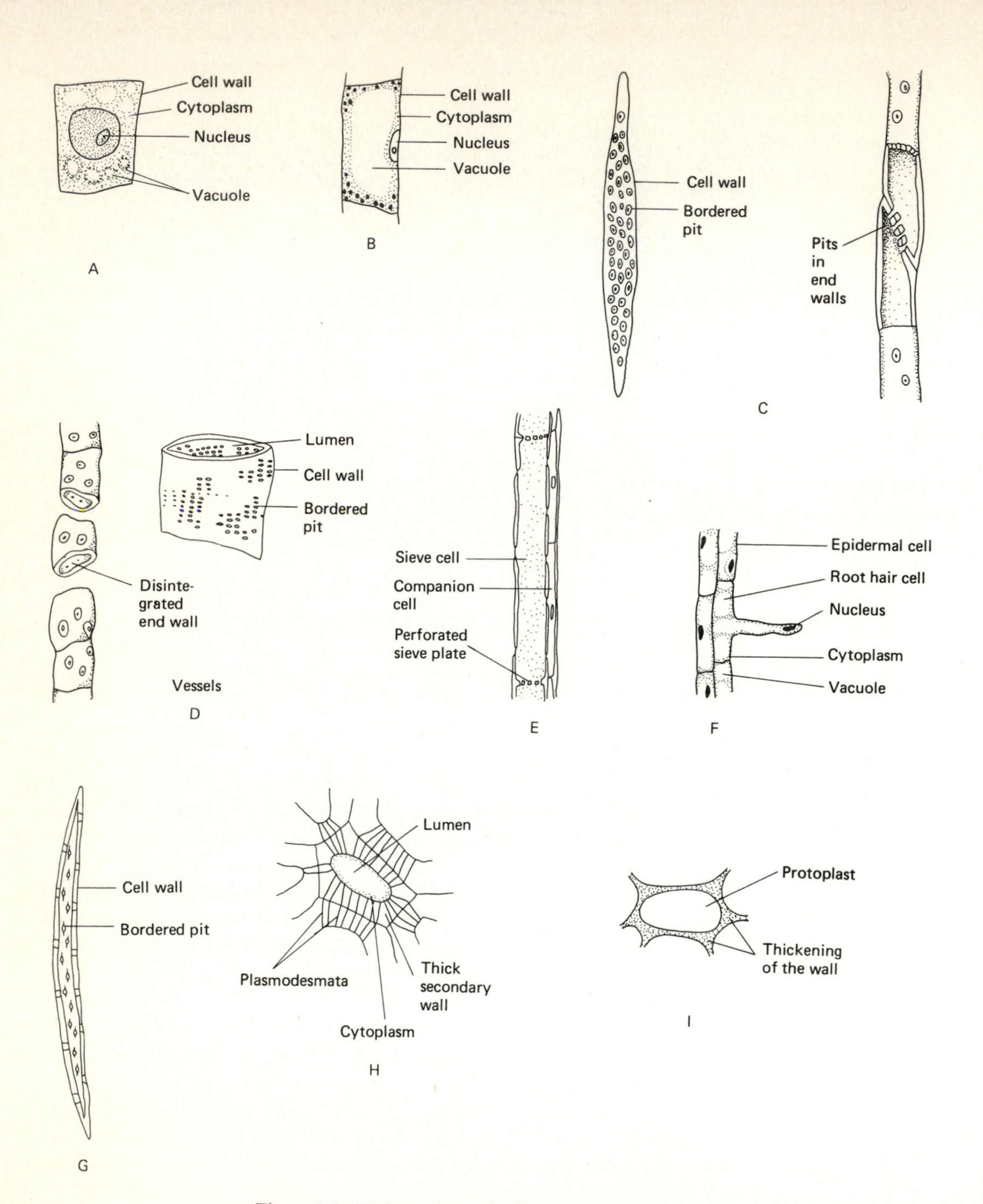

Figure 3-6 Various types of cells in plants (not to scale). (A) Meristematic cell. (B) Parenchyma cell. (C) Two adjacent tracheids showing pits in their end (connecting) walls. (D) Several vessel cells stacked end to end to form a tube. (E) Sieve cell and companion cells. (F) Epidermal cell and root hair cell. (G) Fiber. (H) Stone cell. (I) Collenchyma cell.

being turned "on" and others "off." The nature of a cell thus depends on the complex of genes that is active. The problem then becomes: What turns genes on and off?

The student of the physiology of higher plants can learn how to control certain differentiation phenomena by chemical and physical means. This knowledge, however, is purely empirical; its use is like inserting a key into a door and opening it without knowing anything about the intricate mechanism of the lock. Certainly, future progress in experimental biology will largely depend on some detailed understanding of the phenomena that intervene between the application of the specific agent and the appearance of the altered form.

Tissue Organization:

Roots

The cells on the exterior layer of the young root (*proepidermis*) adopt one of two final forms; they become either flattened *epidermal* cells or *root hair* cells (Fig. 3-6). In some plants, their fate is determined by the cell division that produces them, for although the nuclei of dividing proepidermal cells split equally, the cytoplasm is allocated unequally by cell wall formation. The larger cells produced by this unequal division become ordinary epidermal cells, while the smaller ones ultimately produce root hairs. Root hairs are single epidermal cells with elongated filamentous protuberances (the "hair"), whose great surface area makes them very effective in the absorption of water and minerals. During the rapid growth of such cells, the nucleus is almost invariably found at the tip of the hair, and appears to be the center of great metabolic activity. Root hair cells are relatively short-lived, are present in great numbers, and are rapidly produced as the root tip continues to push its way through the soil. For example, a single rye plant growing in soil in a box about 30 cm × 30 cm × 55 cm was found to produce about 387 *miles* of roots in four months. This figure was obtained by adding up the lengths of all 13 million roots and branch roots as well as some 15 billion root hairs. One can thus easily appreciate how much root hairs do to increase the surface area of root that is in intimate contact with the external medium.

Roots contain no pith, and their central tissues differentiate into the vascular elements (Fig. 3-5). The same pattern occurs even in tissue culture, where it is rare to see tracheids or vessels differentiating on the surface of a mass of callus tissue. Deep within the rapidly growing mass of an undifferentiated tissue culture, however, may be found little whorls of nonfunctional tracheid elements that are morphologically similar to tracheids in functional xylem tissue. We therefore suppose that there is something about the interior of a mass of tissue that favors the xylem type of differentiation. This "something" could be either the suboptimal aerobicity or the nutritional conditions found within a mass of cells or, conversely, a lack of contact with the medium.

The xylem tissue forms the main pathway for the transport of water and minerals. It contains several different types of elongate, dead, conducting cells (Fig. 3-6), in intimate contact with living parenchyma cells. One group of cells, all heavily lignified, includes *tracheids*, *vessels*, and *fibers*, all of which lack living contents at maturity. Vessels and tracheids have a relatively large central space, the *lumen*, through which water is conducted. The lumen is sometimes blocked by inward extension of neighboring parenchyma cells, called *tyloses*, and then can no longer transport water. Fibers resemble tracheids, except that their walls are thicker and their lumen smaller; their major function is to support the organ physically. All these cells contain bordered pits; the vessels alone have large perforations in their end walls. Another type of supporting cell, *collenchyma*, is thickened only at the corners of the cell.

Around root xylem cells are found bundles of *phloem*, as well as the meristematic *pericycle* that gives rise to branch roots, and an *endodermis* that surrounds the entire central vascular cylinder, or *stele*. Unlike stems, whose branches arise from superficial buds at nodes, root branches arise deep within the root tissue in the pericycle layer, and the branch root forces its way to the exterior by growing from the stele through the cortex and epidermis. The phloem is the main pathway through which carbohydrates formed during photosynthesis are transported. Phloem tissue also contains several different types of cells. *Sieve tubes*, the main conducting elements of the phloem, are strings of elongate cells with perforated end plates, through which extend cytoplasmic strands that are continuous with the contents of the tubes themselves. Such sieve cells contain no nuclei at maturity, but the small *companion cells* that generally lie alongside the sieve cell have large nuclei. Numerous phloem parenchyma cells are also in contact with sieve cells.

The *endodermis* is characterized by the *Casparian strip*, a band of waxy, water-impervious material on the radial cell walls. The Casparian strip acts as a barrier that prevents diffusion along the wall between cortex and stele, and forces movement of all materials from the *apoplast* (the spatial continuum external to the plasma membranes) through the differentially permeable membrane of the endodermal cell into the *symplast* (the spatial continuum inside plasma membranes and plasmodesmata). The implications of this change are discussed in chapters 6 and 7.

Between the internal vascular cylinder and the epidermis lies a region of large, thin-walled, loosely packed cells called the *cortex* (Fig. 3-5). These parenchymatous cells, with large nuclei and central vacuoles and abundant leucoplasts, serve mainly as a storage depot for reserve materials such as starch. As a tangentially dividing *cambium* develops between the xylem and the phloem of the central cylinder, and as the root thickens because of the activity of this cambium, the cortex cracks and flakes off at the periphery, the wounds being sealed through the activity of numerous isolated *cork cambia*.

Finally, in an older root, the epidermis and cortex are completely lost, and the new external layer is derived from a now continuous secondary meristem, the *cork cambium*, which gives rise to corky *periderm* tissue at the exterior.

Stems

Cell division and differention in stems resemble, in many ways, corresponding processes in roots. The stem apex of a dicotyledonous plant contains a meristematic zone of cells in rapid division (Fig. 3-7). Behind the cell division area is a region of rapid cell elongation that in turn overlaps and merges with the region of cellular differentiation. In the region of differentiation, one notes the development of epidermal tissues, a central vascular cylinder, and cortical cells between the two. The main anatomical difference between stems and roots is that stems generally have a central *pith* and the primary xylem and phloem are arranged in a different manner. In dicotyledonous plants the xylem surrounds the pith and the phloem surrounds the xylem, while in monocotyledons, vascular bundles containing xylem and phloem may be scattered through the pith–cortex continuum. Although the anatomy of the stem of any plant is genetically determined, the environment can control certain patterns; for example, stems grown in the light generally do not have a readily identifiable endodermis, while stems grown in the dark usually do.

In stems, as in roots, the cambial layer develops between the xylem and

Figure 3-7 Left: the apical meristem of wheat. Right: a median longitudinal section through the apical meristem. The swelling at the left is a floral primordium; the smaller one at the right will form a leaf-like organ, the lemma, which subtends the flower. The larger swellings below are older primordia. (Courtesy of C. Barnard and the Division of Plant Industry, Commonwealth Scientific and Industrial Research Organization, Australia.)

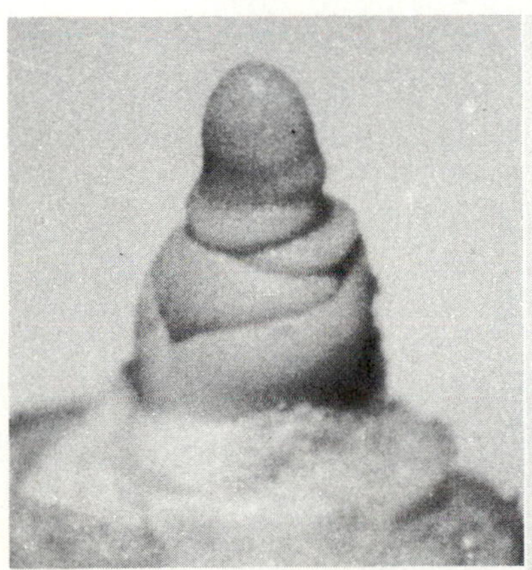

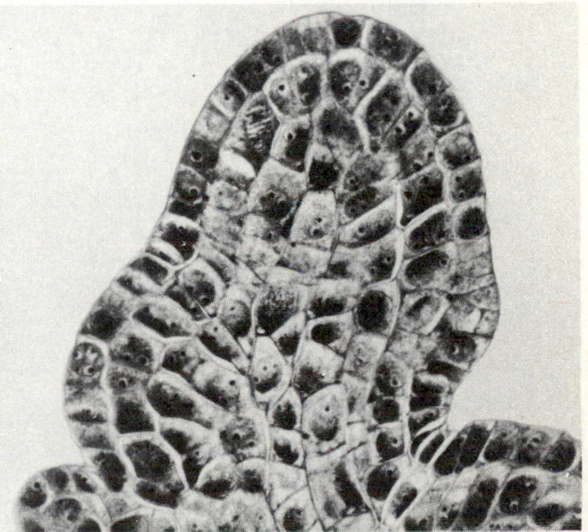

phloem, and, by rapid divisions in both directions, gives rise to cells that differentiate into xylem elements toward the interior and phloem elements toward the exterior. As in roots, the stem thus grows in girth because of an internal expansion. Ultimately, great pressures are created by this internal expansion, and the external layers crack and flake off. As they do so, the plant produces new protective cells under the lost areas. Again, it is a cork cambium that appears, and the cells produced by this cambium have thick walls containing *suberin*, the waterproof material typical of corky tissue. With continued cambial growth and an increase in girth, the successive layers of cork, now part of the bark, are split off and replaced in turn by new cork cells produced from existing or newly formed cork cambium layers. Commercial cork, obtained from the trunk of a species of oak tree, *Quercus suber*, is an economically important product of this process.

The annual rings characteristic of the stems of trees result from the different climatic conditions during different periods of the year. In the spring, when the supply of water is ample and other conditions are favorable, the cambium produces cells that are relatively thin-walled and contain a large central lumen. In the summer and fall, when the water supply tends to be less

Figure 3-8 The annual rings in the wood of a tree are caused by the changing size of tracheids formed by the cambium. In the spring, large tracheids are formed. Thus the ring itself is the line between the summer tracheids marking the end of one year and the spring tracheids marking the beginning of the next year.

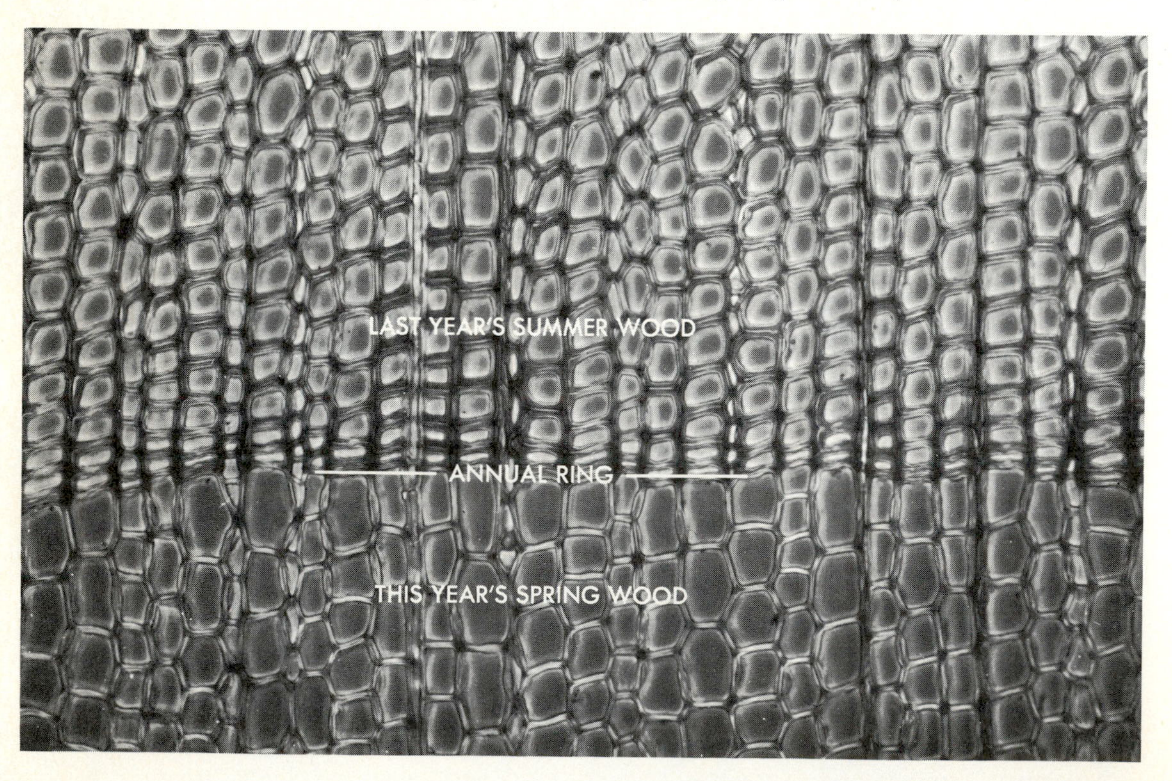

optimal, the newly formed tracheids have thicker walls and a smaller lumen. This regular alternation of spring and summer wood produces the annual ring (Fig. 3-8). Generally, the transition from spring to summer wood in any one year is gradual, whereas the abrupt change from summer wood at the end of one growing season to the spring wood of the following year is abrupt, and provides the line which is counted in assessing the age of a tree.

The regularity of annual rings has enabled us to date trees and also civilizations in which remains of trees have been found. For instance, we know that certain climatic cycles involving variations in rainfall and temperature have occurred in various regions. If a year is especially favorable to growth, a very thick annual ring is produced, while in drought years very small annual rings appear. Thus, the structure of a piece of wood forming the supporting timber of a house in a now extinct civilization could be collated with other materials whose age is known and the civilization dated by this technique. Using this technique, a round table in Winchester Cathedral in England, alleged to have belonged to King Arthur, was found instead to have been made much more recently, in 1538. Although extremely useful, the method is not infallible, because trees under unusual conditions can produce two or more growth rings in one year, and the annual rings of successive years are not always sharply separated. Other techniques, such as radiocarbon dating, can provide additional evidence where needed.

Leaves

At periodic intervals, cells in the stem apex divide to produce clusters of cells that become *leaf primordia*, first visible as small mounds of folded tissue layers. Portions of the meristem are also left behind in the *axil* of the leaf (the angle between leaf and stem), giving rise to buds. These can grow to produce stems with yet other leaves, or under appropriate conditions, flowers. The nature of the bud is frequently controlled by temperature and light conditions, as discussed in Chapter 12.

The first buds that form almost always give rise to stems bearing leaves. In some plants, a single bud develops at a node, whereas in others buds develop in groups of two, three, or more. When the buds are laid down singly, successive leaves are often arranged on the stem in a helical pattern; when the buds occur in pairs, the leaves are usually opposite each other. The pattern varies with the type of plant and is determined genetically.

The arrangement of leaves on the stem is called *phyllotaxy*, sometimes expressed as a numerical fraction. Thus, a phyllotaxy of 2/5 indicates that one would encounter five leaves during two spiral excursions around the circumference of the stem, ending exactly above the first leaf counted.

The outermost cells on the upper and lower sides of the leaf primordium differentiate into epidermal cells whose surfaces are coated with a waxy *cuticle* layer (Fig. 3-6). This secreted coating is relatively impermeable to water, and aids in protecting the leaves from desiccation. The epidermis, especially on the lower side of the leaf, contains a large number of specialized cells, the *guard cells*, whose change of shape controls the opening and closing of stomatal pores (see Chapter 4). Stomatal opening regulates the exchange of water, oxygen, carbon dioxide, and other gases between the interior of the leaf and the atmosphere. Between the upper and lower epidermis lies the photosynthetic *mesophyll* tissue, sometimes called the *chlorenchyma* because the cells contain numerous chloroplasts. Chlorenchyma may be divided into *palisade* tissue, tightly packed cells whose long axes are perpendicular to the leaf surface, and *spongy* mesophyll tissue, loosely packed cells with intercellular gas spaces that communicate directly with the stomatal pore (Fig. 3-9).

Interspersed among the photosynthetic cells of the leaf are numerous veins, composed of xylem and phloem cells that transport water and minerals

Figure 3-9 A diagrammatic cross section of a leaf. The leaf consists of an upper and lower protective layer, the epidermis, between which are the actively photosynthesizing cells. Closest to the upper epidermis is the palisade layer, and below it is the spongy layer. The epidermal layers are frequently perforated by stomatal openings, the aperture of which is controlled by the turgor pressure of the guard cells surrounding them. Intercellular air spaces serve as passageways for the interchange of water, CO_2, and oxygen gases. The small veins containing xylem and phloem serve to conduct water and minerals into and synthesized sugars out of the leaf. A waxy cuticle, largely waterproof, usually covers both lower and upper epidermal cells.

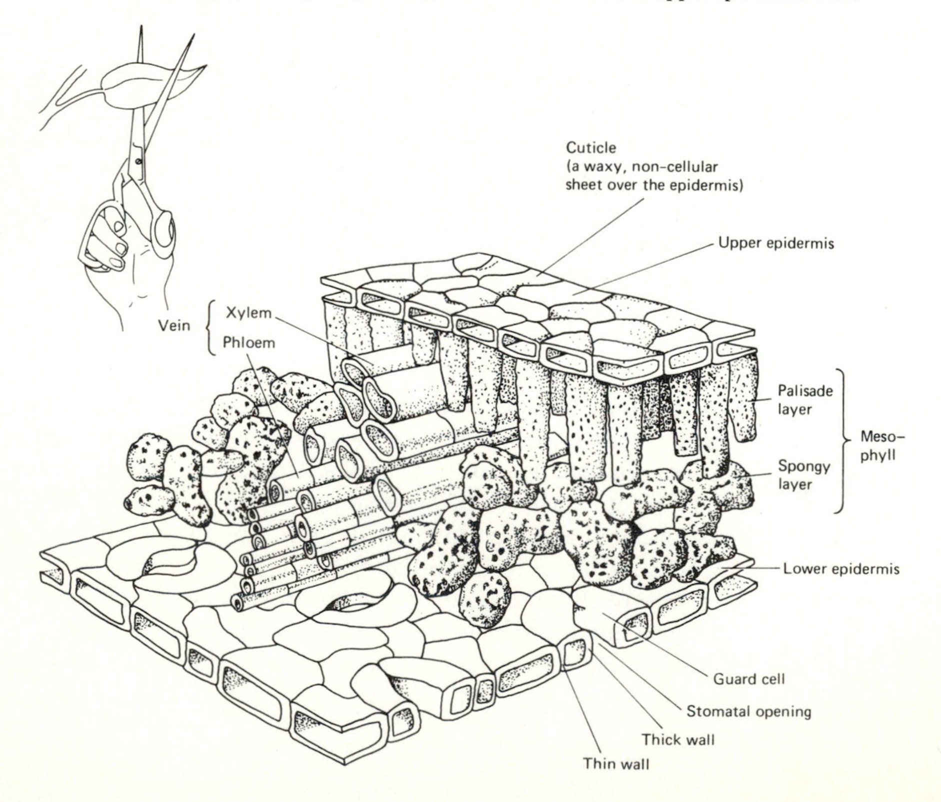

into the leaf and photosynthetic products out of the leaf. These veins are ultimately connected to the leaf midrib, which passes through the petiole and joins the main vascular system of the plant.

The Differentiation of Reproductive Organs

One of the most dramatic examples of the alteration of form occurs when the plant makes a transition from vegetative to reproductive growth. In many angiosperms, roots, stems, and leaves are produced for a long period of time. Then, at some point in its life history, the plant ceases vigorous vegetative growth and begins a series of transformations leading ultimately to the production of reproductive organs. The first morphological differences, noted at the stem apex, involve a change in shape from a narrow cylinder to a flattened dome of tissue. This dome ultimately produces floral primordia, containing *sepals*, *petals*, *stamens*, and *carpels*, instead of leaf primordia. In the stamen and carpel will occur the important process of meiosis, which leads to the production of the haploid cells and the 1n (gametophyte) generation of the plant (Fig. 3-10). In the stamen (the male organ), meiosis produces haploid *microspores* that become the male gametophyte (pollen grain); in the singly- or multiply-carpellate ovary (female organ), meiosis produces a haploid *megaspore* that becomes the female gametophyte (embryo sac). The male and female gametophytes ultimately give rise to sex cells, i.e., the *sperm* nuclei that travel through the pollen tube and fertilize the *egg* in the ovule. With the union of the two sex cells, the 2n (sporophyte) generation is restored, and the life cycle is complete.

Within the last several decades, biologists have accumulated much information about the physiology of reproduction and the nature of the chemical changes that lead to this altered form. We will examine this important process in greater detail in later chapters. At present, let us consider a system that is simpler than the angiosperm, the prothallus of a fern (Fig. 3-11). The fern prothallus is the gametophyte generation, since each of the cells has only one-half the normal chromosome complement of the mature sporophyte fern plant. The haploid generation occupies a large portion of the total life cycle of some algae, as well as mosses and ferns, but lasts for a much shorter time in angiosperms.

The fern gametophyte is formed by repeated mitotic division from a single-celled haploid spore produced by meiosis in sporangia found on leaves. Upon germination, cell division produces either a filamentous structure (in the dark) or a flat plate of cells (in the light). Ultimately, the heart-shaped prothallus consists of several layers of cells bearing *rhizoids*, or rootlike structures, growing downward into the substrate. At some stage in its life cycle, the prothallus will initiate both *antheridia*, the male sex organs, and

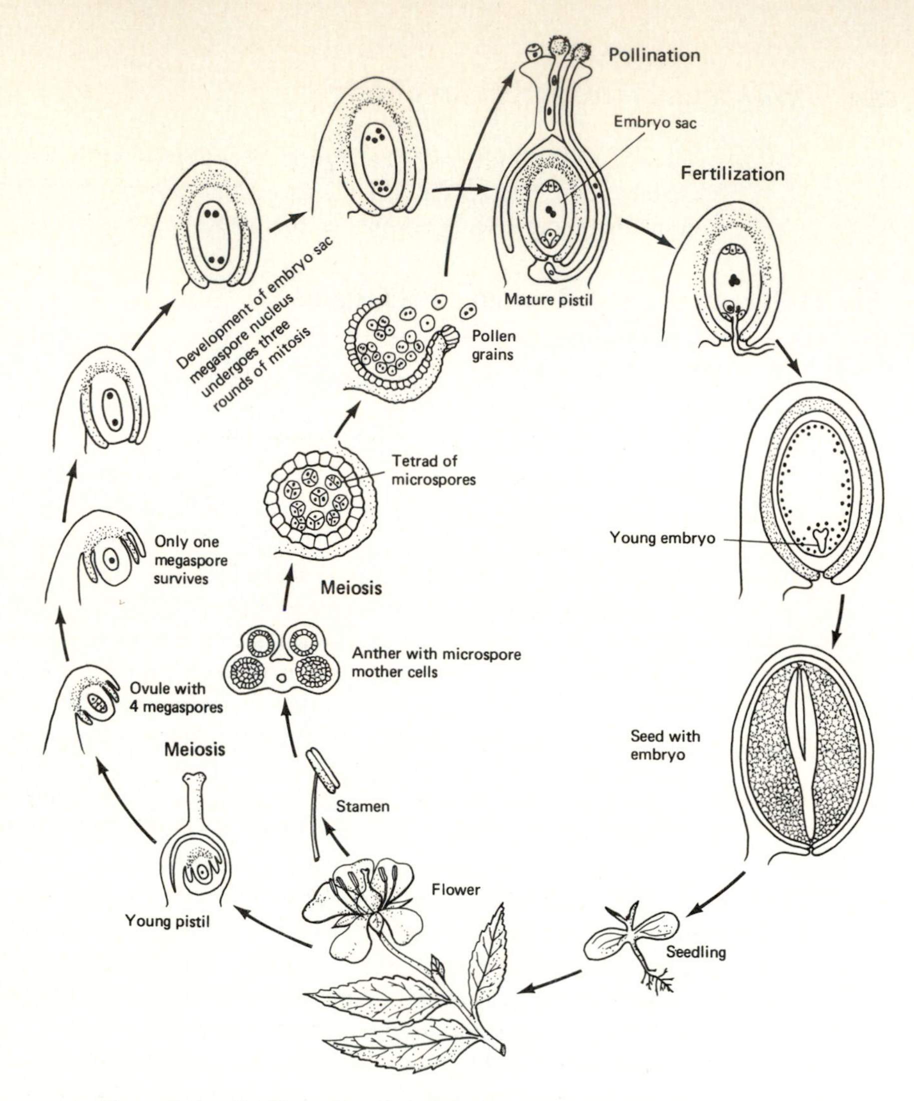

Figure 3-10 The life cycle of an angiosperm.

archegonia, the female sex organs. Within each antheridium will be borne a group of sperms, within each archegonium a single egg.

We may now pose the question: What governs the differentiation of the cells of the prothallus into each of the sex organs? The answer seems to lie in the production of specific antheridium-inducing substances. For example, if spores of the sensitive fern, *Onoclea*, are placed singly in test tubes containing a synthetic medium adequate for growth, the prothallia will develop normally, except that they will never give rise to antheridia. If, however, they are grown in a medium containing many other prothallia, or in the culture filtrate from older prothallia of *Onoclea* or related ferns such as *Pteridium*, antheridia are initiated promptly. Simple dilution experiments show that in the old *Pteridium* medium a substance exists that specifically causes the differentiation of antheridia on the prothallus, without having any marked effect on its growth rate. If large quantities of this material are added to the culture medium, almost every cell of the mature prothallus will give rise to an antheridium.

An interesting fact emerges from a detailed study of this phenomenon; by the time a prothallus has become old enough to produce the antheridium-inducing factor, it is too old to respond to it. Therefore, a single prothallus cannot induce itself to form antheridia. This implies also that the production of antheridia in nature depends on the simultaneous presence of many prothallia of heterogeneous ages. The old ones produce the specific factor that induces antheridial formation in the younger prothallia, but themselves go on to form archegonia directly, without prior antheridial formation. Thus, the successful differentiation of the sex organs, which in turn eventually leads to completion of the sexual reproduction of the plant, is dependent on "together-

Figure 3-11 The life cycle of a fern.

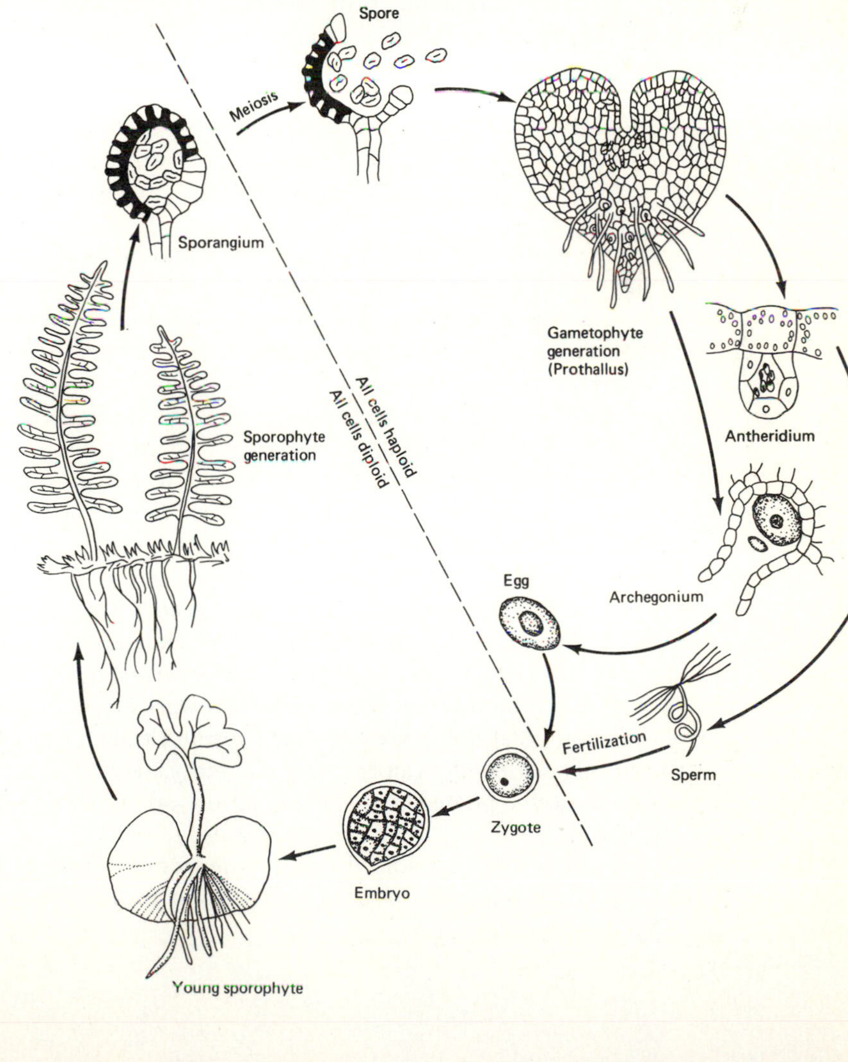

ness." An analysis of various families of ferns reveals that there is some specificity in the production of this substance. Generally, all members of the family *Polypodiaceae* respond to the *Pteridium* factor but do not respond to substances produced by fern prothallia of other families. In the same way, nonpolypodiaceous gametophytes will not respond to the *Pteridium* factor, but will respond instead to substances produced by their own prothallia. Chemical isolation and identification work has shown the effective substances to be related to the *gibberellins*, a class of plant hormones to be discussed later.

There is good evidence that in certain algae, fungi, and higher plants as well, the production of sex organs, their maturation, the release of gametes, and the attraction of the sperm to the egg are all mediated by particular substances produced by the organism at a specific period of its life history. This reinforces the view that development consists of a series of stages, each elicited by a specific chemical or physical event, frequently leading to the production of a mobile chemical stimulus, or *hormone*. At each stage, newly developing cells differentiate to assume forms appropriate for their particular functions. At all times, the harmony and integrity of the entire organism are maintained.

SUMMARY

The mature plant body develops from the embryo as the result of a series of coordinated events, whose broad patterns are programmed by the genes, and whose specific characteristics may be environmentally determined. Growth in length involves the formation of new cells at cell division centers (*meristems*) found at the apices of stems, roots, and buds, as well as the elongation of these cells. Growth in circumference similarly involves the formation and subsequent enlargement of new cells from a lateral meristem, the *cambium*, usually a cylinder of cells between bark and wood. *Differentiation* occurs when initially identical cells formed from meristems assume specific and different characteristics.

Seeds contain an *embryo* plant, stored food in an *endosperm* or in fleshy *cotyledons* of the embryo, and a *seed coat*. *Germination* occurs after the seed imbibes water and initiates cell division and elongation. Growth rate is at first slow (*lag period*), then logarithmic (*log period*), again slow, and finally zero. Plants vary in total life span from a few weeks to more than 3000 years. Even in plants that die quickly the experimental transfer of bits of tissue to artificial chemical media can ensure unlimited growth. Tissues in culture are generally undifferentiated, but differentiation can be stimulated by appropriate chemical substances, including *hormones*.

Young roots are surrounded by an *epidermis* of flattened protective cells and *root hairs*, adapted for absorption. At their center is the *xylem*, with its water-conductive tissue composed of *tracheids* and *vessels*, together with *parenchyma* cells and *fibers*. Around the xylem is the *phloem*, a nutrient-conducting tissue containing *sieve tubes*, *companion cells*, and *parenchyma cells*. Just outside the phloem lies the *pericycle*, the divisions of which give rise to *branch roots* that grow through external tissues to the outside. Around this entire vascular core or *stele* is an *endodermis*, whose water-impermeable *Casparian strip* forces movement of solutions through membranes instead of along wet walls. Between the vascular core and the epidermis lies the parenchymatous *cortex*, frequently a storage tissue. When a cambium is activated between xylem and phloem, the epidermis and part of the cortex cracks away because of inner expansion, and the cracks are sealed by a new meristem, the *cork cambium*, that gives rise to a new protective tissue, the *periderm* (*cork*). Stems have a similar organization except that they have a central parenchymatous pith, they frequently lack an endodermis, and branches arise externally from the meristems, rather than internally, as in roots.

The alternating activity of the cambium at different times of year gives rise to tracheids and vessels of different size. The annual patterns thus produced, called *annual rings*, are useful in dating wood objects, and in inferring information about past climates and archeological events.

Leaves are generally flattened masses of photosynthetic tissue (*mesophyll*) supplied with profusely-branched veins containing xylem and phloem, and surrounded by an *epidermis* coated with a waxy *cuticle*. Numerous *stomata*, pores whose opening is controlled by the turgor of adjacent *guard cells*, permit gaseous exchange between interior and exterior of the leaf.

Reproductive organs (flowers) are not carried throughout the life cycle of the plant, usually being produced in response to seasonally timed environmental signals such as length of day and temperature changes. In the flower, the male (*stamen*) and female (*carpel*) organs produce cells that undergo meiosis. Haploid *microspores* produce the male gametophyte (*pollen*) that gives rise to *sperms;* haploid *megaspores* produce the female gametophyte (*embryo sac*) that gives rise to an *egg*. Fusion of sperm and egg (*fertilization*) produces the *zygote*, restoring the sporophyte generation. In ferns, and probably in flowering plants as well, many stages of the reproductive cycle are under the control of specific chemical messengers, or *hormones*.

SELECTED READINGS

CLOWES, F. A. L., and B. E. JUNIPER. *Plant Cells* (Oxford and Edinburgh: Blackwell, 1968).

ESAU, K. *The Anatomy of Seed Plants*, 2nd ed. (New York: Wiley, 1977).

GUNNING, B. E. S., and M. W. STEER. *Plant Cell Biology: An Ultrastructural Approach* (London: Edward Arnold, 1975).

LEDBETTER, M. C., and K. R. PORTER. *Introduction to the Fine Structure of Plant Cells* (New York and Heidelberg: Springer-Verlag, 1970).

O'BRIEN, T. P., and M. E. MCCULLY *Plant Structure and Development* (New York: Macmillan, 1969).

STEEVES, T. A., and I. M. SUSSEX. *Patterns in Plant Development* (Englewood Cliffs, N. J.: Prentice-Hall, 1972).

QUESTIONS

3-1. In the "grand period of growth" of organs, organisms, and populations, a period of rapid growth is followed by a period of slower growth. Why should growth rate slacken in this manner?

3-2. Discuss the relative roles of cell division and cell elongation in the growth of a seedling.

3-3. The most rapidly elongating portion of a root is a few millimeters behind the tip, although most of the cell divisions are occurring much closer to the tip. What is the reason for this?

3-4. After a grassy lawn has been mowed, grass plants resume their upward growth, whereas dicotyledonous weeds produce new branches from lateral buds. How do you explain this difference in behavior?

3-5. Discuss the various processes involved in the increase in size of a newly produced cell.

3-6. What is meant by *differentiation?* What kinds of influences could possibly operate in causing differences to appear in cells with identical genetic complements?

3-7. What is the probable reason for differentiation into different cell types in plants? Give evidence to support your suggestion.

3-8. A root tip removed from a plant and cultured on a synthetic medium continues to grow as an organized entity with normal anatomy. A fragment of stem below the apex similarly cultured usually gives rise to a more or less disorganized callus. What does this indicate about the nature of the plant's control over differentiation patterns?

3-9. Can differentiated cells dedifferentiate and resume active growth? What does this tell us about the nature of differentiation?

3-10. How may various chemical influences be used to control differentiation patterns in plants? Can the substances themselves be said to produce specific tissue and organ patterns?

3-11. Compare the life cycles of ferns and angiosperms. Comment on the duration and importance of the gametophyte and sporophyte generations in each.

3-12. In what ways is the growth of a root dependent on other parts of the plant?

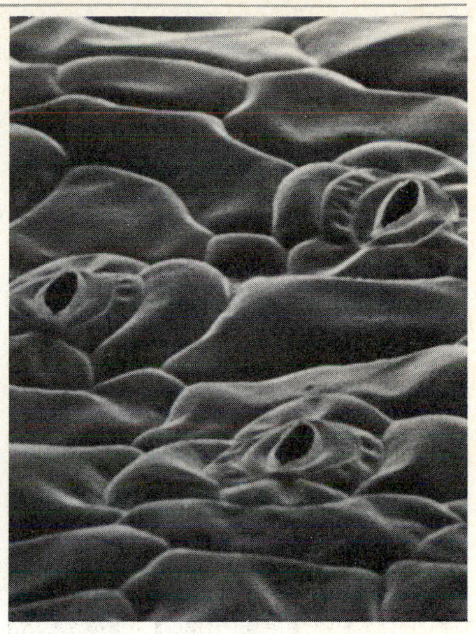

Photosynthesis: the Storage of Energy

Like all higher organisms, the green plant uses sugar and other oxidizable organic molecules as its source of energy, but unlike most organisms, the green plant is *autotrophic* (self-feeding). Plants make their own food, chemically converting atmospheric carbon dioxide to sugar and related substances by means of radiant energy absorbed in the photosynthetic apparatus of the chloroplast. They are thus the primary producers in nature and are independent of any external supply of organic molecules.

Some of the photosynthetically produced sugar molecules are almost immediately converted to large polymeric starch molecules and stored as starch grains in the chloroplast or leucoplast, while others are translocated from the plastids to other parts of the plant. When stored as starch, sugar is temporarily removed from further metabolic transformations, but starch can be broken down again to sugar, which is then readily oxidized to provide energy for future needs. In this chapter, we shall consider the mechanisms employed by the plant in synthesizing sugars from CO_2 and H_2O.

Photosynthesis: An Overview

When light of appropriate wavelengths is absorbed by a chloroplast, carbon dioxide is chemically reduced* to sugar, and gaseous oxygen, equal in volume

**Reduction* consists of the addition of electrons, whereas *oxidation* is the removal of electrons from a compound. Once the electron is transferred, a proton may also follow, the to the CO_2 reduced, is liberated. The direction of these changes is exactly the reverse of those accomplished during the oxidation of foodstuffs in the process of respiration, and, indeed, plants are important in the balance of nature because they restore to the air the oxygen needed for respiration by most organisms. Using the formula (CH_2O) to designate the basic unit of the carbohydrate molecule (six of these units would yield $C_6H_{12}O_6$, or glucose), we can write the equation for photosynthesis as:

$$\underset{\text{Carbon dioxide}}{CO_2} + \underset{\text{Water}}{H_2O} \xrightarrow{\text{Light energy}} \underset{\text{Carbohydrate}}{(CH_2O)} + \underset{\text{Oxygen}}{O_2}$$

Notice that all participants in and products of this reaction contain oxygen, and that the equation as written does not indicate whether the oxygen released in photosynthesis comes from CO_2 or H_2O. For many years biologists believed that light energy splits the CO_2 molecule and transfers a C atom to H_2O to form (CH_2O). However, this assumption was challenged by experiments with photosynthetic microorganisms, whose biochemical pathways are analogous to those of higher plants, yet different. For example, photosynthetic purple bacteria utilize H_2S instead of H_2O for photosynthesis, producing sulfur instead of oxygen as a byproduct:

$$CO_2 + 2H_2S \longrightarrow (CH_2O) + H_2O + 2S$$

The sulfur thus deposited may represent an important natural source of elemental sulfur found in various parts of the earth. The sulfur could only have been derived from H_2S, which must thus have been split during photosynthesis. Similarly, certain algae can be "trained" to use hydrogen gas, H_2, instead of water to reduce CO_2 to (CH_2O), the level of carbohydrate:

$$CO_2 + 2H_2 \longrightarrow (CH_2O) + H_2O$$

In both these schemes, it is clear that light energy is used to split (photolyze) the hydrogen donor, and that the reducing power thereby generated is used to convert CO_2 to (CH_2O).

If any common pathway of photosynthesis exists in different creatures, these studies made it seem likely that light energy splits the H_2O molecule during photosynthesis in higher plants. The essential correctness of this view became apparent when biochemists used H_2O or CO_2 labeled with isotopic oxygen (^{18}O instead of ^{16}O) in studying photosynthesis. They were able to demonstrate that the kind of oxygen released always corresponds with the

---

net result being the addition of hydrogen during reduction and its removal during oxidation. Oxygen is the usual final electron acceptor, or oxidizing agent, but oxidations can occur that do not involve oxygen *per se*.

oxygen present in water, not in CO_2. In fact, the photolysis of water appears to be one key to the entire process of photosynthesis, for it represents the major point at which light energy is made to do chemical work.

Since there are two atoms of oxygen in the oxygen molecule released during photosynthesis in higher plants and since each water molecule contains only one atom of oxygen, at least two water molecules must enter into the reaction to satisfy this requirement. To write a balanced equation truly representing the mechanism of the overall reaction, we must add one molecule of water to each side of the equation. If the H_2O contains ^{18}O, the equation is written:

$$CO_2 + 2H_2{}^{18}O \xrightarrow[\text{energy}]{\text{Light}} (CH_2O) + {}^{18}O_2 + H_2O$$

If CO_2 is labeled with ^{18}O, the equation becomes

$$C^{18}O_2 + 2H_2O \xrightarrow[\text{energy}]{\text{Light}} [CH_2{}^{18}O] + O_2 + H_2{}^{18}O$$

The oxygen evolved is derived from water entering the reaction, and the water molecule formed is different from either of the two split in the photolytic process. Figure 4-1 presents a scheme that may help you visualize the

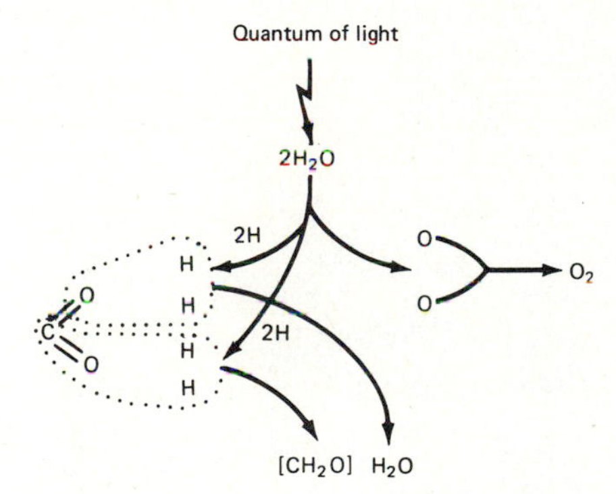

Figure 4-1 Overview of photosynthesis.

basic aspects of the reaction. This scheme shows that light energy is used to cleave the water molecule, releasing oxygen, and that the "hydrogen" (or reducing power) also produced in the photolysis reaction is used in two ways: (1) to reduce CO_2 to photosynthate (CH_2O), and (2) to produce a new water molecule. This is clearly the briefest kind of shorthand, and there are many steps that intervene in each of the simple reactions shown. Some of the steps involve the conversion of light energy to chemical energy, while other parts of the process take place in either light or darkness and are called the "dark reactions" of photosynthesis.

Logistics of Photosynthesis

The CO_2 entering into photosynthesis comes to the green cell of a leaf or stem from the air by way of the *stomata** (singular *stoma*), and a much-branched system of connecting intercellular air canals. The leaf (Fig. 3-9), by far the main photosynthetic organ of the plant, can be visualized as several layers of actively photosynthesizing cells (the *mesophyll*) surrounded by a protective layer (the *epidermis*) and supplied with conducting elements (the *veins*) that are equipped for two-way transport; they carry raw materials to the leaf and photosynthate and other products away from the leaf. These veins branch so profusely that no mesophyll cell lies more than a few cell diameters from a vein.

The stomata that perforate the epidermis (Fig. 4-2) open and close in response to changes in the turgor pressure of two sausage-shaped guard cells that surround them (Fig. 4-3). It is apparent that the guard cells have unusual

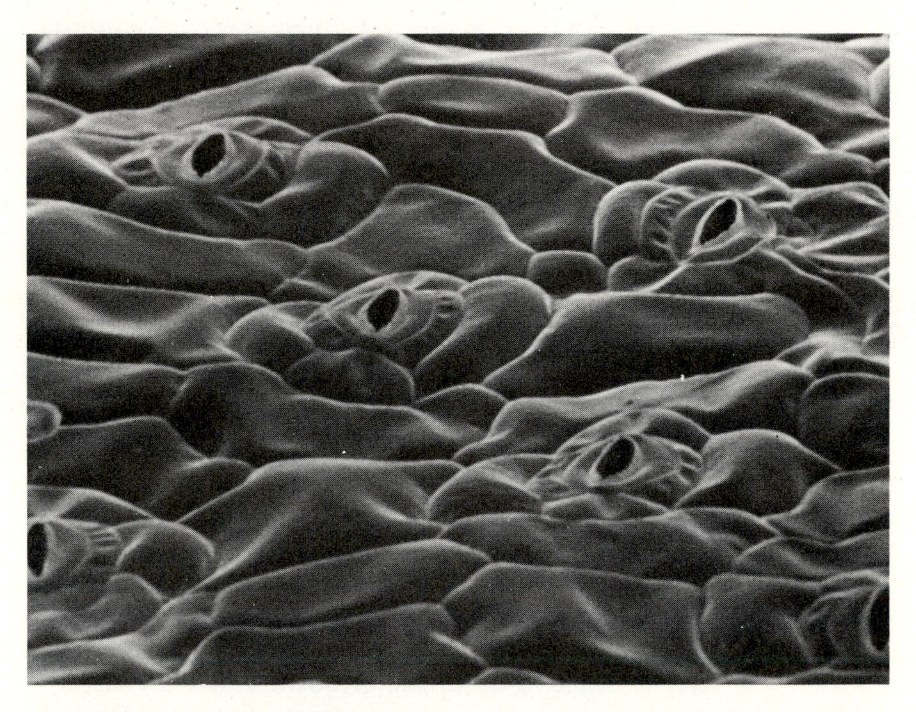

Figure 4-2 Open stomata on the lower epidermis of freeze-dried leaves of *Commelina communis*, seen with the aid of the scanning electron microscope. (X700.) (Courtesy of H. Saxe, Copenhagen University.)

*The terms *stomate* (singular) and *stomates* (plural) are also used.

structural and physiological features enabling them to move in this way. Their cell walls are quite elastic, in contrast to the rigid walls that encase most mature plant cells. In addition, the cellulose microfibrils which form the structural framework of the cell wall are arranged radially, rather than in the more usual longitudinal fashion. Furthermore, the wall is thick and resistant to stretching adjacent to the pore, and thinner and more easily stretched away from the pore. Consequently, when a massive osmotic uptake of water causes the guard cells to become turgid (see Chapter 6), they swell and bulge outward, causing the stomatal pore to open. When they lose water and become flaccid, they shrink, and the pore is effectively closed. It has recently been shown that osmotic water movement, and thus the turgidity of the guard cells, is controlled by massive movements of potassium, chloride, and hydrogen ions between the guard cells and other epidermal cells.

Figure 4-3 Stomatal opening and closure. (From J. Bonner and A. W. Galston, *Principles of Plant Physiology*, San Francisco: W. H. Freeman and Company, 1952.) The photographs on the right are of silicone rubber impressions of closed and open tobacco stomata. (Courtesy of I. Zelitch, Connecticut Agricultural Experiment Station.)

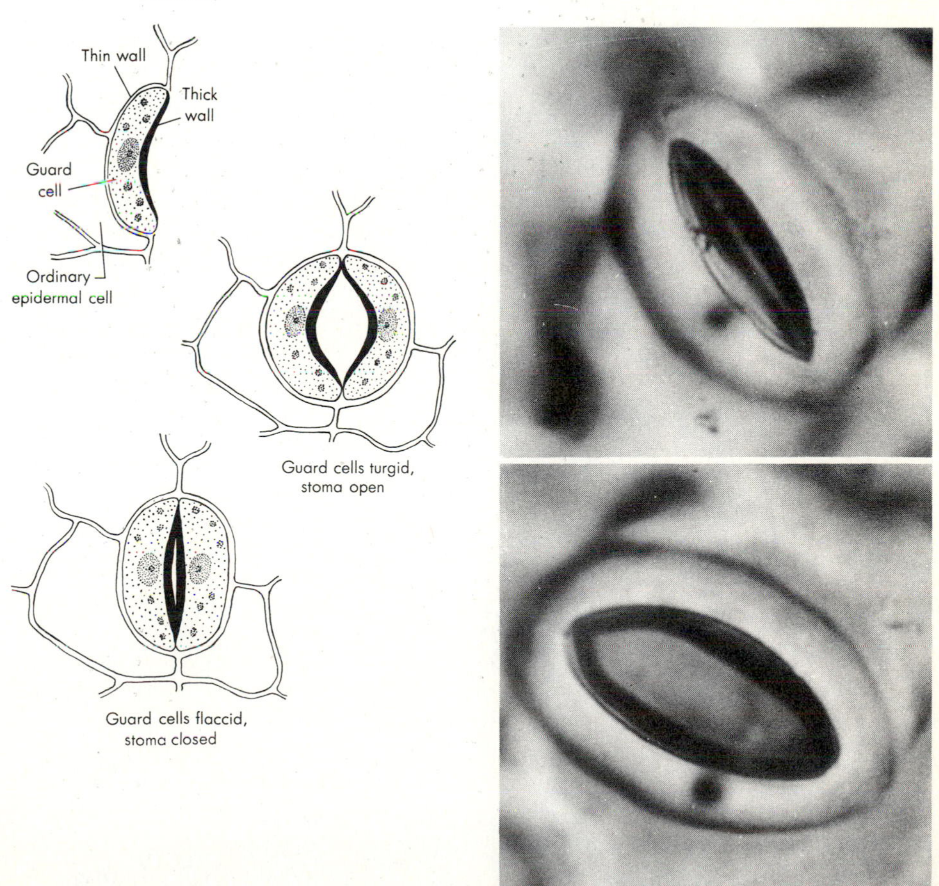

Once the atmospheric CO_2 molecule has passed the stomatal barrier, it enters a substomatal air chamber and connecting air passages, through which it can diffuse throughout the leaf mesophyll. When it reaches the wet surface of a cell, it dissolves in water, becoming hydrated to carbonic acid (H_2CO_3), which is in turn partially neutralized by the cations of the cell to form bicarbonate ions (HCO_3^-). This bicarbonate represents a reservoir of potential CO_2 that might be used for photosynthesis by the cell.

The carbon dioxide upon which we are all dependent exists as a trace gas in the atmosphere, constituting only about three parts per ten thousand (0.03%) of the air. This concentration varies somewhat over the face of the earth; it is higher over urban areas where large quantities of coal, oil, and gasoline are being burned, and lower in rural areas where photosynthesis is proceeding vigorously. It is also low in greenhouses during the sunny part of the day, and some growers artificially elevate the CO_2 content of the atmosphere to raise the photosynthetic rate of plants that are well supplied with light and water. However, this may produce injurious effects in certain sensitive leaves, possibly because high CO_2 levels cause the stomates of some plants to close.

Some students of evolution believe that the CO_2 content of the atmosphere has varied considerably in recent geological times, and may have been responsible for certain changes of vegetation and climate. For example, an elevation in the CO_2 level would not only increase photosynthesis, and thus the amount of plant material on earth, but would also cause a warming of the earth. This would occur because the earth, heated by the sun, normally reradiates a portion of the absorbed energy back into space as infrared radiation. Carbon dioxide absorbs heavily in these infrared wavelengths, thus preventing the complete escape of this heat energy and creating a sort of "greenhouse" over the earth (Fig. 4-4). Warming of the earth through such an effect could lead to partial melting of polar icecaps and glaciers, and

Figure 4-4 The "greenhouse effect" of CO_2 in the earth's atmosphere.

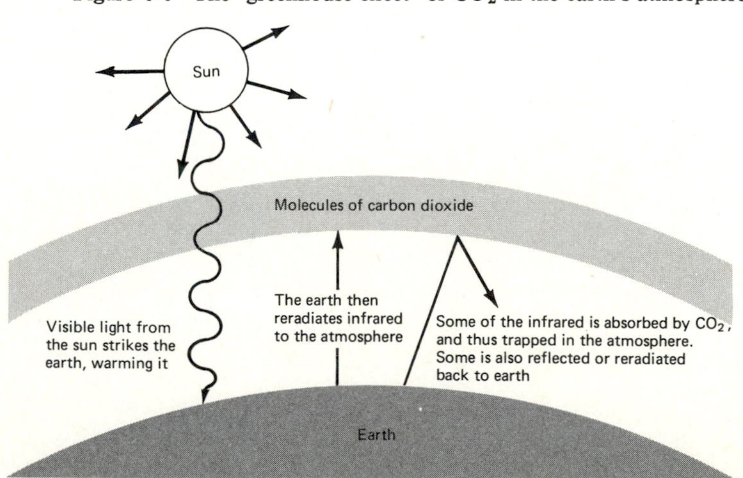

thereby to flooding of the low-lying land areas in which most of the earth's major cities are located. Thus, our prodigious consumption of fossil fuels and the resultant release of extra CO_2 into the atmosphere may have profound consequences for man. This process, however, tends to be self-limiting and perhaps cyclical, in that elevated temperatures and CO_2 levels will result eventually in elevated photosynthesis and a luxuriant growth of plants, as in the Carboniferous Era. This increase of CO_2 fixation in photosynthesis should ultimately lower the atmospheric CO_2 content significantly, resulting in the cooling of the earth and in a reversal of the cycle mentioned above.

The amount of water required in photosynthesis is only a minute fraction of that absorbed and evaporated by plants. For optimal photosynthesis, the leaf must be turgid and the stomata open; thus a suboptimal water supply depresses the photosynthetic rate, although only indirectly, since stomatal closure eventually inhibits the process through limitation of CO_2.

The oxygen liberated in photosynthesis is released to the outside world through the stomata, after passage from the surface of the mesophyll cell into one of the intercellular air spaces that connects with the substomatal cavity. Closure of the stomata because of the loss of guard-cell turgor pressure prevents this gaseous exchange but does not completely prevent photosynthesis or respiration, as the internal oxygen and carbon dioxide will continue to cycle between these two processes. However, photosynthesis, which can proceed optimally at 10–20 times the maximal rate of respiration, is limited, under conditions of stomatal closure, to the magnitude of respiration.

Thus, the maintenance of an optimal rate of photosynthesis in the leaf requires that adequate levels of light energy, water, and carbon dioxide be supplied to the leaf. If any are in short supply, that one in the lowest supply relative to the others will act as the *limiting factor* in photosynthesis (Fig. 4-5). It is also important that the plant maintain an adequate rate of transport of the products of photosynthesis from the leaf, since a "backup" of carbohydrates can inhibit the process. In fact, most plants grow best in alternating light and dark periods, since the products of photosynthesis accumulated in the light are depleted during darkness. Plants of extreme northern and southern latitudes, which have to grow rapidly in the short summer season, are exceptions to this rule and grow best in continuous light.

Almost all plants are "light saturated" at intensities somewhat less than that of bright sunlight, which approximates 10,000 foot-candles. Individual leaves tend to be saturated at 1000 foot-candles or less, depending on the species, but because of the mutual shading of leaves on a plant, it takes several thousand foot-candles to saturate the entire plant (see Chapter 14). Some species, known as "shade-loving" types, can grow in relatively dim light such as that filtering through to the forest floor. In such a location these types can compete successfully, whereas "sun-loving" types that require much higher light intensities for optimal photosynthesis grow poorly.

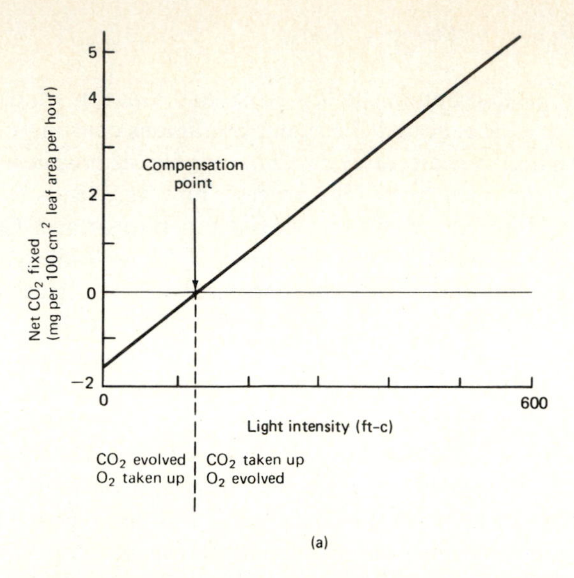

Figure 4-5a The effect of low intensity light on photosynthesis in a typical shade plant. There is no net gas exchange at the compensation point (130 ft-c), for the oxygen evolved during photosynthesis equals that consumed by respiration while the CO_2 fixed by photosynthesis equals that evolved during respiration.

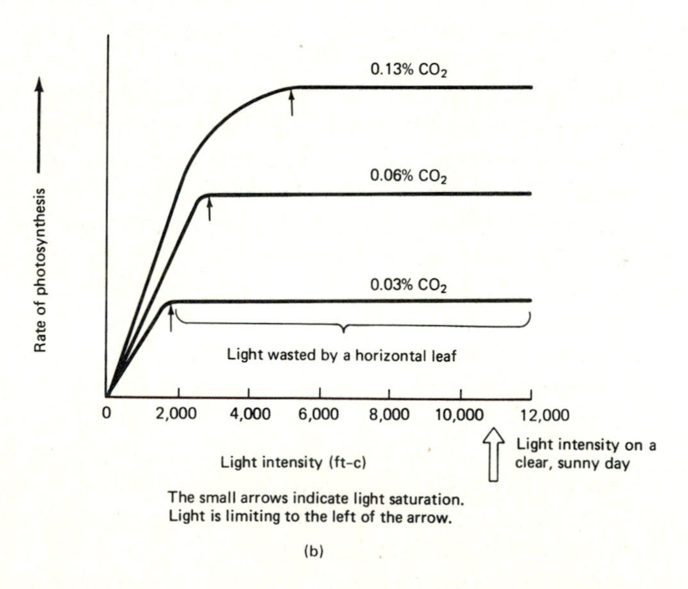

Figure 4-5b The effect of high light intensity on photosynthesis at three different concentrations of CO_2, the two highest of which exceed levels found in most natural environments. The arrows indicate light saturation. Light is rate limiting to the left of the arrow.

The Biochemistry of Photosynthesis

We can best approach the process of photosynthesis by analyzing four major questions: (1) How is light energy "absorbed"? (2) How is it then made available to perform chemical work? (3) Through what pathway is oxygen released from water? (4) Through what pathway is carbon dioxide transformed into sugar? We shall examine each of these questions in turn.

Absorption of radiant energy

Light, to be photochemically effective, must first be absorbed. Those molecules that absorb visible light are called *pigments*. The absorption of a quantum of light (*photon*) by a pigment is a function of the electron distribution pattern in the pigment molecule; the wavelengths absorbed depend on such details as the number and position of double bonds in the molecule and the presence of aromatic rings. In fact, the absorption of a photon changes the electron distribution pattern slightly, and the altered form of the pigment is said to be "activated." Since we understand certain facts concerning the relation of wavelength absorbed to structure of the absorbing molecule, we can deduce the characteristics of a photoreceptor pigment in a particular photochemical reaction from the data relating wavelength to activity.

If monochromatic light of various wavelengths from a filter, prism, or diffraction grating is permitted to fall on a green leaf and the photosynthetic rate is measured at each wavelength, blue light (near 420 nm) and red light (near 670 nm) are found to be maximally effective, and green light (about 500–600 nm) least effective (Fig. 4-6). This *action spectrum*, a plot of relative efficiency versus wavelength, can be understood in terms of the absorption characteristics of chlorophyll, the major pigment of the chloroplast. When extracted from the leaf, chlorophyll heavily absorbs exactly those wavelengths that are most effective in photosynthesis (Fig. 4-7). This similarity between the "absorption spectrum" of chlorophyll and the "action spectrum" for photosynthesis is in fact one of the best proofs we have that chlorophyll is the major receptor pigment in photosynthesis. From the details of the action spectrum, we may also infer that the yellow carotenoid pigments, also present in great quantities in the chloroplast, must also function in light absorption for this process. Since carotenoids cannot operate photosynthetically in the absence of chlorophylls, it is generally assumed that photoactivated carotenoids pass their absorbed energy on to the chlorophylls, which ultimately perform the actual photosynthetic work.

Calculations have shown that only a small fraction of the chlorophyll molecules are actually involved in electron transfer in photosynthesis. The other molecules, termed *antenna* chlorophyll, serve as light gatherers. The passage of energy from carotenoids to chlorophyll, and from one chlorophyll molecule to another, occurs by a process known as *resonance transfer*. In this

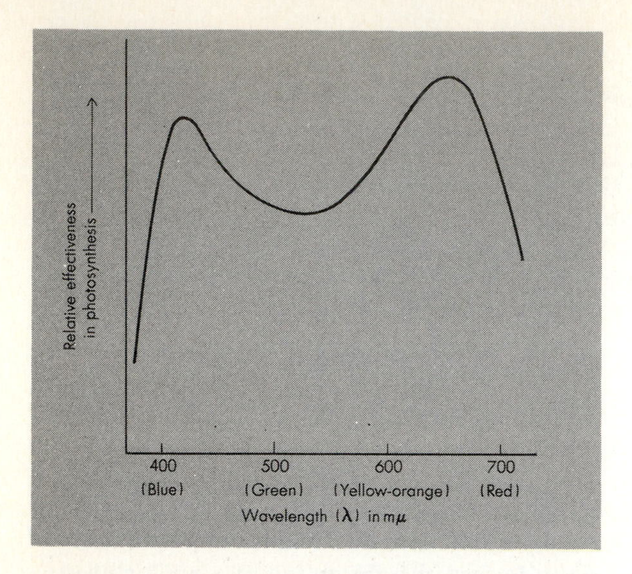

Figure 4-6 An action spectrum for photosynthesis in the green leaf.

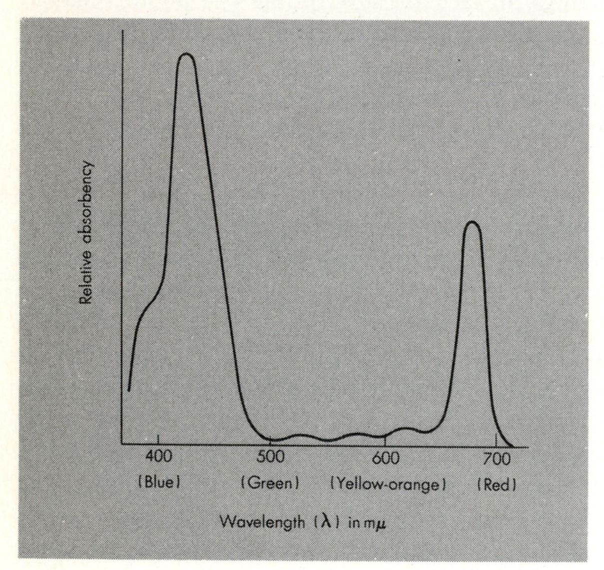

Figure 4-7 The absorption spectrum of chlorophyll *a* in ether.

process, the participating molecules must be closely packed together so that vibrational energy of one is directly communicated to the other. In the grana of the higher plant chloroplast, where the photosynthetic pigments are tightly packed together, as already described in Chapter 2, energy absorbed by any one pigment is easily communicated to certain other pigment molecules.

Purple photosynthetic bacteria contain *bacteriochlorophyll*, a structural variant of chlorophyll which absorbs in the green and infrared regions of the spectrum, where higher green plants do not absorb heavily. Red, brown, and blue-green algae contain, in addition to chlorophyll, large amounts of *phycobilin* pigments (phycoerythrin, phycocyanin, allophycocyanin, and others related to the bile pigments of our bodies), and *carotenoids* such as fucoxanthin and peridinin. These pigments form associations that are the main light-absorbing system in these algae. Phycoerythrin absorbs blue-green wavelengths and therefore appears red, whereas phycocyanin and allophycocyanin absorb most heavily in the yellow and red regions, and appear blue or green (Fig. 4-8). The action spectrum for photosynthesis in these algae differs markedly from that for green leaves (Fig. 4-9).

These algal pigments, attached to proteins, are arranged in structural units called *phycobilisomes*, located on the stroma side of the chloroplast lamellae. Each of the pigments fluoresces, that is, it absorbs energetic photons of particular wavelengths and emits less energetic photons of slightly longer wavelengths through a succession of absorptions and fluorescent re-emissions, eventually passing the light energy to chlorophyll.

The phycobilins, which can comprise as much as 60% of the total protein of certain algae, form an efficient light-harvesting system, although energy transfer among these pigments is less efficient than in the chloroplast of higher plants. In some algae, the ratio of one phycobilin to another varies with the color of the irradiating light, so that an alga changes color in light of different

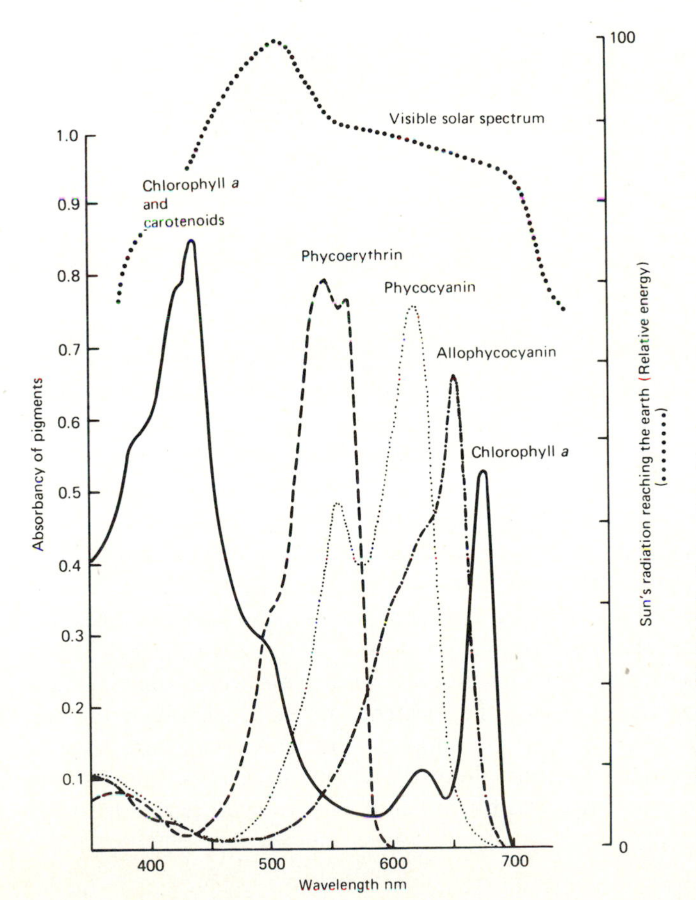

Figure 4-8 Absorption spectra of pigments extracted from a blue-green alga. The major peaks of chlorophyll *a* (—) are at about 435 nm and 675 nm. Phycobiliproteins (phycoerythrin, R-phycocyanin, and allophycocyanin), purified from phycobilisomes, effectively close the gap where chlorophyll absorbs very little. The fluoresence peaks of these pigments closely overlap, forming an integrated transfer system. The visible part of the solar spectrum is shown on top. (From E. Gantt, BioScience 25, 781–788, 1975.)

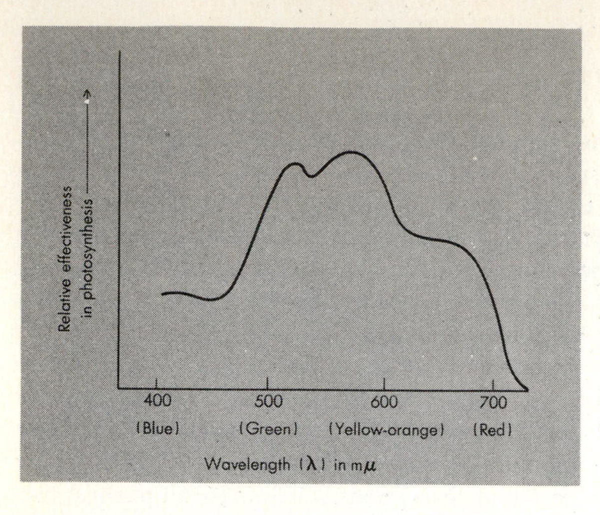

Figure 4-9 Action spectrum for photosynthesis in a red alga, *Porphyra*. (After F. T. Haxo and L. R. Blinks. 1950. J. Gen. Physiol. *33*, 389–422.)

wavelengths, a phenomenon known as *chromatic adaptation*. For example, phycocyanin pigments predominate when the algae are grown in red light, whereas phycoerythrin pigments are most abundant when the light is green. This adaptability enables algae growing at different water depths to absorb sufficient light for photosynthesis, for when light travels through water, some of its energy is absorbed or scattered by water molecules and suspended particles, and it thus changes color with increasing depth. Not all algae are able to adapt chromatically, but some types can nevertheless survive in water of varying depths by synthesizing more pigment molecules as water depth increases.

Transduction of radiant energy to chemical energy

The ultimate photochemical work performed in photosythesis involves the splitting of water or some chemical analog of water, such as H_2S, but before this can happen, the "physical" energy of the trapped photon must somehow be changed or transduced to "chemical" energy. Only a small fraction of the pigment molecules, located at photochemically active centers in the chloroplast, can participate in this transduction step. The energy absorbed by chlorophyll and other photoactivated pigments is transferred to the chlorophyll molecules at such photochemically active centers, called *traps*. As a result, particular electrons gain enough energy to escape from the chlorophyll trap molecule, passing to other nearby molecules called *electron carriers*. The carrier absorbs a fraction of the activation energy, and passes the electron to still another carrier, which repeats the process. In the chloroplast, the various carrier molecules are spatially arranged in a series along or within the chloroplast membrane according to their ability to receive electrons (their "oxidation–reduction potential"). Thus, the electron moves efficiently from one carrier to the next, much as water running downhill moves from one level to another. As each electron travels along the chain of carriers, some of its energy is converted to a chemical form, being used to synthesize adenosine triphosphate (ATP) from adenosine diphosphate (ADP) and inorganic phosphate (P_i) (Fig. 4-10). A considerable amount of energy (8–10 kcal/mole) is stored in the chemical bond between ADP and P_i, and this energy can be transferred during reactions in which this bond is broken. The formation of ATP from visible light energy is known as *photophosphorylation*. ATP, the "energy currency" of living cells, drives many energy-requiring (*endergonic*) reactions in the cell as it is broken down to ADP and either P_i or another phosphorylated compound.

(1) $ADP + P_i + \text{energy} \longrightarrow ATP + H_2O$

energy: (from light or respiration); ATP: (energy trapped in phosphate bond)

(2) *a* $ATP + H_2O + X \longrightarrow ADP + XP$ (some of the energy remains stored in XP)

or

b $ATP + H_2O \longrightarrow ADP + P_i + \text{energy}$ (released as heat or used to drive another reaction)

Several years ago, Robert Emerson discovered that although red light of wavelength greater than 700 nm is relatively ineffective in photosynthesis in higher plants when supplied alone, it becomes quite effective when accompanied by shorter-wavelength red light. This "Emerson enhancement effect" has led to the theory that two different light reactions occur, and that they

Figure 4-10 A, B, C and D are electron carriers of consecutively decreasing (less negative) oxidation-reduction potential. A difference in reduction potential between two consecutive carriers that exceeds 0.16 volt is sufficient to bond ADP + P_i to form ATP.

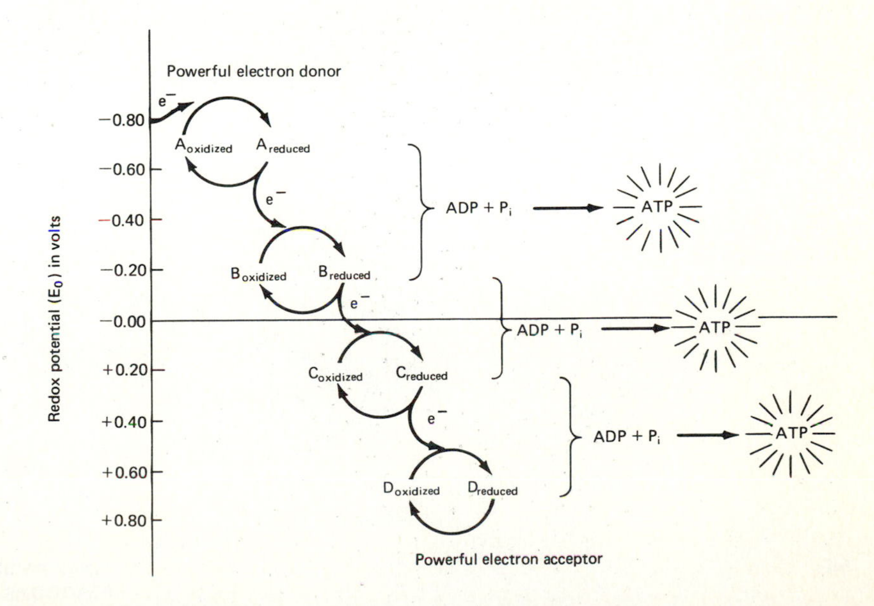

must occur simultaneously if photosynthesis is to proceed optimally. This theory has been validated by the physical separation of two distinct entities, *Photosystem I* and *Photosystem II*, from chloroplasts of higher plants (Fig. 4-11). Each photosystem has its own complex of slightly different chlorophyll molecules and associated electron carriers, and each can carry out only certain photosynthetic reactions. Photosystem I, the only photosystem present in bacteria, evolves no oxygen; since bacteria are the most primitive photoautotrophs, Photosystem I probably predominated early in biotic evolution, when the atmosphere lacked oxygen. The evolution of Photosystem II enabled plants to free molecular oxygen from H_2O, and was thus probably responsible for changing the earth's atmosphere from anaerobic to aerobic.

The chlorophyll trap molecule in Photosystem I is called P_{700}, since 700 nm is the wavelength at which it absorbs light optimally. Absorption of a

Figure 4-11 The "Z" scheme for the light reactions of photosynthesis.

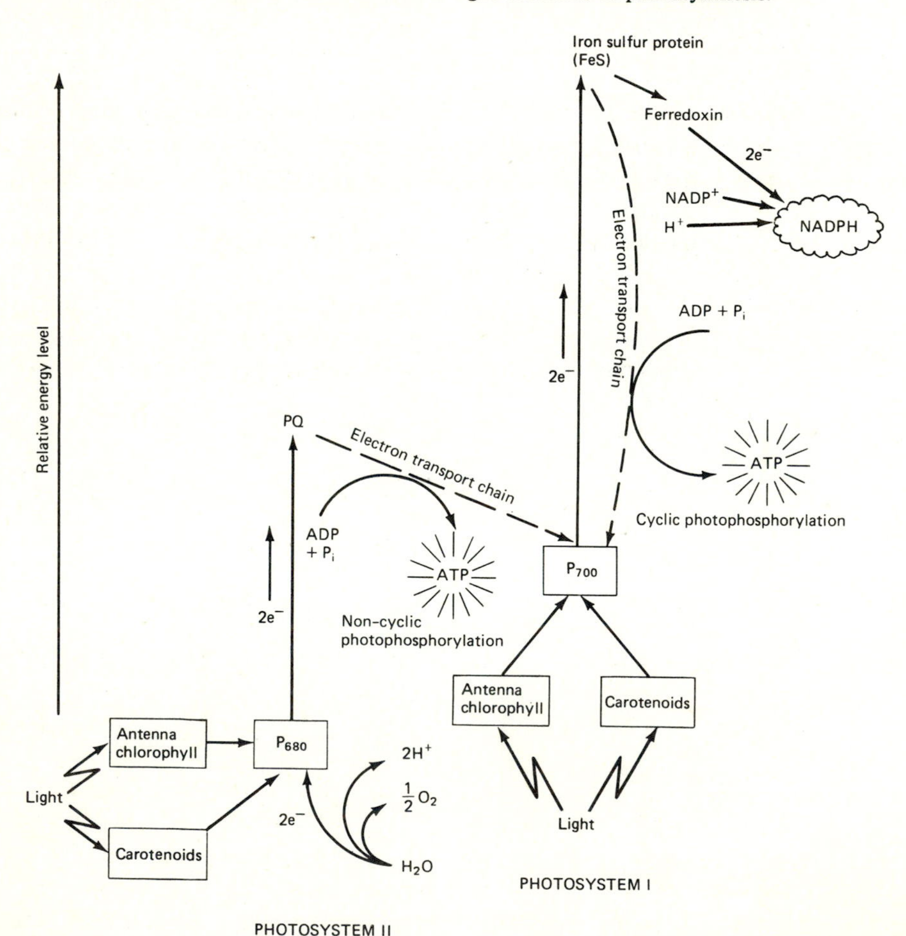

FINAL PRODUCTS OF THE LIGHT REACTIONS = ATP + NADPH

light quantum by P_{700} raises one of its electrons to an excited state, from which it is captured by an iron–sulfur protein (FeS) and then transferred to a carrier called *ferredoxin*. The electron then follows one of two pathways. One pathway (*cyclic photophosphorylation*) involves stepwise transfer of the electron from ferredoxin back to P_{700} through intermediate electron carriers including vitamin B_2-containing flavoproteins and heme-containing cytochromes. En route, the energy of an electron following the cyclic pathway is used to bond ADP to P_i, forming ATP. Oxygen is not evolved during cyclic photophosphorylation, and since there is no gas exchange between the interior of the leaf and the external atmosphere, this type of photosynthesis could theoretically occur even if the stomata were closed. Cyclic photophosphorylation is the only functional photosynthetic pathway in monochromatic light of wavelength greater than 700 nm, since activation of Photosystem II requires shorter wavelengths, but the extent to which cyclic photophosphorylation operates in leaves is still very much in dispute.

Noncyclic photophosphorylation and oxygen release

When the stomata are open and mixed-wavelength light is absorbed by the chloroplast, Photosystem I is coupled to Photosystem II, the combination comprising the *noncyclic photophosphorylation* system. Under these circumstances the activated electrons, ejected by P_{700}, are passed to FeS and then to ferredoxin. At this point, they branch off the cycle and are transferred through different flavoproteins to $NADP^+$, a niacin-containing compound whose reduced form, NADPH, is used to reduce CO_2 down to the level of carbohydrate in subsequent reactions. Now the trap chlorophyll of Photosystem II, P_{680}, is also excited, and ejects an electron, which passes through various carriers (plastoquinone, cytochrome *f*, and plastocyanin) and ultimately reaches P_{700}, filling the "hole" left by the electron photoejected from that system. P_{680} of Photosystem II now fills its "hole" in turn by taking an electron from the OH^- ion produced by the ionization of water. Two of the resulting OH "free radicals" then probably unite to form a peroxide and ultimately decompose with the liberation of oxygen. The oxygen-release reaction requires manganese and chloride ions as cofactors, and can be inhibited by the synthetic herbicide DCMU (dichlorophenyl-dimethyl urea). Death of a plant after DCMU application probably results from the buildup of peroxide or other highly oxidized molecules. Furthermore, electron transport from water is blocked, ATP cannot be formed, and $NADP^+$ cannot be reduced. Thus, DCMU is toxic only to green plants in the light, and seems without effect on other organisms, including man.

The H^+ also produced by the ionization of water remains part of the water medium in the thylakoids of the chloroplast which, we recall, contains H^+ and OH^- ions in addition to water molecules. H^+ buildup on the inner side of the thylakoid membrane does, however, play a role in ATP synthesis (see below). Only electrons ejected from chlorophyll are transferred through the photosystems, although protons (H^+ ions) accompany the electrons through part of the electron transport chain. The free protons in the water medium surrounding the thylakoid membranes are withdrawn to make NADPH when two electrons are finally passed onto $NADP^+$. This entire process, illustrated in Figure 4-11, has been called the "Z scheme."

How ATP is made.

While we have emphasized electron transfer as the basis of ATP synthesis, the newer *chemiosmotic theory* of the Englishman Peter Mitchell, based instead on proton movement, is gaining wide acceptance.

Let us examine ATP synthesis in chloroplasts (Fig. 4-12), where its relation to the "Z scheme" of the light reactions in photosynthesis (see Fig. 4-11) can easily be appreciated. The electron carriers in the photosynthetic electron transport chain are located in the thylakoid membrane. One of the carriers, plastoquinone, transfers protons (H^+ ions) as well as electrons, moving the protons vectorially from the exterior of the thylakoid to the interior. The resultant buildup in concentration of protons inside the thylakoid leads to a large pH gradient across the thylakoid membrane, the interior becoming acid with respect to the exterior. This acidity inside the thylakoid is further enhanced by photolysis, which removes electrons and O_2, leaving behind the protons. The large difference in pH across the thylakoid membrane is a potential source of energy, which is harnessed as protons inside the thylakoid move back out through special channels in stalked knobs on the outside of the thylakoid membrane. These channels contain a proton known as the "coupling factor," or CF_1, that can synthesize ATP. CF_1 is in fact an ATPase, that is, an enzyme that can break down ATP, but, when supplied with appropriate energy, it can also synthesize ATP. This energy is provided by protons flowing across the thylakoid membrane through the stalk and knob, much as water flowing through a turbine stores energy as electricity. The protons flow through the channel as long as their concentration inside the thylakoid exceeds that outside, or, therefore, as long as electrons are passing through the electron transport chain under the influence of light received by chlorophyll. For each two electrons transported, approximately four protons accumulate inside the thylakoid. One ATP molecule is synthesized for each three protons that flow back through the coupling factor.

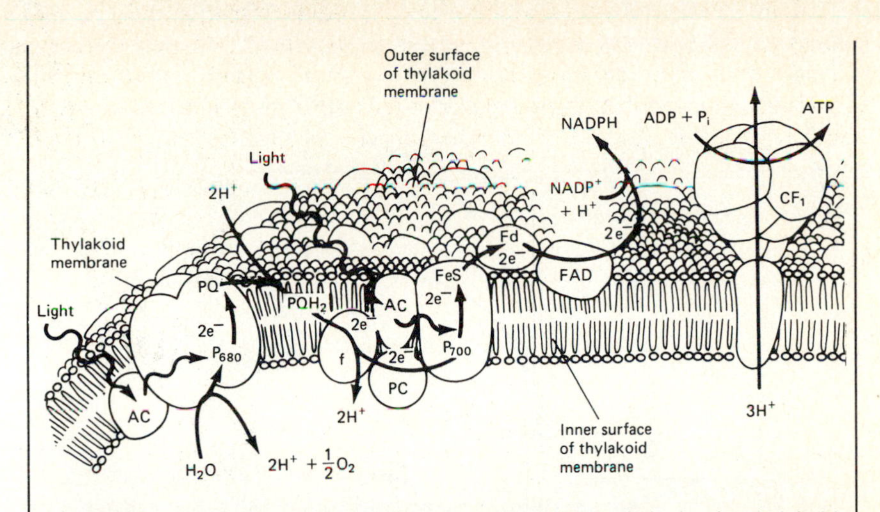

Figure 4-12 A schematic view of the electron transport chain in the thylakoid membrane, showing how it transports protons across the membrane and forms ATP. Radiant energy absorbed by antenna chlorophyll molecules is transferred to P_{680} and P_{700}, which may also absorb light directly. Pairs of electrons are then ejected from excited P_{680} and P_{700} molecules, leaving behind positive regions called *holes*. The electrons ejected from P_{680} are captured by a special form of the electron carrier plastoquinone (PQ). At this point, two protons from the stroma exterior to the thylakoid follow the electrons, thus reducing plastoquinone to PQH_2. The protons are released to the interior of the thylakoid, while the electrons pass to cytochrome *f*(f), then to plastocyanin (PC), and finally to P_{700}, where they fill its hole. The electrons ejected from P_{700} are captured by an iron-sulfur protein (Fes), and are transferred, to ferredoxin (Fd), flavin-adenine dinucleotide (FAD) and finally to $NADP^+$. A proton from the exterior of the thylakoid joins the two electrons and $NADP^+$, to form NADPH. The cycle is completed when H_2O on the inside of the thylakoid is photolyzed to yield 2 electrons, 2 protons, and 1/2 O_2. The electrons fill the hole in P_{680}, the oxygen is released in gaseous form, and the protons in the interior of the thylakoid then flow back to the outside through the coupling factor (CF_1) (on right of diagram). The energy released during this proton movement is used to synthesize ATP from ADP and P_i. (Courtesy of R. E. McCarty and P. C. Hinkle, Cornell University. From a drawing by M. Hinkle.)

The reduction of CO_2 *to sugar*

The overall products of the light reactions of photosynthesis are NADPH and ATP (Fig. 4-13). These substances are then used, respectively, as the reducing power and energy source for the conversion of carbon dioxide to sugar. These steps are commonly termed the *dark reactions* of photosynthesis.

The pathway along which carbon dioxide is reduced to sugar was elucidated more than a quarter-century ago by the use of radioactively labeled carbon, ^{14}C. This isotope, which decays with the emission of a beta-ray, can be detected with a Geiger–Müller-type counter or other radiation detector. The biochemical pathway followed by ^{14}C supplied as labeled carbon dioxide

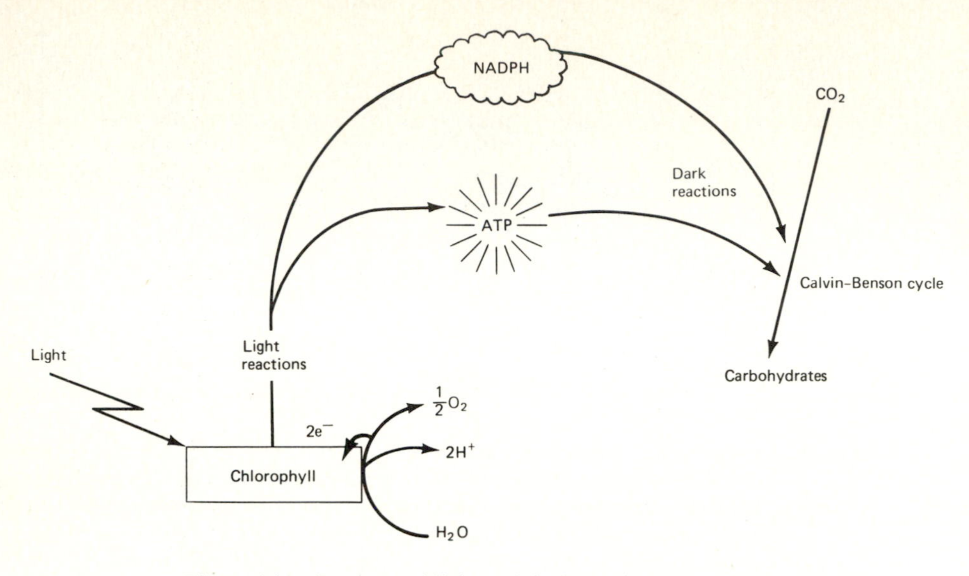

Figure 4-13 Products of light and dark reactions in photosynthesis.

can then be traced by the measurement of radioactivity in various chemical fractions isolated from a photosynthesizing cell at various times after the introduction of the radioactive CO_2. Melvin Calvin and Andrew Benson of the University of California at Berkeley used this general technique to establish the pathway of carbon fixation in the unicellular green alga *Chlorella*. *Chlorella* was allowed to photosynthesize in the presence of radioactive CO_2 and the resulting labeled compounds were extracted with methanol and separated by two-dimensional paper chromatography, first in one direction with one solvent system and then at right angles with a second solvent system. The paper chromatogram was then placed next to X-ray film, which, when developed, showed black spots in the position of any compounds that contained the radioactive carbon. Many labeled compounds were formed quickly, but if the time allowed for photosynthesis was reduced to only 0.5 seconds, then only the 3-carbon phosphorylated compound 3-phosphoglyceric acid (PGA) was formed. PGA is thus the first stable compound formed from CO_2 during photosynthesis.

By stepwise degradation of the isolated radioactive PGA, it can be shown that the terminal carbon of the carboxyl (COOH) group is radioactive (symbolized by the asterisk or ^{14}C) and must thus represent the altered form of the $^{14}CO_2$ that was supplied. One might anticipate that the supplied $^{14}CO_2$ would couple with some 2-carbon fragment to produce the 3-carbon chain of PGA, but this is not the case. Calvin and Benson looked for a compound that accumulated when the $^{14}CO_2$ was removed after a period of photosynthesis, reasoning that it would represent an unused "CO_2 acceptor." Such a compound was found (Fig. 4-14) in the 5-carbon phosphorylated compound

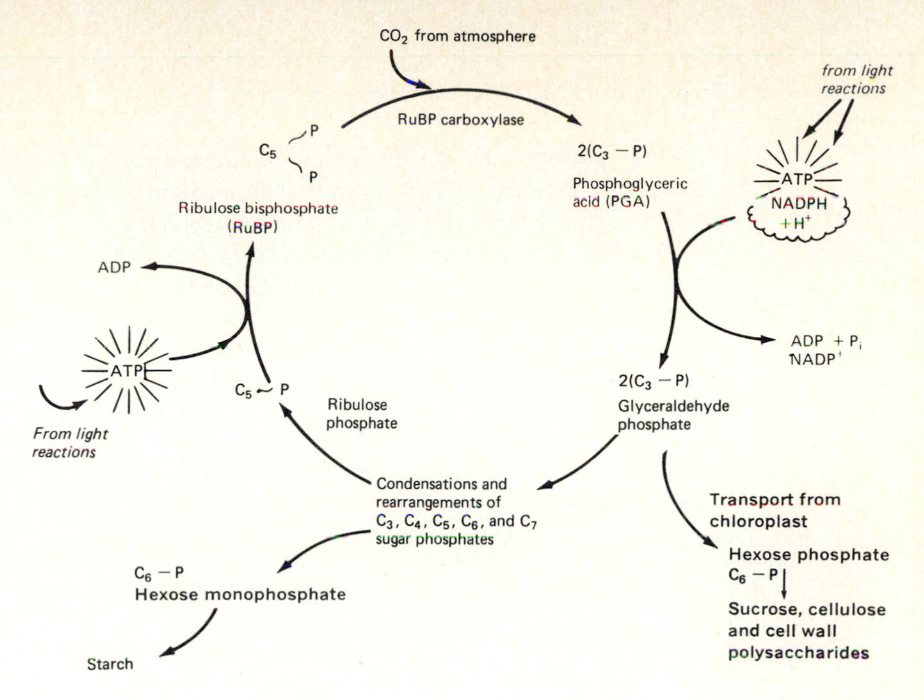

Figure 4-14a Simplified outline of the Calvin–Benson cycle for the dark reactions of photosynthesis.

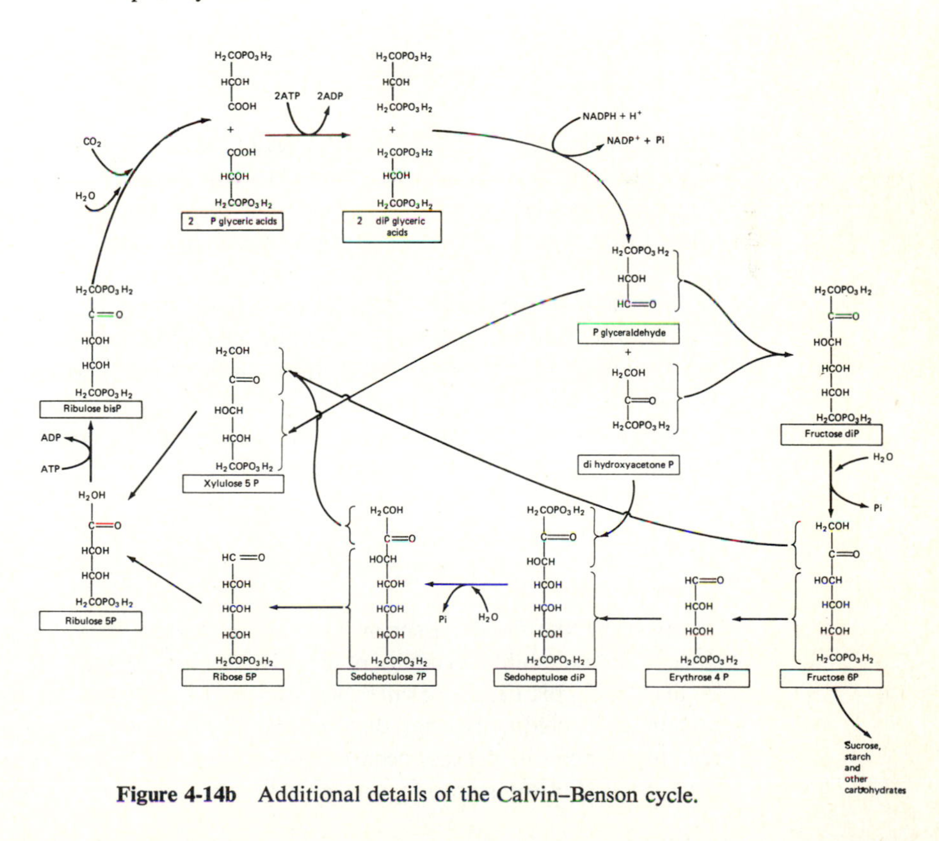

Figure 4-14b Additional details of the Calvin–Benson cycle.

ribulose bisphosphate (RuBP), which on the addition of CO_2 decomposes to two PGA molecules. The enzyme that mediates this reaction, *RuBP-carboxylase*, is the most abundant protein in green tissue.

The phosphoglyceric acid formed from CO_2 is not yet at the reduction level of a carbohydrate, which corresponds to that of an aldehyde (H—C=O with a single bond below C) group, but is rather at the next higher oxidation step of a carboxyl group (HO—C=O with a single bond below C). The reduction to the aldehyde level is accomplished through the reducing power of NADPH and the energy of ATP, both generated by the light reactions of photosynthesis. The final step in the production of a sugar from CO_2 via PGA can be visualized as follows:

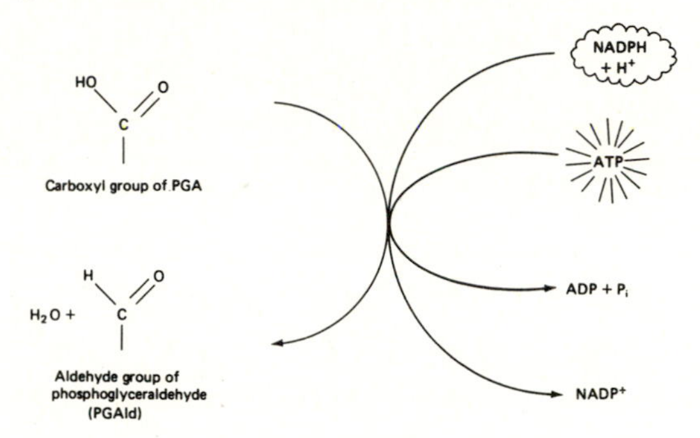

The phosphoglyceraldehyde, a sugar phosphate, has only three carbon atoms, whereas the smallest sugar accumulated to any extent in plants has six carbon atoms. To produce a *hexose*, two PGAld molecules (or simple derivatives thereof) are made to combine "head to head" to yield a hexose biphosphate, which can then be dephosphorylated to form a hexose.

This hexose may then be withdrawn from the cycle for synthesis into sucrose and polysaccharides, or for conversion via a respiratory pathway (see Chapter 5) into the carbon skeletons of all the other organic molecules of the cell. Thus, the sugar produced from CO_2 in photosynthesis is the basic organic molecule used by the cells of higher plants for energy and for structural building blocks required by the cell.

Photorespiration

Plants using the Calvin–Benson pathway exclusively are termed *C_3 plants*, since the first stable product of photosynthesis, PGA, has three carbon atoms. In C_3 plants, a considerable fraction of the photosynthetically fixed carbon is immediately degraded, releasing CO_2 in an oxygen-consuming reaction. Since this process occurs only in the light, it has been called *photorespiration*. Photorespiration was discovered only relatively recently, since

CO_2 evolution during respiration in the light is overwhelmed by CO_2 absorption in photosynthesis. It had been assumed that respiration in the light was equal in magnitude and identical in pathway to that in the dark, but careful measurements of gas exchange immediately after switching the light on and off showed that CO_2 release was greater in the light than in the dark. This additional CO_2 release was due not to an increase in the normal respiratory processes, but to the addition of a completely different pathway, photorespiration, in the light.

Photorespiration occurs because, in the presence of oxygen, the enzyme RuBP carboxylase of the Calvin-Benson cycle can function as RuBP *oxygenase* in addition to its regular function of adding CO_2 to RuBP. In its *oxygenase* capacity, it adds oxygen to RuBP, converting it to one molecule of phosphoglycolate (containing two carbons) and one molecule of PGA, instead of

Figure 4-15 A scheme for carbon metabolism during photorespiration.

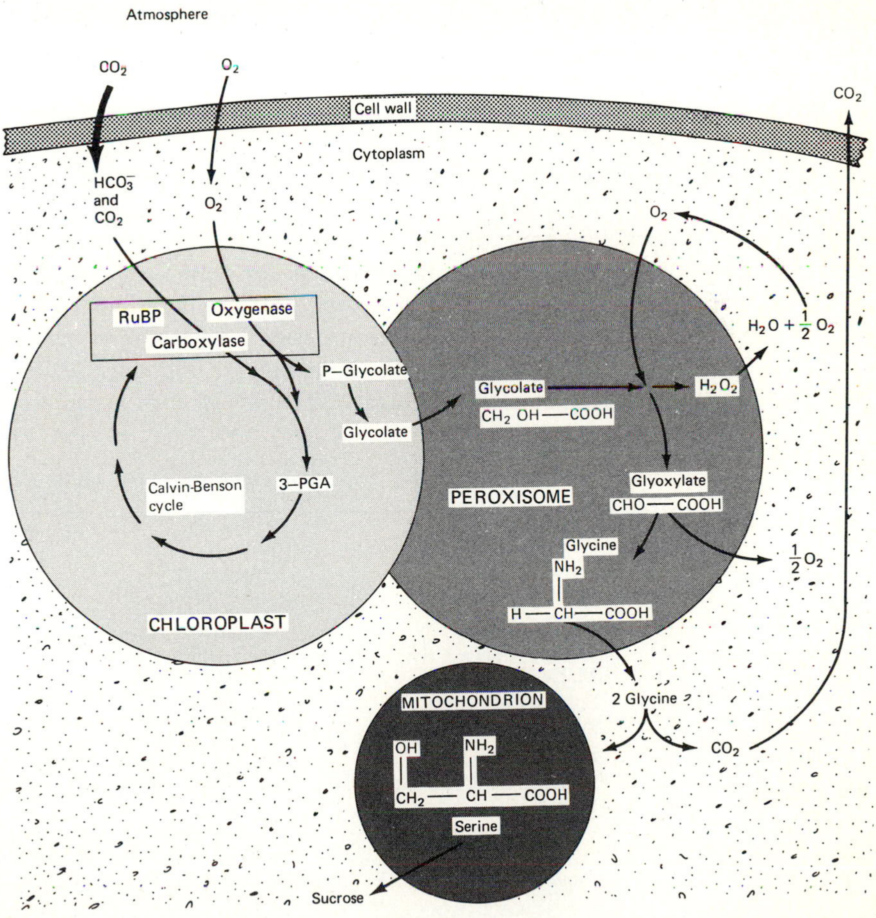

producing two PGA molecules containing three carbons each. Thus, there is no net carbon fixation when this reaction occurs. Phosphoglycolate later loses phosphate to become glycolate, which moves into another membrane-bounded organelle, the *peroxisome* (Fig. 4-15). In the peroxisome, glycolate reacts with oxygen to form glyoxylate and hydrogen peroxide; the latter is immediately broken down to water and oxygen. The glyoxylate is then converted into the amino acid glycine and, outside the peroxisome, two glycines combine to form the amino acid serine (which can be used directly in protein synthesis, or further transformed to glucose), simultaneously liberating CO_2. Thus, a part of the carbon laboriously fixed in the Calvin–Benson cycle is lost before the plant can make any further use of it. The useful function of photorespiration, if any, is not currently understood, but it may have a necessary role in the metabolism or intraorganellar transport of nitrogen compounds, by virtue of the glycolate–glycine transformation. Photorespiration may also have arisen in the context of the early environment of the earth, as photosynthesis evolved. At that time there was presumably no oxygen in the earth's atmosphere, so the production of phosphoglycolate by RuBP carboxylase would not have occurred. As the oxygen produced by photosynthesis accumulated in the atmosphere, phosphoglycolate could start to accumulate, and the evolution of photorespiration would have served to reduce this accumulation.

Photorespiration is not equally vigorous in all plants, and there is considerable variation in the efficiency with which different species fix CO_2 during photosynthesis. The photosynthetic rates of subtropical grasses such as maize, sugarcane, and sorghum, for example, are more than twice as high as those of spinach, wheat, rice, and beans (Table 4-1). The more efficient plants (termed *C_4 plants*, as discussed below) rely on a different pathway for the concentration of CO_2 in bundle sheath cells (called *C_4 metabolism*), which will be discussed shortly. The less efficient group are all C_3 plants, and

TABLE 4-1. Average yields of several leafy crops*

Crop	Type of Photosynthetic Pathway	Calculated crop growth rate (g dry wt per m² ground area per week)
Maize silage	C4	47
Sorghum silage	C4	43
Sugarcane (cane)	C4	50
Spinach	C3	13
Tobacco (leaf plus stalk)	C3	25
Hay	C3	20

*Adapted from I. Zelitch, *Photosynthesis, Photorespiration, and Plant Productivity*, New York: Academic Press, 1971.

they may lose through photorespiration up to 50% of the carbon they assimilate during photosynthesis.

Experimental control of photorespiration poses a challenge to plant physiologists, since it might be possible to double the yield of some plants if this drain on the plant's potential reserves were diminished. Several approaches are being tried. The effect of environmental conditions on photorespiration is being studied with a view to altering these conditions, where possible, to minimize photorespiration. High [CO_2], low [O_2], and low light intensity limit photorespiration, and this is part of the reason that "CO_2 fertilization" increases the growth rate of many plants. Chemical inhibitors of photorespiration without toxicity to the plants or the animals feeding on them are also being sought. Plant breeders are searching for variants or mutants with low rates of photorespiration, and are trying to incorporate this trait into seed stock. However, inhibition of photorespiration in some plants may have deleterious consequences. Recently, scientists studying the growth of soybeans, which have high rates of photorespiration, found that plants exposed to low oxygen levels photorespire less and grow better vegetatively than plants in normal air, but fail to produce harvestable seeds if the oxygen level is below 5%. Thus photorespiration itself, or some of the reactions that accompany it, may be necessary for the orderly progression of the life cycle in some plants. Indeed, it is difficult to understand how photorespiration could have evolved and been maintained unless it had some adaptive value.

C_4 Photosynthesis

For several years, it was believed that RuBP is the initial carbon acceptor, and that 3-carbon compounds are the first stable products formed during photosynthesis in all plants. This view was altered by evidence that 4-carbon organic acids, such as oxaloacetate, malate, and aspartate, are labeled far more rapidly than PGA when certain plants, including maize, sugarcane, and related tropical grasses, are fed $^{14}CO_2$; accordingly, these plants are called *C_4 plants.*

The anatomy of the leaves of many C_4 plants differs from that in C_3 plants. Each vascular bundle is surrounded by a layer of large parenchyma cells, the *bundle sheath,* which is surrounded in turn by a layer of smaller mesophyll cells (Fig. 4-16), an arrangement called *Kranz anatomy.* Chloroplasts in the two types of cells are morphologically distinct; those in the bundle sheath cells have very large starch grains but often lack grana, whereas those in the mesophyll cells contain well defined grana but accumulate very little starch (Fig. 4-17). When the two cell types are experimentally separated and the activity of enzymes in each is examined, marked differences are revealed. The activity of phosphoenolpyruvate (PEP) carboxylase, the enzyme

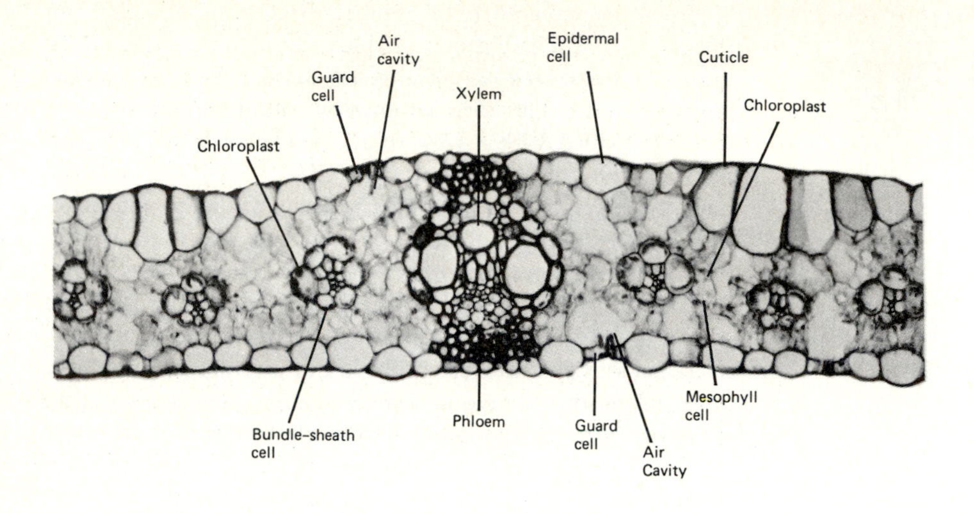

Figure 4-16 Cross section of a leaf in maize, a C_4 plant. Note the ring of tightly packed bundle sheath cells surrounding each vein and the loosely packed mesophyll cells with large interspersed air spaces. (Courtesy of G. Berlyn, Yale University.)

that catalyzes the addition of CO_2 to PEP to form oxaloacetate, is higher in mesophyll than in bundle sheath cells, while the activity of RuBP carboxylase and other enzymes of the Calvin–Benson cycle is higher in bundle sheath than in mesophyll cells. It now appears that the two groups of cells in C_4 plants cooperate in the photosynthetic conversion of carbon dioxide to hexose sugars, followed by the conversion of sugar to starch.

The first reaction takes place in mesophyll cells, where PEP reacts with CO_2 entering from the atmosphere, to form oxaloacetic acid, which is then converted to malic acid in some plants, and to aspartic acid in others. According to one interpretation, malic acid or aspartic acid then diffuses from the mesophyll cells to the bundle sheath cells, where it is decarboxylated to yield CO_2 and a 3-carbon compound. The latter diffuses back to the mesophyll, where PEP is regenerated, and the carboxylation cycle is repeated, utilizing another molecule of CO_2 from the atmosphere. Meanwhile, the CO_2 liberated in the bundle sheath cell enters the Calvin–Benson cycle, reacting with RuBP to form PGA and other intermediates produced in C_3 plants, leading finally

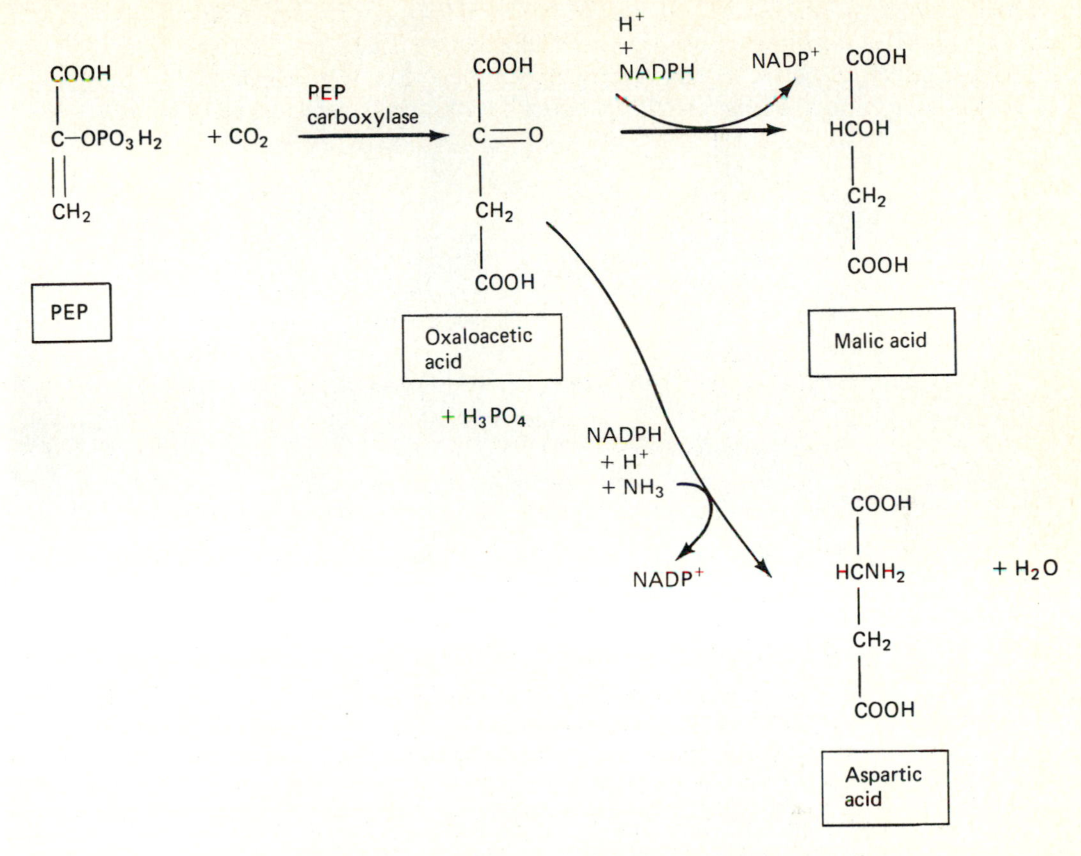

to 6-carbon sugar phosphates. The overall scheme is summarized in Fig. 4-18. Thus, most of the net gain in carbon, even in C_4 plants, occurs via RuBP carboxylase, while the final reaction, involving the conversion of hexose to starch, occurs in the bundle sheath cells.

As noted earlier, C_4 plants are more efficient photosynthetically than C_3 plants, partially because C_4 plants do not photorespire appreciably and thus do not waste the carbon already fixed. This behavior may be related to Kranz anatomy. Photorespiration involves glycolate formation and subsequent breakdown in the presence of oxygen; in C_4 plants the final fixation of the CO_2 by the Calvin–Benson cycle takes place in the bundle sheath cells which are tightly appressed to neighboring cells. It has been suggested that in such compact tissue, with no intercellular spaces, the oxygen level is very low, and that this limits photorespiration. Futhermore, in the absence of intercellular spaces around the bundle sheath cells, CO_2 may also have difficulty reaching the chloroplasts; thus if these cells function alone they cannot be efficient in photosynthesis. The C_4 cells of the loosely compacted mesophyll might then function by collecting and concentrating the CO_2 into the C_4 organic acids

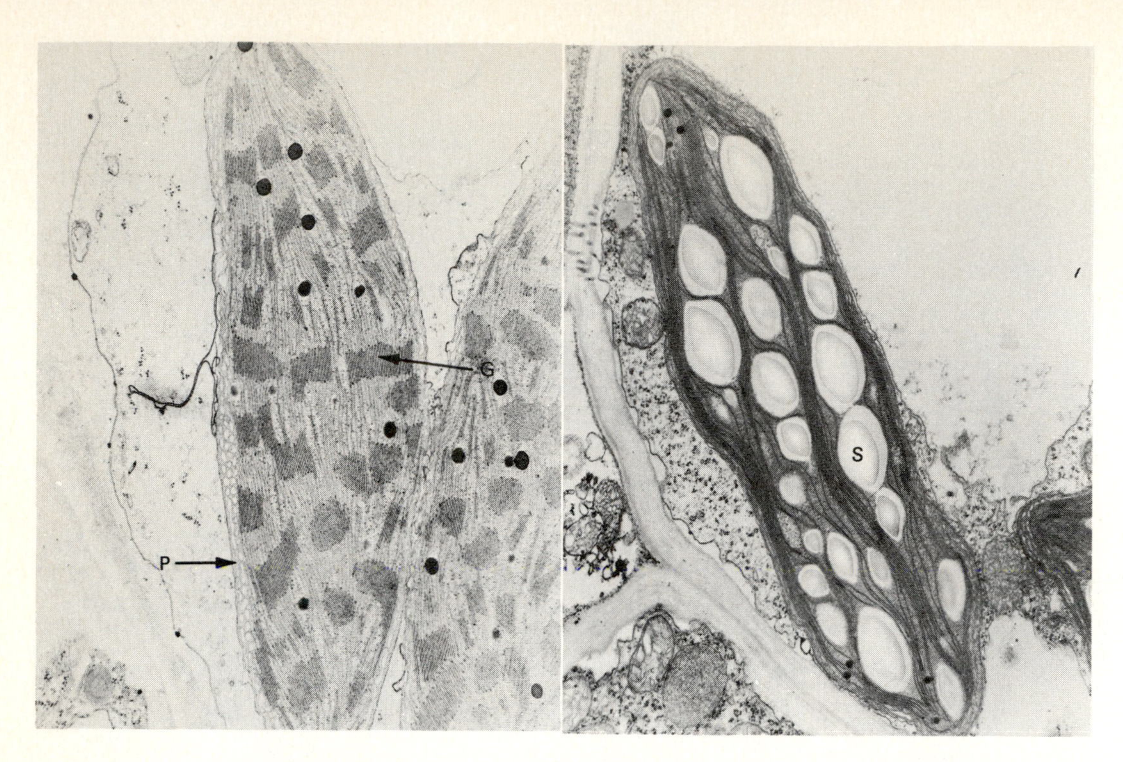

Figure 4-17 Electron micrographs of chloroplasts in a mesophyll cell (× 15,500) (left) and a bundle of sheath cell (× 14,500) (right) in maize, a C_4 plant. Note the numerous large grana stacks (G) and lack of starch grains in the mesophyll plastid; the large starch grains (S) and scarcity of grana stacks in the bundle sheath plastid. Also note the prominent membranous network, called peripheral reticulum (P) in the cytoplasm of the mesophyll cell. (From A. J. Kirchanski, 1975. Amer. J. Botany *62*: 695–705.)

through PEP carboxylase. In fact, the enzyme PEP carboxylase is extremely efficient at this task and will fix CO_2 into organic acids at much lower CO_2 concentrations than will RuBP carboxylase. These organic acids thus formed could then be transported to the chloroplasts of the bundle sheath cells, where the CO_2 is released in high concentration and in a relatively oxygen-free atmosphere, permitting the chloroplasts of the bundle sheath cells to fix the released CO_2 with high efficiency into sugars via the Calvin–Benson cycle. C_4 fixation can thus be considered to be a CO_2 primer pump for the C_3 pathway. The location of the bundle sheath cells also enables them to pass their final products of photosynthesis, such as sucrose, directly into the phloem sieve tubes for transport to other parts of the plant.

This entire explanation, however logical, may not be biological, since even undifferentiated tissue cultures of C_4 plants have recently been shown to retain low rates of photorespiration even though they are derived solely from mesophyll cells and lack the Kranz anatomy. The explanation may thus be more chemical than structural.

The more efficient utilization of CO_2 by C_4 plants enables them to fix CO_2 at lower CO_2 concentrations than can C_3 plants. In fact, if a C_3 and a C_4

TABLE 4-2. Differences between plants having C_4 or C_3 cycles of primary photosynthetic carboxylation.

	C_4 PLANTS	C_3 PLANTS
CO_2 compensation point	0–5 ppm	30–100 ppm
Carboxylation product	Oxaloacetic acid (C_4)	PGA (C_3)
CO_2 acceptor	PEP	RuBP
Photorespiration	Low or absent	High
Effect of O_2 (0% to 50%)	None	Inhibitory
Chloroplasts	One or two kinds	One kind
Leaf veins	Well developed bundle sheath, many chloroplasts	Poorly developed bundle sheath, few chloroplasts
Photosynthetic efficiency	High	Usually lower
Maximum rate of photosynthesis	High	Low to high
Productivity	High	Low to high
Effect of high temperature	Stimulates net CO_2 uptake	Inhibits net CO_2 uptake

*From R. G. S. Bidwell, *Plant Physiology*, New York: Macmillan, 1974.

plant are placed together under a bell jar in a limited CO_2 supply, the C_3 plant will be starved by the C_4 plant and will die. This occurs because the C_3 plant photorespires away some CO_2 which is immediately used by the C_4 plant; thus the C_3 plant "photorespires itself to death." A summary comparison of the characteristics of C_3 and C_4 plants is shown in Table 4-2.

The reasons for the evolution of the C_4 pathway are not fully understood. Many of the C_4 plants are tropical grasses, and as photorespiration is strongly increased by higher temperature, the ability to eliminate photorespiration under such conditions is clearly an advantage. However, until we are certain of the function of photorespiration, it will be difficult to understand the reasons why the C_4 pathway developed, or why some plants do not possess it. In the tropics and deserts, C_3 and C_4 plants exist side by side, with no evidence that either type will take over ecologically.

Crassulacean Acid Metabolism

Succulent plants such as *Cactus*, *Kalanchoë*, and *Sedum*, that grow in arid regions, also incorporate atmospheric CO_2 into 4-carbon compounds. The physiological behaviour of these plants differs, however, from that of other C_4 plants. The stomata of these plants are open in the night and closed during the day, in contrast to the more usual pattern of light-promoted opening and dark closure. Since the evaporative loss of water through open stomata is much lower at night than during the daytime when the temperature is maximum and relative humidity is minimum (see Chapter 6), this pattern is of obvious benefit to plants growing in deserts. These plants absorb CO_2 from the atmosphere during the night and incorporate it into 4-carbon organic acids, mainly malate. PEP is the primary 3-carbon acceptor, as in other C_4

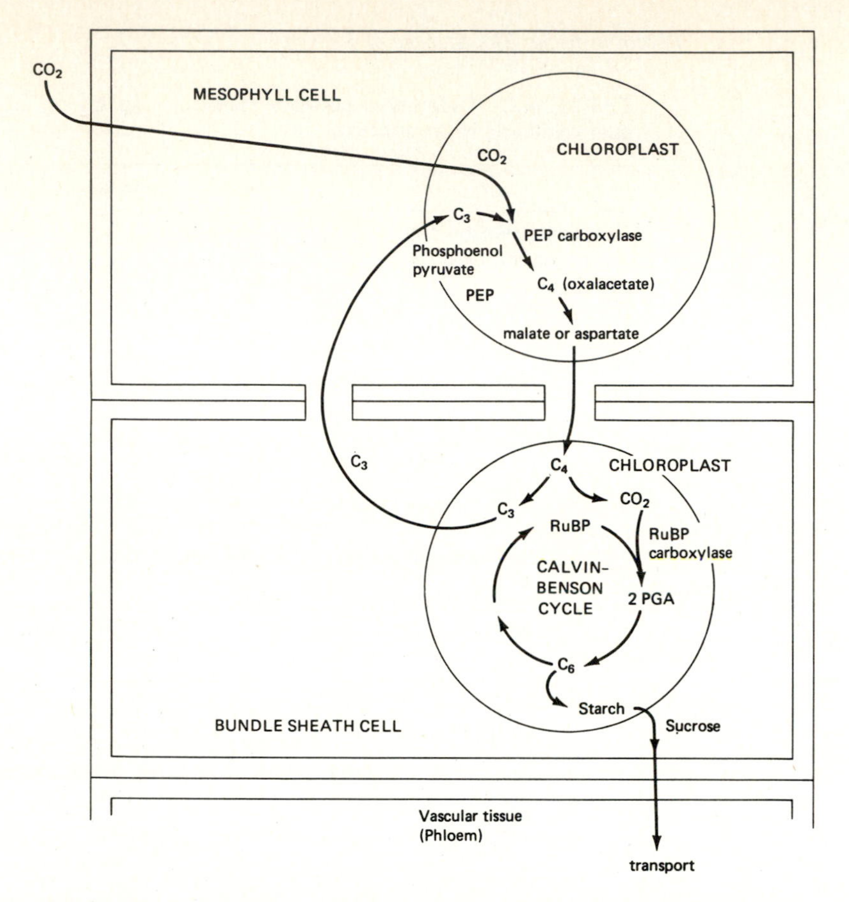

Figure 4-18 A possible scheme for envisioning the cooperative action of mesophyll and bundle sheath cells in C_4 plants. (Adapted from U. Lüttge, Stofftransport der *Pflanzen*, Berlin and New York: Springer-Verlag, 1973.)

plants. In daylight, when chlorophyll is activated by light, malic acid is decarboxylated to yield a 3-carbon compound and CO_2, and the latter is fixed into 6-carbon sugars by the Calvin–Benson cycle. PEP regenerated from the 3-carbon compound serves again as a CO_2 acceptor. This diurnal pattern of acidification in the nighttime and deacidification in the daytime is called *Crassulacean acid metabolism* (CAM), after the Crassulaceae, the family to which many of these succulents belong, and in which the phenomenon was first observed. Thus, in CAM plants, both primary carboxylation and synthesis of 6-carbon sugars take place in the same cells but at different times, while in other C_4 plants, these processes take place at the same time, but may occur in different cells. These relationships are summarized in Fig. 4-19.

The sugar phosphates formed in photosynthesis finally produced are metabolized into starch for storage, into sucrose for transport, or into other essential building blocks via the process of respiration. These processes are dealt with in the next chapter.

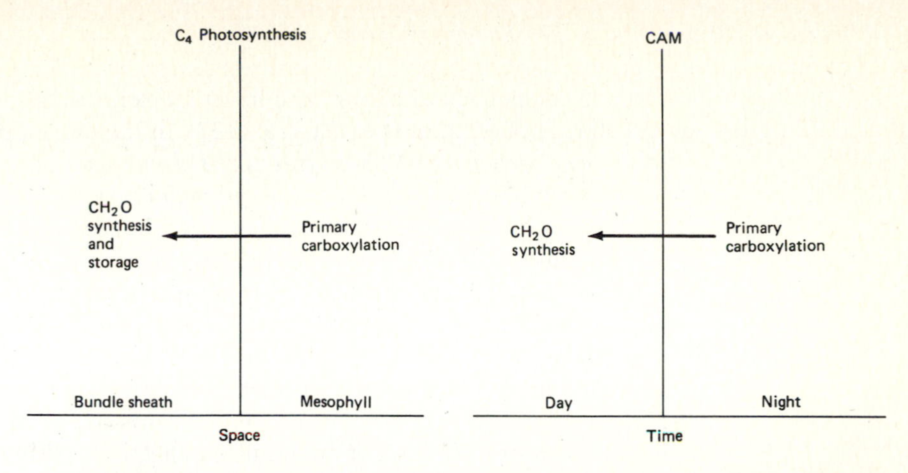

Figure 4-19 A model for the spatial and temporal compartmentation of carbon fixation in C_4 and CAM plants. (Adapted from W. M. Laetsch, 1974, Annual Rev. Pl. Physiol. 25, 27–52.)

SUMMARY

The green plant is an *autotroph*, synthesizing organic molecules through photosynthesis. In this process, carbon dioxide and water are consumed, organic matter and oxygen are formed, and radiant energy is stored as potential energy in the oxidizable organic molecules. The oxygen liberated arises from the water consumed; thus, in photosynthesis, the stable water molecule is cleaved (*photolyzed*) by the radiant energy absorbed in the chloroplast; the resulting electrons are passed through intermediary carriers and used to reduce CO_2 to the level of carbohydrates. Gas exchange in the leaf occurs through the *stomata*, while water delivery and carbohydrate removal occur through xylem and phloem, respectively.

The overall rate of photosynthesis is determined by the single internal or external factor present in relative minimum amount. For most plants, light is usually limiting, but CO_2 becomes limiting during periods of bright sun. When plants are *chlorotic* (low in chlorophyll), the pigment may be limiting. The CO_2 content of the air is about 0.033%, having risen about 10% during the present age of the burning of fossil fuels. Since CO_2 in the atmosphere reduces the escape of heat radiated from the earth, this buildup may cause significant climatic warming.

The *action spectrum* for photosynthesis, relating photosynthetic efficiency to wavelength of light, shows that chlorophyll is the main effective pigment in photosynthesis, although other accessory pigments may help by transferring their absorbed energy to chlorophyll. Two *photosystems* work together in photosynthesis; in both, the initial event following quantum absorption is the ejection of an electron, and its temporary capture by a compound with a high *oxidation–reduction* (*redox*) *potential.* The successive transfer of this electron to compounds with lower and lower redox potentials is accompanied by the storage of some of the energy as *ATP*. In the overall process, termed *photophosphorylation*, the ATP is synthesized *chemiosmotically* when a proton gradient is established across the thylakoid membrane by the vectorial transfer of protons by the carrier plastoquinone, and by protons from photolysed water molecules. When these protons diffuse out through special channels in the thylakoid membrane, their energy is conserved through the synthesis of ATP from ADP and P_i. Some of the electrons liberated during water photolysis terminate in a reduced carrier, *NADPH*. Together with ATP, this compound can reduce CO_2 to the level of carbohydrate. ATP and NADPH are the products of the light reactions in photosynthesis; all the rest is dark-reaction chemistry. The water molecule is effectively cleaved when, after ionization, the hydroxyl ion transfers an electron to a light-excited chlorophyll molecule in Photosystem II that has ejected one of its own electrons; from the resulting hydroxyl free radical oxygen is released.

CO_2 attaches to the chloroplast enzyme called *ribulose bisphosphate* (*RuBP*) *carboxylase*, and there unites with RuBP to produce *phosphoglyceric acid* (*PGA*), a 3-carbon compound which can then be reduced to a 3-carbon sugar by ATP and NADPH. Two such 3-carbon sugars may join to form a 6-carbon sugar. Because RuBP carboxylase also has an affinity for oxygen, some of the RuBP is oxidized rather than carboxylated. The *glycolate* produced is broken down by *photorespiration* with the loss of CO_2. This process, which occurs in C_3 plants producing PGA, is minimized in C_4 plants, where another enzyme, *phosphoenolpyruvate* (*PEP*) *carboxylase*, fixes CO_2 into a 4-carbon acid. The fixed CO_2 is then released to other cells with a lower oxygen environment, where efficient conversion to PGA can occur without appreciable photorespiratory loss. Experimental control of photorespiration may lead to increased efficiency of photosynthesis.

Fleshy plants growing in arid regions open their stomata at night, thus fixing CO_2 at a time when water loss can be minimized. The CO_2 is fixed by *PEP carboxylase* into the 4-carbon compound *malic acid*, from which it is removed in the light and cycled via RuBP carboxylase into PGA and sugars. This pattern is called *Crassulacean acid metabolism* (CAM) after the family of plants in which the process was first discovered.

SELECTED READINGS

Goldsworthy, A. *Photorespiration*. Carolina Biology Reader No. 80 (Burlington, N. C.: Carolina Biological Co., 1976).

Govindjee, ed. *Bioenergetics of Photosynthesis* (New York: Academic Press, 1975).

Gregory, R. P. F. *Biochemistry of Photosynthesis*, 2nd ed. (New York: Wiley, 1977).

HINKLE, P. C., and R. E. MCCARTY. "How Cells Make ATP," *Scientific American* 238 (3): 104–123 (1978).

SMITH, K., ed. *Photochemical and Photobiological Reviews* (New York and London: Plenum Press).

TREBST, A., and M. AVRON, eds. "Photosynthesis I," *Encyclopedia of Plant Physiology*, New Series, Volume 5 (Berlin–Heidelberg–New York: Springer Verlag, 1977).

ZELITCH, I. *Photosynthesis, Photorespiration, and Plant Productivity* (New York: Academic Press, 1971).

See Selected Readings *on page 59 for biochemistry background.*

QUESTIONS

4-1. The sun's radiation covers a broad range of wavelengths from very short-wavelength cosmic rays to long-wavelength radio waves (see Fig. 1-3). Nevertheless both plants and animals absorb only a small fraction of this energy, in roughly the same wavelength region of 400 to 750 nm. How do you account for this?

4-2. The concentration of carbon dioxide in the atmosphere of temperate regions varies in a cyclic manner during the course of a year, with values about $1\frac{1}{2}\%$ higher in winter than in summer. How do you explain this?

4-3. What is the relationship between the light and dark reactions of photosynthesis? What quantitative information about this relationship is revealed in Fig. 4-5b?

4-4. In Chapter 2, we described the chloroplast ultrastructure. Discuss some relationships between chloroplast fine structure and the function of this organelle.

4-5. Why is chlorophyll considered to be the absorbing pigment in photosynthesis when many other light-absorbing pigments are present in a leaf? What is the function of the other pigments?

4-6. What evidence leads us to believe that photosynthesis is composed of a series of discrete reactions, rather than of one step?

4-7. What molecules produced during the light reactions of photosynthesis are utilized in the fixation of CO_2?

4-8. Emerson and Arnold found in 1932 that the photosynthetic yield per unit light could be increased if the light, instead of being administered continuously, were given as brief flashes interrupted by longer dark periods. How can this phenomenon be explained?

4-9. The herbicide DCMU can inhibit photosynthesis without greatly affecting respiration. Explain how this finding may be of use in photosynthesis research.

4-10. If a green leaf is illuminated in the absence of CO_2, it may be observed to fluoresce. The admission of CO_2 to this leaf results in the immediate quenching of fluorescence. What is a possible explanation?

4-11. Trace the movement and metabolic fate of a hydrogen atom of a water molecule from the time it falls on the soil as rain until it ends up in a molecule of starch in the chloroplast of a leaf.

4-12. How do electrons and protons cooperate to make ATP in photosynthesis?

4-13. Plants growing in soils deficient in certain minerals often have low rates of photosynthesis. Discuss some mineral deficiencies that would have this effect.

4-14. Two important enzymes in plants are ribulose bisphosphate (RuBP) carboxylase and phosphoenolpyruvate (PEP) carboxylase. Describe the reaction in which each is involved, their location in leaves, and their functional interrelations.

4-15. What is the raw material fixed in photosynthesis in the Calvin–Benson cycle? At what stage is the material fixed, what is the final product, what provides the energy to drive the process, and for what, chemically, is the energy used?

4-16. Under what conditions is RuBP carboxylase likely also to function as RuBP oxygenase? What would be the likely consequences of such a reaction?

4-17. Why does photorespiration not occur in C_4 plants?

4-18. Does light affect the rate of respiration? Give an explanation for your answer.

4-19. If photorespiration produces amino acids (glycine, serine) why is it described as an inefficient or wasteful process?

4-20. What types of experiments would you perform to determine whether a plant is of the C_3 or C_4 type?

4-21. How does photosynthesis in succulent plants differ from photosynthesis in C_3 or C_4 mesophytic plants?

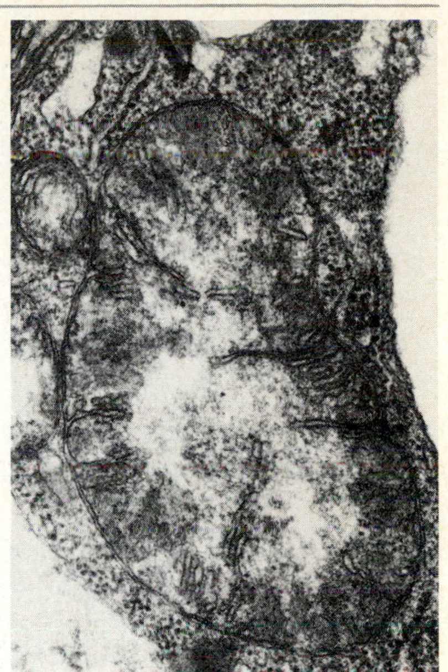

Respiration and Metabolism: Supplying Energy and Building Blocks

In the preceding chapter, we saw how the sun's energy is used during photosynthesis to store energy and carbon in the form of sugar phosphates. From these phosphorylated hexoses are derived not only the basic carbon structures of all the other compounds in the plant but also the energy required to synthesize them. A study of the process of respiration will reveal some of the detailed mechanisms involved in the transformations of the sugar molecule that yield energy and new carbon skeletons.

Energy Conservation and Utilization

Let us first consider how energy is stored, released, and utilized. The chemical energy in organic compounds, a transduced form of solar energy, is stored in their bond structure and is released if the bonds are broken, usually in the process of oxidation. When an organic product such as wood is burned (oxidized) the energy is given off all at once, appearing mainly as heat. For the plant to release such vast amounts of energy all at once would be wasteful, since the energy could not be utilized in orderly, stepwise, constructive processes. Living organisms manage to make use of chemical bond energy by accomplishing the oxidation in many steps, so that the energy is released in small packets which can immediately be used in other processes. The released energy is used to form new energy-rich chemical bonds, often in the form of the "energy currency" molecule ATP discussed previously (Fig. 5-1).

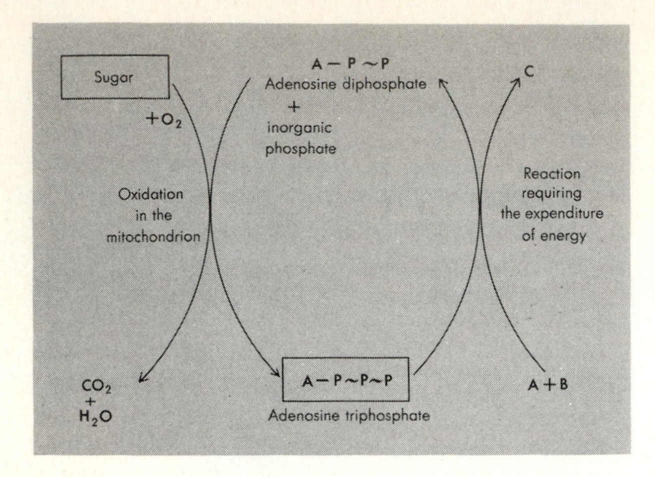

Figure 5-1 In the oxidation of sugar by the cell, energy is stored in the form of special "energy-rich" phosphate bonds in the molecule of adenosine triphosphate (ATP). The ordinary phosphate bond is represented by —P, the energy-rich bonds by ~P. The ATP can then be used to drive reactions requiring the expenditure of energy, such as the union of two small molecules (A, B) to form a larger one (C).

One of the principal forms of stepwise oxidation is the electron transport chain of the mitochondrion. In this process, an electron is passed from one carrier to another, each carrier being at a slightly lower energy level than the former. As the electron moves "downhill" from a partially reduced compound (sugar) to its final union with oxygen, it releases a little of its energy at each stage, and some of this energy is eventually trapped in ATP. In photosynthesis, we considered two electron transport chains, one for cyclic photophosphorylation in Photosystem I and another for noncyclic photophosphorylation between Photosystems II and I. These systems yielded the ATP and reducing power (NADPH) used to fix CO_2 and convert it to the level of a sugar. ATP for most other reactions of the cell, however, is produced through *oxidative phosphorylation;* this process occurs during respiration, and uses another electron transport chain that we shall consider shortly.

Not only must organic molecules be broken down by gradual steps to permit efficient energy capture and utilization, but the synthesis of complex molecules such as proteins, nucleic acids, lipids, or polysaccharides requires multiple steps. Often the simple molecules which will form the larger molecules must be activated so that they contain enough energy to consummate the reaction. Thus, we will usually find that the synthesis of a molecule follows a route that is more intricate than its degradation.

Sucrose and Polysaccharide Synthesis

The hexose phosphate actually produced in the Calvin–Benson cycle is fructose-6-phosphate (F-6-P). This compound is readily converted by specific enzymes to other phosphorylated hexoses such as glucose-6-phosphate (G-6-P) and glucose-1-phosphate (G-1-P), and the reverse process can occur equally easily.

$$\text{F-6-P} \rightleftharpoons \text{G-6-P} \rightleftharpoons \text{G-1-P}$$

These three hexose phosphates are then assembled into chains of sugar molecules used in transport, storage, and synthetic reactions. Before they can be further transformed they must be activated, which usually occurs through their attachment to *nucleotides*, complex ring structures related to the adenylate of ATP. The product of such a union is a sugar–nucleotide. One of the most frequently encountered sugar–nucleotides is uridine diphosphoglucose (UDPG), made from the reaction of uridine triphosphate (UTP) and glucose-1-phosphate (G-1-P). The UTP is itself made indirectly through transfer of a phosphate from ATP to UDP (uridine diphosphate).

$$\text{UDP} + \text{ATP} \longrightarrow \text{UTP} + \text{ADP}$$

$$\text{UTP} + \text{G-1-P} \longrightarrow \text{UDPG} + \underset{\text{(Pyrophosphate)}}{\text{PP}_i}$$

The nucleotides ATP and UTP are present in all cells, for they are used, together with other nucleotides, in the synthesis of DNA and RNA.

Sugars are transported through the plant in the form of sucrose, a disaccharide composed of a glucose molecule attached to a fructose molecule (Fig. 5-2). Sucrose is synthesized through the reaction of UDPG with F-6-P.

$$\text{UDPG} + \text{F-6-P} \longrightarrow \text{GF-6-P} + \text{UDP}$$

$$\text{GF-6-P} \longrightarrow \underset{\text{(Sucrose)}}{\text{GF}} + \text{P}_i$$

This reaction strongly favors the synthesis of sucrose and permits the accumulation of high concentrations of this sugar. Before it can be used further, the sucrose must be broken down; the linkage between the glucose and fructose is hydrolyzed by the enzyme *invertase* to produce free glucose and fructose.

$$\text{GF} \xrightarrow[\text{H}_2\text{O}]{\text{Invertase}} \text{G} + \text{F}$$

This process "wastes" the bond energy between the two sugars; thus the glucose and fructose must again be phosphorylated by ATP before they can

Figure 5-2 The sucrose molecule, a disaccharide.

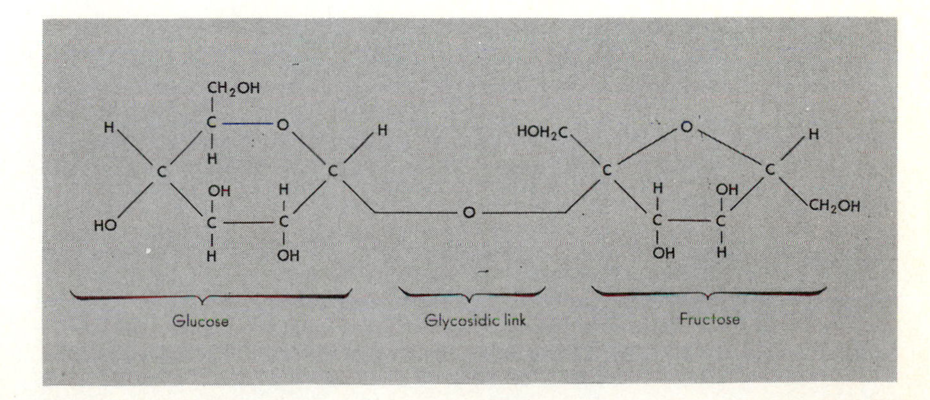

be broken down in respiration or used for the synthesis of polysaccharides. Sucrose synthesis and breakdown illustrate the divergent pathways often used in *anabolic* (building-up) and *catabolic* (breaking-down) reactions.

Starch and cellulose synthesis

Both starch and cellulose are long polymeric chains made exclusively of glucose units, but they differ in the way in which their glucose units are joined. This structural variation makes a considerable difference in the nature of the two glucose polymers (glucans) formed; for example, starch is readily digested by man, while cellulose is not. The main difference is that starch has

Figure 5-3 The structures of starch (left) and cellulose (right). Note that both of these polysaccharides have the same chemical formula, but their oxygen bridges are arranged differently in space. Starch, the plant's main storage carbohydrate, is made up of two distinguishable components: amylose, with its long unbranched chains of glucose, and amylopectin, composed of many short branched chains.

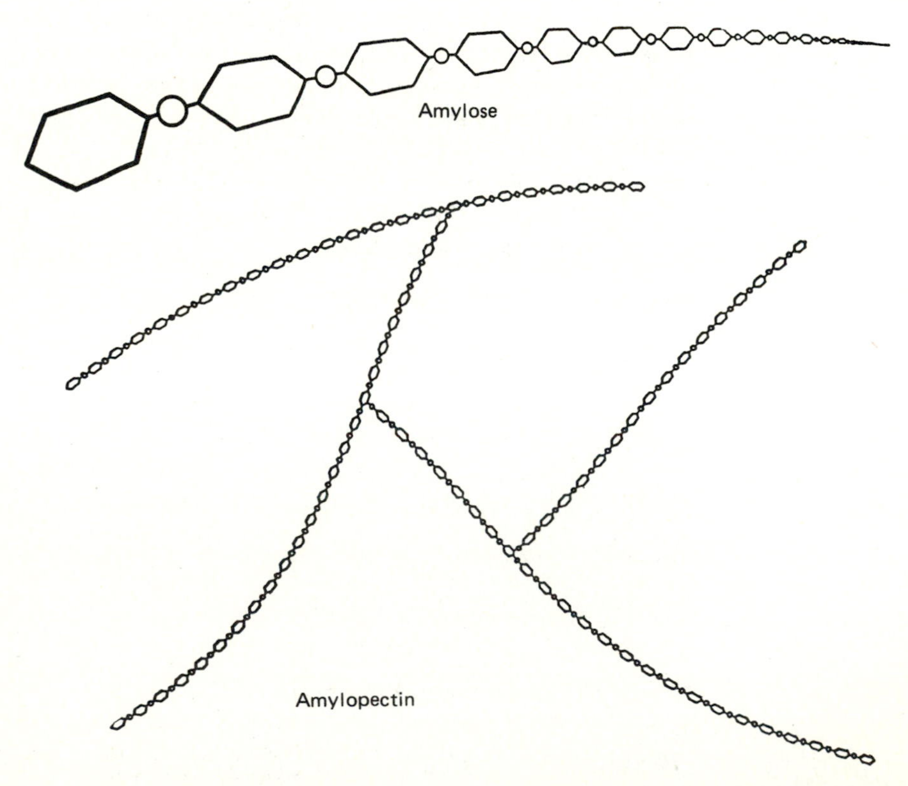

an α-linkage between carbon atoms 1 and 4 of two glucoses, while cellulose has a β-linkage between these two carbons (Fig. 5-3). Starch can be either a linear polymer, composed exclusively of α-1,4- glycosidic linkages, or branched, through the formation of 1,6- linkages as well. The linear form of starch is called *amylose*, the branched form, *amylopectin*. The difference in the linkage alters the spatial arrangement of the glucose chains. *Starch* is the main food-storage polysaccharide. It is insoluble and is laid down layer on layer in *starch grains* in chloroplasts (Fig. 2-17), or in chlorophyll-less leucoplasts in storage tissue of stems, roots or seeds. Sometimes storage cells are

Cellulose, the main component of the primary cell wall, exists in long chains that bond together to form micellar strands, which then bond to form microfibrils. The microfibrils, which are large enough to be seen in the electron microscope, form the warp and woof of the cell wall. (Modified from J. Bonner and A. W. Galston, *Principles of Plant Physiology*, San Francisco: W. H. Freeman and Co., 1952.)

Chemical structure of cellulose

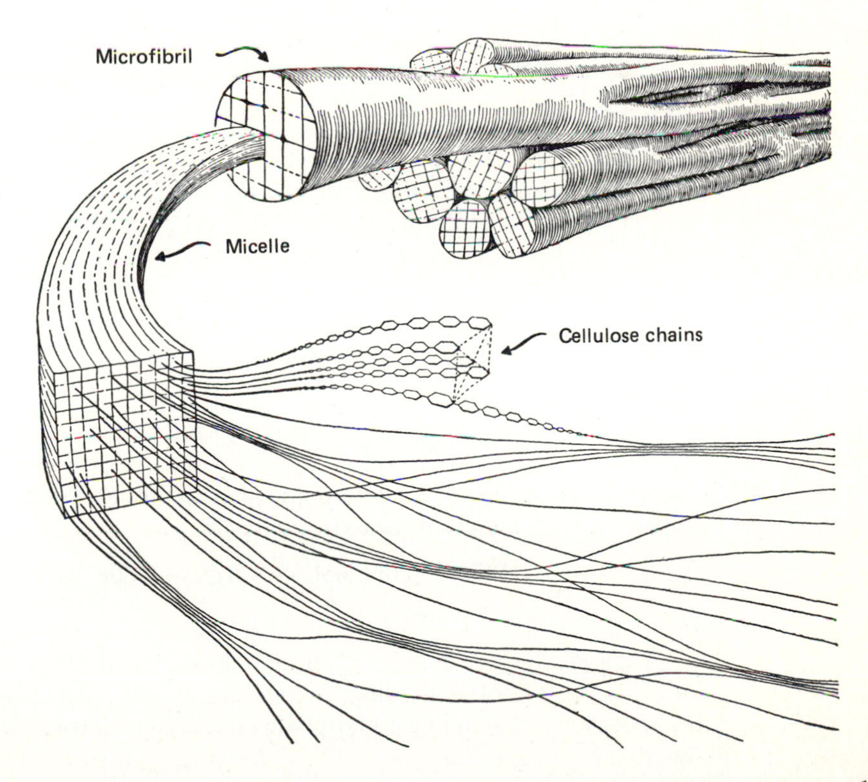

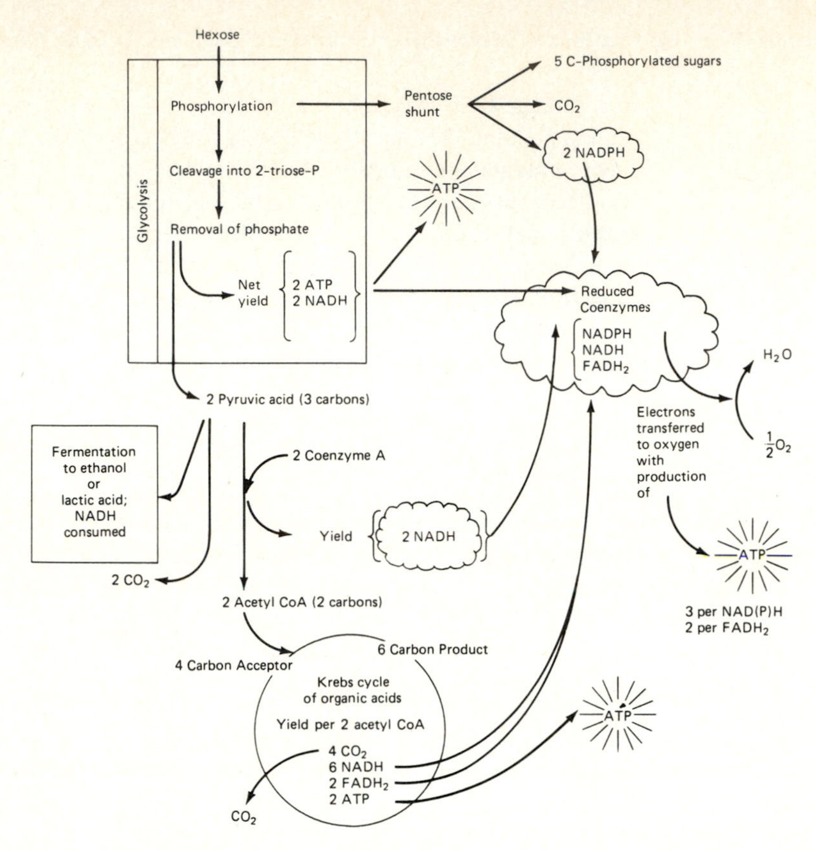

Figure 5-4 The oxidation of glucose to CO_2 and concomitant synthesis of ATP.

completely filled with starch grains, which can be easily visualized as they stain blue-black with iodine. Since the starch is insoluble, it does not, like sucrose or hexoses, create osmotic effects in the cell (see Chapter 6). Its formation in leaf cells during times of intense photosynthesis thus prevents inhibition of photosynthesis by an accumulation of the reaction product. During darkness, the starch is slowly hydrolyzed back into glucose phosphate, which is then changed into sucrose for transport.

Starch synthesis utilizes adenosine diphosphoglucose (ADPG), which is made from ATP and G-1-P:

$$\text{ATP} + \text{G-1-P} \longrightarrow \text{ADPG} + \text{PP}_i$$

The starch molecule is built up one glucose unit at a time by the reaction of ADPG with the preformed glucose chain:

$$\text{ADPG} + [\text{Glucose}]_n \longrightarrow \text{ADP} + \underset{\text{(Starch)}}{[\text{Glucose}]_{n+1}}$$

When the sucrose supply is low, starch is broken down to be reformed into sucrose. The starch is first broken into glucose units, the bond energy being conserved by adding a phosphate onto each to form G-1-P.

$$\underset{\text{(Starch)}}{[\text{Glucose}]_n} + P_i \xrightarrow{\text{Starch phosphorylase}} [\text{Glucose}]_{n-1} + \text{G-1-P}$$

The G-1-P can then enter the sucrose synthesis reaction described above. In seeds and other organs in which massive starch digestion occurs, the starch is broken down to disaccharide units termed *maltose* (G—G) by the enzymes α and β amylase. The maltose is then further degraded to glucose before being resynthesized into sucrose for transport. This second pathway does not conserve the bond energy as does the first, and ATP is needed to reform G-6-P.

$$\underset{\text{(Starch)}}{[\text{Glucose}]_n} \xrightarrow{\alpha\text{-Amylase}} \underset{\text{(Maltose)}}{\text{G-G}} \xrightarrow{\text{Maltase}} \underset{\text{(Glucose)}}{\text{G}}$$

Cellulose, the most abundant carbohydrate on earth, forms the principal component of the primary cell walls. It is formed in a manner analogous to starch except that the glucose donor is another nucleotide sugar, guanosine diphosphoglucose (GDPG), and the linkage between the sugar units is β-instead of α-.

$$\text{GDP} + \text{ATP} \longrightarrow \text{GTP} + \text{ADP}$$

$$\text{GTP} + \text{G-1-P} \longrightarrow \text{GDPG} + \text{PP}_i$$

$$\text{GDPG} + [\text{Glucose}]_n \longrightarrow \text{GDP} + \underset{\text{(Cellulose)}}{[\text{Glucose}]_{n+1}}$$

In some cases, UDPG may also serve as a glucose donor for cellulose.

Except through microbial decay, cellulose is seldom broken down in higher plants. Notable exceptions are in the cells of the leaf *abscission zone* before the leaf is shed and in the digestion of the end walls of xylem vessels. In the abscission zone, the enzyme *cellulase* destroys the wall structure by breaking down the cellulose into individual glucose units. The weakened wall structure then ruptures, and the leaf abscises.

The cellulose microfibrils in the cell wall are joined by a matrix of mixed polysaccharide chains including xyloglucans and arabinogalactans (Fig. 2-26). (Xylose and arabinose are 5-carbon sugars (pentoses); galactose is a hexose related to glucose.) These polysaccharides are also made from sugar–nucleotide precursors, primarily in dictyosome vesicles which fuse with the plasmalemma, thus discharging their contents into the developing cell wall.

All polysaccharide molecules are thus easily interconverted, but synthesis is always via sugar–nucleotides whereas breakdown is more direct.

Respiration

During respiration, in which carbohydrate is oxidized to CO_2 and H_2O, the energy stored in carbohydrate molecules is tapped for the plant's *endergonic* (energy-requiring) activities (Fig. 5-1). At the same time, the many intermediate products which are produced are diverted for use as building blocks for all the other molecules needed by the cell.

The oxidation of hexose sugar to CO_2 and H_2O does not occur in a single step, but involves numerous enzyme-catalyzed reactions and various cellular organelles (Fig. 5-4). Energy is transduced at several of these steps, before it is ultimately used to form ATP from ADP and P_i. Thirty-eight molecules of ATP are formed during the complete oxidation of one molecule of hexose sugar. The overall reaction is:

$$C_6H_{12}O_6 + 6O_2 + 38ADP + 38P_i \longrightarrow 6CO_2 + 6H_2O + 38ATP$$

Glycolysis

Respiration occurs in three stages. The first, an anaerobic series of reactions called *glycolysis*, or the *Embden–Meyerhof–Parnas* (*EMP*) *pathway*, takes place in the cytoplasm, where hexose sugars are cleaved and partially oxidized to the 3-carbon organic acid, pyruvic acid, generally considered in its ionized form, the pyruvate ion. Since hexose

Figure 5-5 Glycolysis and fermentation.

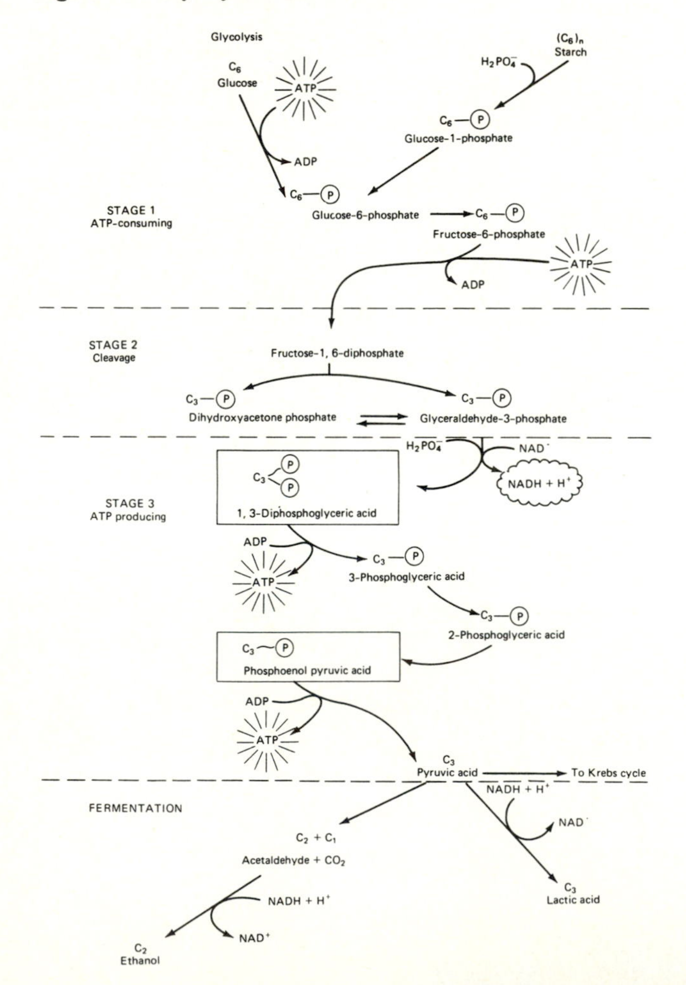

sugars are relatively stable molecules, metabolic energy must be expended to activate the initial reactions in glycolysis. Two molecules of ATP are used in the conversion of hexose to hexose diphosphate; the phosphorylated hexose is then cleaved into two triose phosphates which are in turn oxidized to pyruvate via a series of intermediate reactions. Four molecules of ATP are synthesized in these latter reactions, thus yielding an overall gain of two molecules of ATP during glycolysis. In addition to the direct production of ATP, some of the energy released in glycolysis is used for the reduction of the cofactor nicotinamide adenine dinucleotide (NAD^+)* to its reduced form, NADH. This energy is later converted to ATP by an electron transport chain in oxidative phosphorylation.

If oxygen is available, the pyruvate formed during glycolysis enters the second phase of respiration, in which it is oxidized to CO_2 and H_2O with the formation of additional molecules of ATP. If oxygen is unavailable, pyruvate enters into fermentative reactions (see Fig. 5-5) which do not lead to further significant ATP synthesis. Oxygen deficiency is a common problem for plants growing in poorly drained soil, for when the soil is waterlogged, its oxygen concentration is suboptimal. This limits aerobic respiration, and therefore ATP synthesis, in root cells. Since ATP is used by the plant in taking up minerals from the soil, plants in poorly drained soil often show severe mineral deficiency symptoms (see Chapter 7).

The Krebs cycle

The enzymes that catalyze the oxidation of pyruvate to CO_2 are located in the inner compartment of the mitochondrion (Fig. 5-6). By contrast, the reduced cofactors produced during this process are oxidized by enzymes located on the inner mitochondrial membrane, with the concomitant efficient production of ATP and H_2O. When pyruvate is first split, losing CO_2 (*decarboxylation*) to form acetate, this 2-C fragment is coupled to a substance called *coenzyme-A* (co-A). The resulting compound, acetyl co-A, is fed directly into the *tricarboxylic acid* (*Krebs*) *cycle*, which constitutes the second of the three stages of respiration (Fig. 5-7). The 2-carbon fragment of acetyl co-A is first transferred to an "acceptor" 4-carbon organic acid, forming a 6-carbon compound, *citric acid.* This 6-carbon acid is successively transformed and broken down, one C atom at a time, to the original 4-C acceptor, the two C atoms of the attached acetyl unit being released as CO_2.

At several of the transformation steps, energy is released. This energy

*It takes two electrons and one proton (H^+) to reduce NAD^+ to NADH. These come from two hydrogen atoms (H), the extra proton being released into the medium. Thus: $NAD^+ + 2H \longrightarrow NADH + H^+$

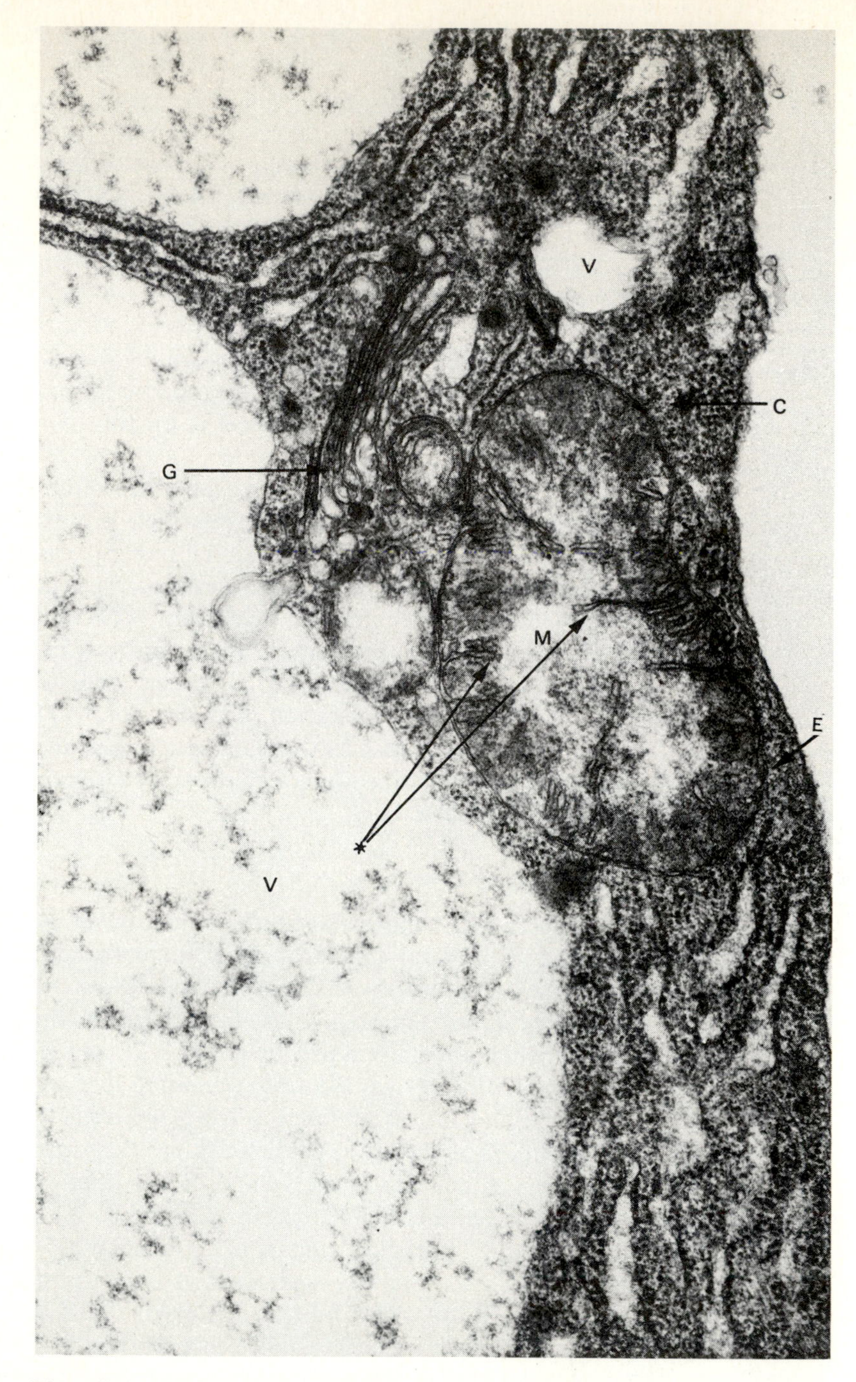

Figure 5-6 Electron micrograph showing a mitochondrion (M) in a cell in a *Phaseolus vulgaris* (bean) cotyledon. Note the outer envelope membrane (E) and the cristae (*) projecting from the inner membrane. (Compare this micrograph to the schematic drawing in Fig. 2-15). V = vacuole; C = cytoplasm; G = Golgi apparatus. (Courtesy of M. Boylan, Yale University.)

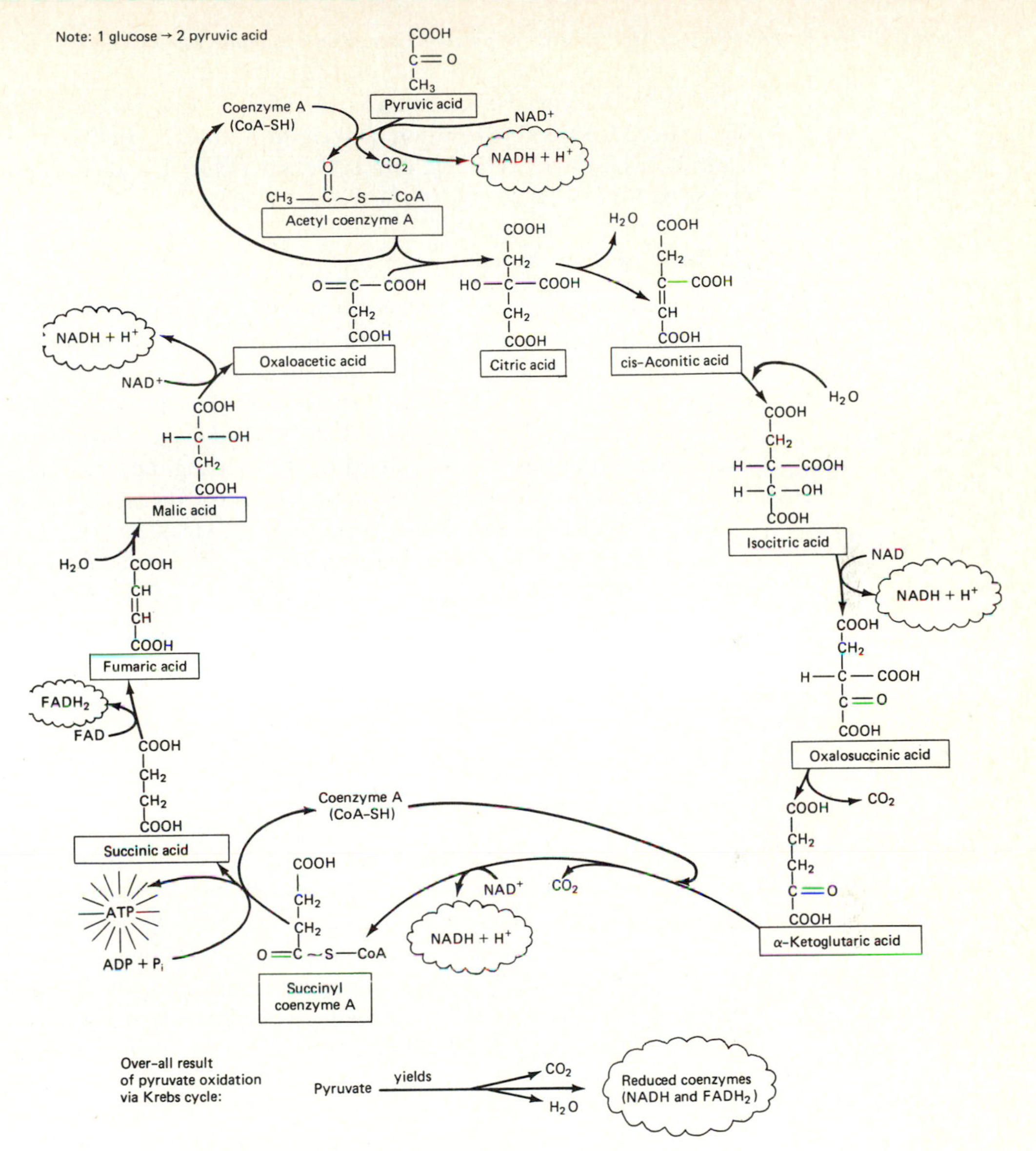

Figure 5-7 The Krebs Cycle.

is either directly used in ATP synthesis in membrane-bound systems or is used to reduce soluble respiratory carriers. Among these carriers are NAD^+, which is reduced to NADH, and the flavin adenine dinucleotide FAD, which is reduced to $FADH_2$. In these reductions, the proton and electron come from the same hydrogen atom in the substrate molecule. In some reductions occurring in the remaining electron transport chain, only electrons are transferred. When needed, H^+ can come from or be released to water in the cell, in which an abundance of H^+ is always available.

The net result of the Krebs cycle, therefore, is that each 2-C acetyl unit formed from 3-C pyruvate is broken down to CO_2. In the process, some electron carriers (NAD^+ and FAD) are reduced, and ATP is also produced.

Oxidative phosphorylation

No free oxygen has thus far been involved in any reaction, yet oxygen consumption is one of the key components of what we term *respiration*. Oxygen enters the picture because much of the energy originally stored in the hexose is now in the form of the reduced carriers NADH and $FADH_2$, from which it is released in the third stage of respiration, when these carriers are again oxidized by transferring their electrons to free oxygen. Since a considerable amount of energy is trapped in these reduced cofactors, it must again be released in steps by passing the electron from the cofactors to a series of spatially oriented protein-bound electron carriers in an electron transport chain. Each of the successive carriers is at a lower level of reduction than its predecessor, and thus contains less energy. The final acceptor for the electron is molecular oxygen, which together with H^+ obtained from the surrounding water medium forms another molecule of water:

$$H \rightleftharpoons H^+ + e^-$$

$$4H^+ + 4e^- + O_2 \longrightarrow 2H_2O$$

The electron transport system (ETS) consists of several carrier compounds which can be in a reduced or oxidized state (Fig. 5-8). These carriers are enzymes, of which various vitamin derivatives are the coenzymes. They are located in the inner membrane of the mitochondrion (Fig. 2-15 and 5-6), much as the electron carriers in photosynthesis are located on the thylakoids of the chloroplast. The final carriers are heme-containing *cytochromes* in which the electron reduces the iron from the Fe^{+++} (ferric) to the Fe^{++} (ferrous) state; from these the electrons pass to oxygen. At several stages of electron transfer in the carrier chain, energy is released and used to synthesize ATP from ADP and inorganic phosphate (Fig. 5-8). It is thought that each NADH which passes its electron to the electron transport chain leads to the production of three ATPs, while each $FADH_2$ produces two ATPs. Since the ATP is produced through the oxidation of each previous carrier, and since the electrons are ultimately passed to oxygen, the process is known as *oxidative phosphorylation*. The complexity of these reactions makes it obvious why such a precise architecture is required in the mitochondrion. Each enzyme must be so placed as to receive one substrate and pass its own finished product on to the next enzyme in line. Blockage at any point in the transfer chain effectively halts electron transport. The

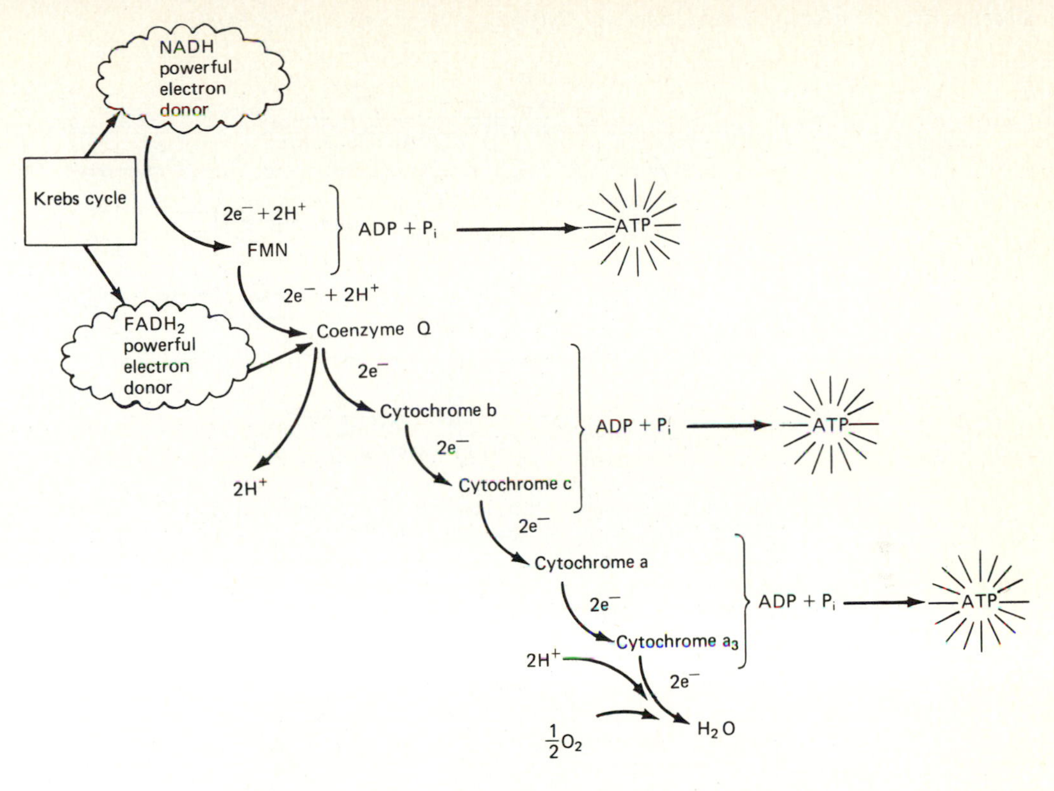

Figure 5-8 Energy transfer through the electron carrier chain in the mitochondrial membrane. Electrons donated by NADPH are passed from one carrier to another, each successive one at a lower energy level, until they finally reduce 1/2 O_2 to H_2O. During these transfers, three molecules of ADP are phosphorylated to yield ATP, the high energy compound used to drive other reactions. The phosphorylation sites are hypothetical, and may actually involve proton movement.

respiratory poisons cyanide and carbon monoxide, for example, act by combining with the iron of cytochromes, thus preventing the $Fe^{+++} \longrightarrow Fe^{++}$ transformation. The affected cell may die unless alternative pathways exist for transferring electrons to oxygen.

In respiration, as in photosynthesis (Chapter 4), the chemiosmotic theory provides a logical explanation of how energy derived from the passage of electrons through the carrier chain is coupled to the synthesis of ATP (Fig. 5-9). The electron carriers are asymmetrically arranged in the inner membrane of the mitochondrion (Fig. 2-15). Some of the carriers transfer protons as well as electrons, leading to a proton gradient across the inner mitochondrial membrane; this is similar to that described across the thylakoid membrane (Fig. 4-12) except that protons are transported outward, making the inside of the mitochondrial membrane more alkaline and the outside more acid (Fig. 5-9). Each pair of electrons that moves through the carrier chain from NADH to $\frac{1}{2}O_2$ leads to the secretion of six H^+ ions through the mito-

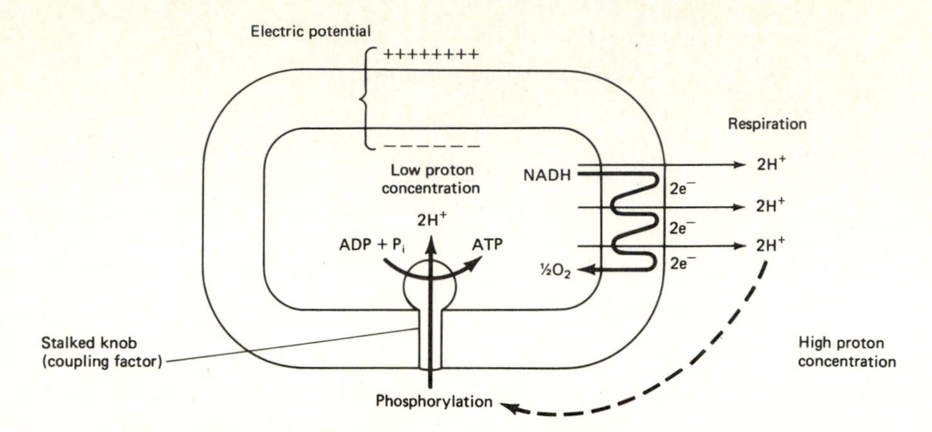

Figure 5-9 Chemiosmotic explanation for the formation of ATP in mitochondria. Three molecules of NAD^+ are reduced to NADH during one rotation of the Krebs cycle. NADH then gives up a pair of electrons, which cross the membrane three times as they move from one carrier to the next, and finally to oxygen, transporting two protons from the inner compartment to intermembrane space each time. This results in a large gradient in pH across the membrane, which drives the protons back into the inner sac, through diffusion channels in the stalked knobs (also called *coupling factor*). The energy of the gradient is used to form ATP from ADP + P_i, the reaction being catalyzed by an ATPase in the coupling factor. (Adapted from Hinkle and McCarthy. 1978. *Scientific American 238*(3): 104–123.)

chondrial membrane. This large pH gradient across the membrane is a potential source of energy, as in photosynthesis. The energy released when the protons move back into the mitochondrion through special channels in the stalked knobs (See Fig. 2-15) is linked by the coupling factor to the synthesis of ATP from ADP and P_i.

The pentose phosphate pathway

In addition to glycolysis, all higher plants also have at their disposal another respiratory pathway, called the *pentose pathway* or the *hexose monophosphate shunt* (Fig. 5-4). The initial substrate is glucose-6-P; this is oxidized by $NADP^+$ to a carboxylic acid called *phosphogluconate*, which in turn is decarboxylated to yield 5-carbon phosphorylated sugars.

$$\text{Glucose-6-P} + NADP^+ \longrightarrow \longrightarrow \text{5-carbon sugars such as ribose-5P} + CO_2 + \text{NADPH}.$$

The 5-carbon phosphorylated sugars can then be incorporated into cell wall polymers (pentosans), metabolized to yield $CO_2 + H_2O$ + ATP, incorporated into the nucleotides of RNA and DNA, or used to form high-energy compounds such as ATP. The pentose pathway is favored when high levels of 5-carbon sugars are required, and also when NADPH rather than NADH is needed as an energy source for synthetic activities.

Building-block production

Whether carbohydrates are first metabolized by the glycolytic or pentose–phosphate pathway, intermediates with varying numbers of carbon units are often removed before complete oxidation to CO_2 and H_2O (Fig. 5-10). In the absence of oxygen, the pyruvic acid can be reduced to lactic acid or the acetaldehyde to ethanol (Fig. 5-5).

The organic acids formed in the Krebs cycle may be coupled with ammonia by *direct amination* or *transamination* to form amino acids and eventually proteins (Fig. 5-10). Other compounds derived from amino acids include *phenols*, *flavonoids*, *anthocyanins*, *lignin* and all formed by a pathway which starts with the deamination of the amino acid *phenylalanine*. Various alkaloids and the plant hormone auxin (indole-3-acetic acid) are synthesized from the amino acid tryptophan.

Figure 5-10 The production of metabolic intermediates during respiratory transformations.

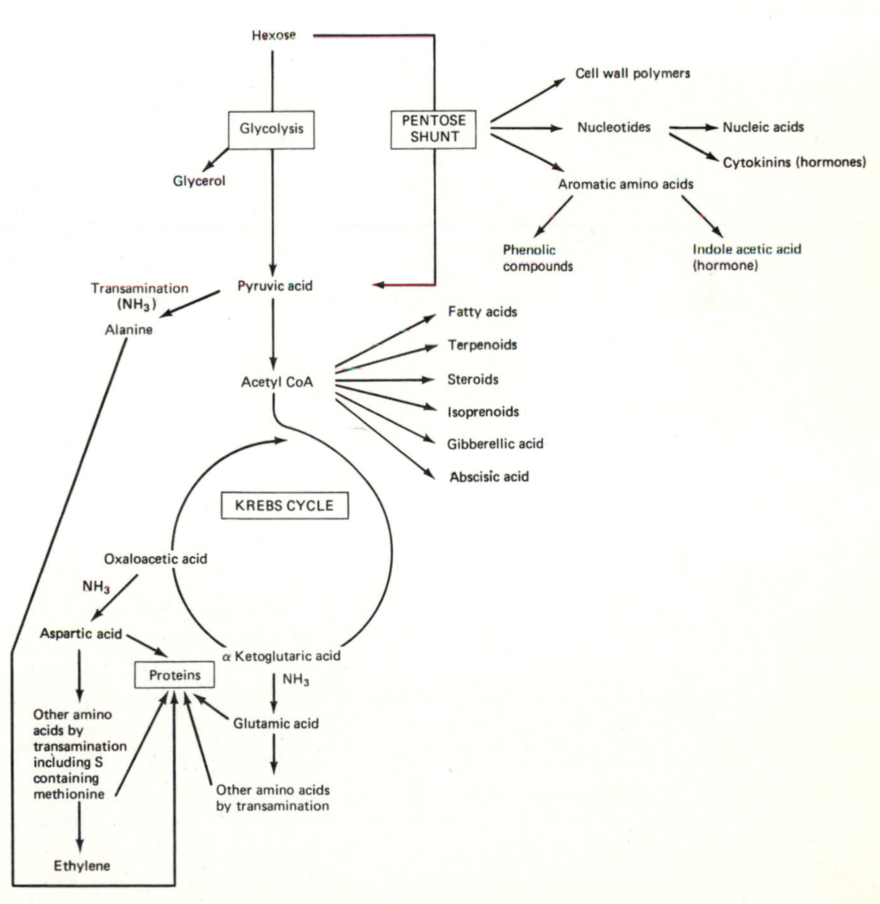

Acetyl co-A, derived from pyruvate, is an important starting point for the synthesis of the fatty acids of *lipids* (see below); for polymers of *isoprene* (C_5H_8) such as rubber; for volatile terpenes; steroids; the hormones gibberellin, abscisic acid, and part of the hormone cytokinin. We will discuss a range of other isoprenoid compounds in Chapter 15.

The two primary functions of respiration are thus the liberation of energy used in metabolism and the formation of building blocks for many other molecules in the cell, processes summarized in Fig. 5-11.

Respiratory quotient

The ratio of CO_2 evolved to O_2 absorbed when a substrate is oxidized to $CO_2 + H_2O$ is called the *respiratory quotient* (RQ). When sugars are the substrate, for example:

$$\underset{\text{Glucose}}{C_6H_{12}O_6} + 6O_2 \longrightarrow 6H_2O + 6CO_2. \quad RQ = 6CO_2/6O_2 = 1.0$$

However when lipids, proteins, or other highly reduced compounds are the substrate, the RQ is less than 1.0. For example:

$$\underset{\text{Stearic acid}}{C_{18}H_{36}O_2} + 26O_2 \longrightarrow 18CO_2 + 18H_2O. \quad RQ = 18CO_2/26O_2 = 0.7$$

Conversely, when compounds less reduced than sugars are the substrate, the RQ is greater than 1.0. For example:

$$\underset{\text{Oxaloacetic acid}}{C_4H_4O_5} + 2.5O_2 \longrightarrow 4CO_2 + 2H_2O. \quad RQ = 4CO_2/2.5O_2 = 1.6$$

The RQ of a tissue, which can be obtained experimentally by simulta-

Figure 5-11 The functions of respiration: a summary.

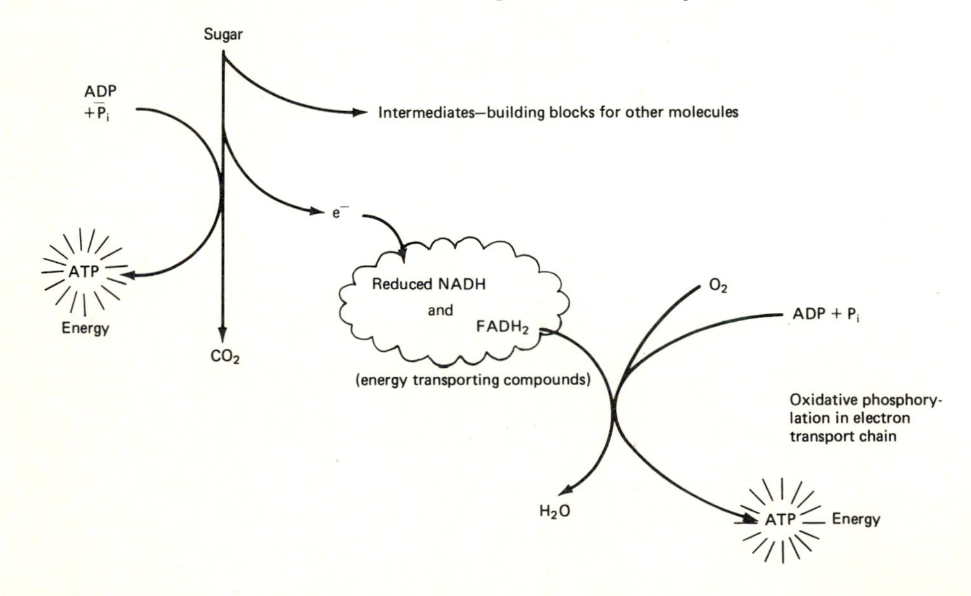

neous measurement of changes in CO_2 and O_2, gives some information about the oxidation state of compounds that are being metabolized. However, such measurements can lead to erroneous interpretations unless the nature of the respiratory process is understood. For example, the partial oxidation of sugars to pyruvic acid utilizes O_2 without any CO_2 evolution, while subsequent decarboxylation to yield acetyl CoA units for the synthesis of fatty acids releases CO_2 without concomitant O_2 uptake. Thus measurements of RQ can yield information about the nature of the substrate only when the substrate is completely oxidized to CO_2 and H_2O.

The Warburg Respirometer

The Warburg constant volume respirometer, designed more than 50 years ago by the German biologist Otto Warburg, is useful for measuring rates of O_2 utpake and CO_2 release by plant tissue. The instrument contains a large number of reaction flasks, each connected to a *manometer* which registers changes in the pressure of the gas in the flask. Small seedlings, seeds, excised plant parts, or cells in culture are placed in the bottom of a flask. The flask with its attached manometer is maintained at constant temperature and volume, so that changes in pressure in the flask will be due to the absorption of gas by the tissue or the evolution of gas from it. Oxygen and carbon dioxide are the two gases that move into and out of plant tissue in greatest quantity; thus pressure changes in the flask are largely due to their uptake and release.

It is important that the temperature and volume of the flasks remain constant, since they both affect pressure by the relationship

$$\frac{P_1V_1}{T_1} = \frac{P_2V_2}{T_2},$$

where P = pressure
V = volume
T = temperature

The temperature of the flasks is kept constant by immersing them in a temperature-controlled bath, while the volume of each flask is kept constant by manual adjustment of the manometer fluid to a fixed reference point before recording pressure. Pressure changes due to small changes in the temperature of the bath are measured in a control flask without plant tissue. This flask is also a monitor of changes in atmospheric pressure.

The respirometer can be used to measure O_2 uptake without interference from CO_2 by placing a strong alkali such as potassium hydroxide (KOH) in the center well of the flask. CO_2 is very soluble at high pH; thus it is absorbed continuously by the alkali, and changes in pressure in the flask are due to oxygen alone. Measurement of CO_2 is somewhat more complicated. The most straightforward method involves use of two flasks, each containing replicate plant tissue, but one containing and the other lacking KOH in the center well. Pressure changes in the flask containing KOH will be due to O_2 uptake, while pressure changes in the flask without KOH will be due to both O_2 uptake and CO_2 release; thus the difference in rates of gas exchange in the two flasks will indicate the rate of CO_2 release.

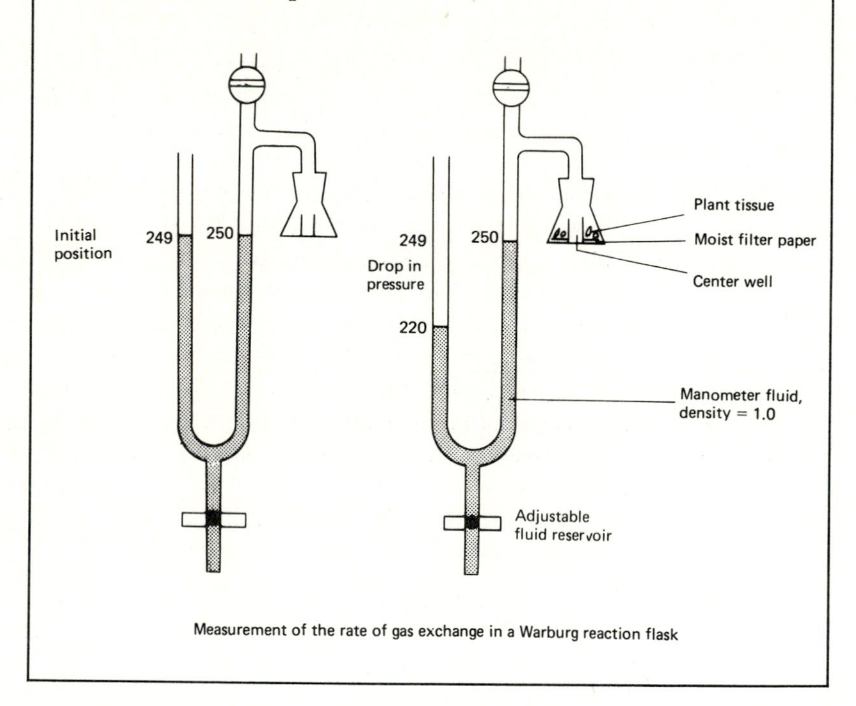

Measurement of the rate of gas exchange in a Warburg reaction flask

Lipid Synthesis and Breakdown

Intermediates of respiration provide the carbon skeletons used in the synthesis of the lipids, which serve both as stored food and as components of the membranes surrounding the cytoplasm and all cell organelles. Lipid molecules such as fats and oils consist of three fatty acids (Fig. 5-12) linked by ester bonds to the 3-carbon alcohol, *glycerol*. The main difference between fats and oils is that at room temperature fats are

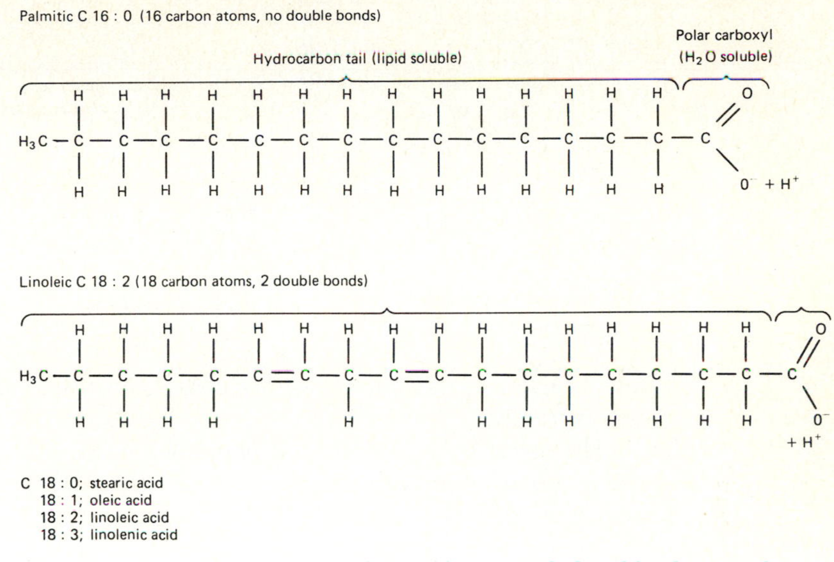

Figure 5-12 Some long chain fatty acids commonly found in plant membranes.

solid and oils are liquid; this difference is in turn due to the length and level of unsaturation (double bonds) of the carbon chains in the fatty acids. Oils tend to have smaller and more unsaturated fatty acids, whereas fats have longer, more saturated fatty acids.

Lipids may be the chief form of stored food, particularly in seeds such as soybean, sunflower, rape, sesame, and cotton. As described above, lipids contain less oxygen than sugars, and are thus a more concentrated energy form because more oxygen must be absorbed to convert them to $CO_2 + H_2O$. Thus, when one gram of sugar is burned, four calories are liberated, while one gram of fat similarly combusted yields nine calories. Lipids are also important in cell structure, since cell and organelle membranes are largely *phospholipid* and protein. Membrane lipids differ from ordinary fats in that one of the three fatty acids has been replaced by phosphorylated serine, choline or ethanolamine (Fig. 5-13). Such a molecule is partly soluble in fat and partly soluble in water and can thus partition itself at an oil–water interface and regulate the uptake of water- or fat-soluble materials. Other important lipids, attached to carbohydrates, are known as *glycolipids;* these can be important as sensors in interactions between cells.

In the form of reserve food, lipids occur as oil droplets in the cytoplasm of the cells. Lipids *per se* are not transported around plants; nevertheless, their carbon skeletons are available for transport upon conversion to sucrose. At some distant destination, fats might be resynthesized from the transported sucrose. Lipids are synthesized from two primary units, glycerol and fatty acids; each of these is formed from

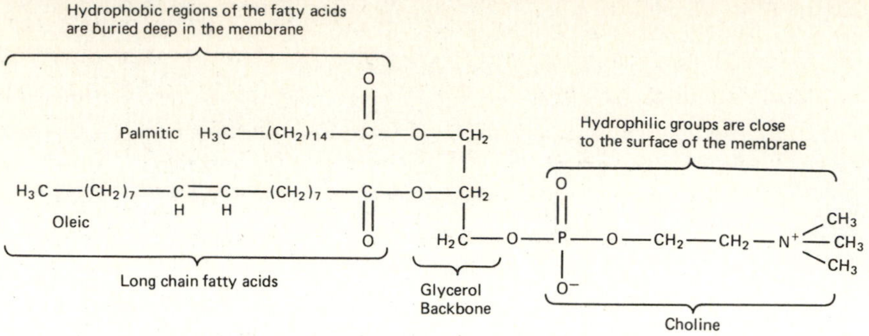

Hydrophobic regions of the fatty acids are buried deep in the membrane

Hydrophilic groups are close to the surface of the membrane

$$\text{Palmitic}\quad H_3C\text{—}(CH_2)_{14}\text{—}\overset{\overset{\large O}{\|}}{C}\text{—}O\text{—}CH_2$$

$$H_3C\text{—}(CH_2)_7\text{—}\underset{H}{C}=\underset{H}{C}\text{—}(CH_2)_7\text{—}\underset{\underset{\large O}{\|}}{C}\text{—}O\text{—}CH_2$$

Oleic

$$H_2C\text{—}O\text{—}\overset{\overset{\large O}{\|}}{\underset{\underset{\large O^-}{|}}{P}}\text{—}O\text{—}CH_2\text{—}CH_2\text{—}N^+(CH_3)_3$$

Long chain fatty acids

Glycerol Backbone

Choline

Figure 5-13 Phosphatidyl choline, a typical phospholipid found in most plant cell membranes.

intermediates in respiration, so that conversion from sucrose to lipid occurs readily.

The glycerol part of the fat molecule is derived as α-glycerophosphate from dihydroxyacetone phosphate (DHAP), a 3-carbon compound formed in glycolysis.

The fatty acids that combine with glycerol to form fats are derived from acetyl co-A by an indirect route. Acetyl co-A first reacts with CO_2 and ATP to produce malonyl co-A, a 3-C compound attached to co-A. The malonyl co-A then loses CO_2 as it donates the residual 2-C acetyl fragment to another acetyl co-A to produce a butyryl 4-C unit attached to co-A. The process is repeated until the final length of fatty acid chain is reached. Since 2-C fragments derived from malonyl co-A are added each time, natural fatty acids always have an even number of carbon atoms. The most common saturated fatty acids in plants are palmitic $[CH_3(CH_2)_{14}COOH]$ and stearic $[CH_3(CH_2)_{16}COOH]$, while the most common unsaturated fatty acids are oleic $[CH_3(CH_2)_7CH{=}CH(CH_2)_7COOH]$, linoleic $[CH_3(CH_2CH{=}CH)_3(CH_2)_7COOH]$, and linolenic $[CH_3\text{—}(CH_2)_3\text{—}(CH_2\text{—}CH{=}CH_2)_2\text{—}(CH_2)_7COOH]$, bearing, respectively, one, two, and three double bonds in the 18-carbon fatty acid chain.

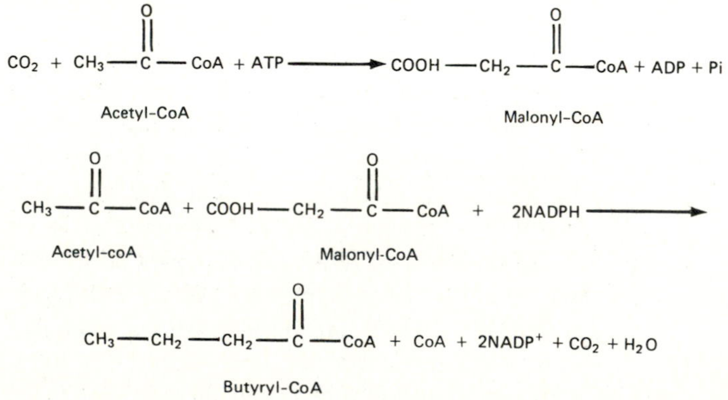

$$CO_2 + \underset{\text{Acetyl-CoA}}{CH_3\text{—}\overset{\overset{O}{\|}}{C}\text{—}CoA} + ATP \longrightarrow \underset{\text{Malonyl-CoA}}{COOH\text{—}CH_2\text{—}\overset{\overset{O}{\|}}{C}\text{—}CoA} + ADP + Pi$$

$$\underset{\text{Acetyl-coA}}{CH_3\text{—}\overset{\overset{O}{\|}}{C}\text{—}CoA} + \underset{\text{Malonyl-CoA}}{COOH\text{—}CH_2\text{—}\overset{\overset{O}{\|}}{C}\text{—}CoA} + 2NADPH \longrightarrow$$

$$\underset{\text{Butyryl-CoA}}{CH_3\text{—}CH_2\text{—}CH_2\text{—}\overset{\overset{O}{\|}}{C}\text{—}CoA} + CoA + 2NADP^+ + CO_2 + H_2O$$

The entire synthetic process occurs while the co-A derivatives of the fatty acids are attached to a special protein called *acyl carrier protein* (ACP), a multienzyme complex.

In phospholipid synthesis, only two fatty acyl co-A's react with glycerophosphate; the third carbon of the glycerophosphate is coupled to a nitrogen-containing compound (choline, ethanolamine, or serine) which is in turn linked to a hydrophilic phosphate group.

When lipids are broken down, glycerol and fatty acids are formed again. The fatty acids are activated by coupling with co-A, then broken down stepwise into 2-C units by enzymes located in the outer membrane of the glyoxysome, a single membrane-bound organelle. They emerge as acetyl co-A, which then enters the glyoxylate cycle (Fig. 5-14). The C_4 compound succinic acid, one of the products released from the glyoxysome, then enters the Krebs cycle in the mitochondrion, producing oxaloacetic acid. The oxaloacetate is then converted to phosphoenol pyruvate in the cytoplasm, and rebuilt into hexose phosphates by the reverse of glycolysis, using energy released during the glyoxylate cycle. The glycerol part of the fat is reoxidized to glyceraldehyde phosphate, and this too can be converted into hexose phosphates, from which sucrose or other metabolites may be synthesized.

By these various pathways, the carbon originally fixed by photo-

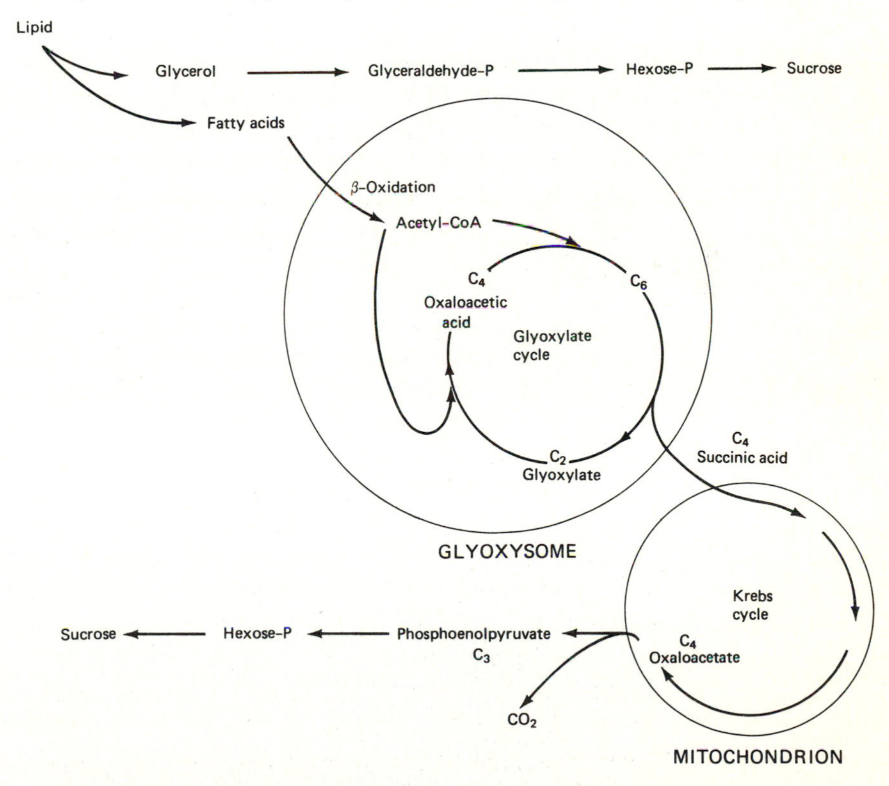

Figure 5-14 Breakdown of lipids to form carbon skeletons for the synthesis of sucrose and other metabolites. Note that some of the reactions occur in glyoxysomes and others in mitochondria and cytoplasm.

synthesis finds its way into all the organic constituents of plants. Thanks to its specific enzymes, the plant is a master chemist, and can even synthesize complex structures still not duplicated by man.

SUMMARY

The chemical energy stored in the sugar molecule is made available to the cell through a series of regulated oxidations, occurring mainly in the *mitochondria*. During such *aerobic respiration*, ATP is synthesized from ADP and P_i; this process is known as *oxidative phosphorylation*. When the energy stored in the bonds of ATP is released for cellular activities, ATP is broken back down to ADP and P_i.

Sugars can be enzymatically transformed to one another, usually through the mediation of ATP or other nucleoside triphosphates (UTP, GTP). *Sucrose*, the major sugar of most plants, is made when UTP reacts with glucose to make UDP-glucose + P_i; the UDPG then reacts with fructose-6-phosphate to form sucrose, P_i, and UDP. The breakdown of sucrose proceeds by a *hydrolytic* route, the enzyme *invertase* catalyzing its reaction with water to form glucose and fructose. Similarly, *starch*, the main storage form of sugar, is built up from phosphorylated derivatives (ADPG), but broken down hydrolytically through action of the enzyme *amylase*. *Cellulose*, the major cell wall component, is built up from GDPG units linked in a β-form rather than an α-form, as in starch. Cellulose is not usually degraded in plants except during leaf abscission, although fungi make *cellulase*, an enzyme that can catalyze cellulose hydrolysis.

Respiration begins when sugar is anaerobically degraded to the 3-carbon compound *pyruvic acid*, a process called *glycolysis*. This pyruvate then loses CO_2 as it joins its remaining two carbon atoms to a 4-carbon acid to form *citric acid*. In the *Krebs*, or *citric acid*, *cycle*, these two carbon atoms are released stepwise as CO_2, while the electrons from the remainder of the molecule are transported to oxygen to produce water in a process which generates ATP. In electron transport, carriers containing the vitamins *niacin* (NAD^+ and $NADP^+$) and *riboflavin* (FMN, FAD), and the iron-containing *heme* group (*cytochromes*) are important. $NADP^+$ can also accept electrons from glucose, oxidizing it to a *carboxylic acid* that loses CO_2 to become a 5-carbon sugar, or *pentose*. This pathway generates ribose, deoxyribose, and other pentoses important in metabolism. Some organic acids in the Krebs cycle can combine with *ammonia* and others can be *transaminated* to form amino acids. The amino acids so formed are used mainly in protein synthesis, but can also be transformed to *alkaloids*, *flavonoids*, and *hormones*. *Acetylcoenzyme-A*, formed when pyruvate is decarboxylated and joined with co-A, is the starting point for the synthesis of *fatty acids*, built up of successive 2-carbon additions. When three fatty acids are joined to glycerol, *fats* are formed; when two fatty acids and one phosphorylated compound are joined to glycerol, *phospholipids* are formed. Phospholipids are important as components of membranes, since their phosphate ends are water soluble, and their fatty acid ends fat soluble. Acetyl co-A is also the starting point for the synthesis of some hormones, terpenes, isoprenoids, and steroids.

SELECTED READINGS

BEEVERS, H. *Respiratory Metabolism in Plants* (Evanston, Ill.: Row, Peterson and Co. 1961).

HASSID, W. Z. "Transformations of sugars in plants," *Ann. Rev. Pl. Physiol.* 18: 253–280 (1967).

HINKLE, P. C., and R. E. MCCARTY. "How Cells Make ATP," *Scientific American* 238 (3): 104–123 (1978).

IKUMA, H. "Electron transport in plant respiration," *Ann. Rev. Pl. Physiol.* 23: 419–436 (1972).

STOCKING, R. C., and U. HEBER, eds. "Transport in Plants. III. Intracellular Interactions and Transport Processes," *Encyclopedia of Plant Physiology*, New Series, Volume 3 (Berlin–Heidelberg–New York: Springer-Verlag, 1976).

See Selected Readings *on page 59 for biochemistry background.*

QUESTIONS

5-1. List three forms of reducing power in cellular processes. How is their energy utilized?

5-2. Discuss the various meanings that can be applied to the term *oxidation*. What have all oxidations in common?

5-3. Explain briefly the function of an electron transport chain.

5-4. Compare and contrast photosynthesis and respiration.

5-5. Discuss the relationships between mitochondrial ultrastructure and function.

5-6. Why is aerobic respiration more efficient than anaerobic respiration?

5-7. What happens to the (1) carbons, (2) oxygen, and (3) hydrogen in the pyruvic acid molecule as it is degraded in respiration?

5-8. Why is respiration a multistep process rather than a direct oxidation of glucose?

5-9. What role does phosphorus play in respiration?

5-10. Describe the similarities and differences in chemical structure, path of synthesis, function, and location in the cell of starch and cellulose. On the basis of chemical structure, explain why cellulose occurs as long fibers, whereas starch occurs as rounded grains.

5-11. Given a molecule of glucose, what reactions would be necessary to produce the following: fructose, sucrose, a fatty acid, aspartic acid, starch, ethyl alcohol?

5-12. Discuss how respiratory metabolism may serve as a central mechanism linking the carbohydrates, organic acids, fats, and proteins.

5-13. Describe the reactions that occur when a hexose phosphate (fructose-6-P) formed in the Calvin–Benson cycle in photosynthesis is (1) converted to starch, (2) then transformed from starch into sucrose for transport. Include the reactions describing the synthesis and degradation of starch and the synthesis of sucrose. Write the required equations, in order, using words or accepted abbreviations (for example, G-6-P = glucose-6-phosphate).

5-14. The breakdown of macromolecules (polysaccharides, lipids, proteins, and nucleic acids) frequently follows a pathway that is different from their synthesis. Using relevant examples, discuss this statement.

5-15. Trace the movement and metabolic pathways of an atom of carbon in atmospheric CO_2 until it ends as a lipid molecule stored in a developing seed.

5-16. From what respiratory intermediate are fatty acids derived, and with what other substance do they combine to form fats and oils?

5-17. Plant oils are most abundant in seeds, and are often localized in the embryo. Of what advantage is this to the energy economy of the plant?

5-18. All fatty acid molecules in plants have even numbers of carbon atoms. Why?

5-19. The fat composition of organisms varies with the temperature to which they are exposed, being in general richer in unsaturated fatty acids the cooler their environment. Is this of any possible benefit to the organism? Through what kind of mechanism might temperature influence the kinds and amounts of fatty acids produced?

5-20. Why should plants have different temperature optima for growth during the light and dark periods of the day?

5-21. Compare and contrast the roles of polysaccharides and lipids in plants.

5-22. Suppose that you were asked to examine a sample of soil from another planet, delivered to earth by a spaceship, to determine whether life existed on the other planet. What tests would you make?

6 Water Economy

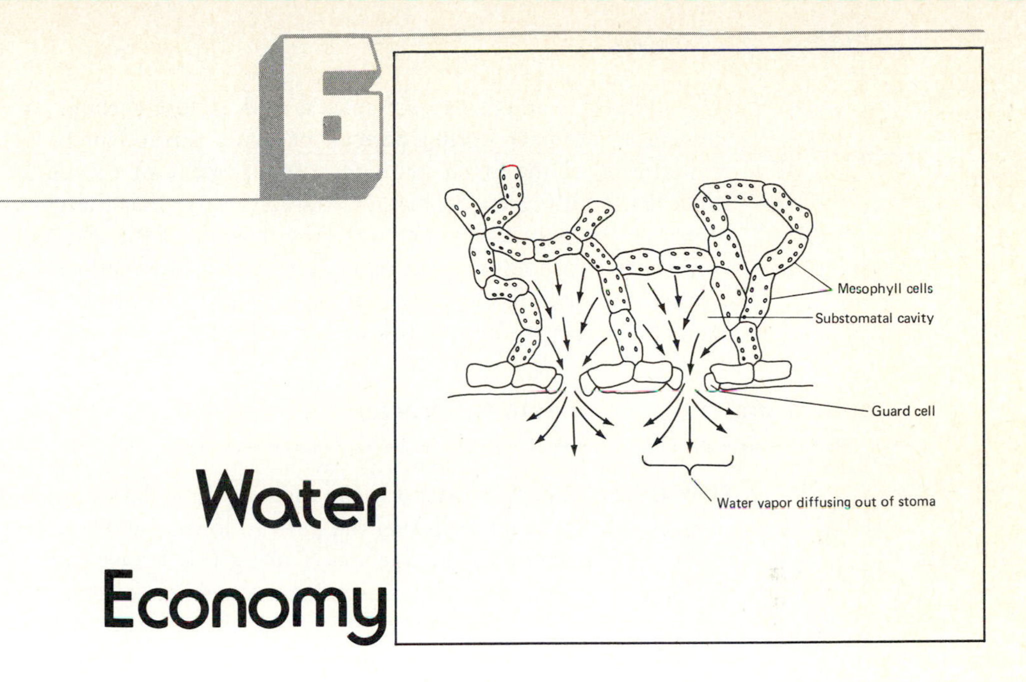

Water is the most abundant component of active plant cells, sometimes constituting 90% or more of their fresh weight. In contrast, dormant seeds may have as little as 15–20% of their weight as water. Most of the water in mature plant cells is found in the large central vacuole occupying 80–90% of the cellular volume. When a plant cell is turgid, its central vacuole presses the cytoplasm firmly against the wall, maintaining the shape of the cell in soft organs such as leaves. Thus, the typical form of herbaceous plants is maintained in part by hydrostatic forces. During periods of water stress, water is withdrawn from the vacuoles, leading to loss of turgidity and consequent wilting. Normally a plant readily recovers from wilting when the water supply is restored; the cells then absorb water, regaining their turgor.

The moisture content of the protoplast, however, must be maintained within fairly rigid limits if cell vigor is to be maintained. Despite large fluctuations in rainfall and the moisture content of the soil, the green plant does manage to maintain a relatively constant internal water content. Such control is accomplished by a reduction, during periods of water stress, of the evaporative loss of water. Plants continually absorb water from the medium in which they grow, and evaporate part of it into the environment about them. *Transpiration*, the evaporation of water by the aerial parts of a plant, is an inevitable consequence of the architecture of the leaf. Structured for efficient photosynthesis, the leaf is usually a large, flat, wet organ, full of pores connected to branched air passages. When exposed to the sun, such an organ loses considerable water. Water evaporates from the surfaces of wet mesophyll cells, diffuses through the air spaces, and is lost through open stomata. Closure of stomata during periods of water stress can reduce the water requirement of the green plant to a few percent of its requirement when stomata are fully open. Stomatal closure, however, has the undesirable effect of preventing gas exchange between the interior of the leaf and the external atmosphere, with consequent reduction in photosynthetic efficiency. In this chapter, we shall discuss how the plant balances its need to conserve water against its other needs, and how it manages its water economy.

Osmotic Uptake of Water Into the Vacuole

Since most of the cell's water is in the vacuole, we will begin our consideration of water transport by analyzing the path followed by water molecules from the cellular exterior to the vacuole. During this transit, water must move through two membranes (the plasmalemma and the tonoplast) as well as the cytoplasm that lies between them. Since very little is known about differences in the water-conducting properties of each of these entities, they are usually treated as a single membrane barrier.

To understand how water moves through this membrane, consider a cell with a central vacuole containing salts, sugars, amino acids, and other solutes, sitting in a tank of distilled water (Fig. 6-1). As we know from the kinetic theory of matter, the molecules of all substances are in constant, rapid, random motion, the velocity of which is dependent on the energy of the molecules. Their average velocity is determined by (and is actually a measure of) the temperature. Since water molecules are small and move through membranes of the cell much more rapidly than do other substances, we may, for simplicity, consider only the movement of the water molecules. These diffuse in all directions, both into and out of the cell and the various organelles. But since the cell's vacuole contains solutes in appreciable quantity, and since the solute molecules dilute the water molecules and exert attractive

Figure 6-1 The generation of an internal pressure against the cell wall is the result of differential rates of movement of water into and out of the cell. The vacuolar solute molecules (black dots) reduce the activity of the water molecules, making the average rate of outward diffusion lower than that of inward diffusion.

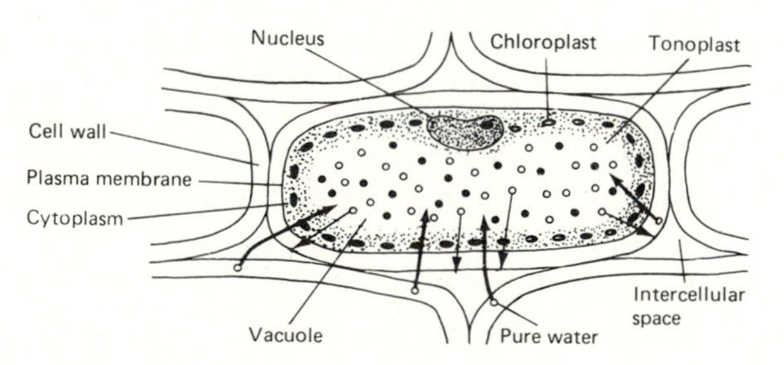

forces on neighboring water molecules, they serve to retard net outward water movement. In a sense, solutes decrease the activity of water molecules in the cell, and as a consequence, the kinetic energy of water inside the vacuole is less than that of the relatively pure water outside the cell. This means that, in any given area of the membrane, more water molecules will bombard and enter the vacuole than will leave in any unit of time. This rapid, unequal, two-way diffusion of water molecules through the membrane ultimately increases the volume of the vacuole, producing *turgor*, a pressing of the cell contents against the more rigid wall.

The movement of water through a differentially permeable membrane is known as *osmosis*, and the solute concentration in the vacuole is a measure of the cell's maximum water-absorbing capacity. The energy level of any molecule, manifested as its diffusion rate, is termed its *chemical potential*, but as we are here concerned only with water, we will use the special term *water potential*. The water potential, ψ (Greek *psi*), expresses the tendency for water to diffuse, to evaporate, or to be adsorbed. Its units are energy per unit volume, which is equivalent to pressure, and its magnitude is usually expressed in atmospheres, or *bars* (1 bar = 0.987 atm). We cannot actually measure the energy of the water molecules in, say, a beaker of water; by convention, we call the ψ of pure water zero at standard temperature and pressure. What we *can* measure are differences in the energy of water molecules in different situations. The lower the energy of the water molecules, the lower will be the water potential; since we started at zero for pure water, ψ will be more negative as the solute content increases. In osmosis, the solute molecules lower the energy of the water molecules so that a solution has a more negative potential than does pure water.

Water molecules always move from a higher to a lower water potential, much as water moves from a higher to a lower energy level when flowing downhill. Since other factors such as pressure also affect the water potential of a solution, we use a special term, *osmotic potential*, ψ_π (Greek *psi* with subscript *pi*), to describe that component of water potential which is due to the presence of dissolved solute. The osmotic potential of a solution is directly related to the concentration of dissolved solute. As the solute concentration is increased, the osmotic potential becomes more negative. If 1 mole (the molecular weight in grams) of an undissociated solute such as sucrose is dissolved in one liter of water (to produce a *molal* solution) the solution will have an osmotic potential of -22.7 bars (at standard temperature and pressure). Less concentrated solutions will have proportionally less negative osmotic potentials.

If pure water is separated from an enclosed solution by a differentially permeable membrane, water will move into the solution and cause the buildup of an actual pressure (osmotic pressure) equal but opposite in sign to the original osmotic potential. A solution has the *potential* to build up such a pressure if placed in such an apparatus, termed an *osmometer* (Fig. 6-2). The

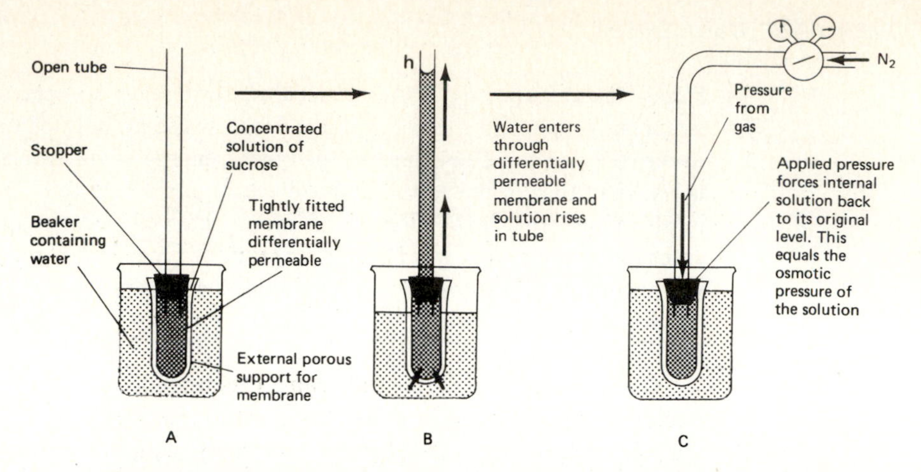

Figure 6-2 The pressure generated by osmotic water movement can be demonstrated by observing the height to which a column of solution will rise (h) when the solution is separated from pure water by a differentially permeable membrane in a leakproof system (A and B). This *osmotic pressure* is measured by determining that pressure that must be applied (using a device such as compressed gas) to prevent water from rising in the tube. (C). The *osmotic potential* of the solution is numerically equal but opposite in sign to the maximum osmotic pressure the solution can generate. The osmotic potential of a solution exists at all times, even if the solution is not actually generating an osmotic pressure.

numerical value of the osmotic potential is that pressure, applied to the solution in an osmometer, which will precisely prevent the net entry of water.

A cell containing a membrane-bounded solution (the vacuole) in a confined space (limited by the cell wall) is really an osmometer. When such a cell is placed in pure water, water enters the cell. At zero wall pressure, the tendency of water to enter the cell is determined by the water potential of the cell (ψ_{cell}) which is initially identical to the osmotic potential (ψ_π) of the solution in the vacuole. But as water enters the vacuole and increases its volume, it dilutes the cell sap and starts to build up a pressure against the cell wall. How long does water continue to enter the vacuole? Theoretically, if the osmotic potential were the only determinant, the net movement inward would never stop. Actually, as the volume of the vacuole increases, pressing the cytoplasm against the cell wall and generating turgor pressure, the wall exerts an equal and opposite back pressure on the cell contents (wall pressure). The term, *pressure potential* (ψ_{p}), usually designates wall pressure, but may also represent turgor pressure (equal in magnitude, opposite in direction). When ψ_{p} is great enough, it prevents a further increase in water content of the vacuole. A dynamic equilibrium is established, and *net* water entry ceases, although water molecules still move rapidly through the membrane in both directions. At that point the positive pressure potential exactly balances the negative osmotic potential so that the cell no longer gains water, and the cell water potential is now zero.

The water potential of a cell at any time depends on the difference between the pressure potential and the osmotic potential. A cell that is permitted to attain maximum turgor in pure water will continue to absorb water until $-\psi_p = \psi_\pi$, and $\psi_{cell} = 0$. This cell will then be unable to absorb additional water from any solution or any other cell. If two cells with different ψ_{cell} are adjacent, water will move from the cell with the higher (less negative) ψ_{cell} through the wall to the cell with the lower (more negative) ψ_{cell}.

Now let us consider the movement of water into and out of the vacuole of a plant cell *in situ* (that is, in its normal place in the organism). The vacuole and its surrounding protoplast are encased in a cell wall through which water freely diffuses. (The wall can be considered rather like filter paper, which is also composed of cellulose.) The cell wall continuum is saturated with water, as long as soil moisture is adequate and transpiration is not excessive. Under these conditions, ψ is higher in the wall region than in the vacuole and there is a net movement of water into the vacuole. However, during periods of water stress there may be a water deficit in the cell walls, and consequently ψ may be lower in this region than in the vacuole. This leads to a net loss of water from the vacuole, so that the cell loses turgor; it can become flaccid and limp as the turgor pressure decreases. If the turgor pressure decreases to zero through extreme water loss, the leaf wilts completely; further water loss results in disruption of the protoplast and death of the cells, although, as we will see below, plants can avoid severe water loss by closing their stomata rapidly in response to water stress. When an adequate water supply is restored, or at night when transpiration virtually ceases and water uptake from the soil catches up with evaporation, water re-enters the cells, provided they are still intact, and turgor is regained.

ψ = water potential
zero for pure water
zero or negative for cells (ψ_{cell})

ψ_π = osmotic potential
always negative

ψ_p = pressure potential
usually positive in living cells
(i.e., cell contents under pressure)
but negative in xylem cells
(under tension)

ψ_{cell} is the net result of the action of ψ_π and ψ_p;

i.e., $\psi_{cell} = \psi_\pi + \psi_p$.

At full turgor $\psi_\pi = -\psi_p$ and $\psi_{cell} = 0$.

At incipient plasmolysis $\psi_p = 0$ and $\psi_{cell} = \psi_\pi$.

Under artificial conditions, a further shrinkage of the cell protoplast can be obtained. If pieces of plant tissue are placed in external solutions more concentrated than their vacuoles, water will continue to leave the cell until the protoplast withdraws from the cell wall and contracts to the middle of the cell. The external solution moves freely through the cell wall, which presents almost no barrier to water movement, to fill the space between the shrunken protoplast and the cell wall. A cell in this state is said to be *plasmolyzed* (Fig. 6-3). Provided that plasmolysis is neither too prolonged nor too severe, the cells will regain their normal turgor when the tissue is replaced in water. Plasmolysis *per se* does not occur in plant cells in the atmosphere, as there is no free solution to enter the space between protoplast and cell wall in cells deficient in water; if water loss became that severe in nature, cell disruption and death would probably ensue.

Knowing the ψ of a cell, one can predict whether it will gain or lose water from any solution. Experimentally, ψ_{cell} can be determined by immersing cells or bits of tissue in an osmotically graded series of solutions of some solute, such as sucrose. The water potential of the solution in which the cell neither gains nor loses weight or volume represents the ψ_{cell} (Fig. 6-4). A

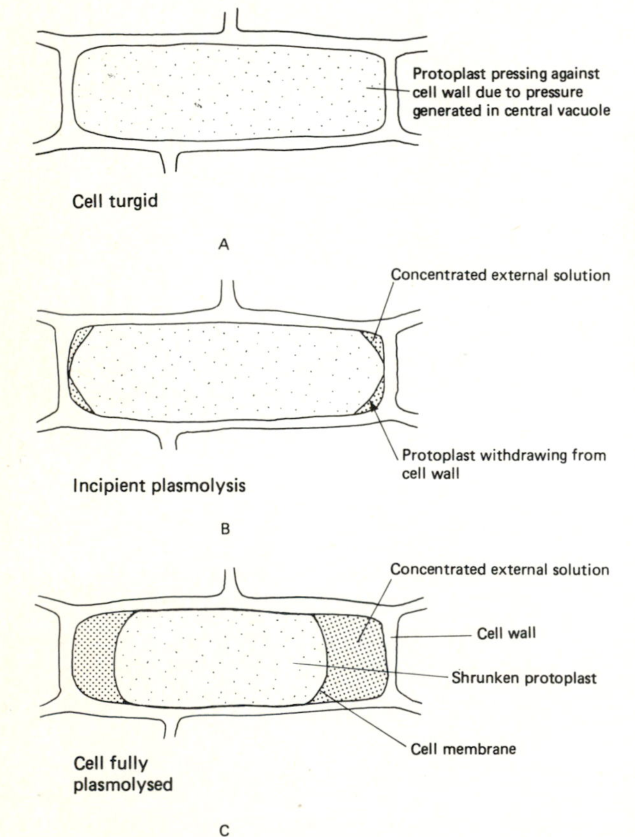

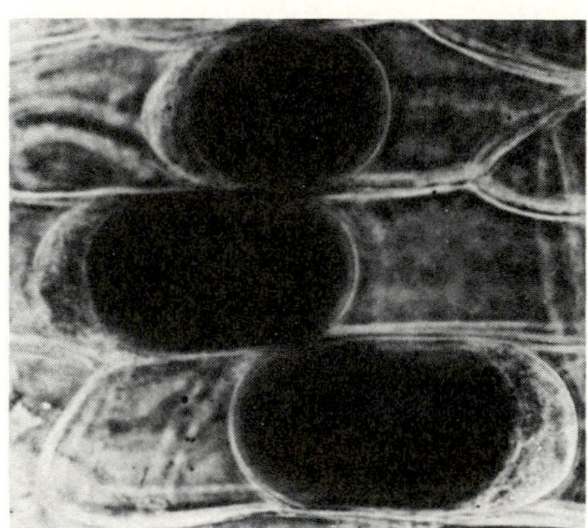

Figure 6-3 The micrograph shows plasmolysis of cells of red onion bulb scale. Note that the protoplast is rounded up in the center of the cell and is withdrawn from walls. (From K. Esau, *Plant Anatomy*, 1st ed. New York: John Wiley & Sons, 1960). The diagram left shows the mechanism whereby plasmolysis is accomplished.

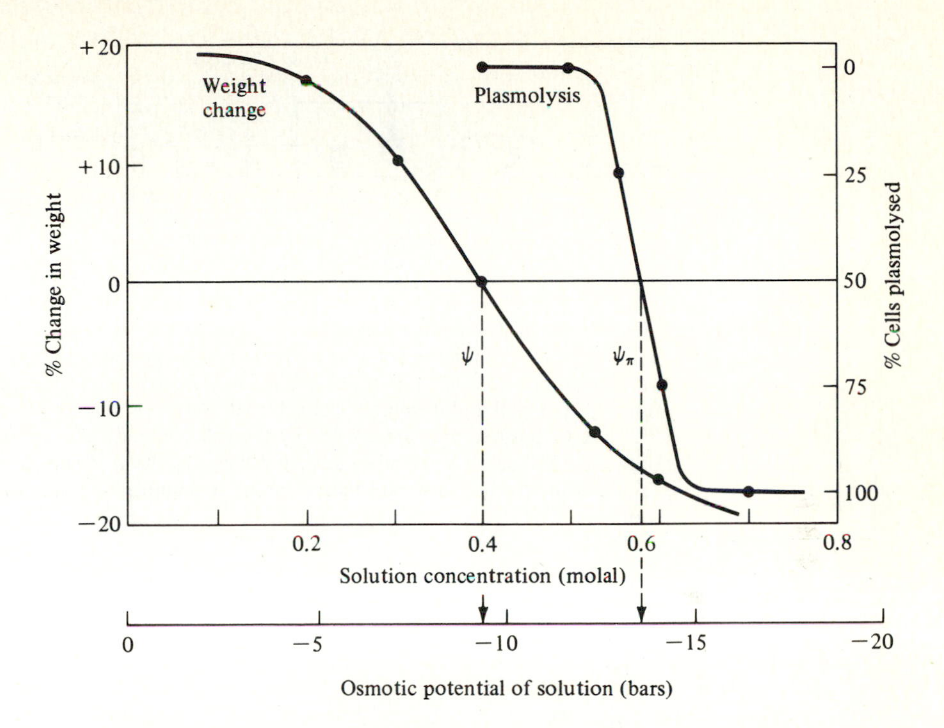

Figure 6-4 The water potential and osmotic potential of a piece of plant tissue e.g. a beetroot, can easily be measured in the laboratory. To determine the water potential, slices of weighed tissue are placed in sucrose solutions of different concentrations for about an hour, and then reweighed. In dilute solutions the tissue will increase in weight; in concentrated solutions it will lose weight. If the change in weight is then plotted against the solution concentration, as in the figure above, the concentration at which the tissue neither gains nor loses weight (i.e. the solution with which the tissue is in water equilibrium), indicates the water potential (ψ) of the tissue.

The osmotic potential of cells in the tissue can be estimated by using a microscope to detect plasmolysis in thin slices of the tissue placed in sucrose solutions of different concentrations. There will be no plasmolysis in dilute solutions, but virtually all the cells will be plasmolyzed in concentrated solutions. The percentage of cells plasmolyzed is then plotted against the solution concentration, as in the figure above. The solution concentration at which 50% of the cells are observed to the plasmolyzed is said to have produced incipient plasmolysis. Since the *pressure potential* (ψ_p) of the average cell = 0 at incipient plasmolysis, the *osmotic potential* (ψ_π) of the cells in the tissue equals the ψ_π of the solution which brings about incipient plasmolysis. The approximate pressure potential of the original tissue can be found by subtraction ($\psi_p = \psi - \psi_\pi$). To be strictly accurate, the change in solute concentration as water is withdrawn to plasmolyze the cells should be taken into account in obtaining both the osmotic potential and pressure potential values.

In the example shown in the figure, the tissue does not change in weight when placed in a 0.4 molal sucrose solution, while incipient plasmolysis occurs in a 0.58 molal solution. The osmotic potentials of these sucrose solutions are shown in the lower two horizontal lines in the figure. (These values are based on measurements made with an osmometer such as the one shown in Fig. 6-2). Since ψ (0.4 molal solution) = −9 bars and ψ_π (0.58 molal) = −13.5 bars, ψ_p (tissue) = 4.5 bars.

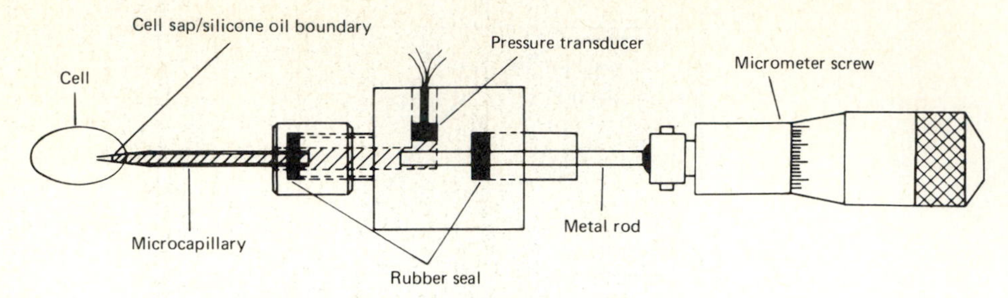

Figure 6-5 a. The micro-pressure probe used to measure the pressure inside large plant cells. A glass micropipet with a very narrow tip contains silicone oil that extends to a reference line near the tip of the pipet. After the pipet tip has been inserted through the wall and membranes of a turgid plant cell, the oil will recede from the tip, due to positive pressure exerted by the cell contents. The micrometer screw is now adjusted so that the oil boundary returns to the reference line. The pressure indicated by the pressure transducer is equal in magnitude to that existing within the cell.

b. A micrograph of the probe tip inserted into an epidermal cell of a halophyte (salt-tolerant species). (Courtesy of D. Huskan, E. Steudle and U. Zimmerman, Institute of Physical Chemistry, Nuclear Research Center, Julich, Germany.)

simple immersion technique can also be used to determine ψ_π, for the osmotic potential of any external solution that produces incipient plasmolysis (zero turgor, that is, $\psi_p = 0$) must be equal to the internal osmotic potential. Incipient plasmolysis occurs when the protoplast barely fills the space inside the cell wall. Thus ψ_{cell} can be determined by microscopic observation of tissues placed in a series of solutions, recording the osmotic potential of the external solution that causes plasmolysis in 50% of the cells. Although old and somewhat primitive, this plasmolytic method of determining the osmotic concentration of a cellular vacuole is still probably the best available. Most other methods require expressing sap from cells, which is likely to alter the vacuolar contents. The pressure potential, ψ_p, has also been measured in the large cells of the filamentous alga *Nitella* by inserting a pressure-recording manometer into one of the cells (Fig. 6-5). However, ψ_p is more difficult to measure in higher plants, and therefore is usually calculated by subtracing ψ_π from ψ_{cell} (Fig. 6-4). In whole plant shoots, a "pressure bomb," to be described later, can be used to measure cellular water potential.

Absorption of Water From the Soil

Although small quantities of water can be absorbed through aerial portions of the plant, almost all the water taken up by the plant is absorbed by the roots. In most plants, the root system is a multibranched network that penetrates a large volume of soil (Fig. 6-6). Water and minerals are absorbed by epidermal cells near the tip; numerous hairs protrude from the epidermis,

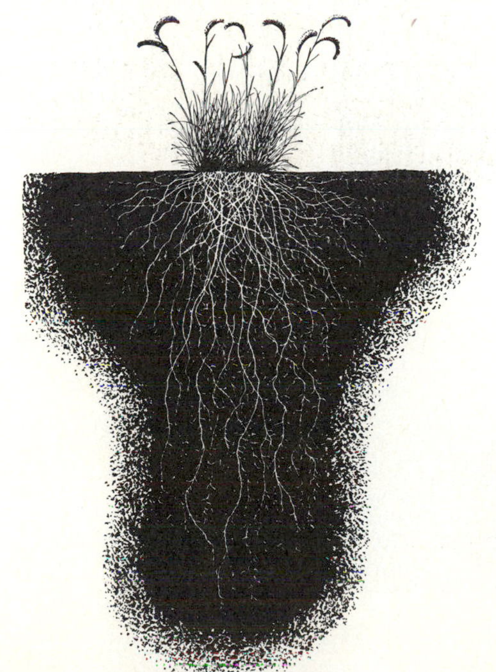

Figure 6-6 The root system of a mature plant of blue grama grass. (Adapted from J. E. Weaver and F. E. Clements, *Plant Ecology*, 2nd ed., New York: McGraw-Hill Book Co., 1938.)

penetrating crevices between soil particles and increasing the absorptive surface manyfold (Fig. 6-7). Water is absorbed entirely by osmotic forces, moving from regions of high ψ in the soil to regions of lower ψ in the roots.

The soil may be considered a sort of water reservoir, which is alternately filled and depleted. After a rain, a front of free water percolates downward through the soil, which is then said to be at *field capacity*. At field capacity, the ψ of the soil is almost zero, and water may be very easily removed from the soil by plant roots, or by mechanical means such as centrifuging or squeezing. As the soil progressively dries, its ψ decreases. When soil ψ is lower than that of root cells, the plant cannot absorb water and wilting occurs. The percentage of water in the soil at this point (the *wilting percentage*) varies widely from soil to soil, being low in coarse sandy soils and relatively high in

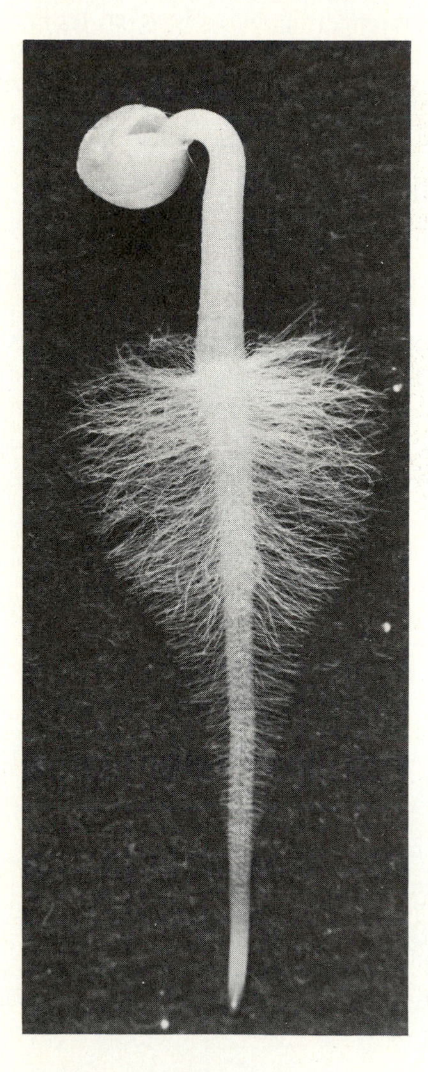

Figure 6-7 The radicle of a mustard seedling. The numerous fine projections are root hairs.

TABLE 6-1. The percent water in different types of soil at field capacity and at the wilting percentage.

	Field Capacity	Wilting Percentage	Percent Available Water
Fine sand	3·3	1·3	2·0
Sandy loam	18·5	10·0	8·5
Silt loam	21·3	10·4	10·9
Clay	28·0	14·5	13·5

finely divided clay soils (Table 6-1). The difference is due to the fact that the finely divided clay particles have a vast surface area which physically absorbs water—so tenaciously that it cannot be removed by roots. High salinity also decreases the ψ of the soil and increases the likelihood of wilting. Irrespective of the type of soil and type of plant, soil water cannot be removed if its ψ is more negative than minus 15 bars. Thus, although clay soils hold more water than do sandy soils, they also retain more water in a tightly bound condition, unavailable to the plant.

Water Movement Through the Plant

Water moves through the plant along a water potential gradient. Water absorbed by root hairs and other epidermal cells diffuses through outer root cells to the centrally located xylem tissue (Fig. 6-8). The main pathway for diffusion in the outer part of the root is through the *apoplastic* cell wall continuum. However, at the endodermis, the cylindrical layer of cells surrounding the vascular tissue, free diffusion through the cell walls is blocked by the water-impermeable suberin layer in the Casparian strip. The water is therefore forced to pass through the membrane and the protoplast of the endodermal cells, which thus form an osmotic barrier between the cortex and stele of the root. In monocotyledons, the inner tangential walls are also suberized, but these walls are penetrated by pits, forming channels through which the water can pass.

Water reaching the xylem travels up these cells to the aboveground portions of the plant. Of the several types of cells in the xylem, the vessels and the tracheids (Figs. 2-6 and 6-9) form the main channels for water movement. Both cell types are well suited for such transport, since they are elongated, lack living contents, and have a large hollow interior that serves as a pipe for water conduction. The lignified secondary walls provide the tensile strength necessary to sustain the large pressure difference that develops as water ascends to the top of tall trees. Vessel cells have large perforations on their end wall and sometimes on their lateral walls as well; joined end-to-end, these cells form long continuous tubes through which water and dissolved minerals can move readily. Although tracheids lack perforations, so that

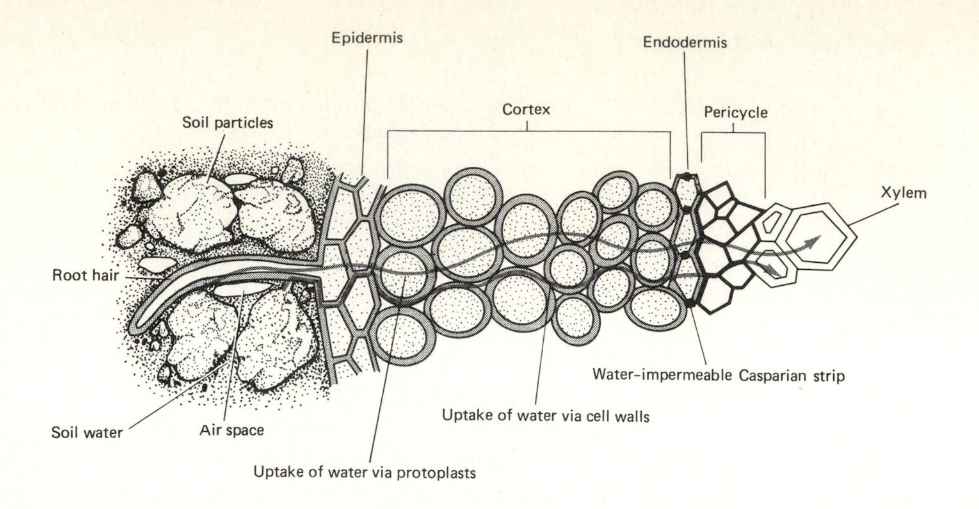

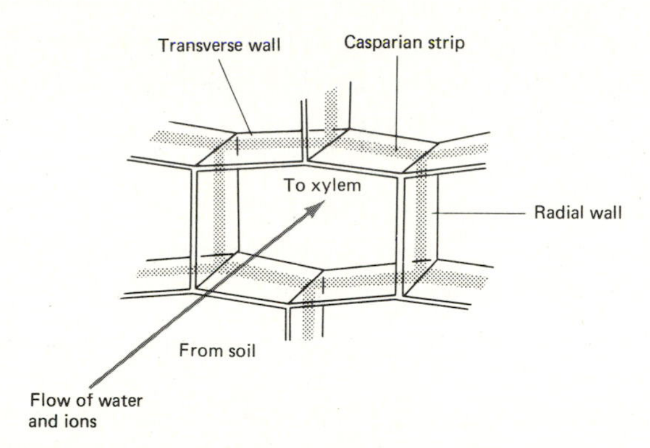

Figure 6-8 Movement of water from the soil into the root. Water can move through either the apoplast or symplast until it reaches the endodermis, where further apoplastic movement is blocked by the Casparian strip. It is this impermeable strip which forces water to move from the apoplast through the membranes of the endodermal cells, as shown in the enlargement below.

water must cross end walls to move from one cell to another, the individual cells are extremely long, and thus form effective channels for water movement. Flowering plants have both tracheids and vessel cells, but more primitive plants generally lack vessels.

Veins containing strands of xylem and phloem tissue ramify throughout the leaf, so that no cell is far from its water supply (Fig. 3-9). Water diffuses from the xylem into the walls of mesophyll cells in the leaf. Thus water in a liquid phase extends all the way from the soil through the root and stem to the mesophyll cells of the leaf. The net movement is always in the direction of decreasing water potential, that is, ψ is highest in the soil, somewhat lower in root cells, and lowest in cells close to the epidermis of the leaf. The low ψ of the latter cells is largely attributed to the evaporation of water into the

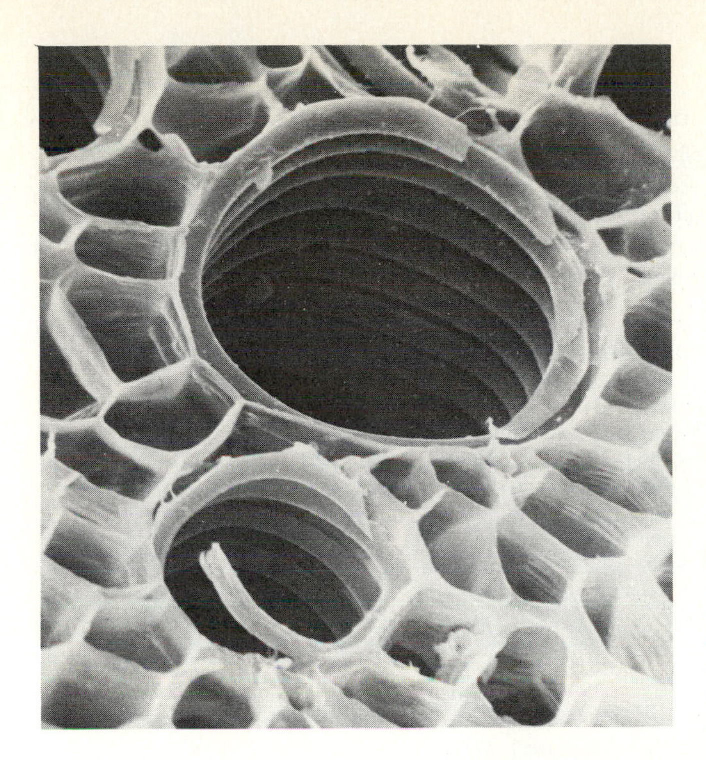

Figure 6-9a Transverse section of the petiole of the castor bean (*Ricinus communis*), showing xylem conduits with spiral thickening. This allows the thickening conduits to extend as the petiole grows. (Larger conduit diamcter ca. 15 μm).

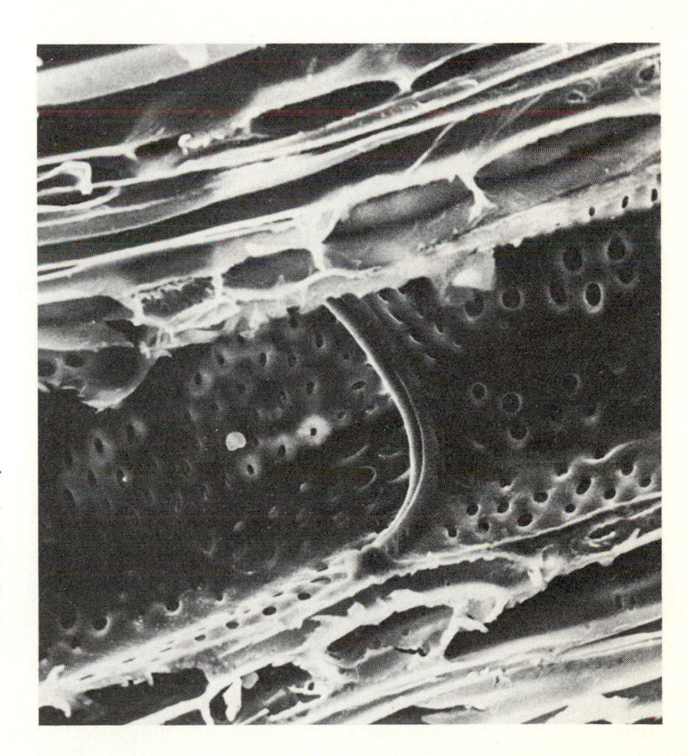

Figure 6-9b Longitudinal section of *Ricinus* stem showing xylem vessel elements and their junction at a perforation plate. Thickening shows elliptical and circular bordered pits closed by pit membranes but lacking a torus, i.e. typical of dicots (Conduit diameter ca. 21 μm). (Scanning electron micrographs courtesy of J. A. Milburn, Glasgow University.)

atmosphere via transpiration, as discussed below. Table 6-2 provides an example of the gradients moving water through the plant from soil to atmosphere.

Most of the water passing from the plant to the atmosphere does so in the gaseous form. There are large intercellular spaces in the mesophyll, and at least one wall of each mesophyll cell adjoins such a space. Evaporation of

TABLE 6-2. Estimated values for water potential (ψ) and water potential difference ($\Delta\psi$) in a hypothetical soil–plant–air system. The plant is a small tree, the soil is well watered, and the relative humidity of the air is approximately 50% at 22°C (ψ = −1000 bars).

	ψ, BARS	$\Delta\psi$, BARS
Soil water	−0.5	
Root	−2	−1.5
Stem	−5	−3
Leaf	−15	−10
Air	−1000	−985

From R. G. S. Bidwell, *Plant Physiology* (New York: Macmillan, 1974).

Figure 6-10 Cross section of a leaf showing an open stoma (S) subtended by an air cavity (AC). Note that large intercellular air spaces permeate the leaf.

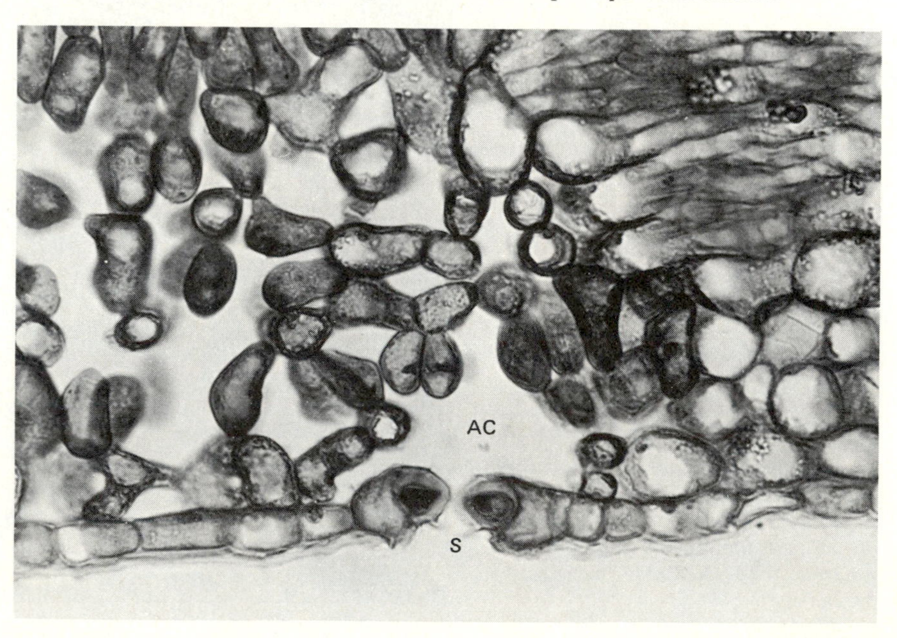

water from the wet walls of the mesophyll cells keeps the air in the intercellular spaces saturated with vapor, some of which is lost to the outside. Since the epidermal cells of most plants are well covered with a waxy, water-impervious cuticle, water vapor loss occurs almost exclusively through the stomata (Fig. 6-10).

Transpiration

Plants vary widely in the number and distribution of their stomata. Those adapted for life in arid regions (*xerophytes*) tend to have fewer stomata per unit area than do mesophytes; their stomata may also be located in sunken areas of a highly cutinized leaf or stem surface, thus minimizing loss of water by restricting air turbulence (Figs. 4-2 and 6-11). In most plants, stomata are

Figure 6-11 Scanning electron micrographs showing stomata on the adaxial (left) and abaxial (right) surfaces of *Distichlis spicata*, a salt marsh grass adapted to a desiccating environment. Stomata on the left (upper view ×85, lower ×860) are in furrows between ridges. During periods of water stress, "bulliform" cells in the furrows collapse and the leaves roll shut, providing protection against desiccation. Stomata on the right are protected against desiccation by the four epidermal cells that partially cover them. (From Hansen et al. 1976. *Amer. J. Botany 63*, 635–650.)

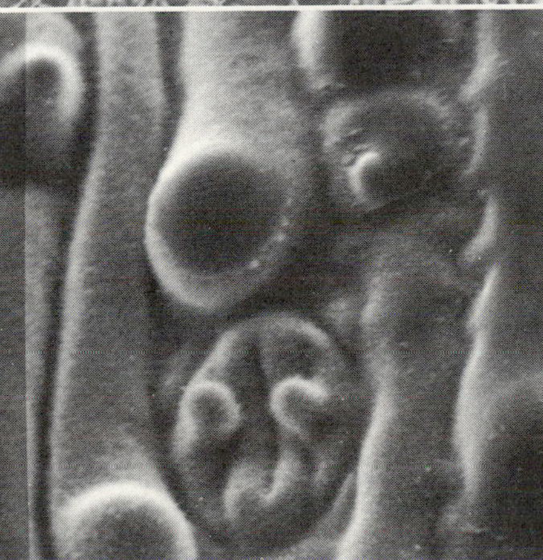

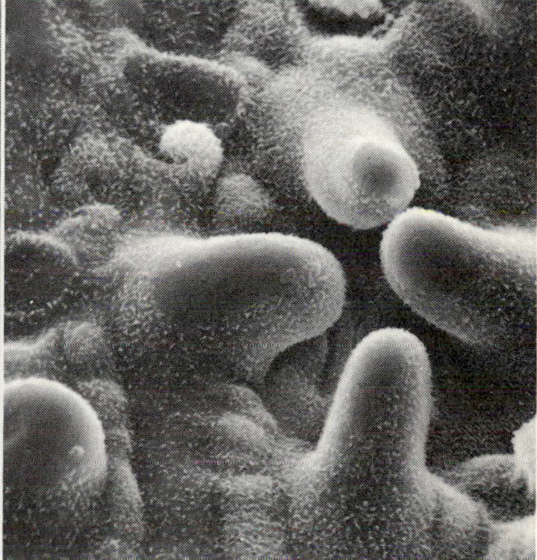

abundant on both surfaces of the leaf; in others, stomata may be restricted to the lower surface. In cucumber there are more than 60,000 stomata per square centimeter of leaf, while in some grasses the number may be fewer than 8,000. Even leaves on a single plant vary widely in the number and distribution of their stomata; "shade" leaves generally have fewer stomata per unit area than do "sun" leaves. In any event, it has been estimated that in a wide variety of plants, the stomata, when fully open, occupy 1-3% of the total area of the leaf and constitute essentially no barrier to the free diffusion of water vapor out of the leaf (Fig. 6-12). This fact accounts for the tremendous quantities of water lost by well-watered plants growing in bright light, in warm temperatures.

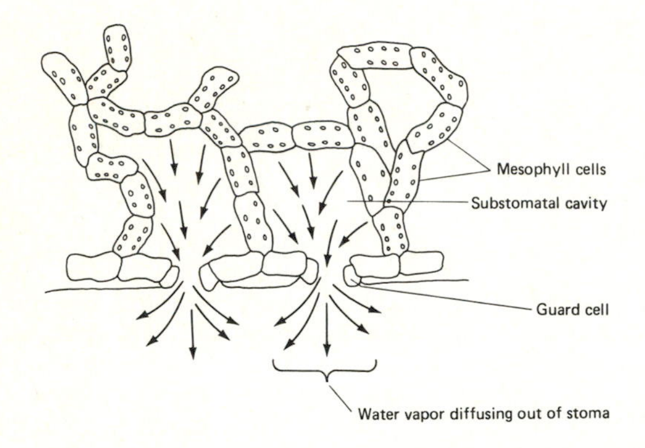

Figure 6-12 The arrows show the path followed by water molecules that diffuse from air spaces in the leaf through open stomata. (Adapted from J. Levitt. *An Introduction to Plant Physiology*, Second Edition, St. Louis, Missouri: Mosby, 1974.)

The rate of *transpiration*, that is, loss of water vapor from the aerial parts of plants, depends upon the size of the stomatal opening, the difference in water potential between air inside the leaf and the atmosphere, and air turbulence. The less water in the atmosphere the lower (more negative) will be the water potential of the atmosphere. (The vapor pressure and relative humidity, which are other measures of the water content of the atmosphere, will also be lower.) When air is moisture saturated, its water potential is zero. As soon as the relative humidity drops only one or two percent, the water potential drops sharply. By the time the relative humidity of the air has decreased to about 50%, the water potential of the atmosphere has dropped to minus several hundred bars. Since the water potential of the leaf cells is seldom below about −20 bars, water will diffuse rapidly from the semisaturated air between the cells, which will have a water potential in equilibrium with the surrounding cells, into the drier atmosphere. As in their movement through the plant, the water molecules leave the plant by moving down a water potential gradient (Table 6-2).

On a sunny day, the temperature inside a leaf can be as much as 10°C higher than that of the surrounding air. This temperature differential enhances the rate of transpiration, since the air inside the leaf is saturated with moisture, and the saturation vapor pressure increases with temperature. Turbulent air also promotes transpiration, since the rapid removal of water vapor from the area surrounding the leaf steepens the diffusion gradient from leaf to atmosphere. Thus during dry, windy, sunny days, and particularly during periods of drought, water is often lost by transpiration more rapidly than it is taken up by the roots. Whenever water loss from leaves exceeds water uptake by the roots for a prolonged period, the plant wilts. On a hot summer day transpiration frequently exceeds water uptake even if there is an adequate supply of water in the soil; under such conditions, leaves of all plants and the stems of herbaceous plants will often wilt slightly in the early afternoon. As the evening progresses, transpiration decreases and the plants start to recover. Overnight, the water deficit of the leaf cells declines as water is taken up by the roots; this continues until the leaf cells are fully turgid, with the plant usually fully recovered by the next morning. Such daily temporary wilting, termed *diurnal wilting*, is quite common, and does not harm the plant except to diminish photosynthesis as the stomata close. If water is unavailable in the soil for prolonged periods, however, a continual wilting is induced which, if prolonged, can lead to the death of the plant.

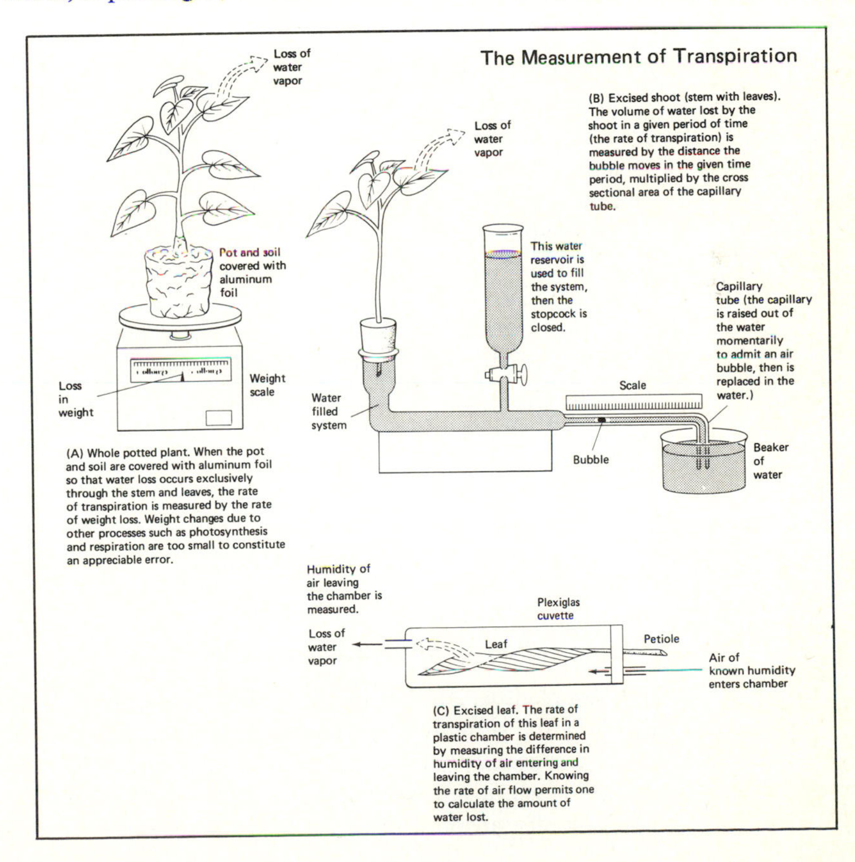

The Regulation of Guard Cell Movement

It has long been known that stomatal aperture is controlled by the turgor of the guard cells surrounding each stoma, as discussed in Chapter 2. Only during the past decade, however, has it become clear that the turgor of these cells depends largely upon their content of potassium salts. During the night, there is a relatively low concentration of solutes in the vacuoles of the guard cell; consequently ψ_π is high, the cells are flaccid, and the stomatal pore is closed. With the onset of the light period, potassium (K^+) ions migrate from neighboring cells into the vacuoles of the guard cells, frequently accompanied by a disappearance of starch and a buildup of malic acid. The resulting large decrease in ψ_π leads to the uptake of water, and since the guard cells have unusually elastic, differentially thickened walls, the cells swell and open the stomatal pore. The subsequent egress of K^+ from the guard cells at the end of the day, or during periods of water stress, leads to cell shrinkage and stomatal closure (Fig. 6-13). The subsidiary cells and other epidermal cells that surround the guard cells function as a reservoir where K^+ ions are stored when stomata are closed. Any changes in the size of the surrounding epidermal cells will be opposite to that of the guard cells and enhance the guard cell opening or closing.

When positively charged ions such as K^+ move across a cell membrane, some additional movement of charged particles must provide for maintenance of electrical neutrality. Either negatively charged anions must move in the same direction as K^+, or cations such as H^+ must move in the opposite direction (see Chapter 7). It has now become clear that chloride (Cl^-) movements play an important role in the regulation of guard cell turgor in some plants, but not in others. In *Zea mays* (corn), approximately 40% of the K^+ ions that move into and out of the guard cells are accompanied by Cl^- ions. Cl^- ion movements in some plants are relatively insignificant, and other ions may function in their place. This substitution can occur even in plants where Cl^- commonly functions. Large-scale migration of H^+ ions across guard cell membranes in a direction opposite to that of K^+ movement probably occurs in all plants. In fact, stomatal opening is correlated with an observed rise in pH, which would occur if H^+ ions left the cell. The source of the H^+ ions could well be organic acids present in the vacuolar solution, since these increase during stomatal opening.

Environmental control of stomatal movement

When a plant is deficient in water, the guard cells tend to become flaccid, thus closing the stomatal aperture and restricting water loss. Until recently, this was assumed to be the plant's primary mechanism to prevent severe wilting; however, it is now evident that plants have a more rapid and efficient way to shut off transpiration. During the initial stage of water stress in many

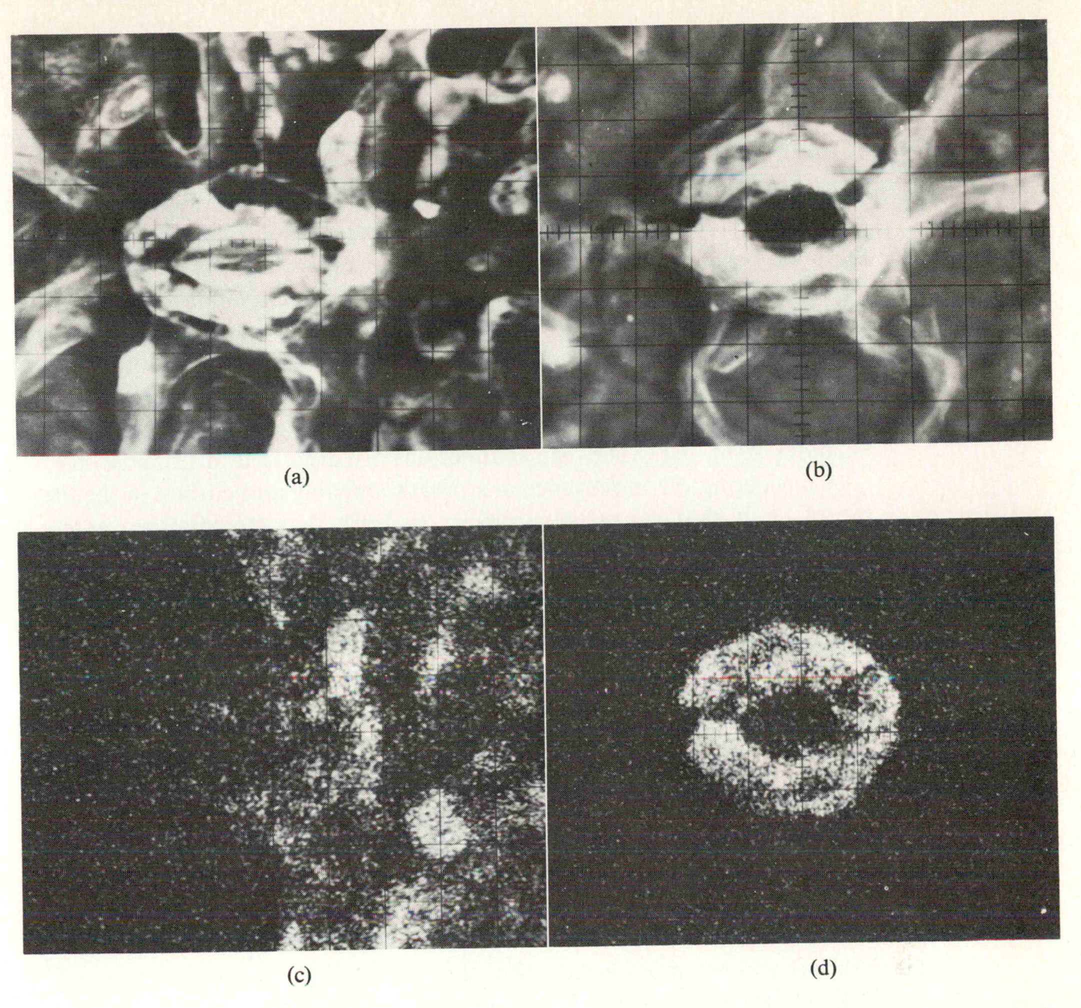

Figure 6-13 The distribution of K (potassium) in cells of the stomatal complex in *Vicia faba* when stomata are (a) closed and (b) open. K was analyzed in peeled strips of epidermal tissue using an electron microprobe. The epidermal strips were frozen rapidly in liquid nitrogen, and then freeze dried. This method was used instead of chemical fixation since K is soluble in most chemical fixatives.

The instrument produces a beam of high energy electrons that, when focussed on the tissue, excite elements in the tissue to emit x-rays. The white dots in the lower photos indicate x-rays emitted by K atoms, while the white regions in the upper photo, due to back-scatter electrons, reveal tissue morphology. Note that K ions stored in cells adjacent to the guard cells when stomata are closed migrate into the guard cells during stomatal opening. (From Humble and Raschke. 1971. *Plant Physiol. 48*, 447–453.)

plants, there is a dramatic increase in the level of the hormone *abscisic acid* (ABA) (see Chapter 10). This somehow leads to K^+ egress from the guard cells, followed by water loss and stomatal closure. Some of the most elegant experiments showing this effect were performed with a *wilty* mutant of tomato, originally obtained by chance after X-irradiation of the seeds of a normal cultivated variety. The mutant wilts rapidly under conditions of only slight water stress, because its stomata always remain open; analysis of hormone levels revealed only one-tenth as much ABA in the mutant as in the parent strain. When the mutant was treated with ABA, its stomata closed and the plant regained turgor more quickly. Obviously ABA, or possibly one of its metabolites, regulates stomatal closure in this cultivar of tomato. It has since become clear that low levels of applied ABA can cause stomatal closure in other plants. In addition, water stress can lead to an increase in the content of endogenous ABA, followed by stomatal closure. Thus, protection from desiccation seems to be one of the important physiological functions of the hormone ABA. Its other regular functions are discussed in Chapter 10.

Considering the importance of stomatal opening and closure in the life of the plant, it should not be surprising to learn that stomatal movement is regulated by other environmental factors in addition to water stress. For example, the CO_2 content of the substomatal cavity is a prime regulator of stomatal opening in many plants. If the CO_2 concentration falls below the 0.03% normally present in the air, the guard cells become turgid and the stomata open. This condition is usually brought about by illumination of the guard cells, which causes photosynthetic activity and an accompanying diminution of the CO_2 content of the surrounding air chambers. Stomatal opening may also be produced experimentally by the removal of CO_2 from the air, whereas stomatal closure can be induced by increasing the CO_2 content. This control by CO_2 helps to explain why stomata are normally open during the day and closed at night.

Although stomatal opening in response to light can be explained in part by reduction in CO_2 levels in the leaf resulting from photosynthesis, light also has another more direct effect. Protoplasts of onion guard cells, which lack chloroplasts, swell when illuminated with blue light, but only in a medium containing K^+ salts. The blue absorbing pigment that promotes K^+ uptake and increased turgor is likely the flavoprotein described in Chapter 11.

The usual daily pattern of fluctuation in transpiration rate is for a sudden increase in water loss to occur at dawn, proceeding to a maximum near noon; this is followed by a temporary decline if the temperature is too high, then by a small rise as the temperature decreases. These changes reflect changes in stomatal aperture. Midday closure is caused partly by the high internal CO_2 concentration at that time; the CO_2 level inside the leaf depends upon the relative rates of respiration and photosynthesis, and the respiration rate increases rapidly with temperature, while photosynthesis is less temperature sensitive. In addition, the water stress in early afternoon may also lead to an increase in ABA, causing stomatal closure.

Stomatal behavior is therefore regulated by the most important variables in the environment: light, temperature, soil moisture content, humidity, and atmospheric CO_2 concentration, which influence internal factors such as

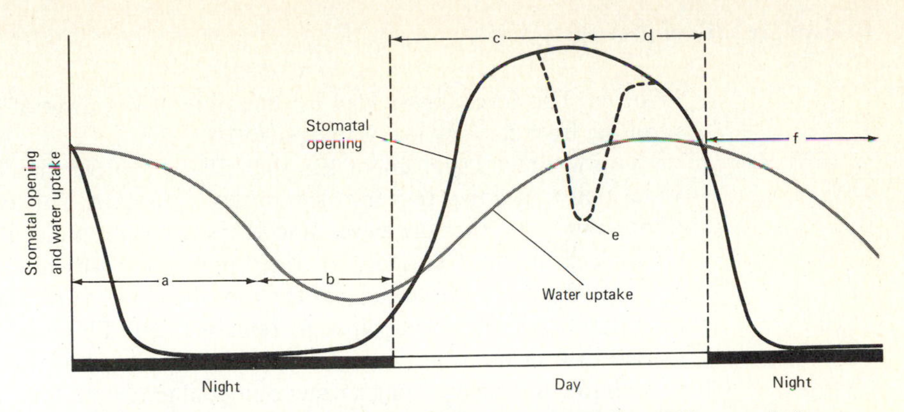

Figure 6-14 Stomatal movement and water uptake during a 24 hour period.
(a) Stomatal closure is the result of lack of light, the accumulation of respiratory CO_2, and the phase of the endogenous rhythm.
(b) Stomata start to open before dawn, for the opening-phase of the endogenous rhythm overrules the other factors. In mesophytes the "night opening" is small, but in Crassulaceae it is much greater because of the CO_2 depletion by dark fixation.
(c) Full opening is due to the direct effect of light and to CO_2 depletion by photosynthesis.
(d) Partial closure in late afternoon can be attributed to the endogenous rhythm passing into its closing-phase, as well as to falling light intensity.
(e) Stomata of some species close at noon if the temperature is high and transpiration exceeds water uptake. Such closure is probably regulated by increase in abscisic acid, which can cause closure during periods of water stress.
(f) Endogenous closure is reinforced by lack of light and by the CO_2 from respiration. (Adapted from T. Mansfield, 1971. *J. Biol. Educ.* 5, 115–123.)
The pattern of the water loss is similar to that of stomatal opening. Water uptake lags behind water loss because of the resistance to water flow through the plant. As a result, a water deficit develops during the day which is eliminated by continued water uptake at night.

water content and ABA concentration of the leaf. In addition, the size of the stomatal pores varies rhythmically, even in the absence of environmental cues. These rhythmic variations are controlled by an internal oscillator, the *biological clock*, discussed in Chapter 12. The combined effects of this rhythm and the environmental conditions on stomatal opening and closure is depicted in Fig. 6-14.

Water Ascent in Tall Trees

The mechanism by which water is transported to the tops of tall trees long puzzled plant physiologists, but is now relatively well understood. Part of the problem is that atmospheric pressure will raise a water column only about 30 feet, but the tallest trees rise to more than 10 times this height; thus, some additional force equivalent to about 10 atmospheres of pressure must be involved. This force is provided by the evaporative power of the atmosphere, resulting from its very low water potential.

You will recall that air of less than 100% relative humidity can have a water potential down to minus several hundred bars. This causes transpirational water loss from the leaves, and keeps the water molecules moving from cell to cell up the tree along a water potential gradient. The evaporative removal of water molecules from the top of the xylem column leads to flow of water along the xylem pipes to replace them. This movement of water caused by transpiration, termed *transpiration pull*, in turn draws water from the soil solution, again along a water potential gradient. Since water loss at the top of the plant leads to a lower water potential there than at the base of the plant, and since water uptake often lags behind water loss because of the resistance of water movement through the cell walls and endodermis of the root, a negative pressure, or *tension*, generally exists in the water columns of the xylem of tall trees. This phenomenon can easily be demonstrated: If a cup of modeling clay filled with ink is appressed to the outside of a tree and a cut is then made into the xylem, the ink will be drawn rapidly into the trunk and will move many meters up the tree. Pressure gauges inserted into tree trunks also show negative pressures, and diurnal changes in tree diameters can be measured, the minima coinciding with the periods of rapid transpiration.

It was originally argued that, if tensions developed in the water columns under these conditions, the water column would cavitate, or form bubbles, leaving spaces filled only with water vapor and thus effectively blocking the upward movement of the water column. Later research has conclusively demonstrated, however, that in a clean glass apparatus, water free of dissolved gases will not cavitate even under a tension of several hundred bars, because of the *cohesion*, the mutual attractive forces, between the water molecules. In the xylem tubes, water molecules are not only tightly bound to one other by cohesive forces, but also, because the xylem cell walls are also extremely hydrophilic, to the wall as well, by adhesive forces. These forces also help to prevent cavitation in the xylem under normal tensions.

This mechanism for water movement in plants—transpirational pull combined with the cohesion of the water molecules in columns under tension—can be demonstrated by sealing a twig in the top of a glass tube of water and placing the base of the tube in mercury. Provided that the glass is clean and that there are no dissolved gases in the water, transpiration by the twig can lift the mercury to a height in excess of 760 mm—the height to which it is normally lifted in a barometer by one atmosphere of air pressure. The system is entirely physical; in no way does it depend on the living properties of plant cells, for transpiration will occur even if collars of stem cells are killed by steam, but not if the leaves are killed. The twig in the apparatus can even be replaced by an evaporating surface made of plaster-of-Paris. This striking demonstration confirms our confidence in the correctness of the *transpiration–cohesion–tension* explanation for the rise of water in tall tress (Fig. 6-15).

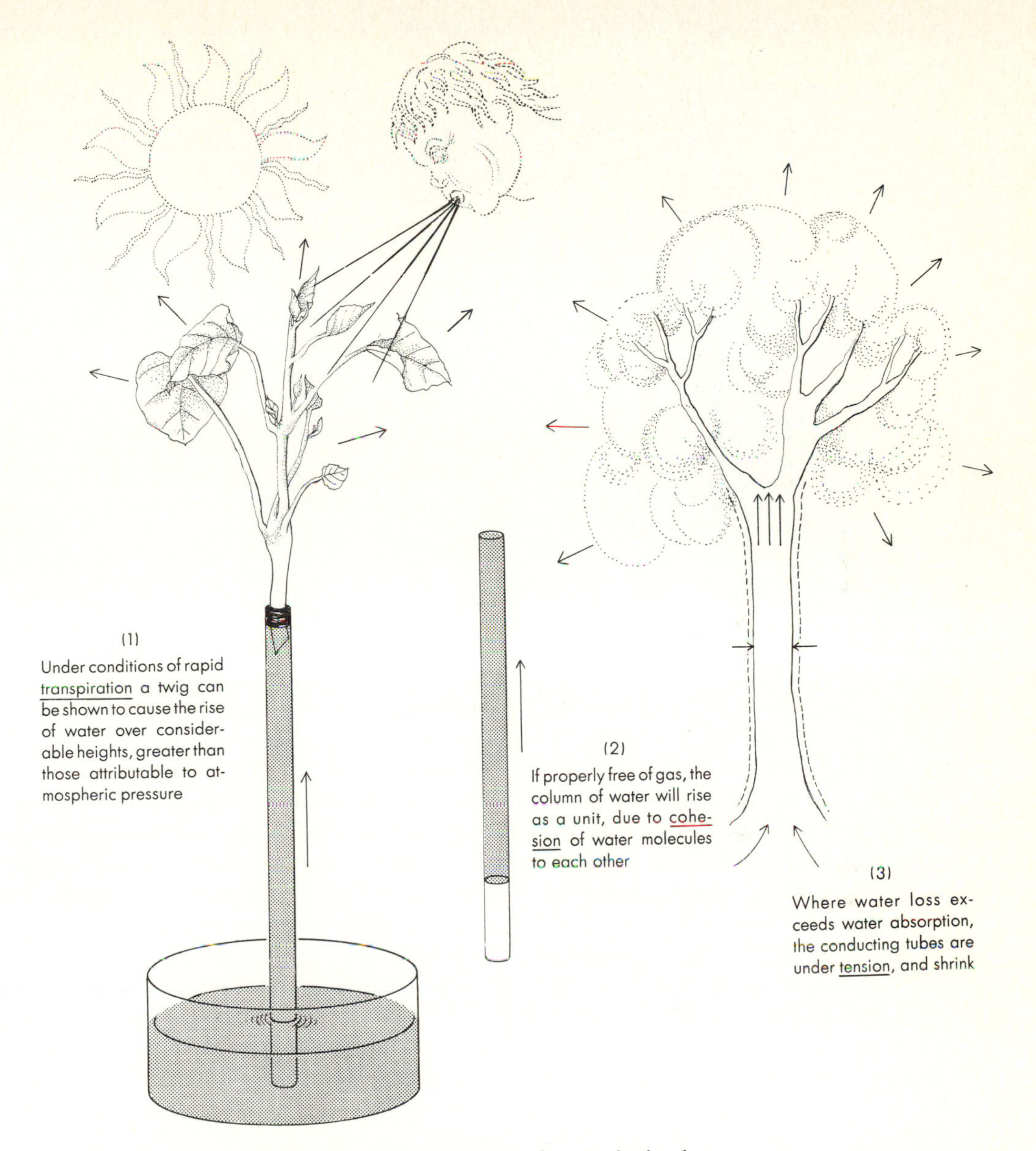

Figure 6-15 The mechanism of water rise in plants.

Water Potential Measurement of Plant Shoots

The existence of tension in the xylem provides a rapid method, adaptable for field studies, for assaying the water potential of a plant shoot. When a stem is cut, this tension is released, and the ψ of the xylem, which is assumed to contain almost pure water, returns to virtually zero. Thus the cells of the

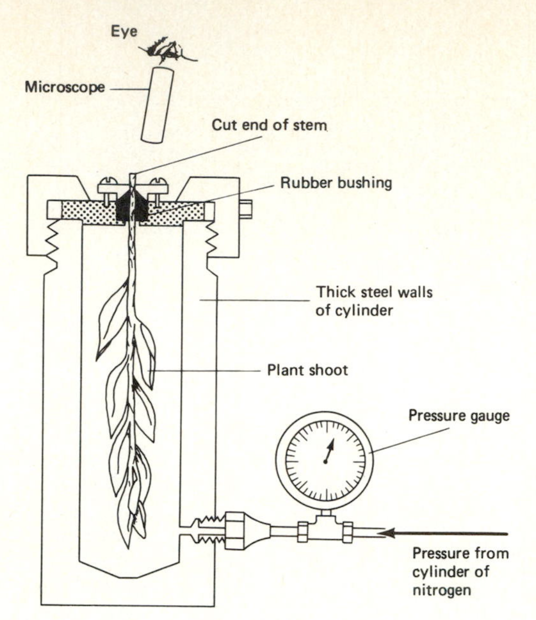

Figure 6-16 The pressure bomb is a good device for measuring the water potential of a shoot and the negative pressure in a stem. When a shoot is cut, tension in the xylem is released, and the surrounding cells can take up water. This causes the water to recede from the cut end. The shoot is placed in the bomb and pressure is applied. The pressure required to force the xylem sap back to the cut surface, although opposite in sign, is numerically equal to the tension previously existing in the xylem. This is the standard research method for measuring the water potential of the shoot. (Adapted from Scholander *et al.* 1966. Plant Physiol. *41*; 529–532.)

shoot, previously at equilibrium with the xylem under tension, are able to take up more water. As the shoot cells take up water from the xylem, the solution in the xylem tubes recedes from the cut surface. Next the cut shoot is placed in a pressurized container (a *pressure bomb*) with the cut end of the stem protruding (Fig. 6-16). A cylinder of nitrogen gas under pressure is connected to the bomb and the pressure outside the shoot increased. By increasing the pressure potential of the cells, this external pressure squeezes water out of the shoot cells back into the xylem. When the solution reaches the cut surface of the xylem cells, as observed through a low-power microscope, the pressure applied in the bomb is considered equal (but opposite) to the tension originally existing in the xylem when the stem was cut. If the xylem solution is assumed to be pure water, then the tension in the xylem was exactly equal to the water potential of the cells of the shoot. (Actually, the xylem solution contains some dilute salts, but since their concentration is very low, the correction needed is very small.) This pressure-bomb procedure provides a rapid estimate of the water status of plants in the field, and is proving very valuable to ecologists and agronomists alike.

Root Pressure and Guttation

While the xylem contents are usually under tension because of transpiration, a positive pressure can develop in the xylem tubes of the root and stem base when rapid transpiration is *not* occurring. Mineral ions actively accumulated by cells in the root are pumped into the xylem, where, because of negligible water movement under nontranspiration conditions, the concentration of salts builds up. This decreases the osmotic potential in the xylem, which then results in the inward movement of water by osmosis.

While water movement through the root tissues into the central cylinder occurs through the cell walls, the water must pass through the membranes and protoplasts of the endodermal cells, because their walls are water-impermeable. The entire ring of endodermal cells thus acts as a single membrane, with a concentrated solution on the xylem side and a weak solution on the soil and cortex side of the root. The entire root thus forms an osmometer, with water diffusing from the soil through the endodermal membrane into the xylem in response to the difference in solute concentration. A pressure thus builds up in the xylem cylinder, much as turgor pressure builds up in a single cell. The impermeable endodermal cell walls also function to prevent the salts pumped into the xylem from leaking back out into the cortex and exterior through the channel formed by the interconnecting cell walls.

When stems of plants in the condition described above are severed just above the ground line, large quantities of fluid are exuded from the cut surface. If a manometer (a pressure-measuring device) is attached to such a cut stem, it will indicate that the roots are producing a pressure (termed *root pressure*) of several atmospheres (Fig. 6-17). In some plants, root pressure

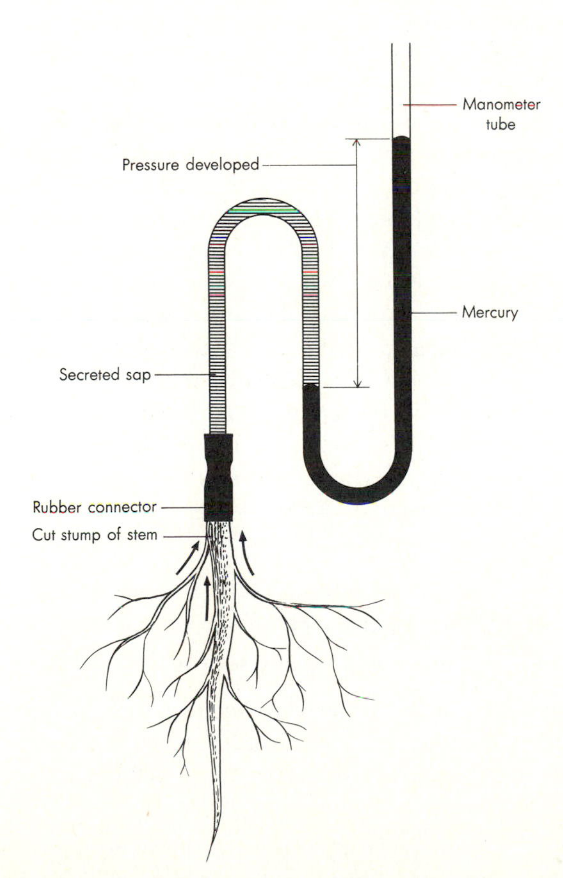

Figure 6-17 Water may move from roots to the stem, causing the development of a *root pressure.*

Figure 6-18 Guttation fluid on the edges of a rose leaf.

causes a small amount of water to leave the plant in liquid form through *hydathodes*, special pores at vein endings in the leaves (Fig. 6-18). This process, called *guttation*, occurs mainly at night, when root pressure is high.

Another phenomenon involving positive xylem pressure, familiar to residents of the northeastern United States and Canada, is maple sap flow. In late winter a sugary solution, derived from carbohydrate stored in the stem, can be tapped from the xylem of several kinds of trees, but most notably the sugar maple (*Acer saccharum*). This flow occurs because of a stem pressure, independent of the root system, which is generally seen on warm days following cold nights. Cold nights lead to the hydrolysis of starch reserves in the xylem parenchyma to sugars, which are then actively transported into the xylem vessels. Warm temperatures cause the release of CO_2 from solution, thus creating a pressure in the xylem. Water and sugar are forced up the trunk under these conditions, which occur just before spring growth starts. This can be shown by severing a tree at the base of the trunk and inserting the trunk into a barrel of water; water is withdrawn from the barrel causing a normal sap flow under "sugaring-off" weather conditions.

Stress and Seasonal Changes in Xylem Transport

The transport of water in plants is believed to be an entirely physical process, the result of pressure gradients. The plant appears to have no active mechanism of ensuring its water supply in times of stress, yet the water supply is maintained through periods of drought and cold. The very high cohesive forces between the water molecules and their adhesion to the hydrophilic xylem cell walls prevents cavitation of the xylem solution in most normal situations. If a plant is put under severe water stress, cavitation starts to occur in individual xylem tubes; this can actually be heard as clicks in a sensitive microphone placed next to the stem. Once cavitation has occurred under water stress, that xylem pipe is lost to the plant as a functional water conduit. As the plant has many such pipes, not all of which cavitate at the same time, the water conducting system as a whole remains functional. When an adequate water supply returns, these pipes transport enough water to the shoot to enable the cambium to form new xylem for the shoot. If all the xylem pipes should ever cavitate, however, the shoot could not survive.

Winter is another period of water stress. As deciduous plants drop their leaves, transpiration—and the xylem tension—are reduced. If the xylem solution freezes, small blockages to water movement can occur, but these generally disappear when thawing occurs. In some trees, the xylem pipes function for only one year, the cambium cells immediately forming new xylem cells as growth renews in the spring.

While root pressure cannot account for water movement in normally transpiring plants, it may function as a refilling mechanism in certain plants. Vines, for example, have large xylem vessels which tend to empty in winter, In spring a large root pressure is built up, which refills the xylem pipes with water; thus a water continuum is formed from the soil to the shoot by the time the leaves unfold. As in other plants, further movement of water to the unfolded leaves would be caused by the transpirational pull created by the negative water potential of the atmosphere.

Adaptation to Water Stress

Since desiccation is always a major threat to plant survival, plants that grow in arid regions have had to develop extraordinary survival mechanisms. In deserts, rainfall is only occasionally heavy enough to wet the ground for even the few weeks needed to establish a newly germinated seedling; seeds of many desert annuals accordingly remain dormant during a light rainfall and germinate only if rainfall is prolonged. Such regulation of germination is accomplished in various ways: Some seeds have particularly thick and impermeable seed coats that resist hydration in the absence of prolonged moist periods, while others contain water-soluble germination inhibitors that leach slowly from the seed during a rain—only a downpour that soaks the soil slowly and thoroughly removes enough inhibitor to permit germination. In other species, the obligatory seed dormancy period is highly variable, and only a small fraction of the viable seed can germinate at any one time. This enhances the possibility that some seed will germinate when conditions for growth and reproduction are optimal. Growth after germination is usually rapid in desert annuals; the reproductive cycle is sometimes completed within just a few weeks, and by the time the environment is again very dry, the plant is again securely perpetuated in the form of drought-resistant dormant seeds.

Desert vegetation is characteristically dwarf, and has restricted leaf area, thereby minimizing the rate of transpiration. Transpiration rates are also reduced by structural adaptations such as small, thick, heavily cutinized leaves, and relatively small numbers of sunken stomata (Fig. 6-11). Most remarkably, the stomata of some xerophytes open at night, when the rate of transpiration is low, and close in the daytime, as discussed in Chapter 4. Since the evaporation of water is a cooling process, one might expect leaf temperature in xerophytes to rise excessively, particularly during hot sunny days, as a consequence of the low rate of transpiration. However the upper surfaces of the leaves of many xerophytes are covered with numerous small hairs, which reflect radiation from the leaf surface, thereby reducing the heat load (Fig. 6-19). Such extreme pubescence seems to be an adaptive mechanism often found on plants in arid regions, and may serve for thermoregulation.

Soil salinity is also a major cause of water stress, since solutes reduce the ψ of the soil solution. Soil salinity is becoming a serious worldwide problem,

Figure 6-19 A cross-section of a leaf of the xerophyte *Encelia farinosa*, seen with aid of the scanning electron microscope. Note the abundance of hairs on both upper and lower surfaces. V = vascular bundle; E = epidermal cell; P = palisade cell; M = mesophyll cell; I = intercellular space. (Courtesy of J. Ehleringer, University of Utah.)

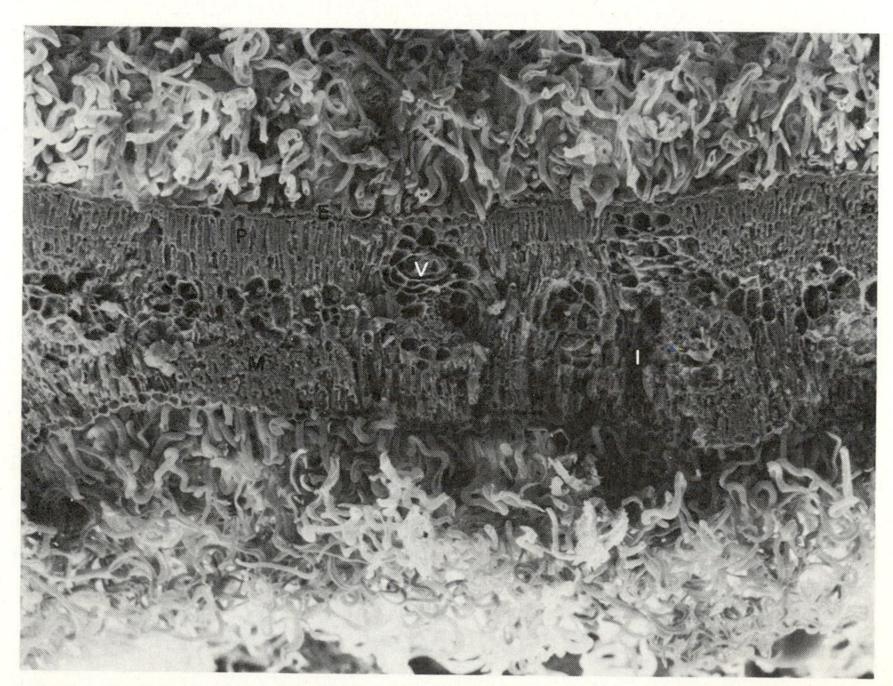

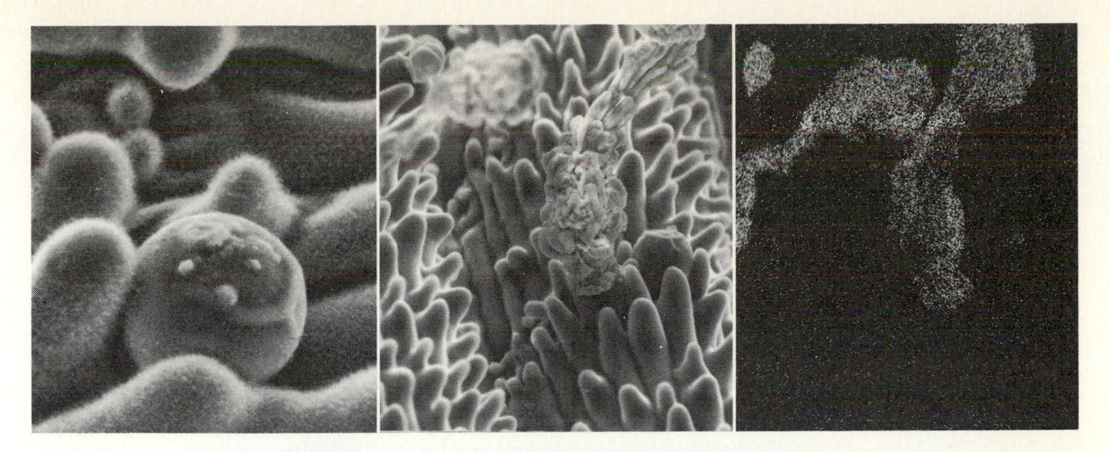

Figure 6-20 Scanning electron micrographs (left and middle) of salt glands on the leaf surface of the salt marsh grass *Distichlis spicata*. The gland on the left has secreted a small amount of salt (×1230) while that in the middle (×500) has an extended salt whisker. An x-ray image (right) shows the distribution of chloride in the salt whisker. The methodology of x-ray analysis is described in the Fig. 6-13 legend. (From Hansen et al. 1976. *Amer. J. Botany 63*, 635–650.)

as more land is irrigated and fertilized more heavily, and as the increase in population leads to increase in soluble waste. Since water is absorbed by plant roots from the soil mainly through osmosis, it will not move into the roots unless $\psi_{\text{root cells}}$ is lower than ψ_{soil}. Thus plants that grow successfully in saline soils have a low internal ψ, accomplished by storing unusually high quantities of solutes in the vacuoles of their root cells. If plants of this type are harvested and removed from the soil after they are full grown, they actually improve the arability of the soil, since they absorb large quantities of solute and thereby decrease soil salinity. Many such plants have special *salt glands* (Fig. 6-20) through which large quantities of salt are actively secreted. In some cases, ions such as sodium and chloride that might inhibit cell metabolism are preferentially secreted, while other ions are retained to maintain an adequate osmotic potential. Botanists are seeking to identify as many high-salt-absorbing plants as possible, so that they can be used in land reclamation and agriculture. Wherever new plants are introduced, the management of their water economy is a prime determinant of their survival, successful productivity, and ultimate value to man.

SUMMARY

Water is the single most abundant chemical compound in plants, often comprising 90% or more of the total weight of the plant body. In addition, vast amounts of water enter and leave the plant for each molecule retained. The absorption of water is osmotic, its transport mainly through xylem, and its loss to the atmosphere mainly by evaporation (*transpiration*) through stomata.

The tendency of water molecules to move is expressed by the *water potential*, ψ. This is a measure of the energy of the water molecules and is expressed in units of pressure; water always moves from a higher to a lower potential. Solutions are said to have an osmotic potential (ψ_π) which is that component of water potential due to solute molecules that lower the energy of the water molecules. Since the ψ of pure water is 0, and the ψ becomes increasingly more negative as the solute content increases, water tends to move toward solutions with greater solute concentrations, provided that the back diffusion of solute is prevented by a membrane.

In a cell, the vacuole accumulates water osmotically, expands, and presses the cell contents against the wall (*turgor pressure*, or ψ_p). When the back pressure of the wall on the cell contents is sufficiently great to oppose the osmotic potential of the cell sap, net water entry will cease because the water potential of the cell is now equal to that of the water outside the cell.

The soil is a water reservoir, saturated at *field capacity* (ψ = approximately 0) and empty at the *permanent wilting percentage* (ψ = approximately − 15 *bars*, or atmospheres). Water enters the root osmotically and moves through apoplast or symplast until it reaches the *endodermis*, at which point the *Casparian strip* compels movement through the symplast. Upward transport of water through the xylem ends in *transpiration*. Stomatal aperture, which controls transpiration, is itself controlled by solute movement into and out of guard cells. Turgor pressure changes in these cells regulate their conformation and thus the stomatal opening. The main solute moved is apparently K^+, balanced by Cl^- or newly formed organic anion such as malate. Under conditions of incipient wilting, a high concentration of the hormone *abscisic acid* may be produced; this causes the guard cells to lose K^+ and other solutes, resulting in stomatal closure. This mechanism can protect the plant against excessive water loss; plants with low ABA levels may not have this protective mechanism intact, and may thus wilt excessively. CO_2 levels in the substomatal air space also control stomatal opening; thus, when light hits the leaf, photosynthesis reduces this level below 0.03% and the stomata open. The converse occurs in the dark.

Water rises in tall trees because of the tension (negative pressure) developed in the xylem by transpiration from the leaves. Because of the high cohesive force between water molecules, the xylem contents are pulled upward as a column. Tensions can be measured in trees during rapid transpiration, and purely inorganic models support the operation of this mechanism.

Water potentials in plant shoots can be measured by *pressure bombs*, in which the application of external gas pressure exactly compensates the internal tension. Thus, that pressure which just serves to express droplets of sap from the xylem is equal and opposite to the internal tension. Under some circumstances, the xylem is under positive pressure as the result of osmotic forces generated by the root (*root pressure*). When root pressure is high, water droplets may be exuded (*guttation*) at *hydathodes*. Maple sap flow is a result of pressures built up in the stem.

Plants adapt to water stress through a variety of structural and chemical modifications, involving reduced leaf area and stomatal number, protected stomatal pores, thickness of cuticle, hairiness, and internal solute level. Plants can be selected, bred, and conditioned for growth under high water stress.

SELECTED READINGS

Dainty, J. "Water Relations of Plant Cells," *Encyclopedia of Plant Physiology*, New Series, Vol. 2 A, U. Lüttge and M.G. Pitman, eds. (Berlin–Heidelberg–New York: Springer-Verlag, 1976).

Kozlowski, T. T., ed. *Water Deficits and Plant Growth*, Vols. 1–5 (New York: Academic Press, 1968–1978).

Kramer, P. J. *Plant and Soil Water Relationships*, 2nd Ed. (New York: McGraw-Hill, 1969).

Meidner, H., and D. W. Sheriff. *Water and Plants* (New York: Wiley, 1976).

Slatyer, R. O. *Plant–Water Relationships* (New York–London: Academic Press, 1967).

QUESTIONS

6-1. List several ways in which water deficit can affect shoot growth. What does each affect? Which is permanent in effect even when sufficient water subsequently becomes available? Why?

6-2. How does the green plant balance its need to conserve water against its other needs?

6-3. What causes water uptake by roots (a) when the plant is transpiring rapidly? (b) when transpiration rate is very low? What is the most likely path of water movement from the soil to the xylem? Does uptake in either (a) or (b) above depend on any particular structural feature of the root? Give reasons for your answer.

6-4. Farmers rarely fertilize their crops during periods of drought, since they have found empirically that they may actually damage their plants by such a practice. Explain why this should be so.

6-5. Drought and soil salinity have somewhat similar effects on water uptake by plants. Explain.

6-6. A plant with roots in pure water may wilt temporarily when salts are added to the water, but will probably regain turgidity after a few hours. Explain.

6-7. Transpiration rate has sometimes been measured by placing anhydrous blue cobalt chloride paper in contact with a leaf, and noting the time required for it to turn pink, indicative of hydrated cobalt chloride. What errors are inherent in this method?

6-8. Why should (a) wind increase transpiration rate? (b) leaf hairs decrease leaf heating in sunlight?

6-9. Trace the path of a water molecule from a raindrop hitting the soil to water vapor being transpired by the leaf of a plant in a plant community. Describe the processes and physical forces operating at each stage.

6-10. Certain bacteria cause wilting of infected plants under conditions in which normal plants remain turgid. Suggest several ways in which such wilting could be brought about.

6-11. Explain why water can be moved to the top of a tall tree while a mechanical lift pump is unable to draw water higher than about 10 meters. What are the conditions driving such movement?

6-12. Water loss by a plant occurs "because of the very negative water potential of the atmosphere." How would you explain this statement (a) in nontechnical language, to a nonbiologist? (b) in terms of free energy? Use diagrams as needed. How does this cause water uptake from the soil?

6-13. How could one measure the velocity of water movement in an intact tree trunk?

6-14. What prevents the columns of water in the xylem from breaking? Why might they be considered liable to break in any case?

6-15. The fungus *Helminthosporium maydis*, cause of the corn blight which destroyed most of the corn crop in the United States in 1970, "starved" the host plants by severely restricting their rate of photosynthesis; this restriction was caused partly by inhibition of potassium uptake. Explain the relationship between potassium uptake and photosynthesis.

6-16. It is thought that the opening of stomata is due to (a) an *uptake of potassium*, (b) which develops *sufficient osmotic force* (c) to create a *pressure inside the guard cells* to open these cells. Describe some experimental methods which could test these three statements.

6-17. Calculate ψ_{cell}, ψ_{π} and ψ_{p} for the cells in a thin slice of plant tissue using the method described in Fig. 6-4. Assume the tissue neither gains nor loses weight when immersed in 0.55 molal sucrose, and that 50% of the cells are plasmolyzed when the tissue is immersed in 0.7 molal sucrose. What errors are inherent in this method?

6-18. If the cell sap of a plant cell has an osmotic potential of -20 bars, what will the cell water potential and cell pressure potential be at (a) incipient plasmolysis, and (b) full turgor, assuming no change in the osmotic potential due to dilution? Use the symbols in question 6-17.

6-19. An algal cell is placed in a hypertonic solution. What happens? How does one describe the cell's condition? How do water potential, osmotic potential, and pressure potential change?

6-20. Explain how the water potential, osmotic potential, and pressure potential of a leaf cell will vary during 24 hours on a summer day.

6-21. When a small cut shoot is enclosed in a pressure bomb with the cut end protruding and sufficient pressure is applied to cause the sap to return to the level of the cut, what does this pressure equal? Explain in terms of water potential what happens when the shoot is cut and what happens in the pressure bomb.

6-22. The following water content percentages were obtained for three soils:

	CLAY	SILT	SAND
Field capacity	38	22	9
Permanent wilting percentage	18	11	3

A similar plant was placed in a pot of each soil and watered until water drained from the bottom of the pot. The plants were then placed in the open air during a dry period and not watered further. Which plant would probably wilt first? Why?

6-23. Embryonic cells are nonvacuolate, while most mature cells have large central vacuoles. How can a vacuole start to develop in the cytoplasm? What causes it to increase in size?

6-24. Explain why the permanent wilting percentage is relatively independent of the kind of plant used in its determination. Why do different soils have different permanent wilting percentages?

6-25. The transpiration of a desert succulent plant is minimal at midday even though that is when evaporation from open water in the same area is maximal. Explain why and how this may occur.

7

Mineral Nutrition

In addition to the water absorbed from the soil and the organic materials produced in photosynthesis, the green plant requires a variety of mineral elements. In the cell, these elements serve a variety of structural functions, and are also involved in the action of specific enzymes that regulate important aspects of cellular metabolism. All mineral elements except one, nitrogen, are ultimately derived from the parent rock giving rise to the soil; nitrogen is obtained from the atmosphere, mainly through the process of *nitrogen fixation*.

Minerals are absorbed from the soil together with water, and are transported upward mainly through the xylem. Since the ratio of the different elements in the xylem sap is quite different from that in the soil, it is clear that mineral uptake through root cells must be a selective process. Mineral elements also tend to be concentrated in cells where they are needed. Such selectivity is regulated by the differentially permeable membranes that surround the protoplast, the vacuole, and the cellular organelles, as well as by localized ion pumps driven by metabolic energy.

Several conditions must be satisfied if the green plant is to acquire adequate quantities of the minerals needed for its growth. First, the minerals must be present in the soil in a form suitable for absorption by cells of the root. Usually, this means that they must be dissolved in the soil solution, but sometimes they may be released from soil particles through the solubilizing activity of roots. Secondly, the soil must be well aerated so that root cells can carry out oxidative phosphorylation, for mineral uptake requires a steady supply of ATP. Finally, the transport system in the plant must function efficiently in delivering minerals to the recipient cells. Before discussing each of these aspects of mineral nutrition, we will first consider the nature and function of each of the mineral elements required by the green plant.

Essential Elements

An *essential element* is one without which the plant cannot complete its life cycle. At present, we know that plants need 16 elements; 4 (C, H, O, and N) are derived ultimately from the CO_2, H_2O, and N_2 of the atmosphere, whereas the remaining 12 (K, Ca, Mg, P, S, Fe, Cu, Mn, Zn, Mo, B, Cl) are derived from the parent rock which gave rise to the soil. Of the soil-derived elements, the last 7, called *micronutrients*, are required in very small quantities; the remainder are *macronutrients*.

Our present knowledge of the mineral requirements for healthy plant growth is based largely on experiments with plants grown in mineral solutions, a technique sometimes called *hydroponics* (Fig. 7-1). If the solution lacks adequate quantities of some essential element, the plant will deteriorate; it will develop deficiency symptoms characteristic of a short supply of that element (Fig. 7-2). A skilled botanist can learn to recognize the deficiency symptoms for each of the elements in a particular plant, and can improve the plant's condition by suitable additions to the soil or culture solution. A more objective procedure is to harvest small bits of the growing plant periodically, and subject them to chemical analysis for the various elements. The results of soil analysis can also be used as a rough guide, but such analysis does not usually indicate the extent to which the nutrients present are available

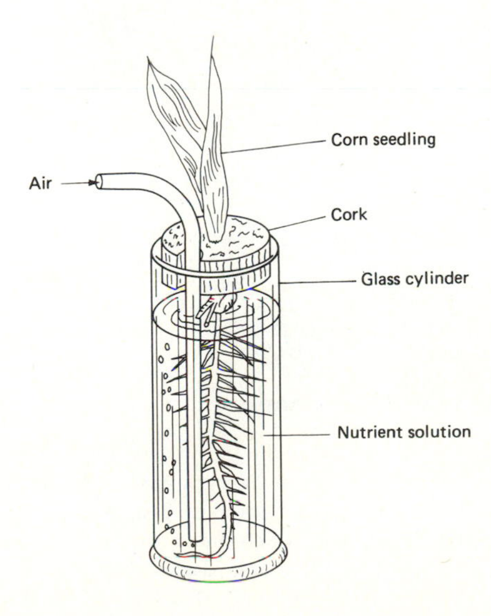

Figure 7-1 Solution culture technique used by early plant physiologists to investigate essentiality of minerals. A seedling is planted in highly washed pure quartz sand in a hard glass, glazed porcelain, or plastic container which is then supplied with a nutrient solution made with distilled water, and containing only the purest salts available. Great care must be taken to exclude organic impurities, microbial contaminants, and dust, all of which may supply traces of mineral-element contaminants. In certain instances, especially with large-seeded plants such as the garden bean, it is necessary in addition to remove the cotyledons, for they may contain sufficient stored quantities of certain elements to eliminate the necessity for an external source.

Figure 7-2 Fifteen-day old bean seedlings, variety Black Valentine, grown in a complete nutrient solution (center), or in a complete nutrient solution for 10 days followed by 5 days in a solution that lacked an essential element. The lower photo shows the first trifoliate leaves. Note the stunted growth of plants deprived of calcium (—Ca), the interveinal chlorosis of plants deprived of iron (—Fe) and the small leaves of plants deprived of nitrogen (—N). (Courtesy of H. Koontz, University of Connecticut.)

to the plants. After plant analysis, any element in short supply can be added to the growth medium, with prompt beneficial effects on plant growth and vigor. Analytical techniques of this type are highly useful to modern agriculture.

One of the first mineral nutrient solutions prepared to support the growth of higher plants, that of the German plant physiologist W. Knop, contained only three salts: calcium nitrate, potassium phosphate, and magnesium sulfate. The six elements furnished by these salts, together with the carbon, hydrogen, and oxygen assimilated in photosynthesis, satisfy the major nutritional needs of higher plants. The exact form in which the major elements are supplied is relatively unimportant, that is, the three cations—K^+, Ca^{++}, and Mg^{++}—may be combined in any way with the three anions—NO_3^-, $SO_4^=$, and $H_2PO_4^-$. In addition, some plants can use ammonium nitrogen (NH_4^+) or organic forms of N as well as or better than nitrate, although most plants prefer the nitrate form.

As commercially available mineral salts became increasingly purer, it became clear that the basic three-salt solution of Knop was not really complete. Plants, in fact, require many other elements, but in quantities much smaller than for the six major elements. These *micronutrient elements* include iron (Fe^{++} or Fe^{+++}), manganese (Mn^{++}), zinc (Zn^{++}), copper (Cu^+ or Cu^{++}), molybdenum (generally as MoO_4^-), boron (as $BO_3^{\equiv}$), and chloride (Cl^-). Some researchers also suspect that the plant may need very small quantities of yet other elements, such as cobalt (Co^{++}), vanadium (generally as $VO_4^{\equiv}$), strontium (Sr^{++}), and iodide (I^-), but have not yet established absolute requirements for these elements. In addition, certain plants have special requirements. For example, diatoms need silicon to build their silica cell walls; as far as we know, most other plants do not require silicon, although horsetails (*Equisetum*) and most cereals do not flourish in its absence. Although silicon cannot be considered a generally essential element, it can certainly be beneficial to plant growth. Similarly, sodium, while not strictly required, seems to aid in achieving high yields of certain fleshy root crops, such as sugar beets.

Functions of the Various Elements in the Plant

The major elements involved in photosynthesis (C, H, O), plus nitrogen, sulfur, and phosphorus, constitute the main building blocks of the plant body. For example, the cell walls that form the plant skeleton are composed almost exclusively of carbohydrates and related compounds containing C, H, and O; proteins, the major organic constituent of the cytoplasm, are composed mainly of C, H, O, and N, with a little sulfur; the nucleic acids found in the nucleus and in some of the cytoplasmic organelles are made of C, H, O, N, and P; and the lipids that are abundant in all membranes are built mainly of C, H, and O, with a little N and P.

Of the 12 elements derived from rock, 4 are used mainly for structural purposes. *Sulfur* is a constituent of several amino acids (cysteine, cystine, and methionine), the structural units ultimately assembled into proteins. Even though plant cells require relatively small quantities of sulfur, almost all of it serves an important structural function, for without the sulfur-containing amino acids, many important proteins of the cell could not be synthesized. Sulfur is also present in *glutathione*, a widely distributed substance believed to play a role in oxidation–reduction reactions because of its ability to change reversibly from the reduced form (—SH, or sulfhydryl) to the oxidized form (—S—S—, or disulfide):

$$\underset{\text{(reduced)}}{\underset{\text{Sulfhydryl form}}{2R\text{—}SH}} \rightleftharpoons \underset{\text{(oxidized)}}{\underset{\text{Disulfide form}}{R\text{—}S\text{—}S\text{—}R}} + 2H$$

The formation of —S—S— bridges between adjacent parts of large protein molecules is important to their ultimate form and stability. Sulfur is also a component of coenzyme A and thiamin (vitamin B_1).

Quantitatively, the main function of *calcium* is its incorporation into the structure of the middle lamella of the cell wall. It forms an insoluble salt when coupled with acidic components of the jellylike *pectins* of the middle lamella. The introduction of calcium into the cell wall, therefore, serves to stiffen a previously semifluid structure. Calcium also plays an important role in regulating the differential permeability of cell membranes. When plants are grown in a calcium-deficient medium, the cell membranes become "leaky" and lose their effectiveness as a barrier to the free diffusion of ions. *Magnesium*, a chemical relative of calcium, is a central part of the chlorophyll molecule, being attached to each of the four pyrrole rings either by direct covalent bonds or by so-called "secondary valences." In cases of magnesium deficiency, the older leaves manifest the yellowing typical of chlorophyll deficiency (*chlorosis*). Magnesium is also known to be a specific cofactor for several enzymes and is needed for nucleic acid stability.

Phosphorus acts mainly as a structural component of the nucleic acids, DNA and RNA, and as part of phospholipids, fatty substances which play an essential role in the structure of the membrane. A deficiency of phosphorus is thus very serious for the cell, for it prevents the formation of new genetic material in the nucleus and cytoplasm, and of new membranes around the surface of the cell and its various organelles. Phosphorus is also critically involved in all energy-transfer steps in the cell, since ATP and its analogs are composed of three phosphates coupled to a nucleoside.

Although phosphorus, magnesium, calcium, and sulfur do have other functions in the cell, the structural roles we have described are quantitatively the most important for these elements.

Potassium is an osmotically active element that is important in regulating the turgidity of plant cells. Most plants contain high levels of potassium, with very little of it fixed into cell structures. It is a characteristically mobile element, and since the membranes of many cells are quite permeable to potassium, large diffusional fluxes through cell membranes commonly occur. The water content of many cells, including those controlling movements, is often related to their internal potassium concentration. This is true of the guard cells whose turgidity controls the opening and closure of the stomatal pores, and the motor cells which regulate diurnal leaf movements (See Fig. 6-13). Potassium is also known to activate several important enzymes, though it has never been isolated as part of an enzyme system.

Chloride is also involved in turgor regulation in some plants, moving together with potassium and thus maintaining electroneutrality. However, chloride levels are rarely as high as potassium levels, and some plants with large potassium-regulated turgor changes contain very little chloride, using organic anions such as malate instead. Thus, it appears that chloride is used in this role when available, but is not absolutely essential for turgor regulation. Chloride is also known to stimulate photosynthetic phosphorylation, but its exact biochemical role in this process has never been exactly defined. It is possible that the minute amounts of Cl^- essential for most plants are involved in this process. The essentiality of chloride was in fact not discovered until relatively recently, when all the air in experimental greenhouses was filtered to remove the small traces of airborne chloride that proved sufficient for healthy growth of most plants.

Of the remaining six elements (Fe, Mn, Cu, Zn, Mo, B), the first five function mainly as essential parts of enzymes in the cell. As previously noted, many important enzymes consist of specific proteins to which are attached special entities called *prosthetic groups*, or *coenzymes*; these groups may consist entirely or partly of metallic elements such as Fe, Cu, Mn, Zn, or Mo.

Iron is a part of many important enzymes, including the respiratory electron carriers called *cytochromes* and the oxidative enzymes *peroxidase* and *catalase*. In all these enzymes, the iron is present in the prosthetic group as heme (an analog of chlorophyll), in which a central iron atom is connected to four pyrrole rings joined into a large cyclic structure. Iron functions in such enzymes by virtue of its reversible oxidation and reduction ($Fe^{+++} + e^- \rightleftharpoons Fe^{++}$); non-heme iron may function in the same way. Copper can also be reversibly oxidized and reduced ($Cu^{++} + e^- \rightleftharpoons Cu^+$), and it is likely that *manganese* present in the enzyme *superoxide dismutase* plays a similar role in various oxidative reactions. Iron is also essential to the enzymes of chlorophyll synthesis, and is a constituent of *ferredoxin*, an electron-transport agent in photosynthesis. Lack of iron causes severe chlorosis in developing leaves, which may appear completely white.

Molybdenum appears to be involved exclusively in the functioning of those enzymes (*nitrate reductase*, *nitrogenase*) involved in the reduction or fixation of nitrogen. When reduced or organic N is fed to plants, the requirement for molybdenum decreases or disappears.

Copper is a part of certain oxidative enzymes, such as *tyrosinase* and *ascorbic oxidase*, which oxidize, respectively, the amino acid tyrosine and vitamin C (ascorbic acid).

Zinc is part of the enzyme *carbonic anhydrase*, which catalyzes the hydration of CO_2 to H_2CO_3. This enzyme may be important in maintaining a storehouse of potential CO_2 for photosynthesis, since H_2CO_3 can readily dissociate to yield bicarbonate (HCO_3^-) or free CO_2. Zinc is also a cofactor in the synthesis of the plant hormone indoleacetic acid from the amino acid tryptophan, and its absence results in stunted plants with poorly developed apical dominance.

Boron deficiency generally results in the death of the meristematic cells, but the exact mechanism by which it acts is obscure. Because boron is known to form complexes with sugars and related molecules, its function in the plant may involve long-distance sugar transport. Some experiments with labeled sugars have supported the view that boron may, under certain conditions, increase the translocation of sugar in the plant, but not all workers in the field accept this generalization. Whether boron is involved in other facets of cell development is as yet undetermined.

In addition to those elements absolutely required for growth, plants contain appreciable quantities of other, nonessential, elements. In certain instances, these nonessential elements may increase growth or vigor. As noted previously, cereals do better in the presence of silicon; thus, wheat plants grown in the absence of silicon are markedly more susceptible to fungus attack than are similar plants grown in the presence of silicon. Beet plants grown in the presence of sodium (Na^+) produce larger and fleshier roots than do those grown in its absence. Despite the beneficial physiological effects of such elements, strictly speaking they cannot be regarded as essential for the plant. Other nonessential elements present in plants may be physiologically inert, or in certain cases may actually be harmful. In fact, some of the microelements, minute quantities of which are essential for growth, become highly toxic if present in excess amounts: these include Mn, Cu, and, in high concentrations, Fe. Boron is an element with an extremely narrow range between deficiency and toxicity, and since plants cannot prevent such a substance from penetrating a membrane simply because it may be harmful, great care must be taken in regulating external levels of this element.

Soil Organic Materials and Plant Growth

When a plant is grown under adequate photosynthetic conditions in an optimal mineral solution, it can develop to maturity vigorously and normally. From such an experiment, it is clear that the green plant is completely autotrophic for all the organic molecules it requires, including vitamins, hormones, amino acids, and miscellaneous other complicated structures. Why, then, are organic fertilizers considered beneficial for plant growth? The answer lies not in the plant itself, but in the nature and structure of soil.

Soil, derived originally from fragmented parent rock, is a highly dynamic and complex medium for plant growth. It includes (a) *rock particles* of various sizes, from coarse sand to finer silt to very fine clay particles; (b) *organic matter*, generally the remains of plant, animal, or microbial cells that have died and are now decomposing; (c) *living organisms* of various kinds, including bacteria, filamentous fungi, algae, protozoa, worms, insects, and even larger animals; (d) a *soil solution*, containing inorganic and organic materials in aqueous solution, generally as a thin film surrounding the rock particles; and (e) a *gas phase*, containing oxygen (which is required for root respiration and active absorption of minerals by root cells), nitrogen, carbon dioxide, and the trace gases of the atmosphere.

The vigorous growth of the plant in soil depends on the proper physical condition of the soil; if the soil particles are too closely packed, the gas phase will be inadequate and the absorption of materials by the aerobically respiring roots will decline because of a lack of oxygen. A soil is said to be in good "tilth" when it has good crumb structure, that is the fine soil particles are cemented together to form larger crumbs, which pack loosely to form a firm, yet well-aerated medium. It is this aspect of soil conditioning that is dependent on organic constituents, for soil granules are cemented into crumbs by mucilage-producing soil microorganisms that consume organic matter. Thus, organic soil additives are required only if the physical condition of the soil demands it; under conditions of optimal growth in completely inorganic media, such as in well-aerated quartz sand watered with a mineral salt solution, organic additives generally produce no additional growth.

Organic residues often contain a wide variety of mineral elements that are freed for use by plants as the organic matter decomposes. While such minerals can be supplied directly in the solution, organic remains provide a source of slow, steady release, and may also add useful amounts of those microelements in which modern fertilizers are often deficient. Thus, the addition of organic matter to soil may be a useful way of recycling organic waste. There is, however, no evidence that organic fertilizers supplied to a soil will improve the nutritive quality of a plant beyond that produced by growth in a well balanced mineral solution. One should remember these facts when considering the spectacular claims of the proponents of "organic gardening." The higher price paid for "organically grown" vegetables does not guarantee their nutritional superiority, although they may well be freer from pesticides than are other produce.

Nitrogen Fixation

Increased efficiency in the fixation of nitrogen is one of the most important goals of biochemists interested in increasing agricultural productivity, since it is the supply of fixed nitrogen which most frequently limits plant growth. Nitrogen, in the form of the stable dinitrogen (N_2) molecule, constitutes 80% of the atmosphere. This molecule must somehow be destabilized and cleaved before fixation (reduction of the nitrogen to ammonium) can occur. The ammonium (NH_4^+) produced by fixation may either be absorbed *per se* by plant roots, or may be absorbed after oxidation to nitrate (NO_3^-) by soil micro-organisms. NO_3^- formation from NH_4^+ occurs so rapidly in most soils that most of the nitrogen taken up by roots is in the form of NO_3^-.

Nitrogen fixation is achieved mainly by certain free-living bacteria that consume the organic matter of the soil (for example, the aerobic form, *Azotobacter*, and the anaerobic form, *Clostridium*). Bacteria of the genus *Rhizobium*, also involved in nitrogen fixation, live in root swellings, or *nodules*, of particular plant species (Fig. 7-3). The host plant is usually a member of the family *Leguminosae*, which includes peas, beans, soybeans, alfalfa, clover, and vetch. Recent studies have shown that nitrogen-fixing bacteria of the genus *Spirillum* surround the roots of a tropical grass, *Digitaria*; this loose association of grass roots and bacteria in the rhizosphere may represent an

Figure 7-3 Nodules on soybean roots. (Courtesy of the Nitragin Company.)

intermediate evolutionary stage between the free-living *Azotobacter* and the nodule-localized *Rhizobium*. The plant "attracts" the bacteria by its secretion of organic molecules; the microorganism in the rhizosphere in turn supplies the plant with fixed nitrogen. Certain blue-green algae (such as *Anabaena* and *Nostoc*) and photosynthetic bacteria (such as *Rhodospirillum*) can fix atmospheric nitrogen by coupling it energetically to photosynthesis. These organisms are the supreme autotrophs of the biological world, although some strains of *Anabaena* will live and fix nitrogen efficiently only when associated with special "pockets" of the water fern, *Azolla*. The reasons for this are not understood.

The association of two organisms for mutual benefit is called *symbiosis*. Since neither the *Rhizobium* nor the host plant alone can fix and reduce atmospheric nitrogen vigorously, the biological complex in the nodule must be regarded as a symbiotic association of bacterium and host. Each type of host plant has its own specific symbiotic *Rhizobium*. Mutual recognition between host and bacterium is accomplished by the binding of a particular protein on the surface of root hair cells (a *lectin*) to a specific bacterium. After binding to the host, the invading organism enters by way of curiously curved root hair cells, which are probably deformed by bacterial secretion of growth hormones of the auxin group (see Chapter 9). Once inside the host cell, the bacterium divides and the resulting progeny change their form to *bacteroids* contained in an *infection thread* running from the tip of the root hair cell wall through the center of the cell (Fig. 7-4). The final result of this invasion is a prodigious overgrowth of the root cells, producing the warty protuberances called *nodules*. Rhizobia can fix nitrogen vigorously only when found in nodules of this type.

The association between *Spirillum* and its host plant is also symbiotic, but is limited to the surface of the roots. This is a looser type of association, since *Spirillum* can be grown apart from its host if supplied with necessary nutrients; accordingly, researchers are investigating the possibility of large-scale cultivation of this bacterium as a source of fixed nitrogen. This bacterium has also been found occasionally in association with corn, which opens the possibility of developing strains valuable in fixing nitrogen in plant species that do not normally possess nitrogen-fixing bacteria.

The fixation of nitrogen is brought about by the Fe-and Mo-containing enzyme *nitrogenase*. Plants with nitrogen-fixing bacteria dependent on this enzyme do not respond to additions of nitrogen-containing fertilizers, since the ammonium (NH_4^+) present in or formed from the added nitrogenous material represses the activity of genes that direct the synthesis of nitrogenase. Efforts to enhance nitrogen fixation have thus included a search for nitrogen-fixing bacteria that lack this "feedback control." The enzyme *glutamine synthetase* appears to be involved in the regulatory mechanism since nitrogenase is synthesized only when glutamine synthetase levels are high. Recently a

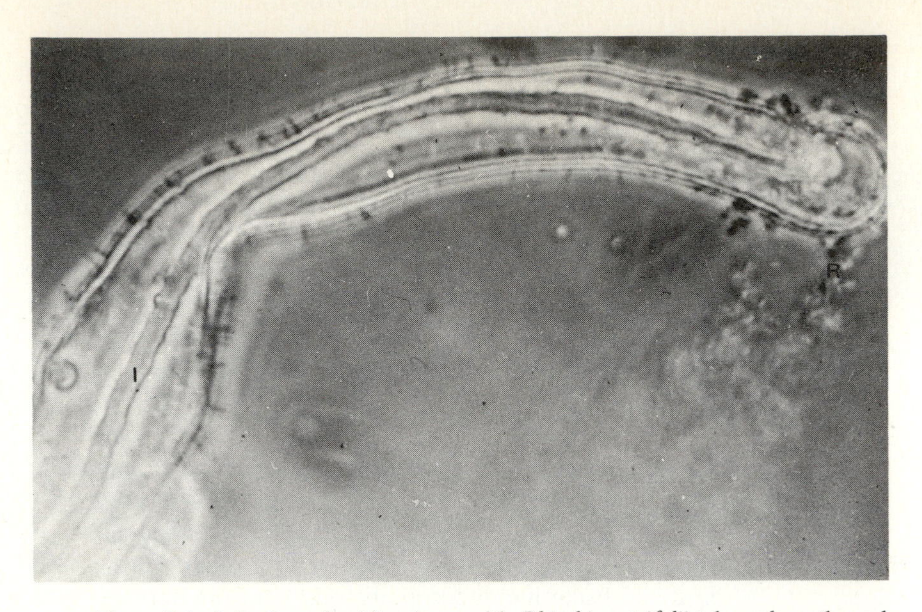

Figure 7-4 Infection of white clover with *Rhizobium trifolii* takes place through root hair cells. Note the massive accumulation of rhizobial cells (R) at the root hair tip and the refractile infection thread (I) inside of the root hair. Single rhizobial cells are attached end-on along the sides of the root hair cell wall. (Courtesy of F. B. Dazzo, Michigan State University.)

mutant bacterium was found which contains high levels of glutamine synthetase; it continues to synthesize nitrogenase even in the presence of NH_4^+. Thus the search for bacteria that can fix N_2 with increased efficiency looks promising.

In addition to nitrogenase, which attaches to and "destabilizes" the N_2 molecule, the reduction of N_2 to NH_3 requires a strong reducing agent and ATP. Ferredoxin, an electron-transport agent that also functions in photosynthesis, serves as the main reducing agent. In symbiotic associations, ATP is provided by the host plant, and the amount of nitrogen fixed is often limited by the rate of photosynthesis. Thus, "fertilizing" a soybean plant with CO_2 can increase its total nitrogen fixation because of the increased photosynthate produced, though this clearly is not a practical procedure in field-grown crops.

All nitrogen fixation systems are poisoned by even slight traces of oxygen, which means that the nitrogenase enzyme must be kept essentially anaerobic, even in aerobic cells. In the legume root nodule, this is accomplished by *leghemoglobin* (LHb), a reddish, iron-containing analog of the animal pigment. Like blood hemoglobin and muscle myoglobin, LHb can form an association with oxygen:

$$LHb + O_2 \rightleftharpoons LHb \cdot O_2$$

This effectively removes the oxygen from the vicinity of the nitrogenase, and ensures optimal rates of nitrogen fixation. The oxygen held on leghemoglobin

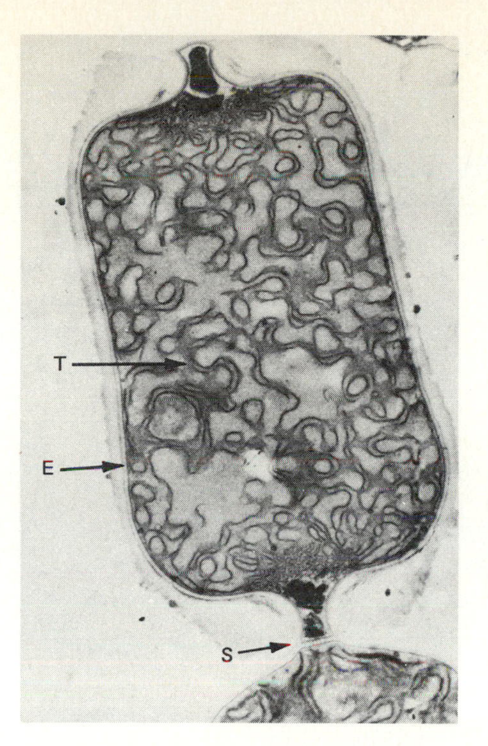

Figure 7-5 An electron micrograph of a heterocyst from the blue-green alga *Anabaena cylindrica*. The heterocyst, enclosed by a thick envelope (E), is separated by a thin septum (S) from an adjacent vegetative cell. Note that the thylakoid membranes (T) are dispersed throughout the cytoplasm, as also occurs in other blue-green algae. (×21,800) (Courtesy of N. Lang, University of California at Davis.)

may also be used for respiratory formation of ATP; since nitrogen fixation requires large quantities of ATP, this could well be important. In general, the redder the nodule, the more active the nitrogen fixation proceeding in it. In the nitrogen-fixing blue-green alga *Nostoc*, the nitrogenase appears to be localized in *heterocysts*, special nonphotosynthetic anaerobic cells (Fig. 7-5). This structural isolation also serves to divert the oxygen released in photosynthesis from the N-fixing system.

Some microorganisms in the soil can oxidize ammonium (NH_4^+) to nitrates (NO_3^-). Most plants preferentially absorb and utilize nitrogen supplied in the form of nitrate (NO_3^-) even though such nitrogen is finally incorporated into plant materials in the form of the highly reduced amino group ($-NH_2$). The enzyme *nitrate reductase* reduces nitrate back to ammonium by means of reduced respiratory carriers such as NADPH. This enzyme may also contain molybdenum at its active center, and in such plants this may be the main metabolic role for molybdenum. The reduction of NO_3^- to NH_4^+ may possibly proceed through such intermediates as hyponitrous acid (HONO) and hydroxylamine (NH_2OH).

Ammonium cannot be allowed to accumulate in substantial quantities, since it is toxic to plant cells. Ammonium is generally converted into amino acids by reacting with *α-ketoglutaric acid* (from the Krebs cycle) to form *glutamic acid* and then, with the addition of another ammonium, the amide *glutamine* (Fig. 7-6). Other amino acids are formed by the enzymatic process *transamination*, in which glutamic acid reacts with other keto-acid precursors of the new

1. Direct amination of organic acids

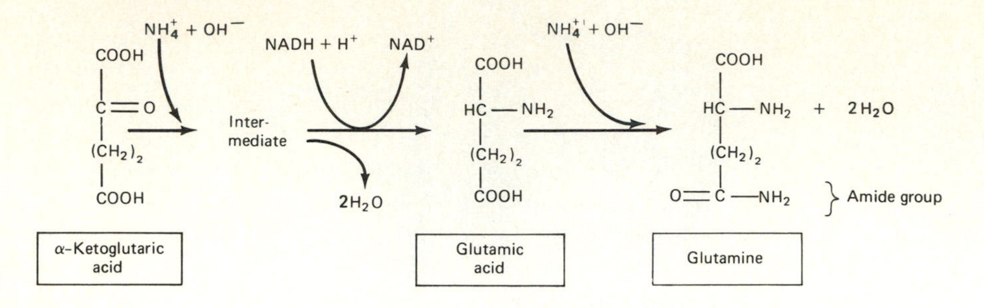

2. Transamination

Glutamic acid + Oxaloacetic acid → α-Ketoglutaric acid + Aspartic acid

Aspartic acid + $NH_4^+ + OH^-$ → Asparagine (amide group) + $2H_2O$

3. Carbon chain transformation

Glutamic acid → γ-Aminobutyric acid + CO_2

Glycine + Formaldehyde derivative → Serine

Figure 7-6 Examples of three methods for the formation of amino acids.

amino acids, transferring its amino group in the process and reverting to α-ketoglutaric acid. *Aspartic acid* is an early product of a transamination reaction involving an oxaloacetic acid receptor. An additional ammonium fixed onto aspartic acid forms the amide *asparagine*. It is mainly in the form of these four compounds—glutamic acid, glutamine, aspartic acid, and asparagine—that fixed nitrogen is transported from the root cells throughout the plant.* Still other amino acids are formed by modification of the carbon skeleton of a pre-existing amino acid.

The overall cycle of nitrogen in nature thus involves passage between a free, gaseous form in the atmosphere and a fixed form in the soil or biological system. In plant cells, absorbed nitrate is reduced back to ammonium, which is then coupled to certain organic acids to form amino acids and ultimately

*In nodulated legumes, substituted urea derivatives (*ureides*) are the main form of nitrogen transported upward in the xylem.

proteins. These materials are ingested by animals and are transformed to animal proteins and nitrogenous waste products such as urea and uric acid. Eventually all animals and plants die, and are decomposed in the soil to simple nitrogenous materials such as ammonium. These materials recycle continuously through biological systems, creating the *nitrogen cycle* (Fig. 7-7). Through the actions of denitrifying bacteria, fixed nitrogen may be returned to the atmosphere as free molecular nitrogen, from which it may again be fixed. Denitrification, an essentially wasteful process, can now be inhibited in the soil by chemical compounds designed specifically for that purpose, an advance which may greatly aid agriculture.

The enzyme nitrogenase, which attaches to the dinitrogen molecule ($N{\equiv}N$) and reduces it to ammonium (NH_4^+), can also attach to acetylene ($HC{\equiv}CH$) and reduce it to ethylene ($HC{=}CH$). Discovery of this activity suggested a method by which the nitrogen-fixing activity of a plant could be assayed directly in the field. A quantity of acetylene gas is brought into the

Figure 7-7 The nitrogen cycle. The nitrogen of the soil, of living creatures, and of the atmosphere is in a state of continuous turnover.

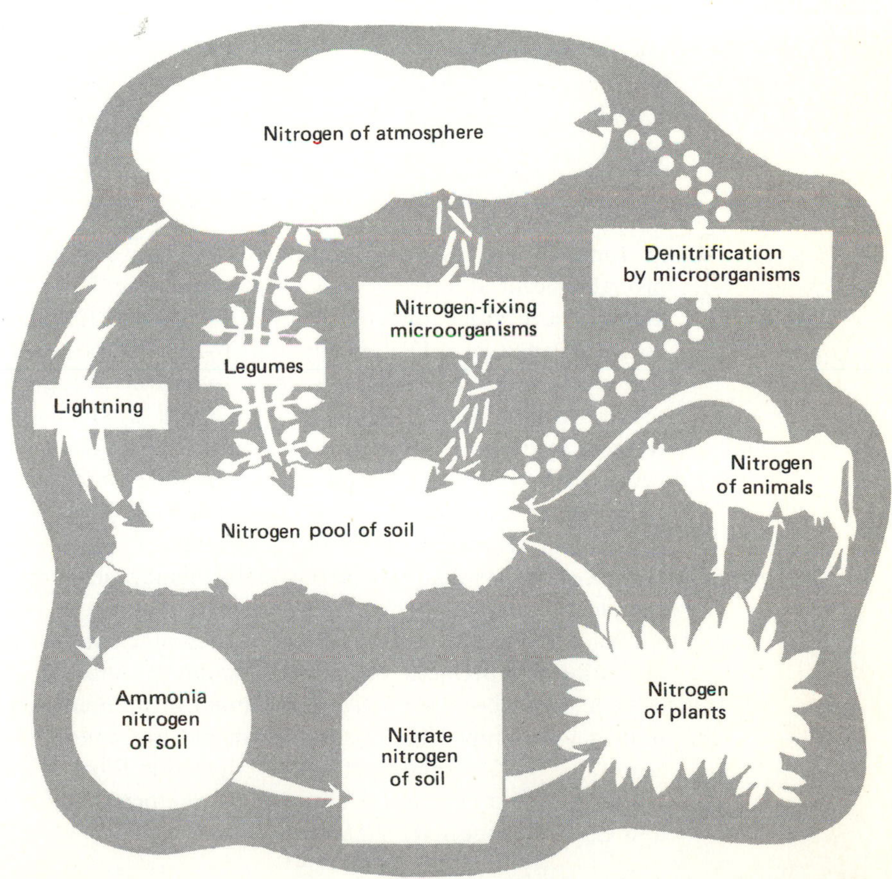

vicinity of plant roots, and after some time is removed; the extent to which it has been transformed to ethylene is an indication of the nitrogen-fixing ability of that plant's roots. Since both acetylene and ethylene are gases, minute quantities can be analyzed by gas chromatography, rapidly yielding accurate and inexpensive data. Using this procedure, plant physiologists have studied the development of nitrogen-fixing ability during the life history of a soybean plant, and have analyzed the physiological factors controlling the efficiency of the process. This could improve future productivity.

The Uptake of Minerals From the Soil and the Movement of Ions Through Cell Membranes

Now that we have discussed the mineral elements required for healthy plant growth, it is appropriate to consider how these substances make their way into the plant from the outside world. Minerals are usually absorbed from the soil by way of the roots. Since they can also be absorbed in small amounts through the leaves, foliar application of some micronutrient fertilizers has become standard agricultural practice. The mineral elements are almost always absorbed in their ionized form; these ions must first traverse the plasmalemma to enter the cytoplasm, and then the membrane surrounding the vacuole or a cellular organelle to enter a compartment within the cytoplasm.

Diffusion

Ions can move through membranes by passive or active processes, which probably occur at different sites in the membrane. Ions diffusing through the membrane in either direction move by virtue of their own kinetic energy, consuming neither ATP nor other energy-rich compounds in the process. The ability of some classes of molecules to move through cell membranes is related to their lipid solubility. Since small lipid-soluble molecules move through membranes more readily than do larger ones, the regions through which such movement occurs are probably best described as small lipid channels traversing the membrane. Although inorganic ions are water soluble rather than fat soluble, they too can move through membranes, and recent evidence suggests that they permeate the membrane through aqueous protein channels appropriately called *permeases*.

Some ions diffuse through membranes much more readily than others, indicating that permeases are somewhat ion specific. The permeability coefficient, P, describes the relative diffusibility of different ions through a membrane under comparable standard gradients. Since most membranes are more permeable to K^+ than to other ions, the value P for K^+ is arbitrarily set at 1.0. In the giant cells of the alga *Nitella*, the permeability constants for Na^+ and Cl^- are, respectively, 0.18 and 0.033.

Polypeptides and proteins with appropriate channeling properties for specific ions have been isolated from bacteria and fungi. These substances, called *ionophores*, can be added to artificial lipid membranes, where they increase the rate of diffusion of a specific ion by as much as one millionfold. Because they disturb the normal pattern of differential permeability, these compounds can act as antibiotics, killing certain kinds of cells; the antibiotic gramicidin, discussed in an earlier chapter, is of this type.

Origin and role of the transmembrane potential in ion transport

Since all ions are charged, their rate of diffusion and their distribution at equilibrium are determined not only by membrane permeability and the differences in their concentration on the two sides of the membrane (*chemical potential*), but also by the *electrical potentials* existing across the membrane. Thus, we speak of ion movement as occurring in response to an *electrochemical potential gradient*. Plant cells typically have a negative potential on the inside of their membranes, resulting in a preferential absorption of cations (+) over anions (−). It is important to consider this potential in more detail.

When different concentrations of freely diffusing ions are separated by a membrane, a voltage called the *transmembrane potential* develops across the membrane. The transmembrane potential can be measured with two microelectrodes connected to a sensitive voltmeter. One electrode, made of glass capillary tubing approximately 1 μm in diameter, is inserted through the cell wall and plasmalemma into the cellular interior, while the other reference electrode is positioned outside the cell (Fig. 7-8). In cells with a large central vacuole, the internal electrode usually penetrates the tonoplast as well as the plasmalemma, so that the measured values indicate the difference in potential between the vacuole and the cellular exterior. Measurements of this type reveal potential differences ranging between 50 and 200 millivolts (mv), the interior of the cell being the more negative.

The membrane potential is due in part to the selective permeability of the cell membrane, for this limits the speed of movement of one ion relative to another. K^+, for example, can diffuse through the membrane much more rapidly than Cl^-. If both K^+ and Cl^- are more concentrated in the cellular interior than in the immediate exterior, the more rapid net diffusion of K^+ ions outward along the concentration gradient would ultimately make the interior of the cell more negative, as an excess of Cl^- would be left behind.

Active transport (pumping) of ions across the membrane is probably the most important regulator of membrane potential. We will discuss some details of this process later, but first we will consider the consequences of the active transport of a *single type of ion* in *only one direction*. Such a process is termed *electrogenic*, since it leads to an accumulation of negative charges on one side of the membrane and positive charges on the other. One of the principal ions involved in the production of the transmembrane potential is

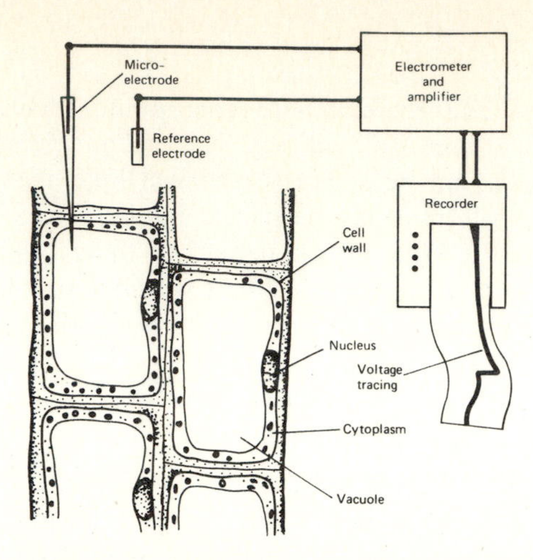

Figure 7-8a Equipment used to measure the transmembrane potential of plant cells. Note that the narrow tipped microelectrode penetrates into the vacuole, while the larger reference electrode is in the solution bathing the tissue. Ions can diffuse freely between the bathing solution and the cell walls. (Courtesy of R. Racusen, University of Maryland.)

Figure 7-8b Electrophysiologist Richard Racusen "demonstrates" the use of microelectrodes to measure the interior potential and external pH of cells in the pulvinus of *Samanea saman*. The pH electrode, composed of glass that is permeable to H^+ ions, is a micro–replicate of the electrode in a conventional pH meter.

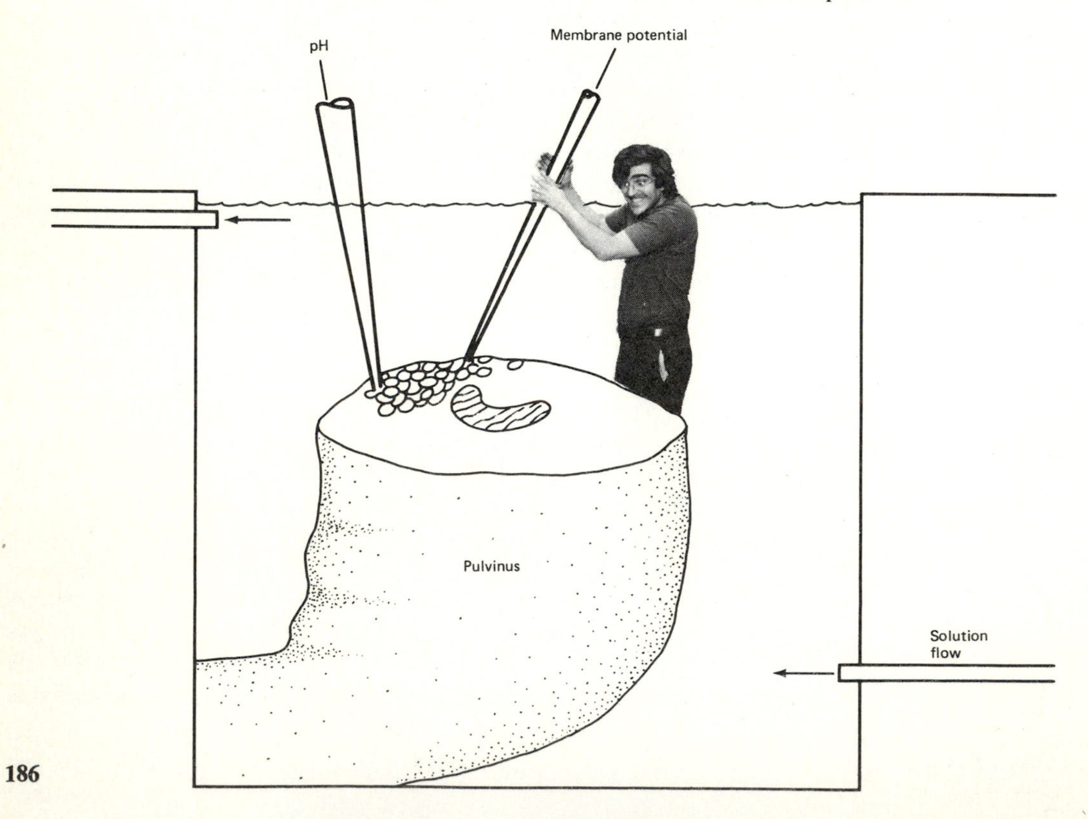

H^+; when H^+ is pumped outward, a negative potential develops on the inside of the cell.

Once generated, the transmembrane potential can in turn affect subsequent ion movements. Let us suppose, for example, that a cell's transmembrane potential is -116 mv, with the interior negative. This negative interior electrical potential would exert an attractive force on the inward diffusion of positively charged ions such as K^+, but would repel negative ions such as Cl^-. The quantitative relationship between the transmembrane potential and the diffusional fluxes of an ion such as K^+ can be calculated by a formula termed the *Nernst equation* (see box below).

Some uncharged molecules such as sucrose move through membranes in conjunction with an ion, usually H^+. This process, called *cotransport*, or *symport*, is particularly important in regulating the movement of sucrose into and out of the phloem (see Chapter 8). H^+–sucrose absorption into ("loading") and secretion from phloem cells ("unloading") probably occur by movement through a permease in the membrane. Sucrose, the uncharged partner, is "pulled" through the permease by H^+, with the direction of net diffusion determined by the electrochemical gradient existing for H^+.

The *Nernst* equation relates electrical potential inside a cell to the distribution of charged ions:

$$E_{(\text{mv})} = \frac{-58}{n} \log_{10} \frac{C_i}{C_o}$$

Where $E =$ the transmembrane potential in millivolts, with the outside electrode grounded.

$n =$ the valence number and charge of the ion.

$C_i =$ concentration (molarity) of the ion inside the cell.

$C_o =$ concentration (molarity) of the ion outside the cell.

Let us suppose that $E = -116$ mv. For monovalent cations such as Na^+ or K^+, $n = 1$ and $\log_{10} \frac{C_i}{C_o} = \frac{-116}{-58} = 2$. Since 2.0 is the $\log_{10}$ of 100, K^+ and Na^+ would diffuse inward, without the expenditure of metabolic energy, until the internal concentration of each was 10^2 or $100 \times$ the external concentration. The diffusion of a monovalent anion such as Cl^-, on the other hand, would lead to an internal concentration of only 10^{-2} or $0.01 \times$ the external concentration.

Typical values of the transmembrane potential and of K^+, Na^+ and Cl^- concentration inside the vacuole of a higher plant cell are shown below. The concentration of Na^+ inside the cell is lower than predicted by the *Nernst* equation, whereas K^+ and Cl^- are higher than predicted. It is apparent that Na^+ is actively transported out of the cell, while Cl^- and K^+ are actively transported into the cell. K^+ is the only element that is close to its *Nernst* value.

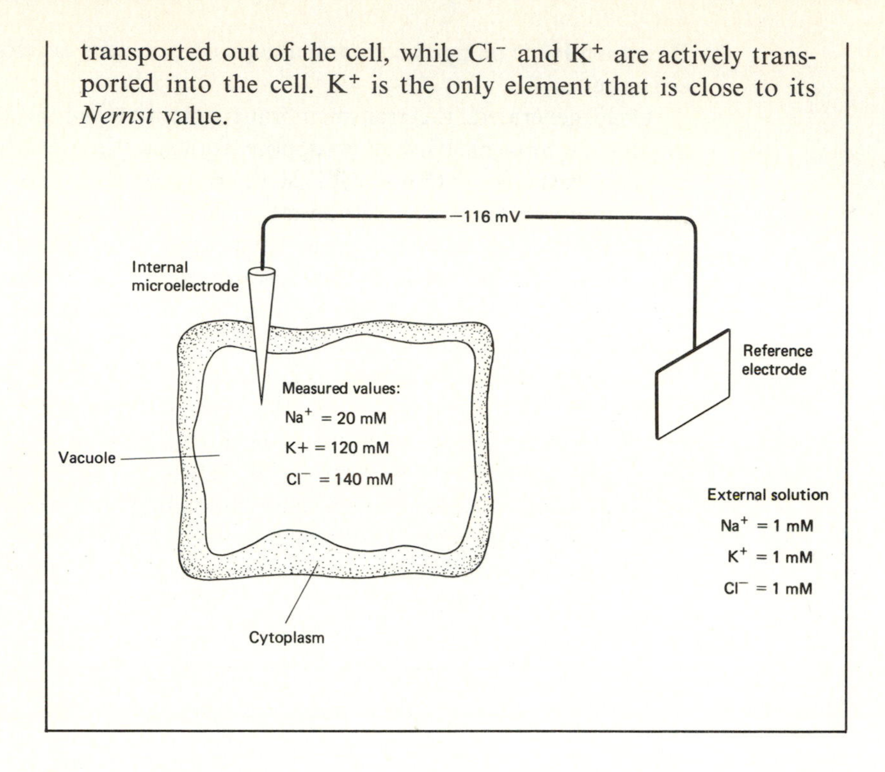

Active ion transport

In addition to diffusing through membranes at specific loci in response to the electrochemical potential gradient, ions also move through *specific* membrane regions called *pumps* by the expenditure of energy, usually in the form of ATP. The distinction between diffusion and pump-mediated transport thus lies in the fact that diffusion can occur only *down* the electrochemical potential gradient, whereas active transport can involve movement *against* that gradient.

When a single ion is actively transported in a single direction, the membrane potential is altered, as already discussed. Active transport, however, need not be electrogenic. The transport of two ions of the same charge (for example, K^+ and H^+) in opposite directions, or of two ions of equal but opposite charge (such as K^+ and Cl^-) in the same direction, is an electrically neutral process, since the balance of anions and cations across the membrane is not changed (Fig. 7-9). In roots, H^+ ions are frequently exchanged for K^+. The K^+ is absorbed from the soil solution or from the surface of soil particles to which it is adsorbed, while the simultaneous export of H^+ ions makes the external medium more acid, and can help break down soil particles. Similarly, the absorption of anions, such as NO_3^-, may be balanced

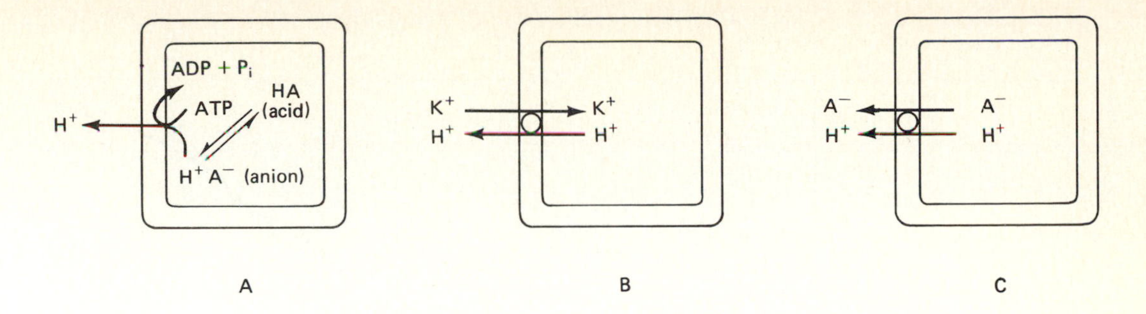

Figure 7-9 Electrogenic and electroneutral ion transport.
(A) Electrogenic secretion of H^+ increases the negativity of the cellular interior with respect to extracellular space.
(B) H^+ secretion is balanced by K^+ uptake
(C) H^+ secretion is balanced by secretion of an anion

by the excretion of bicarbonate (HCO_3^-) or of an organic anion such as malate.

Compounds that inhibit ATP synthesis can be used experimentally to determine whether ions move across a membrane by diffusion or active transport. The simplest experiments of this type are conducted in the dark, so that photophosphorylation by the chloroplasts is eliminated and mitochondria become the chief sites of ATP synthesis. A metabolic poison such as 2, 4-dinitrophenol, sodium azide, sodium cyanide, or *m*-chlorocarbonylcyanide phenylhydrazone (*m*CCCP), all of which interfere with ATP synthesis in mitochondria, is then added. (When such experiments are performed in the light, additional inhibitors, such as DCMU are required to prevent photophosphorylation in chloroplasts.) If a plant is treated with an effective inhibitor of an appropriate concentration, active transport will cease as soon as the plant has consumed its supply of stored ATP. Ionic diffusion will probably continue, but that fraction attributable to the generation of an electrical potential by active processes will be diminished. Similar results can also be produced in the dark by removing oxygen from the atmosphere, since mitochondrial oxidative phosphorylation is very inefficient under anaerobic conditions. The anaerobic environment of roots growing in flooded soils may impair their ability to absorb ions, possibly causing mineral deprivation throughout the plant.

Experiments with excised roots immersed in salt solutions have helped plant physiologists understand how ion pumps function in salt absorption. The rate of uptake of a particular element, usually in the form of a radioactive isotope, is measured as the concentrations of different elements in the solution are varied. A typical experiment with oat roots reveals that the rate of K^+ uptake increases linearly as the K^+ concentration is increased from 0 to 0.02 mM, then increases more slowly as the concentration approaches 0.20

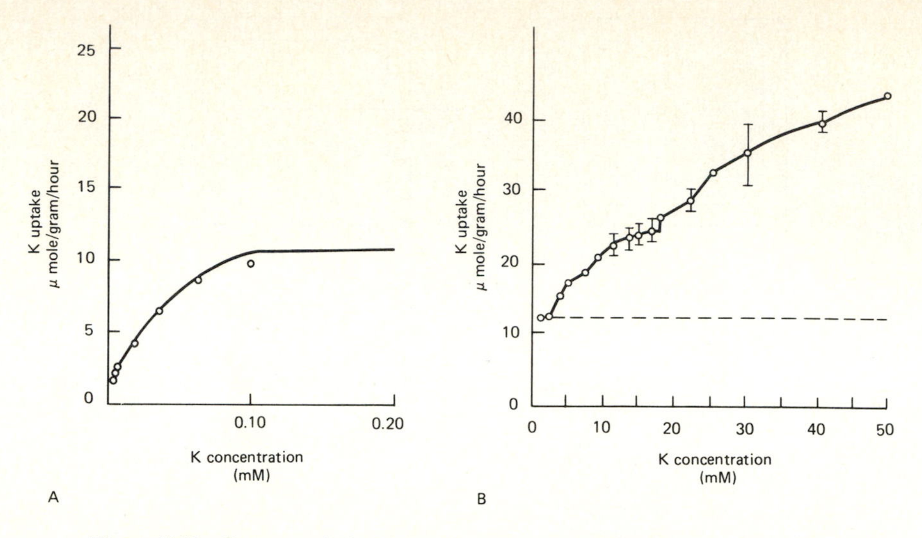

Figure 7-10 Rate, v, of absorption of potassium by barley roots as a function of the concentration of KCl in the solution. The concentration of Ca is 0.5 mM. (A) Low KCl (B) Higher KCl. (From Epstein et al. 1963. *Proc. Nat. Acad. Sci. 49*, 684–692, and Epstein and Rains. 1965. *Proc. Nat. Acad. Sci. 53*, 1320–1324.)

mM (Fig. 7-10a). Most other ions are absorbed more slowly than K^+, but the shapes of their uptake curves are nevertheless similar to that of K^+. It is particularly interesting that these ion uptake curves resemble graphs relating the rate of enzyme-catalyzed reactions to substrate concentration (see Fig. 2-14). It is possible, therefore, that membrane-localized proteins function as

Figure 7-11 The effect of KCl on the hydrolysis of ATP to ADP + P_i was measured in a plasma membrane-enriched fraction from oat roots. The membrane fraction was prepared by the procedure outlined in Fig. 2-1. The reaction mixture contained ATP, $MgSO_4$ and KCl, pH = 6.5. The release of P_i gives a colored product, measured spectrophotometrically. (Adapted from Leonard and Hodges. 1973. *Plant Physiol. 52*, 6–12.)

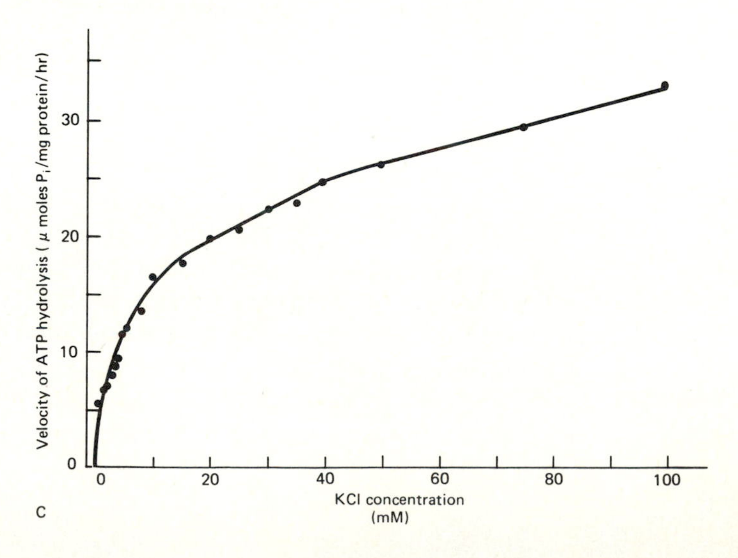

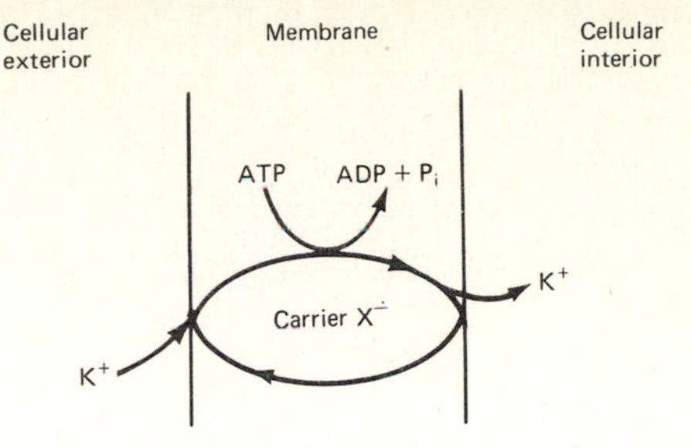

Figure 7-12 "Carrier" hypothesis to explain ion uptake by root cells. A protein, X^-, in the plasmalemma membrane forms a complex with K^+ on the outer side of the membrane. The complex moves K^+ through the membrane and ejects it into the cytoplasm, utilizing the energy provided by the hydrolysis of ATP to ADP + P_i. The carrier then picks up another K^+ ion.

enzymes that facilitate the movement of ions across the membrane, with the rate of absorption of a particular ion reaching its maximum when all the carrier sites for that ion are occupied.

When the external KCl concentration is increased above its first "saturation level," from 1 mM to 50 mM (Fig. 7-10b), the rate of K^+ uptake again increases. However, there is less specificity for K^+ at these higher salt concentrations, since many different ions can alter K^+ uptake. It is not clear whether many different carriers are involved, one with specificity for K^+ and the others with sites that can be occupied by several different cations, or whether a single carrier is responsible for all phases of the K^+ uptake curve.

Recently, an enzyme was found on the plasmalemma of oat root cells that promotes the hydrolysis of ATP to ADP + P_i in the presence of K^+. The activity of this ATPase enzyme is not only enhanced by K^+ in the medium (Fig. 7-11), but the dependence of this effect on concentration is strikingly similar to that of K^+ uptake by intact roots of an oat plant (compare with Fig. 7-10). It seems likely that this enzyme functions as a carrier, using the energy provided by ATP breakdown to transport K^+ ions through the plasmalemma. Although the details of the mechanism of active transport remain largely unknown, available evidence suggests that the membrane protein carrier first binds to its specific ion, then changes its configuration under the influence of ATP. When the carrier finally splits the ATP, its configuration returns to the original state and the ion is released, but to the opposite side of the membrane (Fig. 7-12).

Ion interactions and antagonisms

Some ions interfere with the uptake and transport of other ions (Fig. 7-13). For example, increase in the concentration of Rb^+ in the external solution decreases K^+ uptake, and vice versa; Cl^- and Br^- are also mutually antagonistic. The presence of Na^+, by contrast, has little effect on Rb^+ uptake, but decreases the absorption of Li^+. Presumably, particular groups of ions "compete" for specific membrane-localized carrier sites adapted to match their geometry.

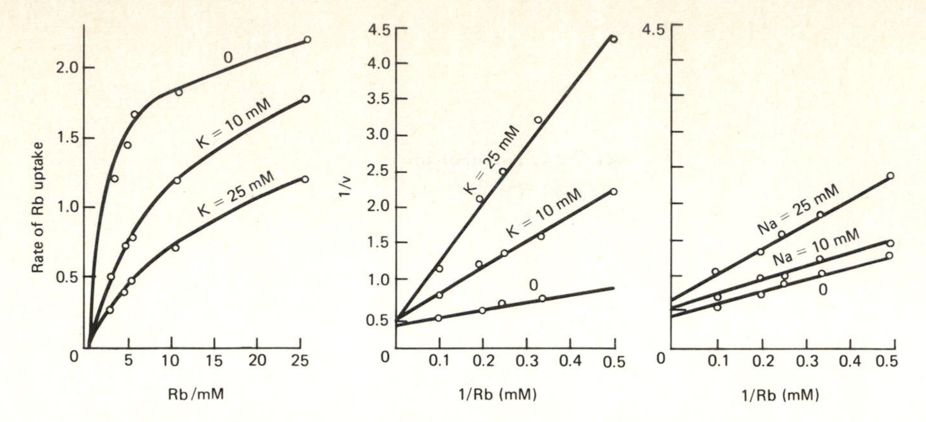

Figure 7-13 Competitive interference of K^+ and Na^+ with Rb^+ uptake. Left: conventional plot. Right: the reciprocal of the rate of Rb^+ uptake is plotted as a function of the reciprocal of the Rb^+ concentration. Double reciprocal plots of this type characteristically yield straight lines in enzyme catalyzed reactions. (See Fig. 2-14b). K^+ is seen to compete with Rb^+ uptake while Na^+ has little effect. (Adapted from Epstein and Hagan. 1952. *Plant Physiol. 27*, 457–474.)

Sometimes the uptake of a certain ion can prevent deleterious effects caused by excess uptake of an other ion. For example, K^+ and other monovalent cations tend to decrease cytoplasmic viscosity and increase membrane fluidity and permeability, while divalent cations such as Ca^{++} act in the reverse manner. Because of such interactions, termed *ion antagonisms*, mixed salt solutions rather than single salts are usually supplied to plants to correct deficiencies of single nutrients.

The Apoplast and the Symplast

Minerals in ionized form, together with water, are absorbed from the soil through root hairs and other epidermal cells near the tip of the root. During their migration through the plant, these absorbed ions may diffuse through either the *apoplast* or the *symplast* (Fig. 7-14). The *apoplast* consists of the wet walls and intercellular spaces of all the cells in the plant body. The walls of adjacent cells are in physical contact, and except for a few specialized regions such as the Casparian strip they form a continuum through which water and ions can diffuse freely without encountering permeability barriers. For this reason, the walls are sometimes called *free space*, although their negative charge can affect relative ion movement.

The plasmalemma that surrounds each protoplast separates the apoplast from the *symplast*. The latter consists of (a) the membrane-bounded cytoplasm of all vacuolated cells, (b) the bridges that connect most cells of the higher plant with their neighboring cells, and (c) the transport cells of the phloem. You will recall that cytoplasmic bridges called *plasmodesmata* penetrate cell walls (Fig. 2-28), thereby permitting molecules to migrate from one protoplast to another without crossing plasmalemma membranes or diffusing through cell walls.

We do not have much precise information regarding the relative utilization of the symplastic and apoplastic pathways by particular molecules or ions; most of our evidence is indirect, and based on inductive reasoning. The walls of many cells involved in a large interchange of metabolites contain numerous plasmodesmata—for example, the walls between mesophyll and bundle sheath cells in some C_4 plants (see Chapter 4). Movement of ions and metabolites in such regions is presumed to occur via the plasmodesmata.

In the aquatic plant *Vallisneria*, both ^{86}Rb (an analog of K^+) and ^{36}Cl are transported through the symplast, for both these radioactive tracers, fed to one end of an excised portion of a leaf floating on water, were recovered at the other end without any loss of label to the bathing solution. Since *Vallisneria* lacks a cuticle and the cell wall continuum is in contact with the bathing solution, it is inferred that ion movements in this floating plant take place exclusively through the symplast.

Less information is available regarding symplastic transport in terrestrial plants. One way to obtain such information would be to determine, by ultrastructural techniques, which molecules are localized in plasmodesmata. Experiments of this type with virus-infected plants reveal that virus particles move through plasmodesmata. The particles of most plant viruses range between 20 and 80 nm in diameter. Since the outer diameter of a plasmo-

Figure 7-14 Schematic drawing showing the apoplast and symplast in cross section of a root. The heavily stipped dark regions (cell walls, intercellular space and the nonliving cells of the xylem) constitute the apoplast, while the lightly stippled regions consisting of the cytoplasm, plasmodesmata, and the transport cells of the phloem constitute the symplast. The vacuole is not part of either system. The Casparian strip creates a discontinuity in the apoplast. Therefore, all ions absorbed by root hairs must cross the plasmalemma membrane of a cell (a, b or c) exterior to the Casparian strip, thereby entering the symplast. After crossing the Casparian strip, ions must cross the plasmalemma again to enter a xylem element and be transported upward to the shoot. (Modified from U. Lüttge, in *Membrane Transport in Plants*, U. Zimmerman and J. Dainty, eds., Berlin: Springer-Verlag, 1974.)

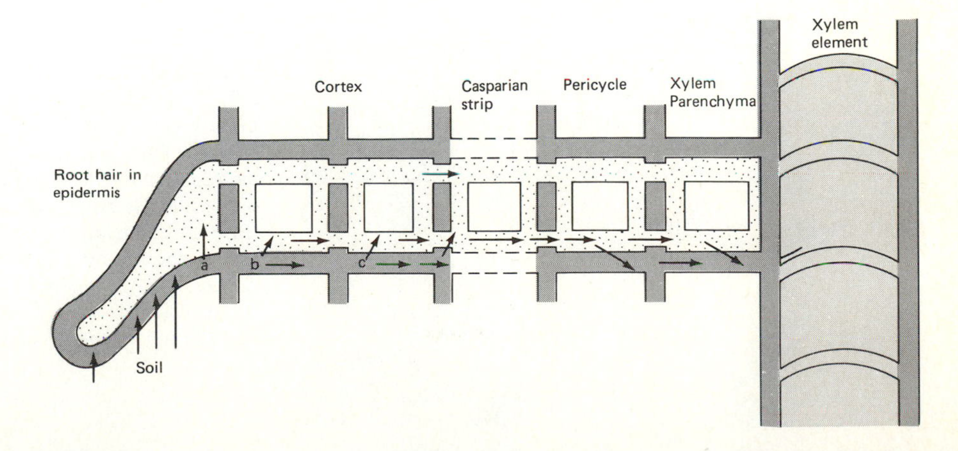

desma falls within this range (see Fig. 2-28), small virus particles might be able to pass through an open plasmodesma without altering its structure. Larger particles, however, appear to modify the size and shape of plasmodesmata between cells of the host plant.

The problem of determining whether soluble molecules move through plasmodesmata presents extraordinary technical difficulties. The fixatives used to prepare plant tissue for examination in the electron microscope can dissolve and move many ions, and are thus unsatisfactory. To circumvent this problem, one can treat the tissue with a reagent that will precipitate certain ions prior to fixation. For example, silver nitrate ($AgNO_3$) can be used to fix the location of the chloride ion, since Ag^+ reacts with Cl^- in the tissue to form the insoluble salt AgCl, which is optically dense when viewed through the electron microscope. Using this procedure, chloride has been detected in plasmodesmata of *Limonium* (Fig. 7-15). Other techniques currently being developed involve rapid freezing of plant tissue at very low temperature to immobilize the ions, which can then be detected and localized by an electron microprobe. In this instrument, the tissue is bombarded by a beam of high-energy electrons. When the elements so activated decay back to their former energy level they emit X-rays; since the frequency of such X-rays is specific for each kind of element, quantitative analysis of the emitted X-rays permits detection of any element. Similar techniques have already been used to trace the movements of potassium and chloride into and out of guard cells during stomatal opening and closure (Fig. 6-13) and into and out of motor cells of mobile leaves (Fig. 12-8). Extension of these methods to the

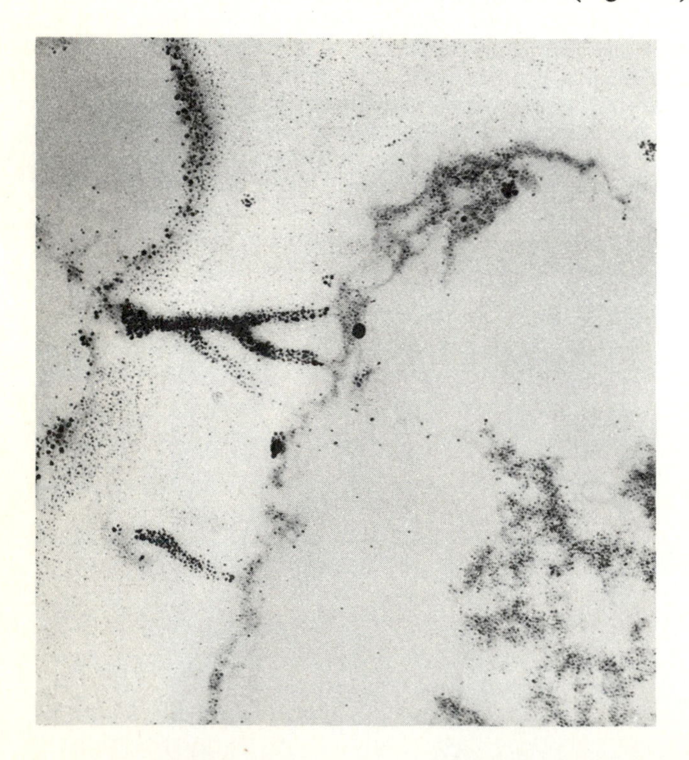

Figure 7-15 Localization of chloride in cells of a salt gland in the halophyte *Limonium vulgare* by a precipitation method. The black dots represent AgCl molecules. Note the heavy concentration of AgCl in the branched plasmodesmata (Pl) that connects a sieve tube cell (St) and a companion cell (Cc). (From H. Ziegler, in *Membrane Transport in Plants*, U. Zimmerman and J. Dainty, eds. Berlin: Springer-Verlag, 1974.)

ultrastructural level should provide quantitative information on ion movements through the apoplast and the symplast.

Little is known about the forces that drive ions through the plasmodesmata. Some plant physiologists believe that plasmodesmata are open pores through which ions readily diffuse, the rate and direction of net movement depending solely upon the size of the pore and the differences in concentration of specific ions in the cells on either side. It is impossible at present to determine whether this interpretation is correct. Plasmodesmata in some plants contain two separate channels (Fig. 2-28): the interior of the central desmotubule, and the cytoplasm lying between the desmotubule and the membrane that borders the plasmodesma. It is not known whether solutes diffuse through both these channels, and the role of each channel remains to be determined.

Until recently, few plant physiologists recognized the importance of symplastic transport, largely because of the difficulty in visualizing plasmodesmata prior to the advent of the electron microscope. Investigators in many laboratories are now focusing their attention on such aspects as ultrastructure, ion localization, and electrical aspects of plasmodesmata, and we should expect important advances in our understanding of symplastic transport to be made in the near future.

Transport of Minerals Through the Plant

Ions enter the plant by diffusing from the soil solution into the cell walls of root hairs and other epidermal cells. They then enter either the symplast or apoplast as a prelude to transport to other parts of the plant. If the ion enters the symplast by crossing the plasmalemma of an outer root cell, its subsequent cell-to-cell movement will occur through plasmodesmata. Alternatively, it may diffuse through the apoplast until it reaches the Casparian strip (Fig. 7-14). Since neither water nor dissolved ions can diffuse through the suberin-impregnated walls that constitute the Casparian strip, all ions must pass through the cytoplasm of the endodermal cells before further transport. Since some ions can move through membrane barriers more readily than can others, the properties of the plasmalemma of cells in the root are important in controlling the exchange of minerals between the soil and above ground portions of the plant. Not only differential permeability, but also active transport of many ions is involved. Negatively charged ions are generally actively transported into the cell, since diffusional movement into the cell is impeded by the internal negative potential. Excessive inward diffusion of certain cations such as Na^+ is counteracted by an active outward transport of these ions, so that they are actively rejected by the plant.

Once ions have entered the cytoplasm of an endodermal cell, they can continue symplastic movement through plasmodesmata to a cell in the pericycle layer, or they can leave the symplast by diffusion or active transport into the xylem or the apoplastic region of the stele. If transport through the stele continues to be symplastic, the ions will ultimately have to leave a protoplast and cross a cell membrane before they can enter the lumen of a dead xylem vessel element. We know that ions are accumulated in the xylem pipes at times of little water movement; since this must involve the movement of ions both into and out of at least one cell, it is clear that the ion-transporting properties of some cell membranes, probably those of the endodermal cells, must differ depending on whether they face the exterior or the interior of the root.

Once ions have entered the xylem, their backward diffusion into the symplast is prevented by the selectivity of the cell membranes, while the apoplastic channel to the soil is blocked by the Casparian strip. The accumulation of ions in the xylem increases the concentration of solutes in the xylem sap and can lead to the development of root pressure, discussed in Chapter 6. Ions in the xylem, ultimately carried upward by transpiration, must be reabsorbed through a plasmalemma before entering the protoplast of a living cell. Whichever path is followed, the plant seems to minimize the number of plasmalemma barriers that must be crossed before a particular ion reaches its final destination.

Mobility of mineral ions

In the root, most ions enter the xylem for transport to the upper parts of the plant, but some ions enter the phloem and are carried with other transported solutes to phloem "sinks" in the growing root apices and in regions of food storage. The newly absorbed ions are freely mobile throughout the plant, but their utilization in the cell may involve their incorporation into some structural molecule. The newly developing tissues require a continuous supply of all essential mineral ions. When these are all available from the soil, no difficulties arise for the plant. If, however, the supply of a mineral ion is deficient, it can sometimes be obtained from the breakdown of preformed molecules in old cells. Thus, N from degraded amino acids and Mg^{++} from degraded chlorophyll are moved from cells in the older parts of the plant to the young, growing cells. These mobile nutrient elements are probably transported from the older tissue to the young growing tissues via the phloem. Removal of these elements from older cells speeds senescence, and causes the appearance of mineral deficiency symptoms in the older parts of the plant.

Some mineral elements become fixed in a cell, and once there cannot be removed. Symptoms of deficiency of such nonmobile elements show up in the youngest tissues. Ca^{++} and $Fe^{++(+)}$ are two elements that are frequently in short supply, causing striking symptoms to develop in growing tissues. When calcium is lacking, the middle lamella for new cells cannot be formed, and since normal membrane permeability also depends on an adequate supply of calcium, the apex ceases to grow and rapidly dies. Iron is needed for chlorophyll synthesis and is a component of ferredoxin and cytochromes; iron-deficient plants frequently show a complete bleaching of the youngest leaves. Whether deficient elements are mobile or nonmobile, healthy, vigorous growth can be restored only by the actual rectification of the deficiency.

SUMMARY

Of the 16 chemical elements required by plants, 4 (C, H, O, N) are acquired from *atmospheric* carbon dioxide, water, and nitrogen in the processes of photosynthesis and nitrogen fixation; the remaining 12 are derived from the *parent rock*, whose decomposition gave rise to the soil in which plants grow. The absorption of soil minerals is usually from solution, but the plant can decompose solid rock particles through its metabolic activities. Ion absorption is an ATP-requiring process, thus demanding an oxygen-containing environment around roots.

Plants can be grown *hydroponically*, that is, with their roots immersed in aerated chemical solutions containing all the required *macronutrient elements* (potassium, calcium, magnesium, nitrogen, sulfur, and phosphorus), and *micronutrient elements* including iron, manganese, copper, zinc, molybdenum, boron, and chloride. In the absence or relative scarcity of any element, typical *deficiency symptoms* appear; these can be corrected by application of salts containing the element in question.

C, H, O, N, S, and P function mainly as components of structural elements such as carbohydrates, proteins, and nucleic acids; calcium is also partly structural, as a component of *calcium pectate* in the intercellular cement of plant tissues; K functions mainly in turgor regulation, while the other elements function mainly as coenzymes or parts of coenzymes, especially those functioning in oxidation–reduction reactions. Sulfur, in the form of *sulfhydryl groups*, can also function in redox reactions, and *phosphorus* serves as a mediator of many biochemical reactions through the energy-rich phosphate bond. *Molybdenum*, involved in nitrate reduction, is not required if the plant is fed reduced nitrogen in the form of ammonia, urea, or amino acids. Plants adequately supplied with minerals are entirely independent of external organic matter, but the soils in which they grow are generally improved by organic addenda. Soil microorganisms produce substances that cement small soil particles into larger *crumbs*; this process ensures adequate soil aeration by preventing excessive packing of small soil particles.

Nitrogen fixation, performed by certain free-living bacteria and blue-green algae, also occurs in root nodules of legumes and other plants infected with bacteria of the genus *Rhizobium*. Fixation involves attachment of the N_2 molecule to the molybdenum-containing enzyme *nitrogenase*, followed by an ATP-mediated reduction to ammonia, using electrons and protons from substrate molecules which become oxidized in the process. The N-fixing system can also reduce acetylene to ethylene, and this is used in the field to assay N-fixation by intact plants. Supplying reduced nitrogen to N-fixing plant–bacterium symbiosis reduces its ability to fix N, as a result of a "feedback repression" mechanism. Nitrogenase is also very sensitive to O_2 and it is thought that the reddish *leghemoglobin* (LHb) of active nodules provides an anaerobic environment by sequestering free oxygen as $LHb \cdot O_2$. Fixed N_2 quickly appears in the amino group of *glutamic acid*, a result of the direct amination of *α-ketoglutaric acid*. Glutamic acid, in turn, can serve to *transaminate* other keto acids to form other amino acids.

Ion movement across membranes occurs partially in response to *electrochemical* gradients, and partially by membrane-localized *pumps*. When movements follow the electrochemical gradient, the ions first attach to particular loci on the membrane (*permeases*). They then move into the cell, as described by the *Nernst* equation, if the combined effect of their concentration gradient from outside to inside and the electrical potential across the membrane provide an inward driving force. Transmembrane potentials are generated in two ways: (1) by diffusion of both anion and cation through the membrane, but at different rates, and (2) by electrogenic transport involving the direct use of energy to pump protons, anions, or cations through membranes against their electrochemical gradients. These two processes always operate to make the interior of the cell negative with respect to the exterior.

Membrane-bounded cells connected by *plasmodesmata* comprise a continuum called the *symplast*, which is separated from the *apoplast* (walls and intercellular space). Ions can move either symplastically or apoplastically, both before and after their passage through the endodermis. Passage between the two systems involves the selective process of permeation through a membrane; since different solutes permeate at different rates, and since even organelles are membrane-bounded, membranes provide a mechanism for compartmentation within biological systems.

SELECTED READINGS

Mineral Nutrition

EPSTEIN, E. *Mineral Nutrition of Plants: Principles and Perspectives* (New York: Wiley, 1972).

HEWITT, E. J., and T. A. SMITH. *Plant Mineral Nutrition* (London: English Universities Press, 1975).

Nitrogen Metabolism

BEEVERS, L. *Nitrogen Metabolism in Plants* (New York: Elsevier, 1976).

BRILL, W. J. "Biological Nitrogen Fixation." *Scientific American* 236 (3): 68–81 (1977).

Ion Uptake

ANDERSON, W. P. *Ion Transport in Plants* (London–New York: Academic Press, 1973).

CLARKSON, D. T. *Ion Transport and Cell Structure in Plants* (New York: McGraw-Hill, 1974).

LÜTTGE, U., and N. HIGINBOTHAM. *Transport in Plants* (Berlin–Heidelberg–New York: Springer-Verlag, 1979).

LÜTTGE, U., and M. G. PITMAN, eds. "Transport in Plants, II. Part A: Cells. Part B: Tissues and Organs," *Encyclopedia of Plant Physiology*, New Series, Vol. 2 (Berlin–Heidelberg–New York: Springer-Verlag, 1976).

ZIMMERMANN, U., and J. DAINTY. *Membrane Transport in Plants* (Berlin–Heidelberg–New York: Springer-Verlag, 1974).

QUESTIONS

7-1. Sixteen elements are now classified as essential nutrients. Do you believe that other elements may be added to this list in the future? If so, why have they not yet been discovered?

7-2. Some nutrient elements are essential because they are an integral part of essential organic molecules within the plant. Name two important organic molecules that contain (a) nitrogen, (b) phosphorus, (c) sulfur. What functions are performed by required elements that do not form part of the structure of organic molecules?

7-3. Describe in detail how you would establish the essentiality or nonessentiality of the element sodium for a higher green plant.

7-4. Discuss several ways for assessing the degree to which a soil is supplying a particular plant with its essential mineral elements.

7-5. Wheat plants grown in a medium low in silicon become highly susceptible to fungus attack, and may die as the result of infection. Would you therefore consider silicon an essential element for wheat? Explain.

7-6. Insufficient iron in the soil causes chlorosis between the veins of young leaves, whereas insufficient nitrogen causes a generalized yellowing of old leaves. Why do iron and nitrogen deficiencies affect tissues of different ages?

7-7. A soil rich in calcium phosphate may actually supply too little phosphorus for optimal plant growth. Explain.

7-8. The "little-leaf" disease of peaches, now known to be caused by a *zinc* deficiency, was originally cured in the field by the application of large quantities of commercial *iron* sulfate. Explain.

7-9. Addition of nitrogen fertilizer to pea plants growing in a nitrogen-deficient soil may not improve crop growth significantly. Why?

7-10. The reduction of nitrate to ammonia in the green alga *Chlorella* is greatly accelerated by light. What is a possible mechanism for this effect?

7-11. Most plants grow best with nitrate as a nitrogen source, but certain plants seem to prefer the ammonium ion. What can you suggest as a possible biochemical reason for this difference in behavior?

7-12. Trace the biochemical pathways followed by a molecule of nitrogen and a molecule of carbon dioxide from the atmosphere until they appear in an amino acid in a plant.

7-13. Lichens are often observed growing on bare rock. What is their source of mineral nutrients? What does this indicate about the role of lichens in soil formation?

7-14. In what ways does soil organic matter contribute to enhanced agricultural productivity?

7-15. Why may cultivation of land previously in grass lead to erosion, floods, and dust storms?

7-16. Trace the pathway taken by a potassium ion as it moves from the soil to a leaf. What forces are involved in its movement?

7-17. The rapidly dividing cells of an apical meristem are separated from differentiated vascular elements by a region of undifferentiated elongating cells. How do the meristematic cells obtain the water and nutrients essential for their great metabolic activity?

7-18. The Casparian strip in endodermal cells may be of importance in the uptake of salts into the xylem of the root and also in the uptake of water under conditions of positive root pressure. Explain what the Casparian strip does and how it functions to produce the above effects.

7-19. Substances may enter the cell in several different ways. (a) By what mechanisms and (b) under what driving force would the following occur:
1) rapid entry of a nonionized non-lipid-soluble molecule such as sucrose;
2) entry of a small lipid-soluble molecule;
3) "passive" entry of a positively charged ion;
4) "active" entry of a negatively charged ion?

7-20. If a plant cell with a negative internal potential is placed in a solution containing cations at the same concentration as the cations in the cell, how will the cations move, if they do, and why?

7-21. What processes are involved in active ion uptake? Name two criteria useful in distinguishing active from passive uptake.

7-22. Ions can move from cell to cell either through plasmodesmata, or by crossing a plasmalemma, diffusing through free space, and crossing another plasmalemma. What unique function does each pathway serve?

Translocation: the Redistribution of Nutrients

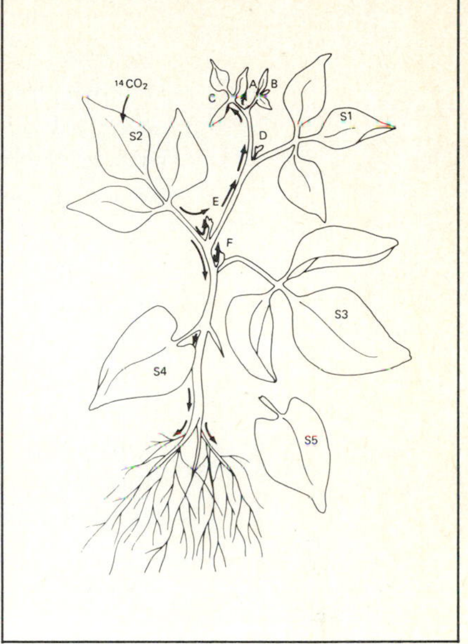

The carbon dioxide fixed in chloroplasts of leaf cells through the process of photosynthesis appears eventually in the form of a sugar. As the disaccharide sucrose, these sugar molecules are available for movement to the rest of the plant. Although photosynthetic products may be stored temporarily as starch in chloroplasts and leucoplasts whenever the rate of photosynthesis exceeds the rate at which its products are removed to the rest of the plant, starch is insoluble in water, and therefore cannot be transported as such. Only when starch is transformed into sucrose, and to a lesser extent other sugars or sugar alcohols (*sorbitol*), can carbohydrate redistribution take place.

While tracheids and vessels of the wood form the plant's pipeline for the distribution of water and some solutes from the soil to the shoot, another tubular system, the phloem, acts as the system through which the sucrose made in leaves is distributed to all the other parts of the plant. Early experiments on trees, in which phloem is found in the bark, revealed the location and importance of the sugar-distributing system. If a ring of bark completely encircling the trunk is removed from a tree, the leaves remain quite healthy temporarily, as their water is supplied through the xylem in the wood. The roots, however, starve to death because of energy and carbon deprivation, while the sugar they need accumulates in the swollen bark above the point where the trunk has been girdled. Clearly, the products of photosynthesis are transported from leaves to roots in the bark. Chemical analysis provided further evidence of the unique function of the bark by showing that it had a high content of sugars, while the xylem sap contained mainly mineral salts.

Bark is the common name given to the tough outer tissues of a woody stem which can be separated from the hard inner woody cylinder at the weak, actively dividing layer of cells comprising the vascular cambium. On its outer surface, the bark is covered by layers of dead, protective *cork* cells, produced by a cork cambium lying immediately below them. As the exterior bark is split off by the pressure of the enlarging trunk, this cork cambium produces more cork cells toward its exterior, sealing the gap made by the split. The inner side of the bark consists of the phloem tissue, containing the sieve tubes, composed of thin-walled, elongated, living cells joined end to end (see Fig. 3-6). In addition to these tubes, phloem contains a considerable number of fibers and parenchyma cells interspersed among the sieve tubes. In young stems and in many herbaceous plants, the phloem is found mainly on the outer side of each vascular bundle, although additional locations may be found in some species.

The Direction of Movement in the Phloem

The sieve tubes contain not only sucrose but also lesser amounts of other sugars, amino acids, hormones, and mineral elements. While xylem contents generally flow exclusively from the root to the leaves, phloem contents move either upward or downward from the *source* of photosynthesis to any point, termed a *sink*, at which photosynthetic products are utilized. Sugars are needed to support growth in the roots, shoot tips, and young leaves; they are also transported to roots and shoots for storage, or to seeds and fruits to be transformed into materials for growth of the next generation. In some species, special organs of food storage such as tubers, rhizomes, or bulbs may act as sinks. When the sucrose reaches such fleshy storage tissues it is usually transformed into insoluble starch, thus preventing the buildup of high concentrations of osmotically active solutes. At some later time, starch in these tissues is reconverted to sucrose and exported to a new sink.

The "source-to-sink" movement of sugars in the phloem has been demonstrated by the use of radioactive tracers. Radioactive CO_2 or sucrose is applied to a leaf, and after a period of transport the plant is harvested, pressed, dried, and placed against x-ray film. Wherever the radioactive ^{14}C is present, a black image is produced on the x-ray film; in this way, the parts of the plant that receive sucrose from the treated leaf can readily be visualized (Fig. 8-1). In general, each sink is supplied by its nearest available source; thus the uppermost photosynthesizing leaves supply the growing bud and youngest leaves, the lower leaves supply the root, and the leaves nearest to a fruit will supply that fruit. In a perennial plant which makes most of its growth early in the season, all the leaves supply sucrose to the widely dispersed storage tissues later in the year, ensuring that a large reserve will be available for the following season. Clearly, the direction of movement in the phloem is not fixed, as it tends to be in the xylem. In the lower parts of the stem it

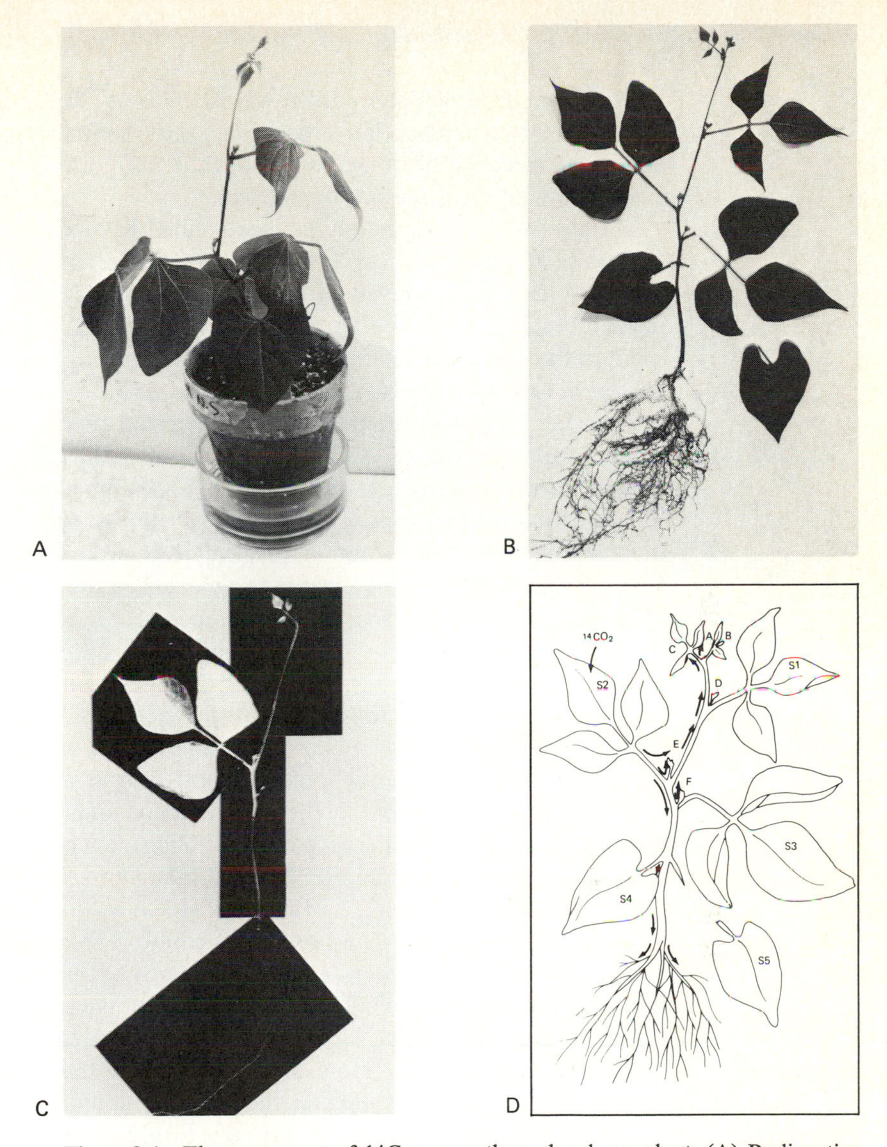

Figure 8-1 The movement of ^{14}C sucrose through a bean plant. (A) Radioactive sucrose was synthesized in the youngest fully mature leaf (S_2) from $^{14}CO_2$ supplied for 2 hours in the light. (B) The plant was then dismembered, dried, mounted on a piece of paper, and placed next to x-ray film for a few days. (C) After the film was developed and printed, radioactivity shows up as white. A small portion of the treated leaf was exposed to the film for a shorter period of time to reveal the higher concentration of ^{14}C-sucrose in the veins. (D) Movement from the source (the treated leaf) to the various sinks is indicated by arrows. The internode below the source leaf, and youngest developing leaf (c) on the same side of the plant are strong sinks for photosynthetic products from the treated leaf. The youngest leaflet (d) is not labeled as it receives its sucrose from source leaf 1. The older leaves are also unlabeled as they are exporting rather than importing sucrose. While the treated leaf contributes some material to the roots, most of the photosynthate translocated to the roots comes from the lowermost photosynthesizing leaves (S_3, S_4, S_5). (Courtesy of D. R. Geiger, University of Dayton, who performed this experiment especially for this book.)

certainly tends to be downward towards the roots, but in other parts of the stem it depends on the relative positions of the source and the sink. Furthermore, the direction of transport can change both during the life of the plant and during different seasons of the year.

Studies with appropriately labeled solutes have shown that transport in the phloem may occur in two directions simultaneously. This apparent bidirectionality is the result of unidirectional flow in separate but adjacent sieve tubes connected to different sources and sinks. The direction of movement in the phloem is determined solely by the position of the sink relative to its source of supply. When sucrose moves from source to sink, other materials are carried along with the transported water and sucrose. Thus, growth substances applied to the leaves are swept along with the sucrose solution stream in the phloem toward the sucrose sink. The importance of this process is illustrated by the fact that some growth substances applied to a leaf kept in darkness are not moved out of the leaf, but become mobile if sucrose is supplied to the leaf or if the leaf is transferred to the light.

An understanding of source-to-sink movement in the phloem is important in analyzing the effectiveness and use of applied systemic pesticides (insecticides, fungicides, herbicides). Systemic pesticides are those which are mobile within the plant, being transported in the xylem, phloem, or both. Xylem-mobile pesticides will not move out of the leaf to which they are applied, becoming distributed throughout the plant only if applied to the soil and taken up through the roots. If the insecticide or fungicide is phloem-mobile, it will work much better if sprayed on the foliage. Such a protectant will move out of the leaves with the sucrose stream, making its way to the newly developing shoots and providing a continuing shield as the shoots grow. Phloem transport also ensures a continuous distribution from the maturing leaves back to the developing shoots, as long as the protectants are not broken down by the plant. For a herbicide to be maximally effective on a weed with a vigorous system of vegetatively spreading stolons or rhizomes it must also be phloem-mobile. Such an herbicide is carried from the site of application on the leaves to the sucrose sinks in the growing points of the stolons or rhizomes, and thus accomplishes their destruction. In fact, one of the chief problems confronting herbicide chemists is not that of discovering a toxic chemical, but rather of finding one that is translocated to the growing zones of stolons and rhizomes before it destroys the translocation system of the phloem.

A wide diversity of substances, including mineral elements, nitrogenous compounds, and plant hormones, are circulated around the plant through a combination of xylem and phloem routes (Fig. 8-2). Mineral ions, for instance, are initially absorbed from the soil and transported to the shoot mainly through the xylem. As the leaves senesce, some mobile elements (for example, K^+, $H_2PO_4^-$, Mg^{++}) are withdrawn and move in the sucrose transport stream of the phloem to a sink. In times of nutrient deficiency these elements can again be redistributed to the growing shoot, which can then remain

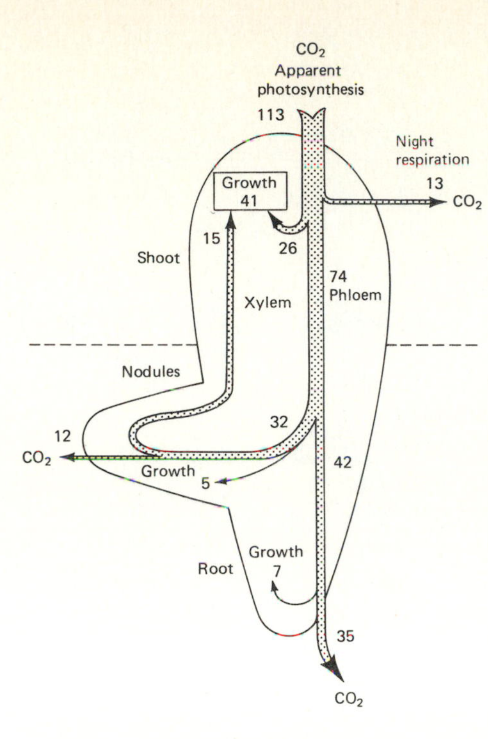

Figure 8-2 Flow sheet diagram of carbon in a nodulated pea plant, prior to fruiting. The numbers are the percent of carbon compounds moving to each sink per 100 units net gain from the atmosphere. Note that some of the fixed CO_2 is translocated to the nodules in the root, then returns to the shoot as nitrogen containing compounds. When fruits appear, they will become the principal sinks for photosynthetic products. (From J. S. Pate, *Crop Physiology—Some Case Histories* L. T. Evans ed., Cambridge, England University Press, 1975.)

relatively healthy while the older leaves become depleted and display deficiency symptoms.

Phloem Structure

Before we can attempt to understand how sucrose is transported in the phloem we must examine the structure of the sieve tubes. These thin-walled, elongated cells are joined end to end to form a continuous pipe. Where the end walls of adjacent cells come together, they are perforated by numerous *sieve pores*, for which the end walls are termed *sieve plates* (Fig. 8-3). The sieve tube cells, unlike xylem vessel cells, are living, but they do not resemble a typical living cell. They have no nucleus, but do retain some other organelles and a plasma membrane, which lines the outer surface of the cell and the sieve pores. This plasmalemma is important in retaining sugars in the sieve tubes and is demonstrably functional, since sieve tube cells can be plasmolyzed. The tonoplast, like the nucleus and many of the organelles, has disappeared. Special phloem protein microfilaments termed *P protein* have been detected through the electron microscope, but no one yet knows the exact structure or function of these proteins.

The ultrastructure of sieve tubes has been difficult to study because when sections are cut, the release of pressure on the contents of the tubes causes all the contents to rush to one end, thus blocking the pores. For this reason, it appeared that the pores were normally blocked, which made it very difficult

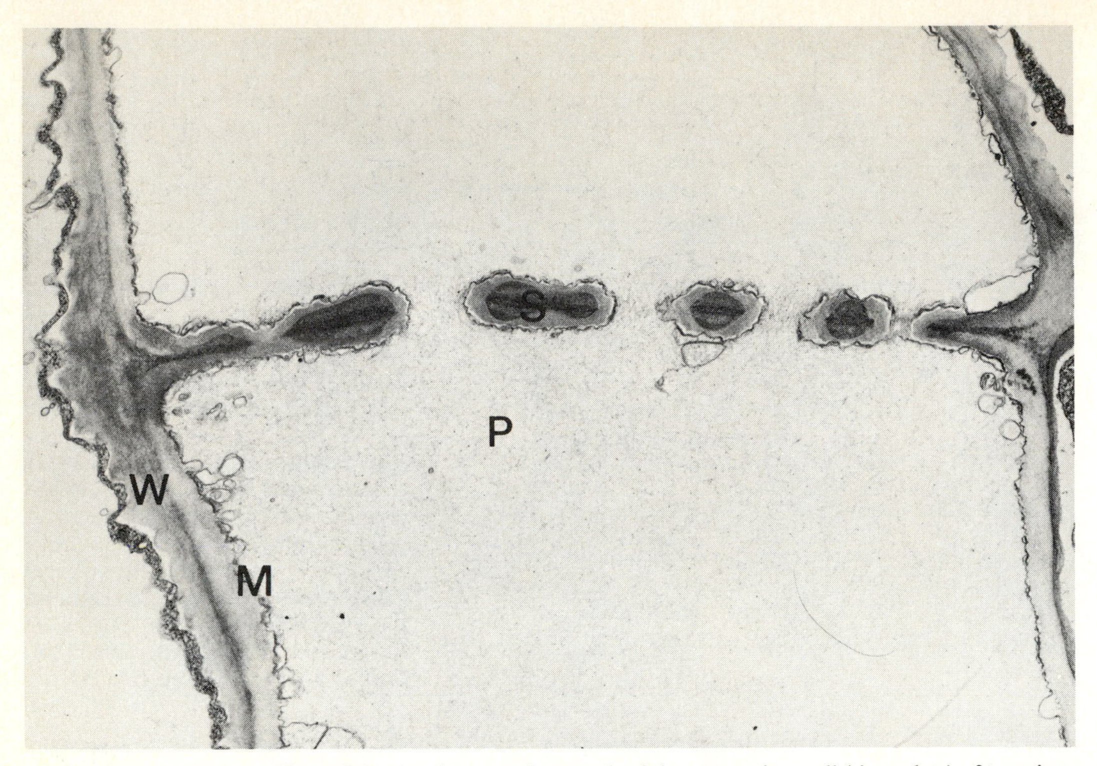

Figure 8-3 An electron micrograph of the connecting wall (sieve plate) of two sieve tube elements in the phloem of *Cucurbita maxima* (×8000). Note the disperse contents with only fine protein microfilaments (P) filling the bulk of the sieve tube, and the open sieve pores except for the continuation of the 'P protein'. W = cell wall, M = sieve tube plasmalemma membrane. (Courtesy of M. V. Parthasarathy, Cornell University.)

to understand how they could function in transport. Newer, more sophisticated techniques, in which the phloem can be fixed for microscopy *in situ* prior to cutting, have shown that the sieve tube contents do not usually block the sieve plate, although occasionally deposits of *callose*, an amorphous glucose polymer, are found in the sieve pores. In most rapidly growing plants, such deposits have also been shown to result from injury, or to occur normally at the end of the functional life of the sieve tube. In most plants, sieve tubes are short lived, and are continuously replaced by new sieve tubes produced by divisions of the cambium. Callose is probably functional in plants that retain their sieve tubes through a dormant period. In some trees (for example, *Tilia americana*) whose sieve pores become blocked by callose deposits at the end of the growing cycle, the callose plugs are hydrolyzed in the spring, with the hydrolytic products providing a supply of substrates used in the resurgence of growth. The unplugging permits a renewal of phloem transport to sinks.

Some of the parenchyma cells surrounding the sieve tubes have special characteristics. Derived together with the sieve tube from the final division of a sieve-tube-forming cell, such parenchyma cells are called *companion cells*. Companion cells have large nuclei; dense cytoplasm; abundant ribosomes, mitochondria, and endoplasmic reticulum; and may perform special secretory functions during transport. They probably also provide some of the energy and informational RNA needed for maintenance of the sieve tubes.

Characteristics of Phloem Transport

Sugars move through the phloem as a concentrated solution, generally in the range of 7–25% (0.2–0.7M). There was originally some argument over whether the water serving as solvent actually moved at the same rate as the sugars. The movement of ^{14}C-labeled sugars was relatively rapid, but the movement of tritiated (^{3}H-labeled) water did not keep pace with the sugars. It is now clear that the sugars are restricted to the sieve tubes, but that the labeled water molecules are free to diffuse through the plasmalemma; thus the random motion of the very small number of tritiated water molecules led to their diffusion out of the sieve tube, giving the impression that water was not moving with the sugars. That the entire solution actually moves as a unit was shown by an elegant experiment in which bundles of sieve tubes pulled out of a *Heracleum* petiole were left attached at each end, so that they were still functional. Heat was applied to a small spot on the bundles through the use of a small light beam, and accurate thermocouples were then used to measure the movement of the warmed material further down the bundle. This method demonstrated that the heated solution moved along the bundle at about 35 cm/hr, consistent with observed rates of sucrose transport.

Sugars are often transported in the phloem over distances of many meters and at rates of up to 100 cm/hr. These distances and speeds are far too great for diffusion to be a significant mechanism of transport. Efficient phloem transport requires living sieve tubes capable of active metabolism. Thus, phloem transport is inhibited by stem girdling, which kills all the living cells in a small ring around the stem. This allows water to reach the leaves through the unaffected xylem tubes, but sugars cannot get through the phloem of the girdled area. In the same manner, respiratory inhibitors can prevent the transport of photosynthetic assimilates.

Phloem transport has proved difficult to investigate, for experiments which in any way disrupt the delicate pressure balance in the sieve tube complex lead to aberrant results. One of the few ways in which the contents and characteristics of phloem have been successfully investigated is through the use of aphids. These insects have the unique ability to locate and penetrate a single sieve tube with their stylets when feeding on a plant (Fig. 8-4a and b). Once they have punctured the sieve tube they no longer have to exert any effort, as they are force fed and inflated by the pressure in the sieve tube. The nature of the sieve tube contents and of the phloem transport process can

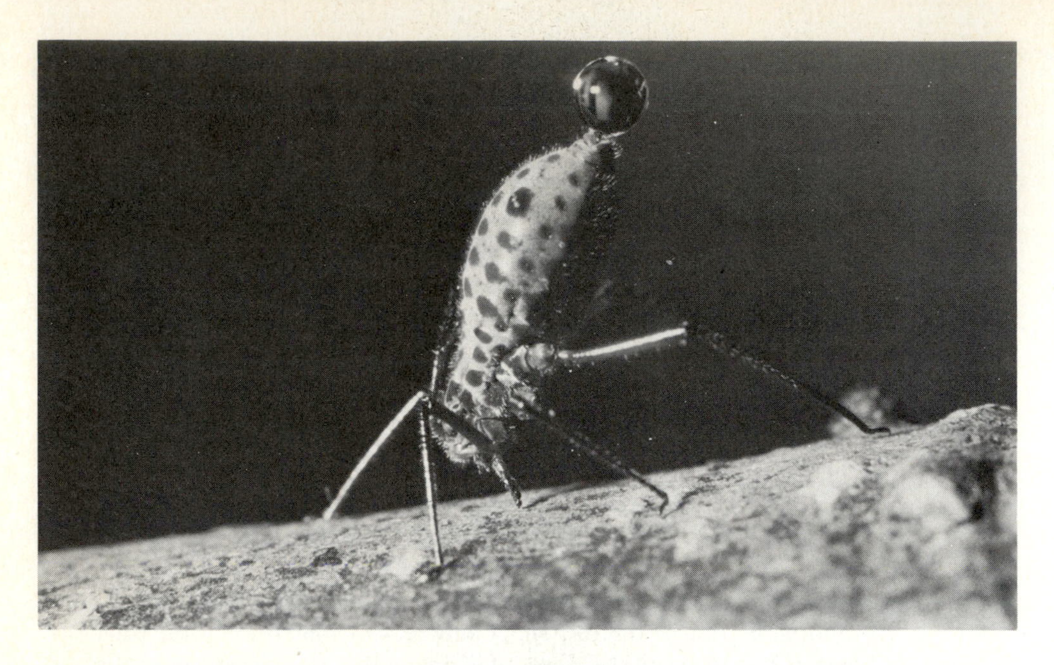

Figure 8-4a Photograph of an aphid on the surface of a plant. Note the aphid's stylet penetrating the surface and a drop of "honeydew" (phloem sap after passing through the aphid's intestine) at the tip of the abdomen.

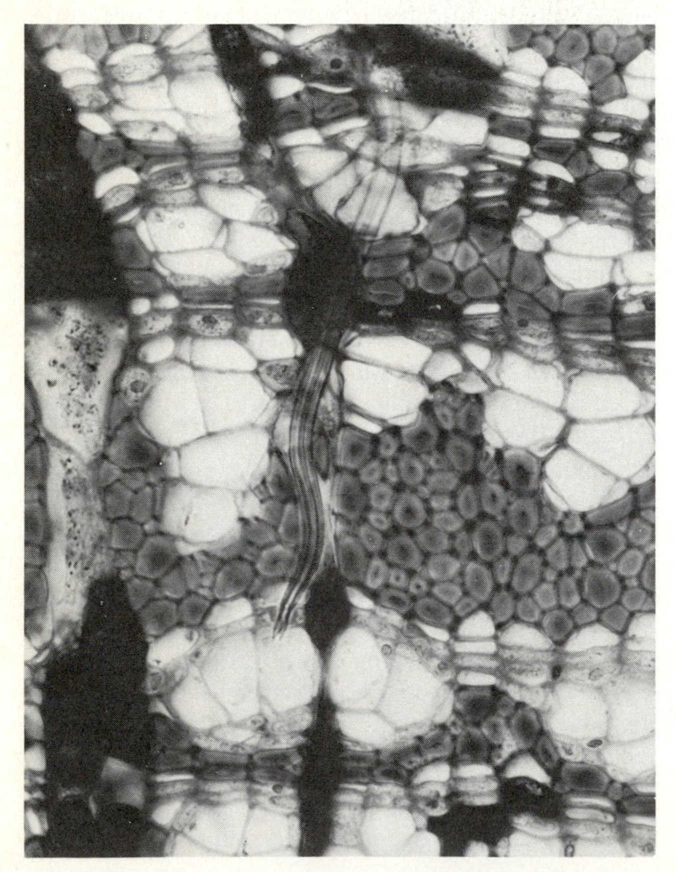

Figure 8-4b The stylet of the aphid penetrates through the outer tissue of the stem to enter a single sieve tube (bottom center). (Both photographs courtesy of M. H. Zimmermann, Harvard University.)

therefore be determined by using the aphid as a tap. The body is removed, but the stylet is left embedded in the sieve tube as a microcannula through which the phloem sap flows under the force of the pressure in the sieve tube. With this technique it can be shown that concentration and pressure gradients exist in the phloem, with the greater sucrose concentration and hydrostatic pressure occurring near the source and lower sucrose concentration and pressure near the sink.

Despite the contiguity of phloem sieve tubes under pressure and xylem vessels under tension, the two systems are in aqueous equilibrium—that is, they have the same water potential. The highly negative water potential of the concentrated phloem sap counteracts the effect of the positive hydrostatic pressure in the sieve tubes. The sieve tubes thus have a net negative water potential approximately equal to that of the xylem vessels, which contain a very dilute solution (slightly negative water potential) and are under tension (Fig. 8-5a).

TABLE 8-1. Analyses of vessel sap of a pear tree and phloem exudate of *Robinia pseudoacacia*

Concentration in sap, mg/liter		
Substance	Xylem	Phloem
Ca^{++}	85	720
Mg^{++}	24	380
K^{+}	60	950
SO_4^{--}	32	—
PO_4^{---}	25	—
Sugars*	—	200,000
N, organic	—	425
N, inorganic	—	135

From R. G. S. Bidwell, *Plant Physiology* (New York: Macmillan, 1974): Data recalculated from F. G. Anderssen: *Plant Physiol.*, **4**:*459–476* (*1929*), and C. A. Moose: *Plant Physiol.*, **13**:*365–380* (*1939*).

**Mostly sucrose.*

The Mechanism of Phloem Transport

Over the years, researchers have advanced many theories on the mechanism of phloem transport. Most of these can be dismissed because they were based on experiments that were later shown to be producing artifacts; with others the data can be satisfactorily explained by a mechanism other than that proposed by the author. Only one theory currently accommodates almost all the known facts, and the small discrepancies which remain will probably be

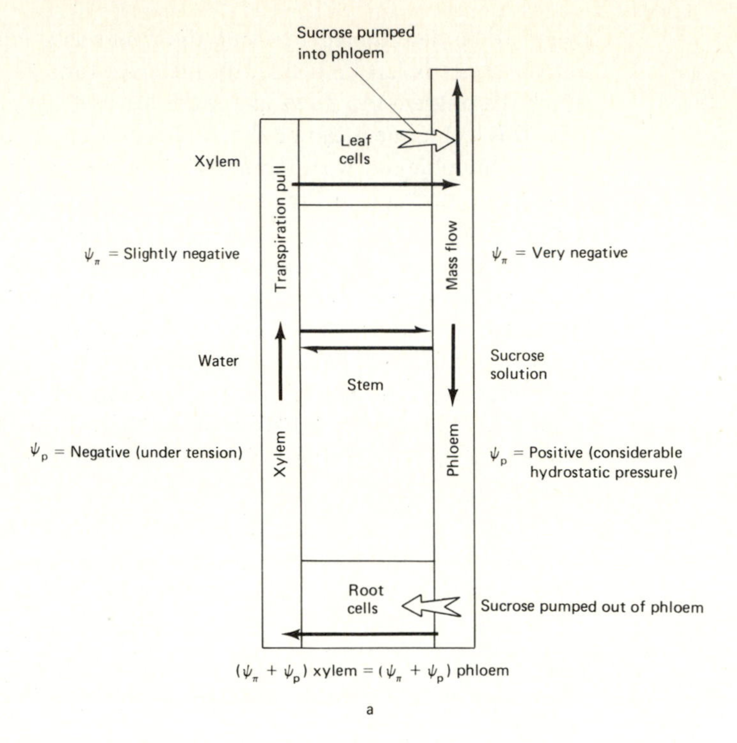

Figure 8-5a A diagram showing how the mass flow hypothesis could function in the plant.

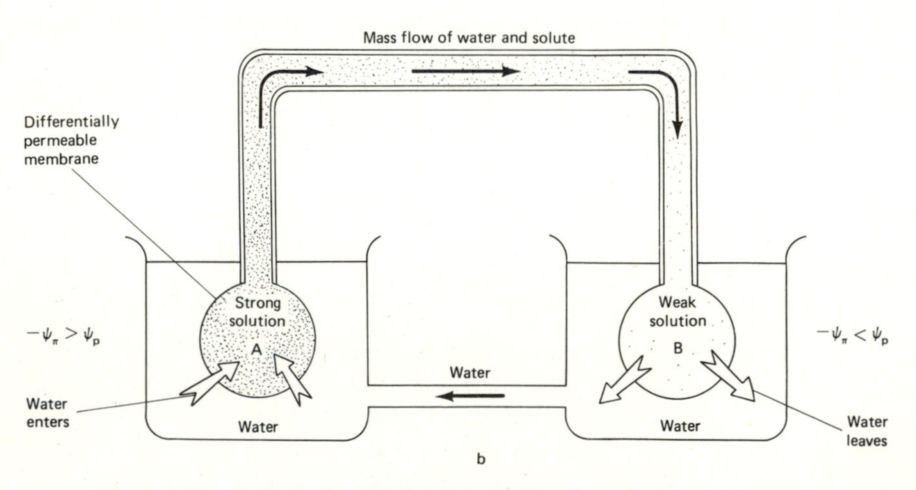

Figure 8-5b A physical model for the mass flow hypothesis.

explained in the future by minor modification of the basic theory. For simplicity, therefore, we will describe only the *mass flow* (or *Münch pressure flow*) *theory*, initially proposed by the German plant physiologist Karl Münch.

Mass flow in the phloem is envisioned as depending entirely on the simple principles of osmosis, which can best be demonstrated by considering a purely physical analog system. Two rigid spheres constructed of differentially permeable membrane are immersed in water and connected by an impermeable tube (Fig. 8-5b). One sphere is initially filled with a strong sucrose solution (*A*) and the other with a weak sucrose solution (*B*); an intermediate solution then fills the tube. Water will enter both spheres by osmosis and pressure will build up in the system. As the pressure increases in the more concentrated solution *A*, it will be transmitted through the tube to the less concentrated solution *B*. If the transmitted pressure from *A* exceeds the pressure built up in *B*, it will force water to leave rather than enter *B*. Since water is now entering *A* and leaving *B*, a *mass flow* will develop from *A* to *B*, carrying the sucrose with it. Flow continues until the concentrations in *A* and *B* are equalized, at which point the entire system will be in pressure equilibrium, and net directional flow of the solution will cease. If, however, we continue to add sucrose to *A* and remove sucrose from *B*, the flow of sucrose solution from *A* to *B* will continue. If now the water removed from *B* can find its way back through the external solution to *A*, a continuously operating closed system will have been established. To depict this, the solutions surrounding *A* and *B* are joined by tubing in the figure.

In the plant, sucrose produced by photosynthesis is actively pumped into the sieve tubes of the small veins of the leaf in a process called *phloem loading*. This process decreases the water potential of the sieve tube, resulting in entry of water by osmosis. Pressure then builds up, forcing the solution along the pressure gradient on its way toward the sink. At the sink, the sucrose is actively removed, raising the water potential of the sieve tubes; the water then diffuses out along the water potential gradient caused mainly by the pressure in the sieve tube. Water returns readily to the source through the xylem.

This theory fits well with the major facts known about phloem: The phloem is under positive pressure; pressure and osmotic gradients exist from source to sink; the contents flow in bulk through the sieve tubes; and the sieve pores are open to permit this flow. The one remaining problem is that more active metabolism is known to be required than would appear necessary for the mere maintenance of the plasmalemma of the sieve tubes. This dilemma may be resolved by the fact that metabolism is required for three separate processes: sucrose loading into the phloem in the leaves, sucrose removal at the sink, and retention of the sucrose inside the sieve tube membrane. Differential operation of these three processes can lead to different directions of movement of the contents of sieve tubes in adjoining vertical files.

Phloem Loading

In the leaves, sucrose, the primary mobile product of photosynthesis, has to be moved into the sieve tubes against a concentration gradient. Characteristically, leaf veins branch repeatedly until at their terminus they are only a few vessels and sieve tubes thick. At this point they are very close to the mesophyll cells that are actively engaged in photosynthesis. In sugar beet (*Beta vulgaris*), sucrose is pumped into these terminal tubes directly from the surrounding mesophyll cell walls and small intercellular spaces into which it has been moved from the mesophyll cells (Fig. 8-6). Transport into the phloem is selective for sucrose, and involves active metabolism, probably a cotransport of sucrose and hydrogen (H^+) through a specific permease on the

Figure 8-6 Autoradiographs showing the loading of ^{14}C labeled sucrose into the small veins of a leaf of sugar beet after the leaf received a 3 min pulse of $^{14}CO_2$ ($\times$ 1/2). The labeled material shows up as white. The leaves were taken at different times after termination of the CO_2 pulse: A) 1.5 min, B) 5 min, C) 8 min, D) 12 min, E) 20 min, F) 100 min. Note that initially the fixed products are all over the leaf; then they are loaded into the veins, and finally transported away. In G) the leaf was labeled for 180 min with ^{14}C sucrose. (Courtesy of D. R. Geiger, University of Dayton.)

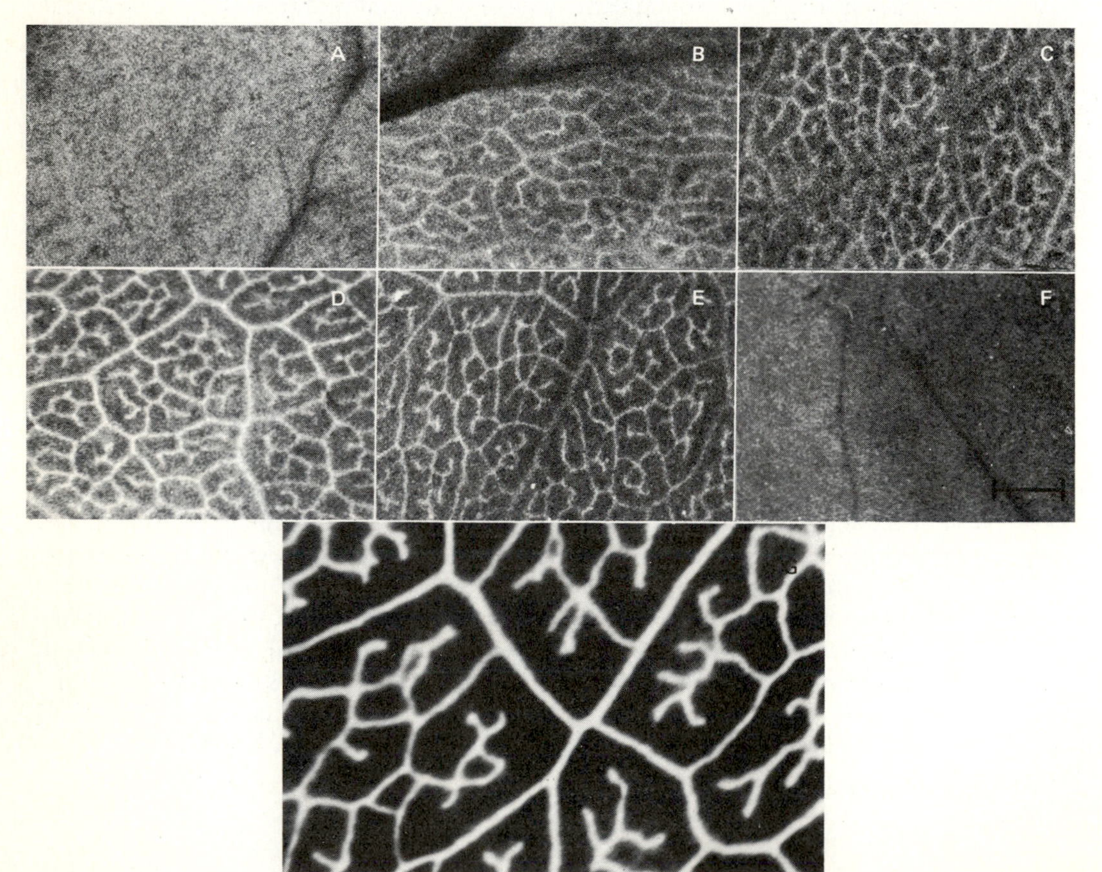

sieve tube plasmalemma, in response to pH and electrochemical gradients. In loading, energy is later used to pump H^+ ions out of the sieve tube by means of an ATP-utilizing H^+ carrier. This enzyme uses the energy released in ATP hydrolysis to move the H^+ out across the membrane. The H^+ can then diffuse back into the cell together with sucrose, utilizing the specific proton–sucrose permease, and moving down an electrochemical gradient for H^+.

Sucrose Retention in Transport

During transport in the phloem there is a highly selective retention of sucrose. Most other molecules can freely diffuse into and out of the sieve tubes and are in equilibrium with surrounding molecules of the same type. The retention of sucrose in the sieve tubes occurs against a considerable concentration gradient and requires the constant expenditure of metabolic energy. Sieve tubes are thought to contain considerable amounts of ATPase, which might be involved in this process. From their dense appearance and abundant mitochondria, the companion cells are believed to be highly active metabolically, and one of their functions could be to provide the energy needed to maintain the membrane selectivity of the adjacent sieve tube.

Despite the high energy requirements for maintenance of the system, transport in the sieve tube is itself not dependent on metabolism. The speed of transport varies with the demand of the sink; thus metabolism is required for the removal of sugars from the phloem, and this process, combined with metabolic loading, provides the energy to keep the system going. Mass flow itself then occurs from source to sink along purely physical gradients.

The mass-flow theory can therefore explain the major features of phloem transport. The only structural feature whose role is not yet understood is the P protein. It has been suggested that these microfilaments provide surfaces for interfacial flow, allowing the solution in the phloem to move rapidly, but as yet we have no experimental evidence to support this suggestion. While it appears that additional mechanisms may be operating, and that the mass-flow theory may have to be slightly modified in the future, it is our best current description of photosynthate transport through phloem sieve tubes.

SUMMARY

The *photosynthate* made in the leaves and the water and minerals absorbed by roots are required by all plant cells. Their movement, or *translocation*, around the plant occurs in the specialized transport elements of the phloem and xylem, respectively. Whereas xylem transport is mainly from root to stem, phloem transport can be either upward or downward, with each individual row of sieve tube elements conducting in a single direction.

Flow in the phloem is from *source* to *sink*, the latter being any region in which translocated materials are removed from the flowing stream. Growing regions and developing storage organs are common sinks, while leaves and storage organs whose resources are being mobilized are the usual sources.

Sieve tube cells lack a nucleus, but have abundant cytoplasm connected to that of neighboring cells through plasmodesmata and the numerous pores of the *sieve plate.* Sieve tube cells also contain strands of filamentous proteins, designated *P protein,* which sometimes traverse sieve pores and extend from cell to cell. Sieve pores are sometimes plugged by the glucan *callose,* especially during dormancy, and can be unplugged when dormancy is broken. Densely cytoplasmic *companion cells* lie alongside sieve tubes; although their function is unknown they probably supply energy to the sieve tubes. Through the use of *aphid stylets,* which penetrate individual sieve elements, researchers have learned that the translocated phloem contents, containing 0.2 to 0.7M sucrose plus other solutes, generally move as a unit at about 35 cm/hr, although rates of up to 100 cm/hr are known.

Transport through the phloem is generally agreed to occur via *mass flow.* The high hydrostatic pressure caused by water movement into sugar-rich regions of highly negative water potential causes a mass flow toward regions of lower pressure; there, sugar removal guarantees the continued existence of the gradient, and thus of the flow. Crucial steps in the maintenance of the system are the *loading* and *unloading* of solutes into and from the sieve cells. Loading is believed to involve *cotransport* of sucrose and H^+ ions through specific permeases in response to pH and electrochemical gradients. Absorbed H^+ is later resecreted through an ATP-utilizing proton transporter. Sucrose retention in sieve tubes, against high osmotic gradients, may involve similar active processes.

SELECTED READINGS

CRAFTS, A. S., and C. E. CRISP. *Phloem Transport in Plants* (San Francisco: Freeman, 1971).

ZIMMERMANN, M. H., and J. A. MILBURN. "Transport in Plants, I. Phloem Transport," *Encyclopedia of Plant Physiology*, New Series, Vol. 1 (Berlin–Heidelberg–New York: Springer-Verlag, 1975).

QUESTIONS

8-1. In a stem, the xylem and phloem are adjacent, yet water does not show a large net movement from one to another, despite the fact that the osmotic potential of xylem is only slightly negative (about −1 bar), while the osmotic potential of the phloem is much more negative (about −20 bars). Explain why there is little net water movement. What do these facts tell us about the mechanisms by which these tissues perform their function?

8-2. In what ways does the transport of inorganic materials differ from the transport of organic materials in plants? In what ways are they similar?

8-3. Phloem transport is considered to result from the same physical processes involved in osmosis, yet transport slows measurably or even ceases if metabolic inhibitors are applied to either the leaf or stem. How do you explain this?

8-4. What governs the direction of flow in the phloem?

8-5. The growth of a fruit, such as an apple, depends largely on the amount of photosynthate transported to it. How could you increase the size of an individual immature apple relative to other apples on the same tree?

8-6. Radioactive phosphorus (^{32}P) was fed to either the leaves or roots of a woody plant, and the radioactivity of the bark and wood halfway up the stem was later measured. The following results were obtained: (cpm = counts per minute).

	Root feeding	Leaf feeding
Bark	90 cpm	1360 cpm
Wood	1120 cpm	8 cpm

What can you conclude from the different distributions of ^{32}P following the two feeding methods? Where would most of the ^{32}P end up following each feeding method? What determines the direction and rate of movement in each?

8-7. Explain how the components of a lipid in a seed are transported to the growing tip of the stem after germination of the seed. Describe all the processes involved, in order.

8-8. Relatively low night temperatures appear to accelerate both net translocation and growth in tomato plants. Suggest a possible reason for the apparent negative temperature coefficient for translocation.

8-9. A deficiency in boron sometimes leads to a poor crop of fruit in apple trees. What is a possible explanation involving translocation?

8-10. What are the probable transport mechanisms in large algae, which lack well organized vascular systems?

9 Hormonal Control of Rate and Direction of Growth

In addition to the water and minerals absorbed from the soil, and the photosynthetically produced carbohydrate needed for energy and protoplasmic building blocks, the plant cell requires certain other chemical substances for optimal growth. Among these are organic materials called *hormones*. Hormones are generally needed in only infinitesimally small quantities, and in most instances are synthesized in adequate amounts by the plant itself. However, by appropriate experimental techniques, one can deplete the hormone supply in a plant, organ, or tissue, demonstrate their existence and function, and deduce a great deal about their nature and mode of action.

A hormone is a substance produced in small quantities in one part of an organism, then transported to another part of the organism where it produces some special effect. The distance over which it is transported may be relatively great, for example, from leaf to bud; it may be smaller, as from apical meristem to subjacent cells; or even minute, as from one organelle to another in the same cell. The definitive criterion is migration from site of synthesis to site of action, that is, action as a "chemical messenger." Several major classes of growth-regulatory hormones have been shown to exist in higher plants. These are *auxin*, *gibberellins*, *cytokinins*, *abscisic acid*, and *ethylene*, to which we now turn our attention. Gibberellins and cytokinins occur as groups of related, similarly acting molecules, while each of the other three classes is represented naturally by only one compound. Because of the complexity of their differences in chemistry and physiology, we will discuss these hormones as if they acted individually on cellular processes. But this is an artificial restriction—several or even all of these growth substances may be present in a cell or tissue at any given time and the way in which the tissue grows or develops is the result of the presence and interaction of all these compounds.

How it all Began

Hormones and their synthetic analogs have found many uses in agriculture as controllers of flowering, fruiting, ripening, dormancy, and abscission; as promoters of rooting; and as selective herbicides. It is paradoxical that so important a series of practical discoveries should have come from a series of investigations in which no practical goal was visualized. The existence of plant growth hormones was inferred, about 1880, from the experiments of Charles Darwin and his son Francis, who were investigating the curvature of grass seedlings toward light. By placing small, opaque, capillary glass shields over the coleoptiles of these seedlings, the Darwins were able to demonstrate that although only the extreme tip of this organ is able to perceive the dim light active in eliciting curvature, a region some millimeters below the tip does the curving. Darwin hypothesized in his book *The Power of Movement in Plants*, published in 1881, that some stimulus passed from the tip down to the growing zone and there exerted a specific growth effect. This idea spurred various investigators to conduct further experiments, and about 50 years later a young Dutch graduate student, Frits Went, was able to confirm that grass coleoptile tips do, in fact, produce significant quantities of a diffusible substance that controls the growth of the subjacent zones. This substance, called *auxin*, became the prototypic plant hormone, and formed the basis for modern chemical agriculture.

This bit of history shows that in the search for knowledge the most impractical of investigations can sometimes produce the most economically important results. Had Darwin been told by some supervisor to develop a weed killer, he undoubtedly would not have performed his perceptive experiments leading to the discovery of auxin. In the organization of research and the allocation of research funds, therefore, opportunity should be given for the unfettered mind to operate without restraint.

Auxin

Auxin is a substance produced by the growing apical regions of stems, including the young leaves. It migrates from the apex to the zone of elongation, where it is specifically required for the cellular elongation process (Fig. 9-1). If the tip of a rapidly growing stem is removed, the growth in the region below the cut will slow down very quickly, and within several hours or days,

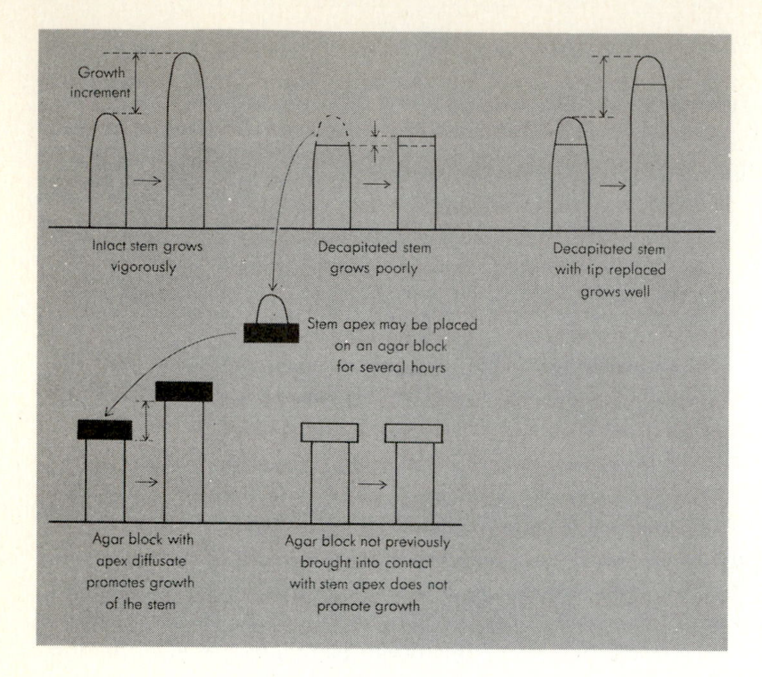

Figure 9-1 Experiments to demonstrate the production of growth-promoting substances by the stem apex.

depending on the type of plant, it will come to a complete halt. If the removed tip is immediately replaced, growth of the stem continues almost normally, showing that some influence emanating from the tip is conducted across the wound to the growing cells. If the tip is placed on a block of gelatin or agar for several hours and the block alone is then transferred to the cut stump of a decapitated stem, the block will partially substitute for the tip in facilitating growth of the subjacent regions. From this experiment, we deduce that a substance, auxin, moves from the tip to the block and from the block down to the base. The natural auxin compound in plants has been shown to be the simple material indole-3-acetic acid (IAA) (Fig. 9-2), synthesized in the plant by enzymatic conversion of the amino acid tryptophan (Fig. 9-3). Auxin promotes growth by facilitating cell elongation through effects on components of the cell wall.

The activity of IAA may be shown as well in pieces of tissue removed from a plant. If the growing portions of a stem, such as that of a pea plant, are excised and placed in petri dishes containing sucrose and a mineral salt solution, growth will be very slow. However, if small quantities of IAA are

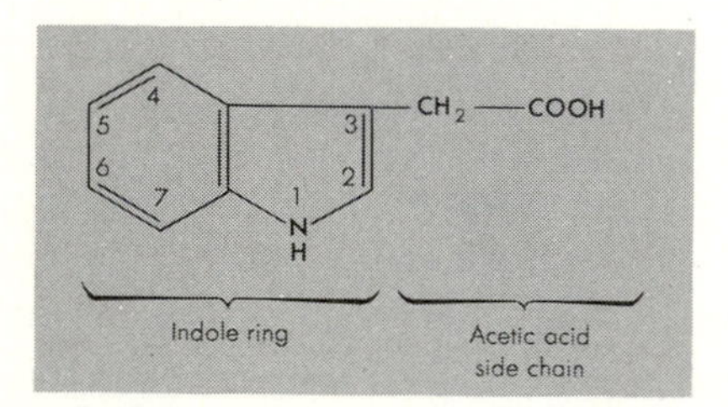

Figure 9-2 Indole-3-acetic acid (IAA), a naturally occurring plant growth hormone.

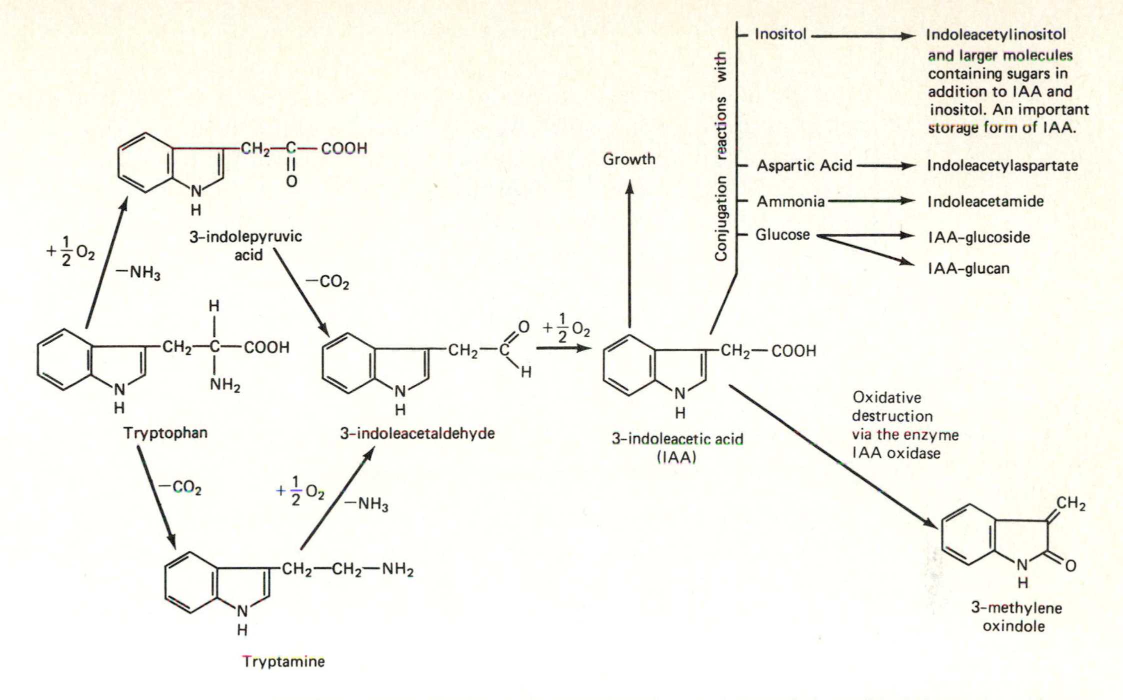

Figure 9-3 The formation, conjugation and degradation of indole acetic acid.

added, growth will be greatly enhanced, the effect being directly proportional, within certain limits, to the logarithm of the concentration of the auxin added. Usually an optimum concentration is attained, beyond which growth again becomes somewhat slower, and ultimately growth may be inhibited completely (Fig. 9-4).

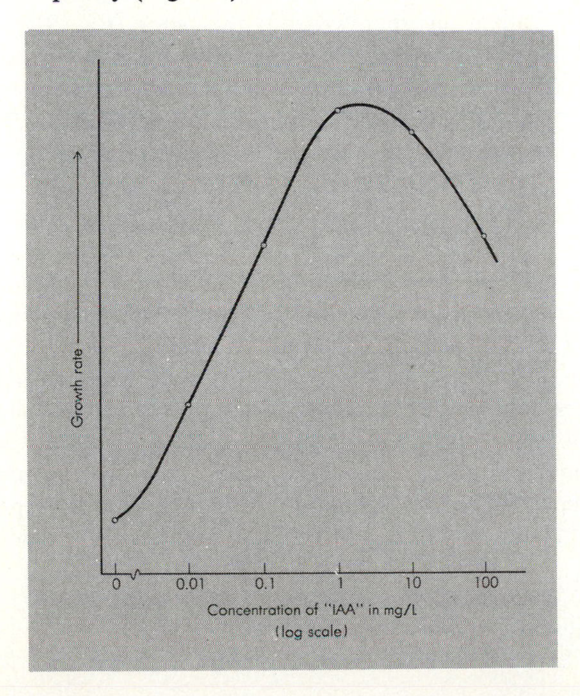

Figure 9-4 A dose-response curve for the effect of IAA on the growth of pea stem sections. (From Galston and Hand. 1949. *Amer. J. Bot. 36*, 85–94.)

Usually, the application of auxin to intact stems produces no extra increment of growth, from which we may conclude that a stem is normally saturated with auxin produced by its own tip. In roots, on the contrary, the application of auxin usually inhibits growth, and in some instances the removal of the tip enhances growth. We may therefore deduce that the root normally operates under conditions of more than adequate auxin supply. The growth of the root is promoted by concentrations of auxin lower than those that promote the growth of the stem, and is inhibited by concentrations lower than those that inhibit the growth of the stem. In all ways, roots are more sensitive to auxins than are stems.

The measurement of hormone levels in tissue

The amount of auxin in any plant tissue may be determined by extraction with a solvent, followed by application of the extract to some tissue that will respond quantitatively to the auxin contained in it. Normally, the tissue is placed in diethyl ether at a temperature near 0° C and gently shaken for a period of 2–4 hours. This ether extract is then concentrated and, when reduced to a small volume, is incorporated into an agar block that is then placed asymmetrically on the decapitated stump of the auxin-sensitive organ. Traditionally, the leaf sheath, or *coleoptile*, of dark-grown oat plants has been used. In this plant, the asymmetrically placed auxin enhances the growth of only the tissue directly below it. The unequal growth of the two sides of the coleoptile causes a curvature of the organ that is directly proportional to the amount of auxin incorporated into the block (Fig. 9-5). Thus, to determine quantitatively the amount of auxin in an unknown organ, the extract is made and applied, the curvature is measured, and this curvature is compared with curvatures produced by known quantities of auxin in another experimental series. This technique of using the response of an organism to measure the amount of a chemical in an extract is called a *bioassay*.

Bioassays, relying on the induction of a certain level of a response normally produced by the hormone, are available for all plant hormones. Today they are generally being supplanted by chemical and physical techniques, mainly because bioassays, while simple to perform, are subject to a wide variety of external influences. One common failing, for example, is that inhibitory compounds are often extracted from tissue along with the hormone, and the mixture induces a smaller growth response than should be produced by the amount of hormone actually present; therefore, an erroneous estimate is obtained of the amount of

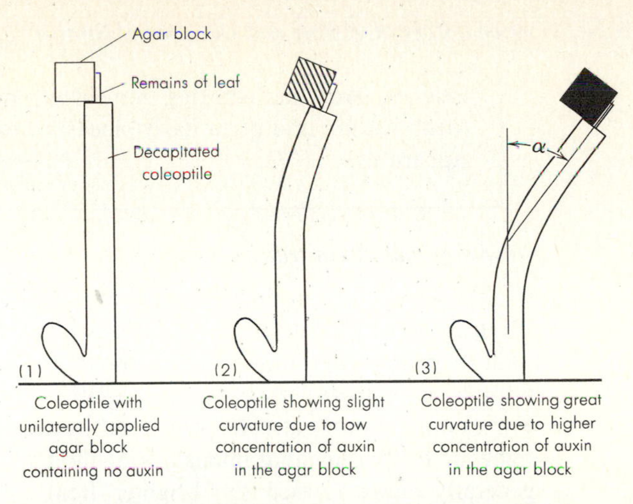

Figure 9-5 The determination of auxin concentration by means of the oat coleoptile curvature test. The angle α is proportional to the auxin content of the agar block.

hormone present. Most modern methods rely initially on purifying the hormone in the extract by a chromatographic method involving adsorption on paper, thin layers of silica gel, or columns of synthetic resin beads. The most sensitive methods of all use gas chromatography, in which a volatile derivative of the hormone is separated from impurities by flow in a gas, at an elevated temperature, through a column of material that differentially absorbs the hormone and the impurities. Another modern method is high-pressure liquid chromatography, in which a pressurized solution of the hormone plus impurities is passed through a column filled with a material that differentially adsorbs the hormone and impurities. Since the hormone and the impurities take different amounts of time to pass through the column in all these methods, the hormone is purified and can be measured directly. In both gas and high-pressure liquid chromatography, the hormone must be detected and positively identified after purification. A physical detector of some sort does this as the compounds come off the column. IAA, for example, can be detected and measured by fluorescence at a specific wavelength when a solution is illuminated with a specific wavelength of ultraviolet light. Gibberellins are identified by passing the compounds as they come off the gas chromatography column directly into a mass spectrometer. This complicated apparatus is used to fragment each molecule (by electron impact or ionization) into many molecular ions that are separated according to their ratio of mass to charge. Since the patterns of fragments (molecular ions) recorded are characteristic of particular starting molecules, one can obtain positive identification and quantitative measure of the hormone with this apparatus.

Tropisms and auxin transport

The curvatures produced by the unilateral application of auxin to plants bring to mind the curvature of various plant organs toward or away from light or gravity. In fact, we now know that such curvatures (called *tropisms*) are related to the asymmetrical distribution of auxin in the organ involved. For example, if an oat coleoptile is subjected to a low intensity of unilateral light, or to two unequal intensities of light from different directions, it will generally curve toward the brighter light (*phototropism*). This curvature occurs because growth an the side near the light is somewhat depressed by the light, while growth on the side away from the light is accelerated (Fig. 9-6). If the tip of a unilaterally exposed coleoptile is removed and the amount of auxin in its dark and light halves assessed by the curvature test described above, the side away from the light will contain about twice as much auxin as the side toward the light, but darkened whole tips contain no more total auxin than illuminated tips. Physiologists have therefore concluded that light acts in producing curvature by affecting the *lateral distribution* of auxin in the organ; this differential auxin concentration in turn elicits differential growth rates on the two sides of the organ, inducing curvature.

Similarly, a stem laid prostrate (Fig. 9-7) will, after some time, accumulate more auxin on the lower surface than on the upper; this results in accelerated growth below and an ultimate curvature upward (*geotropism*) (Fig. 9-8). The growth downward of a prostrate root is at least partly a consequence of the different auxin sensitivity of the root. In the prostrate root, as in the prostrate stem, auxin also accumulates below, but since in the normal root, auxin concentration is already optimal, or supraoptimal, the greater concentration of auxin on the lower side depresses growth, and thus causes a

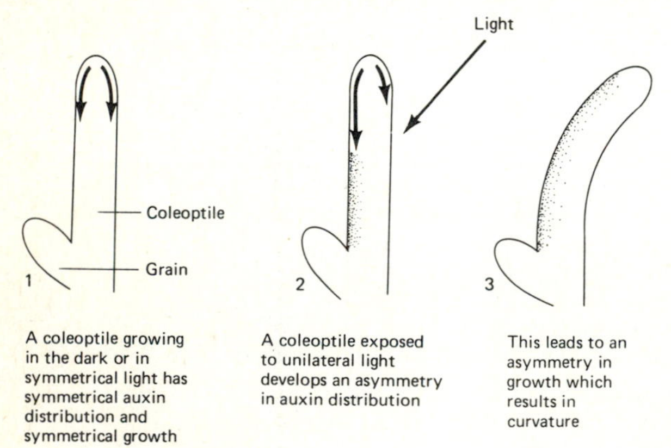

Figure 9-6 An explanation of phototropism based on auxin distribution.

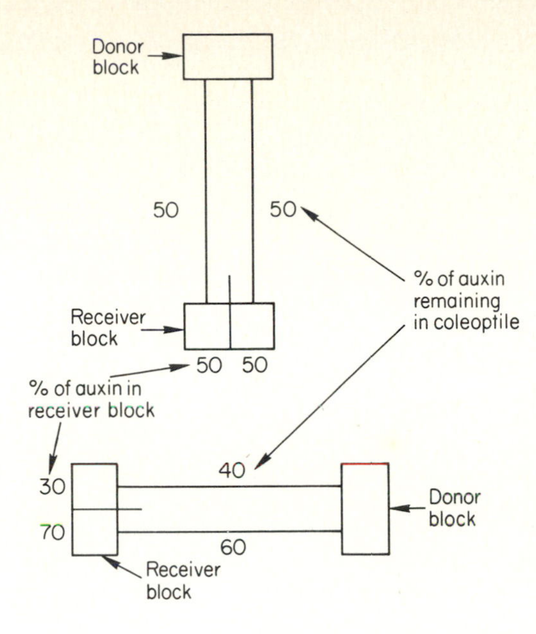

Figure 9-7 The corn coleoptile shown in its normal upright orientation (above) was tipped on its side (below). This geotropic stimulation produces a redistribution of symmetrically applied IAA. As a result of the higher concentrations on the lower side, the coleoptile grows more rapidly on this side and thus turns upward. (Data from Gillespie and Thimann. 1963. *Plant Physiol. 38*: 214–225.)

downward curvature. Recent investigations indicate that other hormones are more important in root tropisms. It has been suggested, for example, that the inhibitory hormone abscisic acid may also accumulate at the lower side of a horizontal root and so contribute to the depressed growth of the cells there. We shall discuss this phenomenon in a later section.

Plant physiologists are not completely certain as to the plant's mechanisms for detecting the unilateral light and gravitational stimuli. The action spec-

Figure 9-8 The response of a shoot of a coleus plant to being tipped on the side. The photos were taken on tipping the shoot (a) and at 10 (b) 12 (c) and 18 (d) hours afterwards. Note the overshoot in c which is then corrected by the plant. (Courtesy of C. R. Granger, Washington University.)

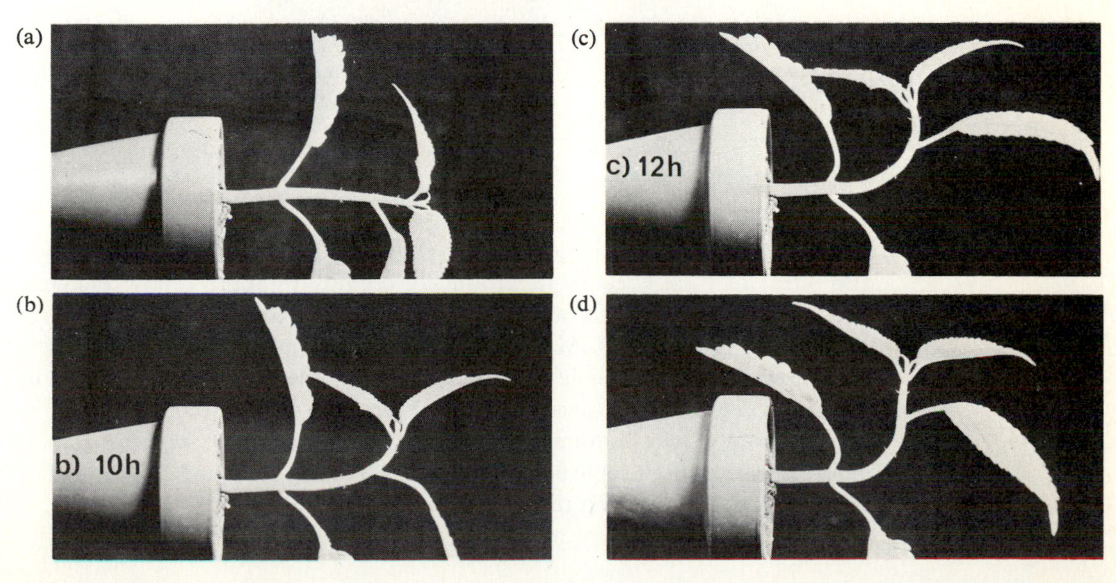

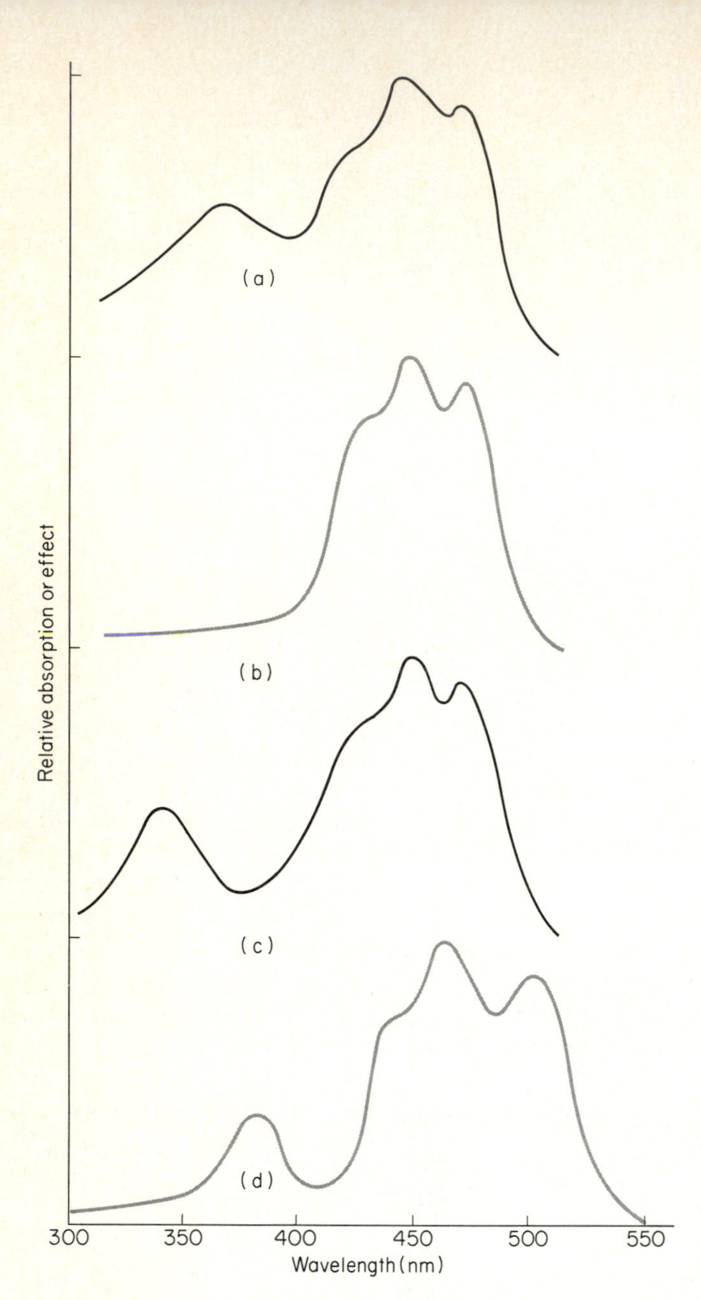

Figure 9-9 Comparison of (a) the action spectrum for phototropism with absorption spectra of (b) *trans-β* carotene, (c) *Cis-β*-carotene unknown in nature, and (d) riboflavin in castor oil. (From Galston. 1967. *Amer. Sci. 55*: 144–160.)

trum for phototropism indicates that only blue light and some portions of the ultraviolet are effective in producing curvature. Some yellow pigment, probably related either to carotene or to riboflavin, could be involved as the photoreceptor (Fig. 9-9). Many other reactions in plants and fungi show a similar wavelength dependence. In some fungi, the blue light absorbed by a flavoprotein enzyme causes the reduction of a cytochrome that can be detected in a spectrophotometer. This reaction is currently regarded as possibly central to phototropic perception and linked to subsequent physiological response by means not yet thoroughly understood.

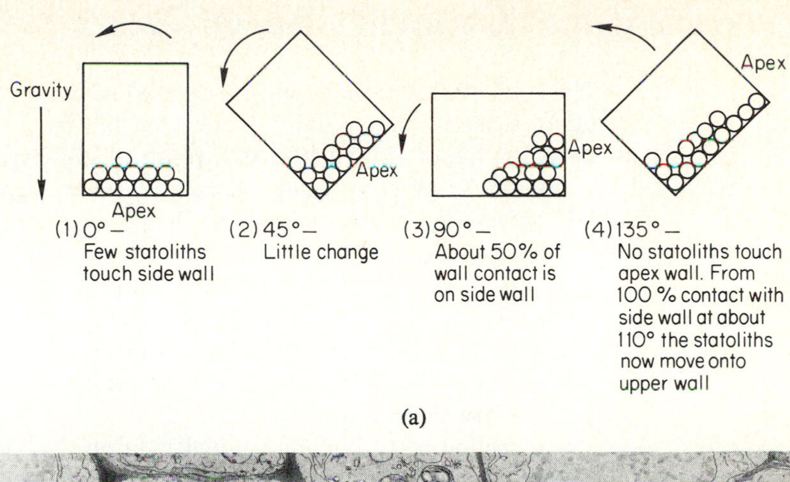

(a)

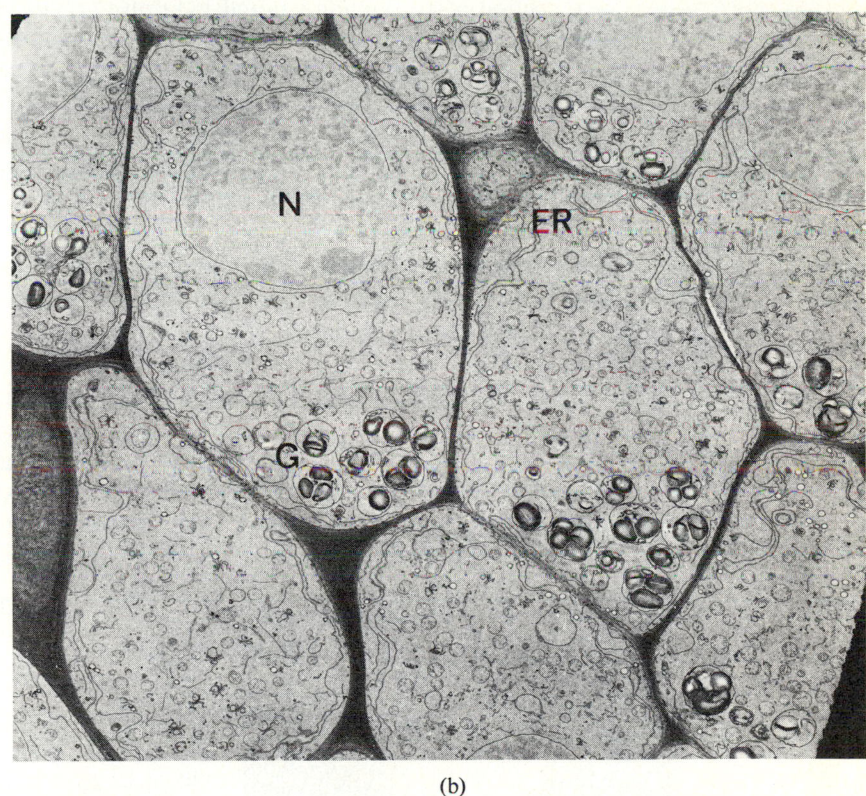

(b)

Figure 9-10a If we represent a root cap cell by a box filled with balls symbolizing starch grains, then it can be seen how they fall as the root is tipped from the vertical. The maximum number of balls touching the lateral wall occurs at 110° from the vertical–exactly the same angle as for maximum geotropic response.

Figure 9-10b In an actual root cap cell as seen in the electron microscope, we can see an effect similar to that in the box in (a). The starch grains (G) have fallen to the lower side of the cell, while the endoplasmic reticulum (ER) and nucleus (N) have been displaced towards the top of the cells. Such changes might be detected by the cell membrane and translated into hormonal signals which control the direction of growth of the root. (Courtesy of B. E. Juniper, Oxford University.)

The gravitational sensing (*graviperception*) mechanism appears to involve *statoliths*, especially dense starch grains that fall from one wall to another as the cell is tipped (Fig. 9-10). As a result of gravity or centrifugal force, they come to lie asymmetrically against cell walls, signalling in some way that the orientation of the stem or root has been altered. In addition, some other organelles in the cells, such as the endoplasmic reticulum, rise to the upper side of cells when the plant's axis is tipped to one side. Thus, other signals besides those from the shifting starch statoliths may be involved in graviperception. Whatever the sensing organelle, the stimulus seems to be transduced to a chemical message by proximity of the organelle to the peripheral plasmalemma.

The detection of both unidirectional light and gravity occurs at the tip of the root and the stem. In a root, the cap (Fig. 9-11), which is rich in starch grains, is the sensitive region, whereas in the stem, the extreme tip, including the youngest leaves, perceives the stimulus. The zone that responds to the tropistic stimulus is, however, some distance (from a few millimeters to several centimeters) behind the detector cells at the tip. Thus the auxin has to migrate from the tip to the growing zone below it. It does so by moving in the parenchyma cells of a coleoptile or in the pith parenchyma and youngest undifferentiated vascular cells of a stem at a rate approximating 1 cm/hr at normal temperatures. This transport is unusual, however, in that it occurs only in one net direction, that is, away from the tip (Fig. 9-12). The application of a block of auxin-containing agar to the morphological apex of a stem, therefore, results in the rapid passage of auxin into the tissue, from which it may be collected by extraction or by diffusion into a basal block. However, if a similar block is applied to the base of the same piece of tissue, auxin will not move through the base into the apex and out the end into a receptor block. Unidirectional transport of auxin in stems and coleoptiles is referred to as *polar* transport.

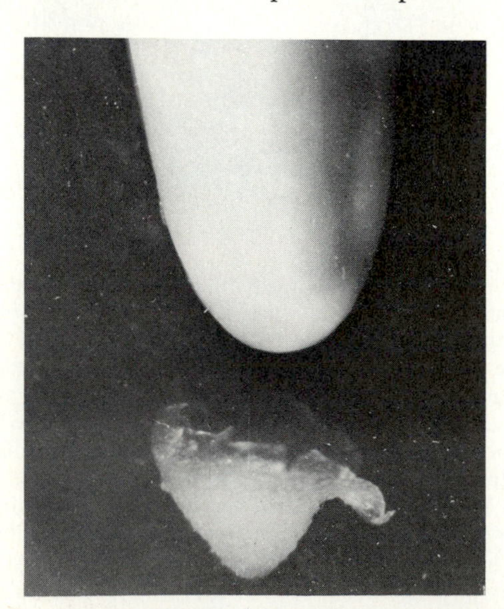

Figure 9-11 The root cap is a protective group of continually-replaced cells fitting snugly over the tip of the root. Here we can see the root cap after it has been pulled off the end of the root. The root cap helps the root to penetrate the soil. Another function of the root cap is the detection of gravity. The starch statoliths and the production of the root growth inhibitor involved in geotropism both occur in the root cap. (Courtesy of Dr. B. E. Juniper, Oxford University.)

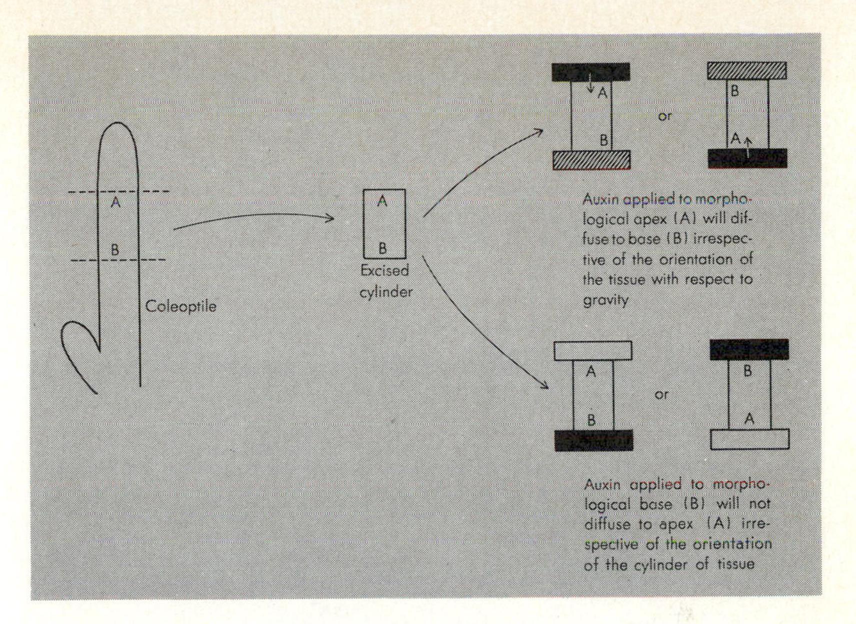

Figure 9-12 The flow of auxin through coleoptile tissue is unidirectional.

The physiological basis of polar auxin transport is not completely understood, but the mechanism probably involves some of the following steps (Fig. 9-13): Auxin diffuses from the base of one cell into the top of the underlying cell in the form of the nonionized indoleacetic acid molecule. Since the cytoplasm is nearly neutral, while the cell wall is slightly acid, the indoleacetic acid outside the cell is less ionized than that inside the cell, and thus a diffusion gradient exists for nonionized IAA to enter the cell. The ionized IAA in the cell is insoluble in lipid and therefore cannot diffuse out, while nonionized IAA continues to flow in. Thus, IAA accumulates in the cell. It has been suggested that ionized IAA can leave the cell by means of specific carrier proteins located in the cell membrane at the lower end of the cell; these are able to pick up the ionized IAA inside the cell and transport it to the outside. As there is a high concentration of ionized IAA in the cell and a lower one outside, the carrier proteins essentially act as turnstiles through which ionized IAA can flow out, to be picked up by the next cell. Since the hypothetical auxin carriers are located only at the bottom of the cell, auxin can flow only from apex to base. Thus, the basis of polarity is assumed to be structural, at the molecular level. Polar transport also needs active energy metabolism, if for no other reason than to maintain a higher pH in the cytoplasm than outside. This can occur through active pumping of hydrogen ions out of the cells, using ATP as a source of energy.

We must now knit together our discussion of tropisms, although there are still many gaps in our understanding. We do not know how the receptors of unilateral light or gravitational stimuli are able to affect the lateral distribution of auxin, although the unequal distribution of auxin on the two sides of a stimulated stem or root appears to result mainly from transverse migra-

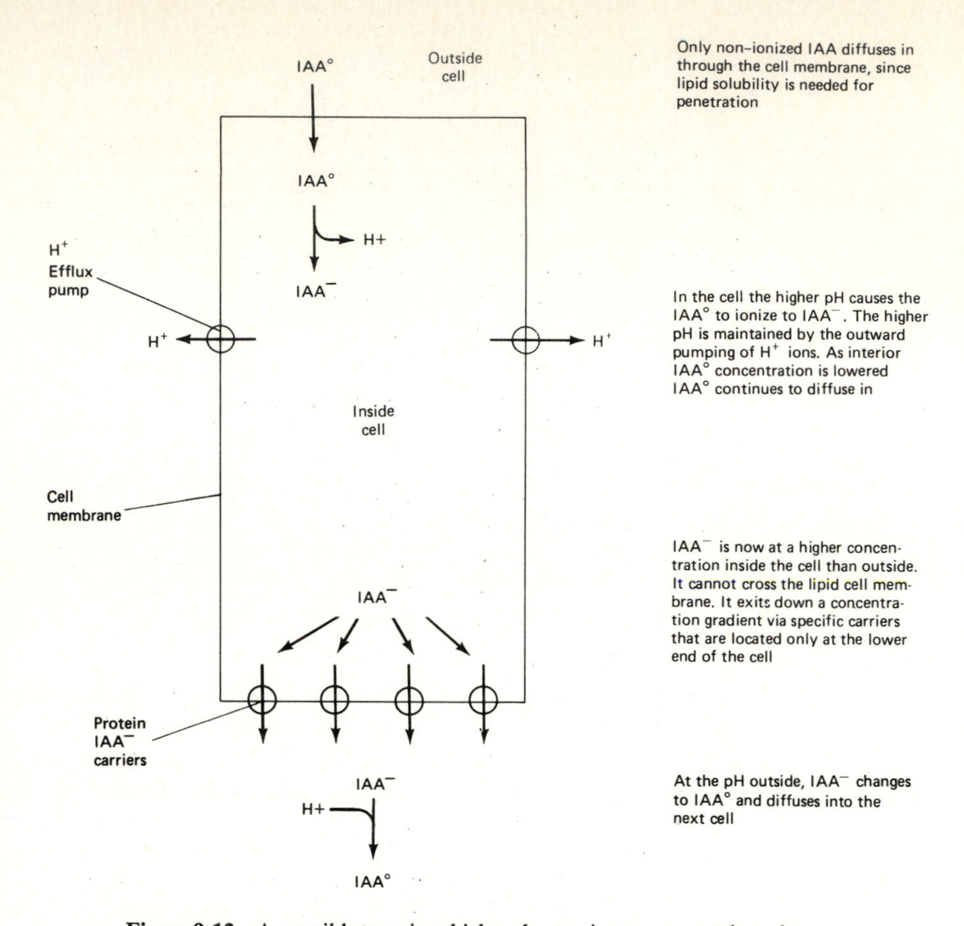

Figure 9-13 A possible way in which polar auxin transport takes place.

tion. For example, symmetrically applied ^{14}C-labeled IAA accumulates in the shaded longitudinal half of a phototropically stimulated organ and in the bottom half of a geotropically stimulated organ. One possibility for explaining this effect is that basipetal transport (transport to the morphological base) is inhibited on the stimulated side, so that there is continuous diffusion from the light to the shaded side during stem phototropism and from the upward to the downward side during stem geotropism.

The root presents us with a somewhat different situation. The root cap is the gravity-sensing region and, by pulling the root cap off, one can show that it is also the source of a potent growth inhibitor that can become asymmetrically distributed, leading to growth curvatures. Experiments on segments of root from behind the growing zone show that the movement of IAA in such segments is *toward* the root tip, not away from it; in other words, there is unipolar transport of IAA, extending all the way from stem tip to root tip. Most of the IAA in roots in fact derives from the stem tip, and is transported down the stem all the way into the root tip. It also accumulates at the points where new branch roots are developing, there promoting cell division. Roots do produce very small amounts of auxin, most probably in the root cap, and this auxin can migrate 2–3 millimeters back to the growing zone of the root, though little further. If the root is on its side, the auxin from the tip is redistributed toward the lower side, there possibly inhibiting growth and resulting in downward curvature. Another component of the root cap inhibitor is probably abscisic acid (see Chapter 10). Recent investigations indicate that it is more important than auxin in root tropisms.

Transportability of synthetic auxins used to control plant growth and development is related to their activity. In general, the synthetic compounds active in promoting growth are also polarly transported, which means that some part of the auxin structure that is responsible for its action is also responsible for its binding to the specific sites on the transport protein. If we measure the auxin-controlled growth of a particular stem we always find that the growth rate correlates not with the *total* auxin content, but with the *diffusible* auxin that is transported out of the stem if it is cut and placed with its lower cut surface on a block of agar. This shows us that not all the auxin in the cell is acting to stimulate growth. Probably, growth is controlled by the auxin in a particular part of the cell, such as in the cytoplasm. It is also this auxin which is available for transport. However, there may be considerable auxin in other places in the cell, such as the vacuole, where it is available for neither transport nor growth. Yet this immobile auxin is included in the measurement if we extract the tissue with solvents, giving an incorrect picture of the actual auxin status of the cells being studied. This phenomenon of *compartmentation* is important throughout physiology—the location of a compound or enzyme is frequently more important than the total amount present. The same almost certainly applies to all the other hormones.

The amount of active auxin present in any part of the plant will depend on several factors: how much is synthesized in the shoot tip; how much is transported; how it is compartmentalized in the cells; and finally how much is broken down or metabolized in other ways. Auxin is known to be broken down by the enzyme *IAA oxidase* to produce an inactive product, 3-methylene oxindole (Fig. 9-3). Some tissues in the plant, especially in the root, are extremely active in degrading auxin in this way. There are also various conjugation reactions of auxin with other molecules, such as aspartic acid and inositol. These products generally arise only when extraordinarily high levels of IAA are fed to tissue, thus they are assumed to be storage or detoxification products.

In cereal grains, where virtually no free IAA can be found, considerable amounts of IAA appear when the tissue is treated with dilute alkali. This IAA occurs in the form of glycosides, that is, combinations of IAA with sugar derivatives. On germination, these substances are transported to the coleoptile tip, where IAA is released and made available for basipetal polar transport, as previously described.

Other effects of auxin

In addition to exerting control over cell elongation, auxin can also initiate or promote cell division. For example, when normal cells are excised from stems or roots and grown in tissue culture in chemically defined media, cell division is dependent on auxin, either formed by the cells or present in the medium. Similarly, the initiation of cambial activity in trees in the spring is controlled in part by auxin diffusing downward from developing buds. In roots and stems, the initiation of adventitious or branch roots from the pericycle region is under the partial control of auxin. In these mitosis-inducing activities, auxin works together with another group of plant hormones, the *cytokinins*, to be discussed later.

Auxin not only controls the initiation of cambial activity but may also

Figure 9-14 If a shoot tip is implanted into the top of an undifferentiated block of cultured callus tissue (a) nodules of vascular tissue containing xylem develop. This effect can be shown to be related to auxin produced by the shoot since, if a block of agar containing auxin is used instead of a shoot, (b) then a similar development of vascular tissue results. In a stem apex, xylem develops below each leaf primordium. These primordia are the main sites of auxin production.

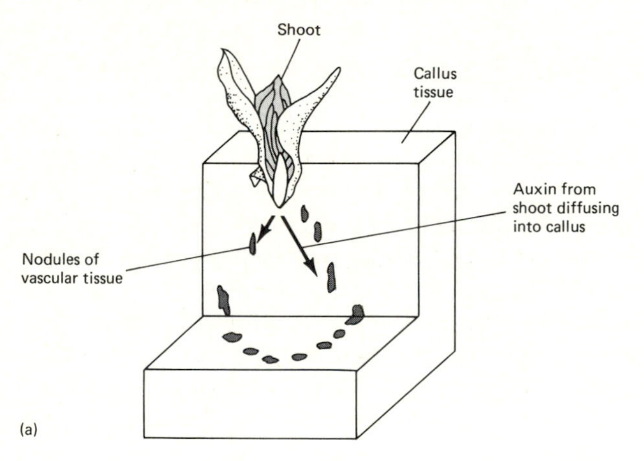

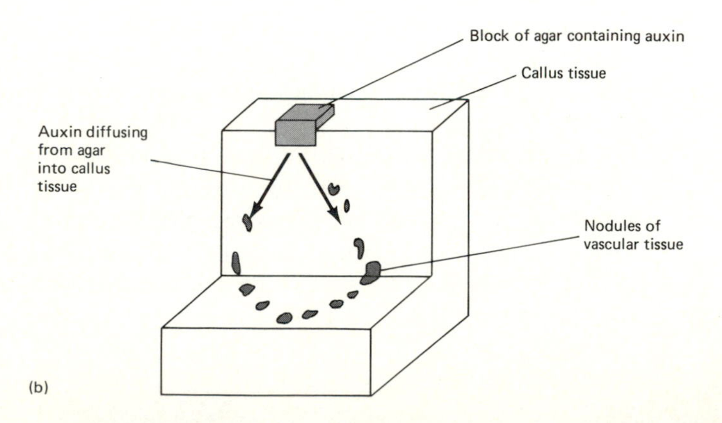

determine the nature of the cells differentiating from the cambial products. The presence of auxin on the xylem side of the cambium, mainly in the young differentiating xylem elements in which it is transported from the stem tip, promotes the development of cambial derivatives on that side of the cambium into xylem cells (Fig. 9-14). On the outer side of the cambium, high sugar concentrations and gibberellins in the mature phloem cause the outer cambial derivatives to develop into phloem cells. As we shall see later, gibberellins are produced mainly by the young expanded leaves and would therefore tend to be found in the mature phloem together with photosynthetically produced sugar.

If a shoot is excised, polar transport of the auxin produced by the tip causes its accumulation at the base of the stem. Similarly, any auxin applied to the shoot also accumulates at the base. There, after some time, the accumulated auxin will lead to the production of a swelling or *callus* containing many parenchymatous cells produced by division of cambial cells at the base of the stem. Such callus tissue is usually undifferentiated, but may contain vascular elements in random orientation. Frequently, adventitious roots develop in profusion following activation of cambial cells by auxin. This effect of auxin is widely used in horticulture for the propagation of desirable plants by the rooting of cuttings (see Chapter 14).

The prodigious growth of the ovary into a fruit is another auxin controlled phenomenon. Normally, following pollination and fertilization the ovary wall which ultimately forms, the fruit is stimulated to tremendous growth through an increase in both cell number and cell size. That auxin plays a role in this process is indicated by the dramatic increase in auxin levels of ovaries and fruit during such development, and by the fact that the application of exogenous auxins at the appropriate time can enhance the process, even substituting in some species for normal pollination. If no pollen reaches the pistil, the development of the ovary can be stimulated by the application of fairly large quantities of synthetic auxins to produce "artificial," or *parthenocarpic*, fruits (Fig. 9-15). Beta-naphthoxyacetic acid, for example, can be sprayed or daubed onto a tomato ovary to produce a fairly typical fruit that is large, red, and tasty, but lacking in viable seeds. Since the pollen itself contains little auxin, its function must be in part to activate the formation of auxin, possibly by carrying into the ovary some stimulator of the enzyme that makes auxin from its precursors.

So far, we have considered the promotive action of auxin in cell division, elongation, and differentiation. But auxin can also exercise an inhibitory or restraining action on the growth of plant parts, as it does in influencing the competition between the various buds of a shoot. In plants showing extreme *apical dominance*, only the terminal bud will grow, the lower buds being repressed. If the terminal bud is excised, one or several of the buds below it will start to grow, one of these usually becoming dominant. But if following

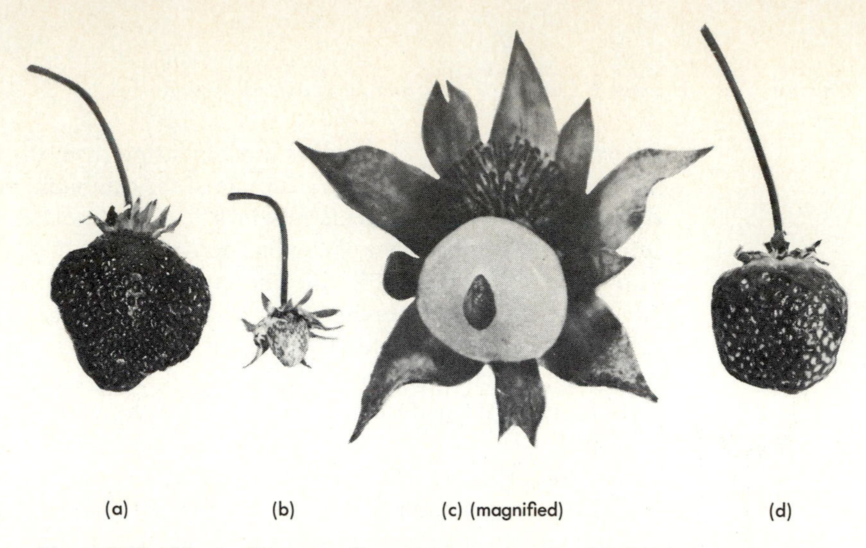

Figure 9-15 The development of strawberry receptacles ("fruit") is dependent upon auxin produced by the achenes seeds. (a) Normal fruit. (b) Fruit of same age with all achenes removed. (c) Fruit with all achenes removed except one. The receptacle develops only under this achene. (d) Fruit with all the achenes removed and replaced by lanolin paste containing 100 ppm β-naphthoxyacetic acid. (From Nitsch. 1950. *Amer. J. Bot. 37*: 211–215.)

excision of the terminal bud some auxin is applied to the cut surface, the lateral buds do not develop, implying that auxin coming from the apical bud inhibits the development of the lateral buds. Recent evidence indicates that the "decision" on inhibition versus growth of lateral buds involves a competitive action of at least two growth hormones—auxin, coming from the apex, and cytokinins, probably coming from the root system. Thus, even in a system inhibited by terminally produced auxin, the localized application of cytokinins to a repressed bud may cause it to grow selectively. Once the lateral bud has overcome the inhibition, applications of auxin are no longer inhibitory and may, in fact, enhance lateral bud growth.

How auxin works

As we have seen, auxin can produce a multitude of effects on plants, controlling such processes as cell division, cell elongation, and the direction of development of cells during cell differentiation. It can produce many physical, chemical, and physiological changes within cells, some of them extremely rapidly; for example, the rate of cytoplasmic streaming in excised oat coleoptile sections increases within a few minutes after auxin application. Although this rapid response may be an early manifestation of auxin action, it is difficult to relate it directly to cell elongation, which must at some point involve basic changes in the cell wall. It is unlikely that any single "master reaction" involving auxin can account for all of its effects, so plant physiologists have investigated several different processes. We now have a reasonable idea as to how auxin controls cell elongation, and are beginning to find some clues regarding its control of other aspects of development.

Unlike animal cells, plant cell protoplasts are enclosed in a semirigid "wooden" box, the cell wall. No matter what changes occur inside the box, the cell cannot enlarge unless the walls can be extended. An analogous situation in animals is the growth of an arthropod such as a lobster, whose entire body is encased in a rigid exoskeleton made largely of chitin. The animal can grow only when the old rigid exoskeleton is shed (*molted*), and the new layer underneath is temporarily soft and stretchable. Plant cells cannot cast off their cell walls as lobsters cast off their exoskeletons; the only way they can increase in volume is through modification of the existing cell wall to make it more extensible. This is the process in which auxin has been shown to exert considerable influence.

In plant cells, growth is governed by two factors: the extensibility of the cell wall and the turgor pressure of the cell contents pushing on the cell wall. We can represent this situation with a partly inflated balloon, in which an increase in volume is proportional to the internal pressure, and inversely proportional to the resistance offered by the balloon wall. The balloon will increase in size if we increase the internal pressure by pumping more air into it; this is analogous to increasing the turgor pressure inside the cell. Alternatively, if we could make the wall more stretchable by applying a rubber-softening chemical, the balloon would also increase in size, without any increase in the original internal pressure. An hour or two after auxin is applied to a stem or coleoptile, its cells have increased in size much more than those of controls without auxin. The expansion is mainly in length rather than in width, both because the spiral arrangement of the cellulose microfibrils in the cell wall allows elongation in preference to radial growth of the cell, and because auxin increases the extensibility of the cell wall. We can measure this by boiling a stem or coleoptile segment to kill it (thus destroying the turgor pressure in the cell), then clamping the two ends of the segment and measuring the force needed to stretch it. This force will be inversely related to the extensibility of the wall. Wall extensibility can be resolved into two components, *plastic* (irreversible deformation) and *elastic* (reversible deformation). If the plant segments are treated with auxin prior to being killed and measured, less force is required to stretch them irreversibly a given amount than is needed for controls without auxin. This auxin-induced increase in the plastic extensibility of the cell wall can occur only with living cells; auxin has no effect if it is applied directly to the cell walls of the dead stem or coleoptile segments.

How can we explain the effect of auxin on cell wall loosening? For example, could it be due to an increase in the amount of the cell wall digesting enzyme, *cellulase*, in the cell? This enzyme does increase dramatically in activity several hours after auxin application, and might digest some of the cellulose microfibrils, making the wall easier to stretch. Although such an idea is feasible, the activity of cellulase apparently starts to increase only *after* auxin has already started to influence growth. Clearly, any acceptable theory of auxin action must involve a process that starts *before* the auxin-stimulated increase in growth rate.

To measure accurately the earliest time after auxin application that an increase in growth rate can be detected, scientists have used the apparatus shown in Fig. 9-16. With this equipment it is possible to measure the time for any changes in section length to within a few seconds' accuracy. There is no increase in the growth rate for about 10 to 15 minutes after the addition of auxin; thereafter growth increases rapidly, achieving a maximum rate within the next 15 minutes. Thus a period of 10 to 15 minutes after auxin application is required to produce the changes in metabolism that ultimately increase growth, and these changes must be maximized in about 30 minutes. Therefore any changes that cannot be detected within an hour or more cannot be part of the *initial* action of auxin in inducing growth. However, they could be part

Figure 9-16 An apparatus for recording rapid changes in growth, with a typical growth tracing resulting from the addition of IAA to a buffer at about pH 6.5.

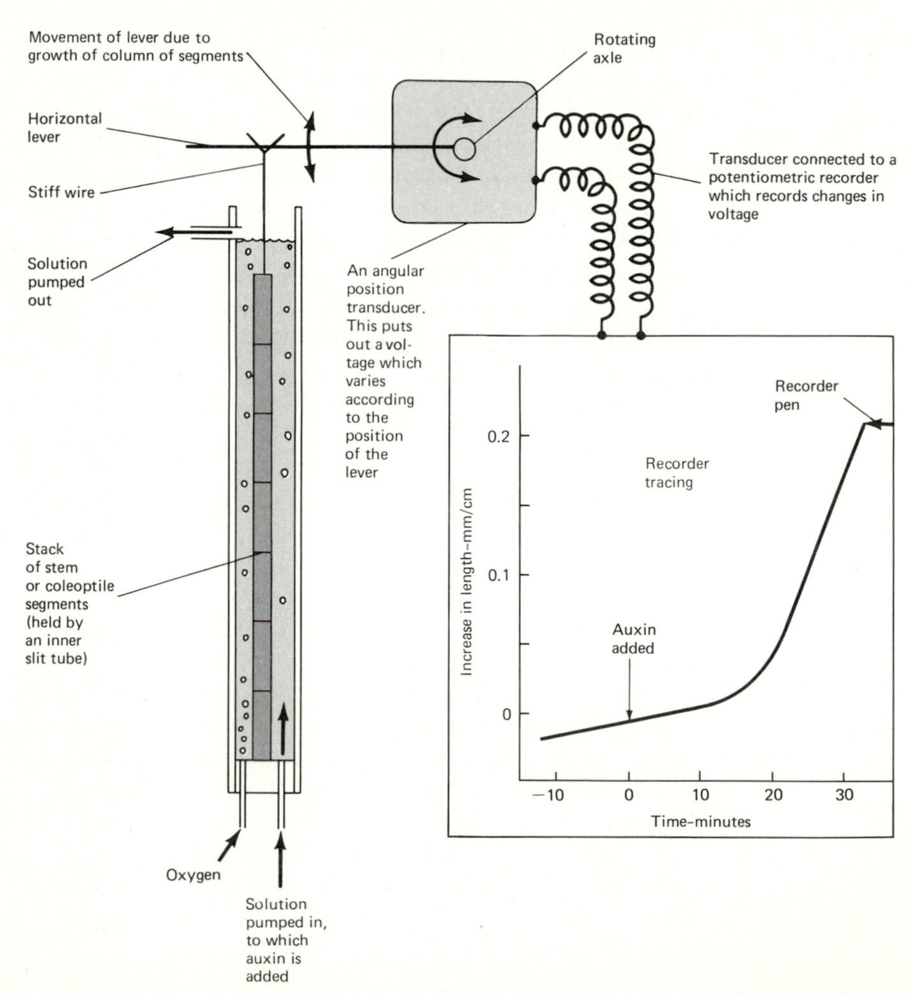

of the mechanisms that need to be switched on later for sustained auxin-induced growth.

What change occurs in so short a period that it can be related to auxin-induced cell extension? The answer probably lies in a phenomenon first noted in the 1930s but largely overlooked, then rediscovered and described in the early 1970s: Acid conditions (that is, low pH) were found to favor cell extension. If stem or coleoptile segments are placed in the apparatus shown in Fig. 9-16, and then surrounded by an acid solution, their growth is increased. If one peels the epidermis from the segments to facilitate penetration of the acid, a pH of about 5 produces optimal increase in growth and extensibility of the cell walls. If killed segments are stretched mechanically, the force required for a particular extension is less if the segments are surrounded by a solution at pH 3.5 rather than pH 7. Thus, the acid effect on extension does not require the presence of metabolically active cells but is a direct consequence of hydrogen ions acting on the components of the wall. The relation between the acid effect and auxin was suggested by the similarity of their effects on growth, except that (a) the acid effect is on the walls whereas auxin acts on the cells, and (b) the acid effect lasts for only a short time whereas the auxin effect lasts for at least several hours. The two can be connected simply: If auxin causes the cell walls to become acid, the acid could then cause wall loosening. This mechanism explains only the initial growth promotion, since the wall soon runs out of material for continued stretching; for a continued growth effect, a second effect of auxin is needed, one involving synthesis of new wall. It is this latter process that is lacking when wall extension is induced by acid alone.

Metabolic pumps for numerous substances such as mineral ions, sucrose, hormones, and hydrogen ions are known to exist in cell membranes and in the membranes of the various cell organelles (Chapter 7). According to one currently popular theory, auxin stimulates the action of a pump in the plasmalemma that moves hydrogen ions (H^+) out of the cytoplasm into the wall. Such pumps are proteins that derive the energy for their work by breaking down ATP into ADP and inorganic phosphate; the resynthesis of ATP depends of course on metabolism. The theory supposes that the H^+ pump functions only when IAA, acting as an *effector* (activator), is bound to it (Fig. 9-17). Such a binding is presumably reversible, and IAA is free to bind or leave depending on its concentration in the surrounding cytoplasm. This model proved to be essentially correct when peeled coleoptile segments in a solution were found to acidify the surrounding solution only if IAA was added to the medium. Thus, while some difficulties remain to be resolved and some details filled in, the theory of an auxin-activated proton (H^+) pump fits most of the data, and is a reasonable explanation for auxin action on cell elongation.

But we are not yet finished. *Why* does acidification promote wall elongation? To answer this, we must return to the structure of the plant cell wall

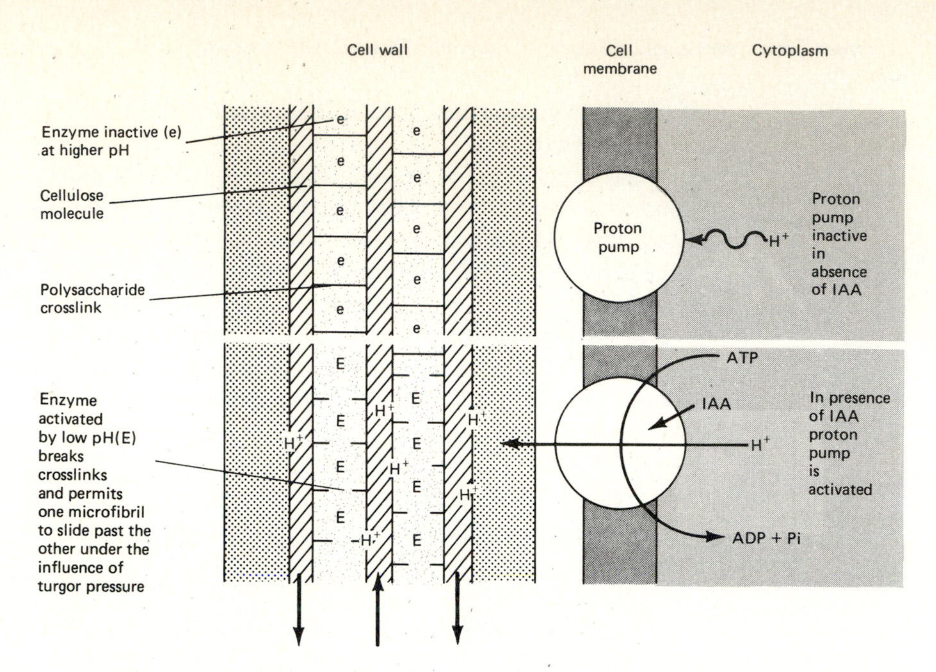

Figure 9-17 The proton pump theory to explain the action of IAA in inducing wall extensibility and cell elongation.

(Fig. 2-26). The wall consists of long molecules of cellulose joined to other cellulose chains by polysaccharide bridges composed of a mixture of sugars such as galactose, arabinose, and xylose. These polysaccharide bridges form part of the wall that used to be called *hemicellulose*. Because the long cellulose molecules are bound together by these crosslinks, the cell wall is rigid. When the wall stretches irreversibly, as during growth, the cellulose chains have to slide past one another, and the only way this can be accomplished is by breaking the crosslinks attaching one cellulose chain to the next. When the process of wall stretching is completed the crosslinks reform, fixing the cellulose chains in their new positions (Fig. 9-18). The cell has now grown. For the acid in the wall to increase the stretchability of the wall, it must obviously break the crosslinks. It has been hypothesized that there exists in the wall a pH-dependent enzyme that breaks the crosslinks. This enzyme has no effect at pH 6–7 but is fully active around pH 5. When H^+ ions are pumped into the cell wall under the influence of the auxin inside the cell, the enzyme is activated, breaking the crosslinks; the cellulose chains are then free to slide over each other. With the turgor pressure of the cell contents pulling the cellulose chains past each other, wall extension and cell growth are accomplished. We do not yet know the nature of this hypothetical enzyme, whose existence is inferred because certain chemicals which ordinarily prevent enzyme action are able to inhibit acid-induced extension of isolated cell walls.

Acid-induced growth proceeds for only a short period in the absence of auxin, while it continues for many hours in its presence. What is the long lived process activated by auxin? If one adds an inhibitor of protein synthesis (such as cycloheximide) to auxin-stimulated stem segments, they cease growth about 15 minutes after the inhibitor stops protein synthesis. When the inhibitor is added long before the auxin is added, the auxin still stimulates growth, but only for about the same 15 minutes. We can therefore conclude that while the initial growth stimulation caused by the outward pumping of H^+ ions under the influence of auxin is independent of protein synthesis, continued auxin-induced growth requires protein synthesis. One of the required proteins appears to be the pumping protein itself, since in the presence of cycloheximide, auxin-induced acidification of the exterior medium ceases. This pump protein must therefore be rather short-lived in the presence

Figure 9-18 How wall elongation probably occurs under the influence of auxin.

1
Cellulose molecules
Polysaccharide crosslinks
Original wall structure

2
Crosslinks broken –wall loosened

3
Stretching force
Under influence of turgor pressure cellulose molecules slide past each other. Wall stretches

4
Crosslinks reform to fix wall in its newly stretched position

5
More wall material synthesized. Wall now elongated but still its original thickness. Process is then repeated.

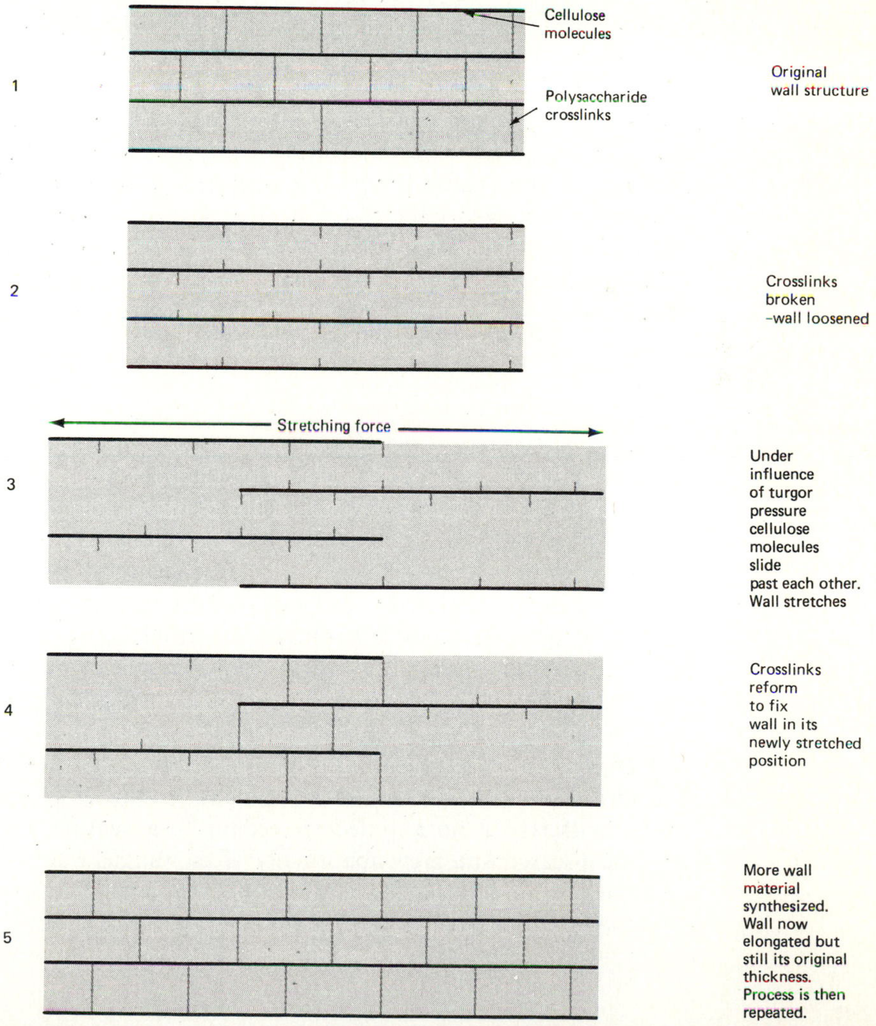

of auxin, and must require continual synthesis. Other processes are required as well for continued growth. The polysaccharide crosslinks must be reformed if the elongated state of the cell wall is to be fixed and made rigid. Then, more cellulose and crosslinking polysaccharide have to be synthesized to prevent the wall from becoming continually thinner, as it is known to remain at approximately the same thickness during cell growth. It is thus interesting that auxin has been found to enhance the activity of cellulose-synthesizing enzymes located on the cell membrane and in the dictyosome vesicles known to be involved in wall synthesis.

For continued cell division, elongation, and differentiation under the influence of auxin, continued synthesis of new proteins is necessary. During cell growth, proteins are needed to construct the additional cytoplasmic components of the cell, and during differentiation, new enzymes are needed to function in the new processes characteristic of the differentiated state. Auxin has been shown to trigger an increase in the rate of protein synthesis by increasing the synthesis of messenger RNA in the nucleus. Together with other hormones and regulators, it probably also alters the type of messenger RNA being produced, and thus changes the types of enzymes in the cell, so that more enzymes specific for cell growth and differentiation are synthesized. In addition, more ribosomal RNA is also made to accommodate the increased requirements for protein synthesis. We do not yet know which enzymes synthesized under the influence of auxin induce continued growth and differentiation. Such problems are difficult to solve, but eventually we will understand more about the details of growth and differentiation and will gain better ideas on where to look for the answers.

The Gibberellins

Another group of important plant growth hormones, discovered through a series of chance occurrences and shrewd observations, is the *gibberellins*. In the last decade of the nineteenth century, Japanese rice farmers noticed the appearance of extraordinarily elongated seedlings in their paddies. They watched these seedlings closely, for the alert farmer usually considers any large-sized plant as a possible source of breeding stock for the improvement of general vigor. These tall seedlings, however, never lived to maturity and only rarely flowered. The disease was aptly named *Bakanae* or "foolish seedling" disease. In 1926 a Japanese botanist, Kurosawa, discovered that these seedlings were infected with a fungus later named *Gibberella fujikuroi*, a member of the *Ascomycete*, or sac fungus, group. If spores of the fungus were transferred from an infected seedling to a healthy plant, the recipient became diseased and grew abnormally. If the fungus was grown on an artificial medium in a flask, the nutrient medium accumulated some substance that, when transferred to a receptor plant, produced the overgrowth symptoms typical of the "foolish seedling" disease. This substance was named *gibberellin*, after the fungus that produces it.

In the 1930s, Japanese physiologists and chemists succeeded in isolating from the *Gibberella* growth medium several substances, some inhibitory and some growth promoting. The structural formula they finally proposed for the growth promoter, gibberellin, was somewhat in error, but nonetheless they had correctly determined the general nature of the substance and had produced crystals that, when applied to test plants, produced the typical hyperelongation symptoms of the Bakanae disease. This information was published in Japanese in many papers before 1939, but unfortunately, World War II intervened and diverted the attention of most scientists to military matters. The exciting story of gibberellin remained unknown in the Western world until about 1950, when several groups of workers, both in England and in the United States, uncovered these old papers and attacked the problem again.

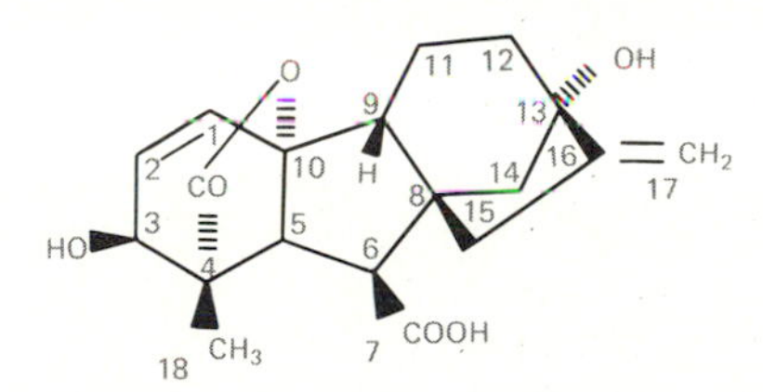

Figure 9-19 The structural formula of gibberellic acid or GA_3. There are over 50 different gibberellins differing mainly in the number and position of additional groups such as —OH. In order to describe these differences, each carbon in the main skeleton has been assigned a number and then each group is given the number of the carbon to which it is attached. (◀ lines indicate that the bonds are directed above the plane of the paper, ▪▪▪▪▪ behind that plane, and —— in the plane of the paper.)

By 1955, British workers had confirmed the original Japanese observation and had also isolated a substance they named *gibberellic acid* (Fig. 9-19), which differed slightly from the material isolated by the Japanese. Soon many other molecules with a similar basic structure were discovered both in fungi and in the uninfected tissues of higher plants. It is clear that the gibberellins are a family of compounds, now numbering over 50, which all have the same basic gibbane ring, each with various modifications and chemical groups (for example—OH) in different parts of the molecule. Gibberellic acid, when applied to some plants, produces tremendous hyperelongation of stems, and in some instances also results in a diminution of leaf area. Its most dramatic effect is probably its prompt stimulation of flower stalk elongation (*bolting*) and in many cases of flowering in long-day plants (discussed in Chapters 11 and 12) (Fig. 9-20). Gibberellin promotes bolting by increasing both the number of cell divisions in particular locations and the elongation of cells

Figure 9-20 One of the main effects of gibberellin is to induce stem elongation, particularly that of the flower stalk in plants that normally need long days to induce elongation and flowering. This bolting is sometimes, but not always, followed by flowering. The cabbage plants on the right were treated with gibberellic acid, those on the left were untreated. (Courtesy of S. H. Wittwer and M. J. Bukovac, Michigan State University.)

produced by such divisions (Fig. 9-21). Where only small amounts of gibberellins are applied, bolting occurs but floral primordia may not be differentiated; higher quantities of gibberellin generally produce not only bolting, but flowering as well. Some data are available to support the contention that in long-day plants the administration of a long-day inductive treatment stimulates gibberellin production, which then causes the morphogenetic responses. In other instances gibberellin activity in apices of plants declines as bolting proceeds, indicating that gibberellin may be used up during bolting. Gibberellin is ineffective in causing flowering in short-day plants, and in fact appears

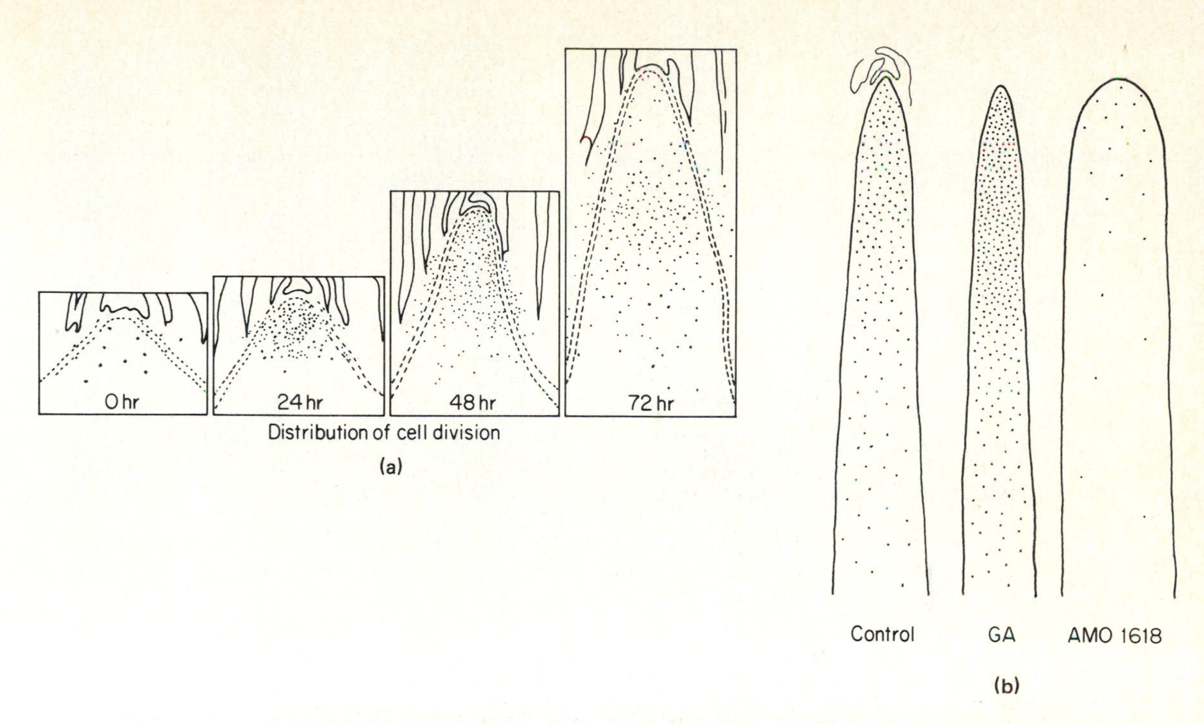

Figure 9-21 Gibberellin applications to rosette plants produce increases in both cell division and cell elongation. (a). This can readily be seen in longitudinal sections through the axis of *Samolus parviflorus* in which division increases following application of GA (Each dot represents one mitotic figure in a 64-micron-thick slice.) (b). By the use of AMO 1618, an antigibberellin growth retarder, the meristematic activity in stem tips of *Chrysanthemum morifolium* can be seen to be directly due to GA. The heavy lines show the vascular boundaries within which each dot represents one mitotic figure in a 60-micron-thick slice. (From Sachs. 1965. *Ann. Rev. Plant Physiol. 16*: 73–96, and Sachs et al. 1959. *Amer. J. Bot. 46*: 376–384.)

to have the opposite effect. What hormonal influence controls flowering in short-day plants, or in those long-day plants which elongate but fail to flower in response to gibberellic acid, is uncertain. In the latter case it is possible that some other gibberellin, not gibberellic acid, is involved in the induction of flowering.

Not all plants respond to gibberellin, however. In surveying the results of numerous tests, plant physiologists have noted a correlation between the original size of a plant and the degree to which it responds. For example, if we compare dwarf peas with tall peas, or dwarf corn with tall corn, we note that the application of gibberellin to the dwarf causes it to assume the tall appearance, but application of gibberellin to the tall form produces little or no effect (Fig. 9-22). Since in many instances the difference between dwarfness and tallness resides in a single gene, the attractive hypothesis is suggested that dwarfness is actually caused by the inability of the plant to produce enough gibberellin to satisfy its major needs. The application of gibberellin to certain genetic dwarfs would thus produce the tall form. Such dwarf plants

Figure 9-22 (A) Effect of gibberellic acid on normal and dwarf corn. Left to right, normal control, normal plus gibberellin, dwarf control, dwarf plus gibberellin. (From B. O. Phinney and C. A. West, "Gibberellins and the Growth of Flowering Plants," in *Developing Cell Systems and Their Control*, D. Rudnick. The Ronald Press, 1960.) (B) Dwarf pea treated with water (left), and gibberellic acid (right). (Courtesy A. Lang, Michigan State University.)

made to look tall by treatment with gibberellin would still, of course, have the dwarf genotype and when bred would yield dwarf progeny. Organisms whose phenotypes are altered by chemical or physical treatments to make them resemble a different genotype are called *phenocopies*.

Some tall plants have indeed been found to contain more total gibberellin than dwarf plants, although this is not true in all cases. However, when separated chromatographically, the gibberellins of dwarf and tall plants frequently show some qualitative differences. In such instances, dwarfism may be due to the presence of gibberellins of a type less active in growth promotion. It is clear that not all cases of dwarfism are associated with abnormalities in gibberellins.

Gibberellin may also act, either alone or in combination with auxin, to induce the formation of parthenocarpic fruits. An example of this is the apple, whose parthenocarpic development had long been sought but never achieved by the use of auxin alone. Now, combination auxin–gibberellin sprays produce the desired results. A rise in the natural gibberellin content in apple seed is also correlated with the maximum period of seed growth, indicating possible gibberellin control of seed development as well as of the ovary wall. The latter aspect of gibberellin function has found an application in agriculture in the production of seedless grapes.

Not only do gibberellins probably play a role in seed development, but they are also involved in seed germination in two different ways. First, they are probably involved in breaking the dormancy of seeds, as can be easily

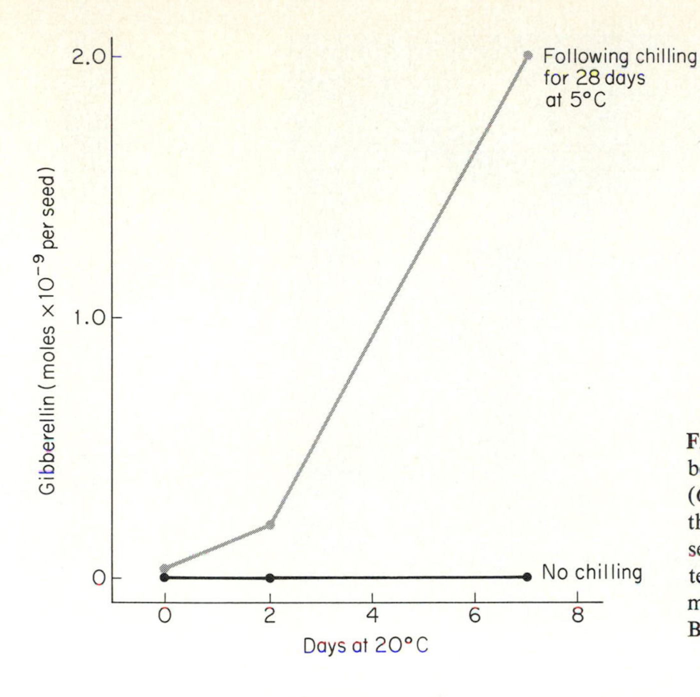

Figure 9-23 Some seeds require chilling before they will germinate. In hazel (*Corylus avellana*), pre-chilling triggers the production of gibberellin when the seeds are returned to the germination temperature, which then promotes germination. (From data of Ross and Bradbeer. 1968. *Nature 220*: 85–86.)

demonstrated by applying gibberellic acid to dormant seeds, which then germinate. Gibberellin will also substitute for light or chilling in some seeds which normally require these stimuli to break dormancy; in nature the dormancy is probably broken by a rise in the natural gibberellin content (Fig. 9-23). Secondly, in cereal grains it is gibberellin which controls the mobilization of the reserve food material in the endosperm. The seeds of many cereals, such as barley, contain reserve starch which is rapidly hydrolyzed at the onset of germination. If barley seeds containing embryos are soaked, starch hydrolysis begins rapidly. If the embryos are removed prior to soaking, no starch hydrolysis occurs in the embryonectomized seeds. If gibberellin is supplied to such embryoless seeds, starch hydrolysis proceeds (Fig. 9-24). It thus appears that the embryo, shortly after imbibition of the seeds, normally produces gibberellins that activate the starch hydrolysis process by a separate mechanism, to be examined more closely in the next section. The two actions of gibberellin in seed germination are quite distinct, in that the dormancy-breaking process occurs in the embryo before the mobilization of the food reserves. The latter process is started by the action of gibberellin on the *aleurone* layer which surrounds the endosperm (Fig. 9-25).

There appear to be two ways in which gibberellins control particular physiological processes. The first is simply the synthesis of gibberellin by the plant, and the consequent initiation of the gibberellin-dependent process. The second possibility is more subtle. Remember that there are about 50 gibberellins, differing in relative activity depending on the particular process being affected. As the gibberellins all have a similar molecular structure,

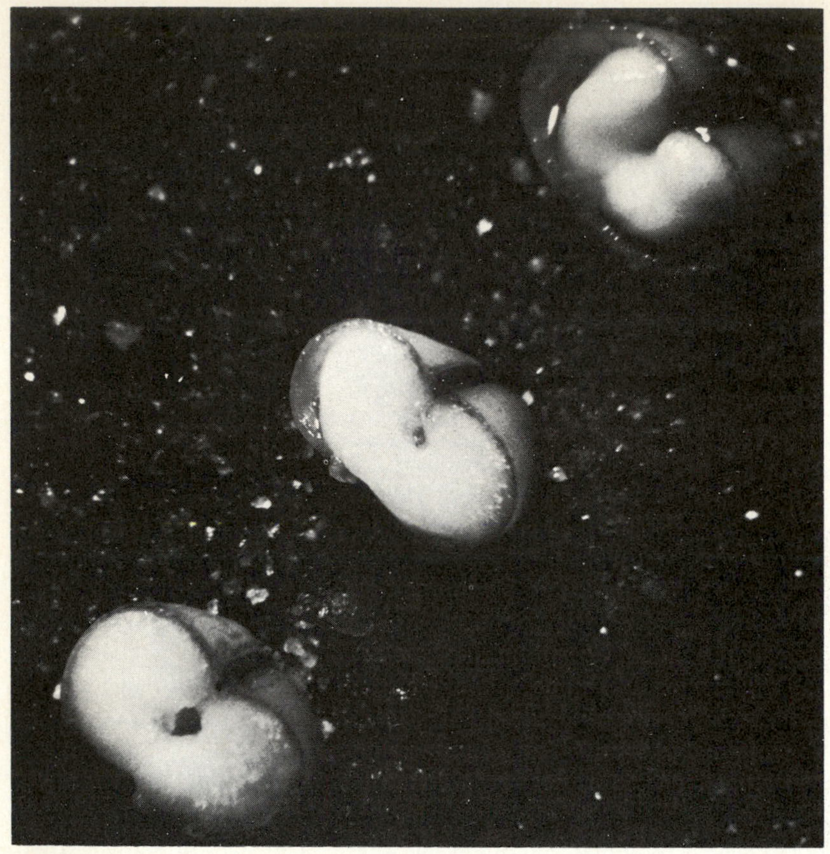

Figure 9-24 Three sterile halves of barley grain without embryo. To the open surface of each was added (left to right) either 0.5 microliter of water, gibberellic acid at a concentration of 1 part per billion, or gibberellic acid at a concentration of 100 parts per billion. The photograph, taken 48 hours later, shows that digestion of the starch-filled storage tissue is already taking place in the grains treated with gibberellin. The hormone gibberellin promotes production and secretion of the enzymes that cause hydrolysis of the storage material. (Courtesy of J. E. Varner, Washington University.)

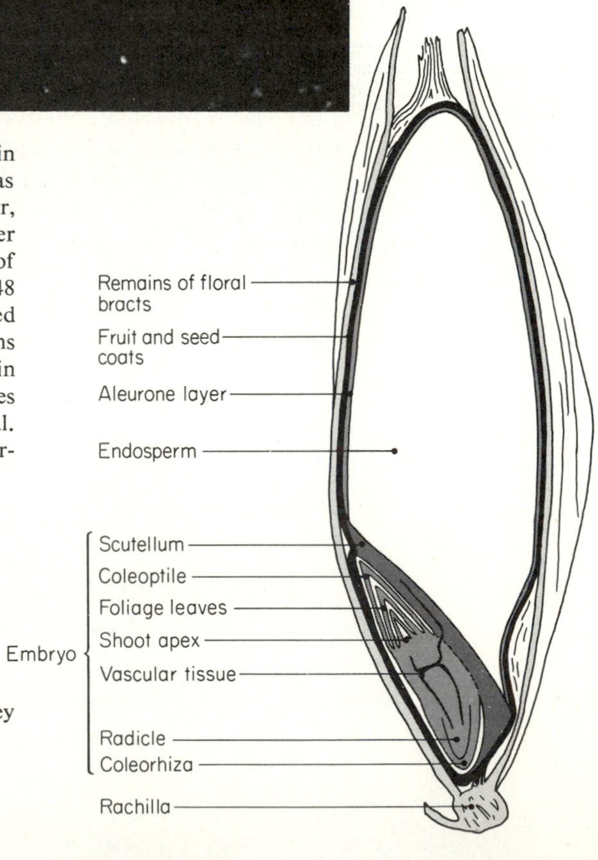

Remains of floral bracts
Fruit and seed coats
Aleurone layer
Endosperm
Embryo
Scutellum
Coleoptile
Foliage leaves
Shoot apex
Vascular tissue
Radicle
Coleorhiza
Rachilla

Figure 9-25 Longitudinal section of a barley grain.

any one gibberellin can be changed into another by only slight modifications of the molecular structure—for example, adding an $-OH$ group in one or more of several positions. Thus a plant could control an internal process by metabolizing an inactive gibberellin into an active one, or vice versa.

Gibberellins, like carotenoid pigments, steroids, and rubber, are isoprenoid compounds formed from acetyl co-A during respiratory metabolism (see Chapter 5). The pathway of gibberellin biosynthesis, which was elucidated by feeding ^{14}C-labeled compounds to plants, is:

$$\text{Acetate} \longrightarrow \text{mevalonate} \longrightarrow \text{numerous rearrangements} \longrightarrow \text{kaurene (a diterpene)} \longrightarrow \text{gibberellin}$$

The identification of diterpenes as intermediates in gibberellin biosynthesis is further confirmed by the fact that AMO-1618, a growth retardant that acts as a gibberellin antagonist, prevents the formation of the diterpene from its precursor.

How gibberellin works

With gibberellin, as with auxin, we have the problem of explaining how very minute quantities of such a substance can control numerous and varied morphogenetic responses, including seed germination, cell division, cell elongation, and floral initiation. Only one phenomenon has been analyzed in detail—the induction of starch hydrolysis in embryonectomized barley seeds.

We now know that the control of starch digestion by gibberellin involves the regulation of enzyme production and release. The application of gibberellin to embryonectomized barley seeds results in the appearance and secretion of *amylase* (Fig. 9-26), as well as other enzymes. This enzyme causes the hydrolysis of the starch (Latin name, *amylum*) contained in the endosperm of the barley grain. By removing the aleurone layer, one can show that enzyme formation occurs in that tissue. The aleurone, therefore, produces and secretes hydrolytic enzymes for digesting the food reserves in the endosperm, and it is these aleurone cells that are the "target cells" that respond to gibberellin. This system provides an example of organ-specific growth regulation, in that gibberellin, the key to the food reserves, it made in the embryo, which also contains the only tissues in the seed capable of growth.

How does gibberellin cause the appearance of α-amylase activity? First, it is clear that the enzyme is newly formed from its constituent amino acids and is not simply an activated form of a previously synthesized inactive stored protein. This was shown when the addition of radioactively labelled amino acids to the barley grains or aleurone layers incubated with gibberellin resulted in the incorporation of radioactivity into the protein. This incorporation was prevented by inhibitors of protein synthesis such as cycloheximide. The point in the protein synthesis process at which gibberellin acts is indicated by the fact that inhibitors of DNA-dependent RNA synthesis (for example,

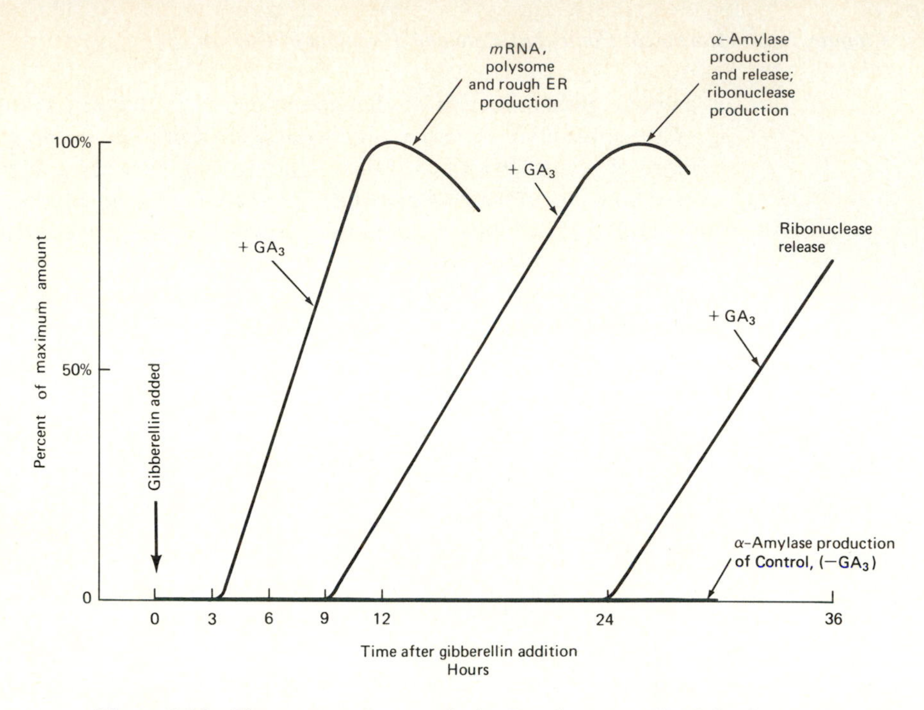

Figure 9-26 The course of events in barley aleurone cells following application of gibberellin.

actinomycin D) also prevent amylase synthesis. Thus, we infer that gibberellin must be involved in the production of messenger RNA molecules on the DNA template, acting as a derepressor of genes that code for the hydrolytic enzymes, and allowing for the production of these enzymes.

The difficulty in proving the production of this specific messenger RNA results from the very small amount produced, and the lack of techniques to distinguish it from other m-RNAs. The latter problem has recently been resolved by the discovery that m-RNA has a chain of adenine residues at one end of the molecule. Since adenine hydrogen-bonds to uridine, this property has permitted m-RNA to be separated by using a column of bound polyuridine with which the adenine will associate. The messenger RNA adheres to the column, while all other RNA passes through; later, by changing the solution in the column, one can wash off and assay the m-RNA. This method has shown that, if gibberellin is present, labeled nucleosides are incorporated into messenger RNA by the nuclei of aleurone cells, starting about four hours after the addition of the gibberellin. This is several hours before the appearance of the α-amylase. In addition, the appearance of α-amylase is inhibited by the early addition of the inhibitor *cordycepin*, which is considered to be specific for preventing the completion of m-RNA molecules. The cordycepin effect decreases the later it is added, so that it no longer inhibits if added about 12 hours after the gibberellin. The gibberellin-induced synthesis of α-amylase messenger RNA must therefore be complete by that time.

The specific nature of the newly synthesized m-RNA has been conclusively established by a combination of elegant techniques. When the isolated m-RNA was supplied to an *in vitro* protein-synthesizing system containing ribosomes, t-RNA, and the necessary enzymes and amino acids, the resultant protein was shown, by a combined immunochemical and electrophoresis technique, to be identical with true α-amylase!

At about the same time that the messenger RNA appears, there is a pronounced increase in the number of polysomes and rough endoplasmic reticulum in the aleurone cells. Such changes are often typical of cells manufacturing enzymes for secretion, and indeed, gibberellin seems to function in promoting the secretion, as well as the synthesis, of enzymes. Gibberellin has been shown to initiate the formation of other hydrolases, notably protease and ribonuclease in addition to α-amylase; thus, one hormone appears to initiate a series of events that soon renders all the food reserves of the seed available to the young plant. Gibberellin also promotes the secretion of all these enzymes from the aleurone cells into the endosperm. Alpha-amylase synthesis and release starts about nine hours after gibberellin is added (Fig. 9-26), while ribonuclease is synthesized at about the same time, but not released until more than 24 hours after gibberellin application. The enzymes break the stored food reserves down into soluble products which are then transported to the growing apices of the plant, to be used for energy and materials needed for the production of new cells.

If gibberellin can derepress certain genes in aleurone cells, it is not surprising that it can also affect cell division and differentiation in other parts of the plant by "turning on" certain genes. Which genes are turned on almost certainly depends on the nature of the cells. Not much work has been done with the initiation or control of cell elongation by gibberellin. In fully grown oat plants, gibberellin is responsible for the considerable elongation of the internodes of the stem prior to flowering. In the absence of auxin this has been shown to be due entirely to cell elongation, though in the natural situation, with some auxin present, cell division also takes place in the node. Initial results indicate that gibberellin induces the elongation of the cells through acidification of the cell walls in a manner similar to that described earlier for auxin. These gibberellin-sensitive cells do not, however, respond to auxin. The difference probably results from the presence of different hormone-receptors in these cell types.

Cytokinins

We have previously mentioned that cell division in cultured plant tissue is dependent on auxin and other cell-division factors termed *cytokinins*. The same controls probably operate to regulate cell division and growth in other parts of the growing plant, such as lateral buds.

The discovery of cytokinins stemmed from attempts to induce the division of particular plant cells in culture. One attempt, in the early 1940s, involved extracting the unfertilized egg from the ovule, and culturing it on a defined medium in an attempt to produce haploid embryos. This was never achieved, but it was found possible to promote growth of very young normal embryos through the use of coconut milk, the liquid endosperm of the huge coconut seed. Since no combination of then-known growth factors could duplicate this result, it was clear that coconut milk must contain some substance or substances responsible for this unique effect. Several researchers therefore initiated attempts to isolate and characterize the hypothetical cell-division factors from coconut milk.

The discovery of cytokinins emerged from the attempts of Skoog and his colleagues at the University of Wisconsin to induce the division of mature tobacco pith cells. In the intact plant, these cells enlarge but never divide, yet when explanted to a medium containing auxin and coconut milk, they initiate division. Testing various compounds, Skoog's team found adenine to have some cell-division-promoting activity in the presence of auxin. Subsequently nucleic acids, in which adenine occurs as a component, were also examined, and one old sample of herring sperm DNA was found to yield results quantitatively superior to those produced by adenine. Later it was found that any sample of DNA, or even adenosine alone, could be converted to an extremely active material merely by autoclaving it under appropriate conditions. The new substance formed under these abnormal conditions was isolated and identified as 6-furfuryladenine. Because of its great activity in inducing cell division (cytokinesis) in tobacco pith when applied together with auxin, it was named *kinetin.*

Once the nature of this substance was discovered, researchers were able to look for other similar materials in plants. Fruitlets and developing cereal grains proved to be rich sources of activity. Eventually a substance with a related structure was isolated from developing maize (*Zea mays*) grains, and was called *zeatin* (Fig. 9-27). This, or the closely related compound *isopentenyl adenine*, have now been found in all plants examined. These compounds, generically called *cytokinins*, are usually found attached to ribose (*ribosides*) or to ribose and phosphate (*ribotides*). The growth-promoting activity of coconut milk has been found to be due to a mixture of many compounds, of which the cytokinin component has been identified as zeatin riboside. Synthetic cytokinins such as kinetin and benzyladenine are widely used in research, since the more active natural cytokinins are more difficult to obtain.

Cytokinin-auxin interaction

Cytokinins, when present with auxin, are extremely active in inducing division of plant cells placed in tissue culture; they sometimes also promote the development of buds in the resultant callus. The differentiation in tissue culture depends mainly on the relative concentrations of the growth substances present. High molar ratios of cytokinin to auxin lead to the formation

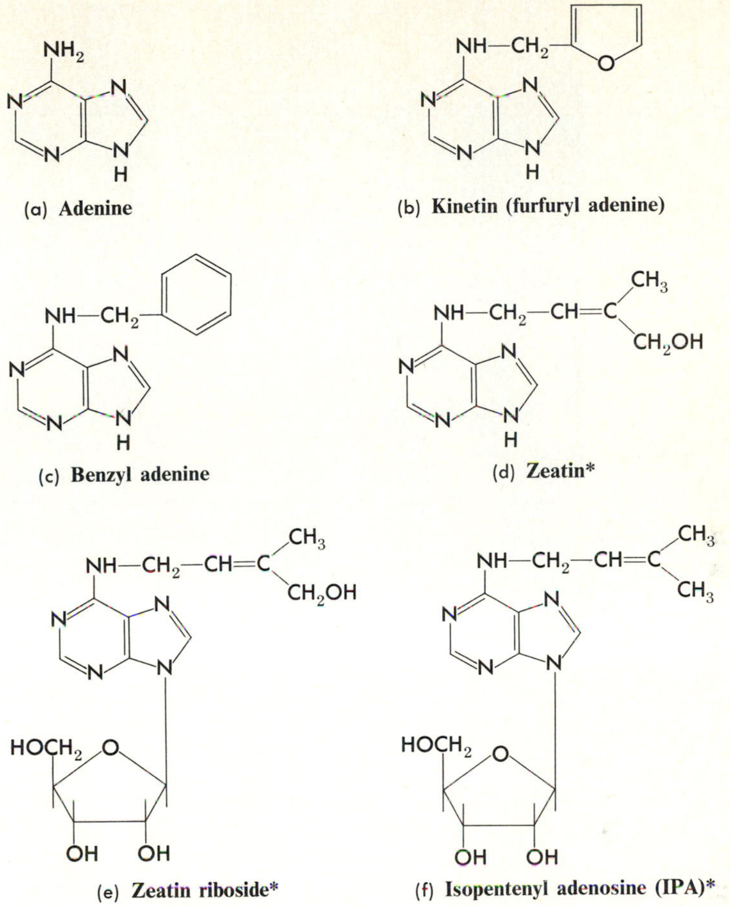

(a) **Adenine**

(b) **Kinetin (furfuryl adenine)**

(c) **Benzyl adenine**

(d) **Zeatin***

(e) **Zeatin riboside***

(f) **Isopentenyl adenosine (IPA)***

Figure 9-27 The structural formulae of various adenine derivatives with cytokinin activity. The starred compounds are natural cytokinins.

of buds; at roughly equal concentrations of auxin and cytokinin, undifferentiated callus growth is favored, and when the ratio of auxin to kinetin is high, root growth tends to be initiated (Fig. 9-28). In typical cell-elongation systems such as oat coleoptile and pea epicotyl, kinetin tends to inhibit auxin-stimulated longitudinal growth and to promote transverse growth.

Discovery of the interaction of kinetin with auxin necessitated a systematic investigation of other physiological systems, such as apical dominance, known to be influenced by auxin. Here it was found that local application of kinetin to repressed buds causes a release from inhibition, and consequent growth (Fig. 9-29). The inhibition of lateral bud growth by auxin has in fact been shown to be due to the induction of ethylene production in the lateral bud (see Chapter 10), whereas application of cytokinin inhibits the production of ethylene. The expansion of young leaves is also promoted by cytokinins.

In mature plants, one of the main sites of synthesis of cytokinins is the root system; from there the cytokinins travel to the shoots through the xylem.

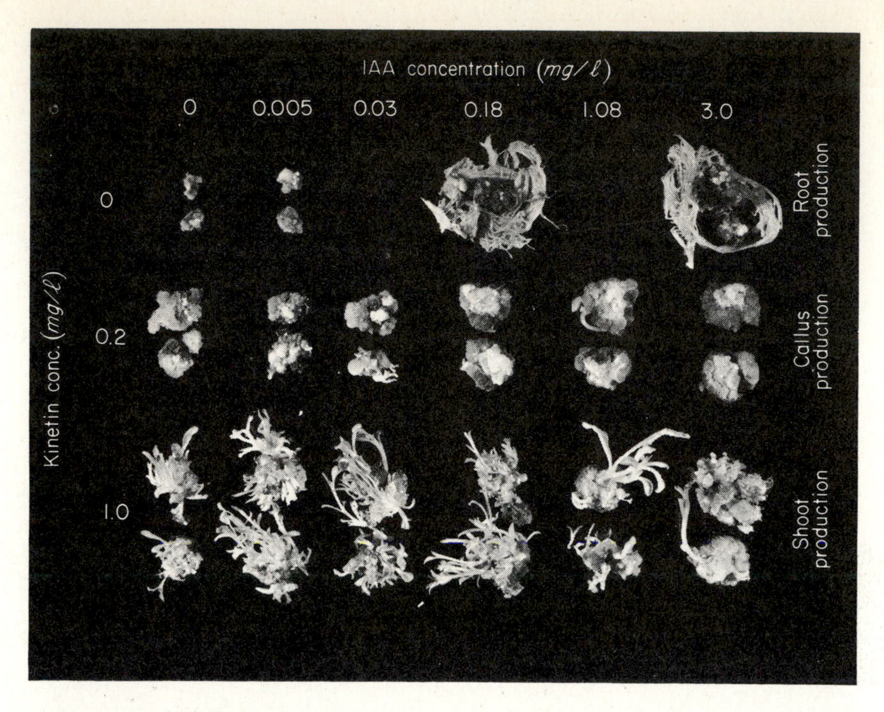

Figure 9-28 The interactions of kinetin and IAA on callus production and tissue differentiation in tobacco tissue culture. Maximum tissue production requires the presence of auxin and kinetin. High ratios of auxin/kinetin produce roots; high ratios of kinetin/auxin produce shoots; at equal concentrations callus predominates. (From Skoog and Miller. 1957. *Symp. Soc. Exp. Biol. 11*: 118–131.)

Figure 9-29 The release of lateral buds from apical dominance by kinetin. The plant on the right was treated 3 days previously with a local application of 330 ppm kinetin to the bud. (From Sachs and Thimann. 1967. Amer. J. Bot. *54*: 136–144.)

The lower down on the plant that the lateral buds are situated, the farther they are from the auxin source in the stem apex, and the closer to the cytokinin source in the roots. Thus, as the apex of a plant showing apical dominance grows away from the lateral buds, the buds start to "break" as the influence of the cytokinin overcomes that of the auxin. This auxin–cytokinin antago-

Figure 9-30 A "witches'-broom" on a willow tree resulting from the growth of many lateral buds caused by cytokinin production by an invading plant pathogen (compare regular branches upper right). (Courtesy of W. A. Sinclair, Cornell University.)

nism also accounts for the formation of "witches' brooms," which result from a complete removal of the inhibition normally placed on dormant buds separated from the apex by only a few short internodes (Fig. 9-30). This happens naturally when the fungi that induce witches' brooms invade the plant and produce substances with cytokinin activity.

Cytokinins explain another phenomenon that had long puzzled plant physiologists. It had often been observed that when leaves, such as those of tobacco, are removed from the plant, their protein content diminishes rapidly, while soluble nitrogen rises. This massive breakdown of protein was thought to account, at least in part, for the short life span of many cut plants and plant parts, especially leaves. It was fortuitously discovered that the incorporation of kinetin into the nutrient solution bathing the petiole of excised *Xanthium* leaves caused a prolonged retention of the green color of the leaf. Thus cytokinins have an antisenescence effect, later shown to be produced by their maintenance of the levels of protein and nucleic acid, probably by decreasing their rate of breakdown, and also by maintaining the integrity of cell membranes. Application of cytokinin to leaves on intact plants has also been shown to retard their senescence (Fig. 9-31), and it is

Figure 9-31 An important function of cytokinins in plants is the prevention of the processes of senescence. The lowermost primary leaves on a bean (*Phaseolus vulgaris*) plant usually turn yellow and abscise as the later leaves develop (plant on the left). Painting the cytokinin benzyladenine on the leaves of the plant on the right has prevented senescence and the leaves have maintained their green color. (Courtesy of R. A. Fletcher, University of Guelph, Canada.)

probable that cytokinin must be continuously supplied in the water arriving in the leaves from the roots to prevent leaves from senescing.

How cytokinins work

In the absence of cytokinin, DNA synthesis continues as long as auxin is present, but cells do not divide. Cytokinin is required for some process that occurs after DNA replication is complete, but before actual mitosis begins. This process may be the production of proteins required for cell division, perhaps those involved in the mitotic apparatus. The appearance of different proteins has in fact been noted when cytokinins are added, and this occurs even when synthesis of new messenger RNA is blocked by inhibitors. It would therefore seem likely that cytokinins control the translation of preformed m-RNA's specific for some aspect of cell division.

Another clue as to how cytokinins function may lie in their structure: They are related to the components of nucleic acids. A cytokinin can in fact be found as one of the bases next to the anticodon in several transfer RNA molecules, and in this position is probably involved in the binding of the transfer RNA to the ribosome during protein synthesis (Fig. 2-10). This is not how cytokinins function as hormones, however, since when cytokinins are applied to plant tissue they do not become incorporated as such into transfer RNA. Rather, they are degraded to adenine, which, after incorporation into the polynucleotide chain, is modified by the addition of an isopentenyl side chain. Nevertheless their relationship with the structure of nucleic acids, together with their observed effects in the plant, does indicate a possible role for cytokinins in protein synthesis. Cytokinin molecules do bind loosely to ribosomes, the bodies on which proteins are synthesized in the cell; the binding site is actually a protein which is itself bound to the ribosome. The cytokinin and binding protein may be part of an initiation factor for specific protein synthesis on the ribosomes. Whether this is the way cytokinins are able to control cell division and perhaps other aspects of cell metabolism remains to be discovered.

As with the other hormones, cytokinin probably acts in different ways in producing its various effects. Consider, for example, the action of cytokinin in the prevention of leaf senescence. In the presence of cytokinin the levels of protein in leaves remain higher than if cytokinins are absent; in this case, cytokinin inhibits protein breakdown rather than enhances protein synthesis. As already mentioned, cytokinin has also been shown to maintain the integrity of cell membranes. How it carries out these functions is currently unknown, but the mechanism should become evident as we learn more about how hormones function at the molecular level.

SUMMARY

Hormones are organic molecules that control and integrate the plant's activities. Produced in minute quantities in one organ or tissue, they move to another organ or tissue, and by virtue of specific reactions in the target tissue, control such processes as growth, development, and differentiation. Five classes of plant hormones are currently recognized: *auxins*, *gibberellins*, *cytokinins*, *abscisic acid*, and *ethylene*.

Auxin, discovered through early experiments on *phototropism*, controls the rate of cellular elongation and, through differential effects on the two sides of a cylindrical organ, the direction of growth as well. *Indole*-3-*acetic acid* (IAA) produced from the amino acid tryptophan, is the native auxin. It is synthesized in stem tips and migrates *polarly* basipetally, facilitating cellular elongation below the apex, repressing the growth of lateral buds, and stimulating mitotic activity in the pericycle, which gives rise to branch roots. It may also be metabolically inactivated by the enzyme *IAA oxidase* or sequestered as conjugates. Its effect on cells increases proportionally with the logarithm of its concentration up to some optimal point, after which its effect diminishes and may even become inhibitory. Roots are more sensitive to auxin than are stems.

Auxin concentration in tissue may be measured by *bioassay* or by direct chemical or physical techniques after auxin extraction and purification. Such analyses have shown that in *tropistic curvature*, light and gravity cause a *lateral displacement* of auxin from one side of the organ to the other. Gravitational perception is correlated with displacement of starch *statoliths*, while phototropism results from the differential absorption of blue light, probably by *flavoproteins*. The promotion of cell elongation is probably related to auxin-promoted *proton secretion* into the wall area; the resulting higher hydrogen-ion concentration causes more active enzymatic breakage of bonds that crosslink cellulosic microfibrils. Chemical analogs of IAA, as well as IAA itself, can be used commercially to *initiate roots* on cuttings, promote *parthenocarpy*, *inhibit sprouting* of tuber buds, control leaf and fruit *abscission*, and *selectively kill* unwanted plants.

Gibberellins are a group of isoprenoid plant substances that control *elongation*, *flowering*, and aspects of *seed germination* in some plants. Some genetic dwarf plants treated with gibberellic acid (GA) grow rapidly, becoming *phenocopies* of the genetically tall strains, while the tall plants show little or no extra growth in response to GA. Most *rosette* plants bolt and flower when treated with GA; certain *biennials* that need a cold treatment in order to flower can use GA as a substitute for low temperature. In seeds, gibberellin produced by the embryo facilitates the mobilization of stored food reserves, by increasing synthesis and secretion of hydrolytic enzymes. In this action, it appears to act at the gene level, controlling the synthesis of m-RNA molecules specific for hydrolytic enzymes.

Cytokinins are adenine molecules modified by the substitution of isopentenyl or modified isopentenyl groups on the 6-amino group. Synthesized in the roots, cytokinins migrate upward and interact in various ways with auxins. Cell division is promoted by median molar ratios of the two hormones, while high cytokinin/auxin ratios tend to overcome lateral inhibition of buds and even cause bud initiation in some tissue cultures. In contrast, low cytokinin/auxin ratios facilitate root initiation on the plant or callus culture. Cytokinins tend to oppose the action of auxin on cell elongation, and to inhibit the senescence phenomena that occur in detached leaves by maintaining adequate levels of RNA, protein, and chlorophyll synthesis. Their role in facilitating cell divisions is not yet understood, although they are known to bind to ribosomes, which may then permit the formation of a protein essential for cell division.

SELECTED READINGS

Audus, L. J. *Plant Growth Substances*, Vol. I. *Chemistry and Physiology* (London: Leonard Hill Books, Ltd., 1972).

Galston, A. W., and P. J. Davies. *Control Mechanisms in Plant Development* (Englewood Cliffs, N. J.: Prentice-Hall, 1970).

Leopold, A. C., and P. E. Kriedemann. *Plant Growth and Development*, 2nd ed. (New York: McGraw-Hill, 1975).

STEWARD, F. C. *Plant Physiology* Vol. VI B. *Physiology of Development—The Hormones* (New York: Academic Press, 1972).

THIMANN, K. V. *Hormone Action in the Whole Life of Plants* (Amherst, Mass.: University of Massachusetts, Press, 1977).

WAREING, P. F., and I. D. J. PHILLIPS. *The Control of Growth and Differentiation in Plants*, 2nd ed. (New York: Pergamon Press, 1978).

QUESTIONS

9-1 It has been calculated that 1 molecule of auxin can affect the deposition of 1.5×10^5 glucose molecules in the form of cellulose of the cell wall. What does this signify concerning the mode of action of auxin?

9-2 Discuss various ways by which the auxin level of tissues is regulated. Why is such regulation important?

9-3 Several hours after excision of a 5-mm-long oat coleoptile tip, the new apical region acquires the ability to produce auxin, an ability which it lacked in the intact coleoptile. How can one explain this?

9-4 How is auxin involved in tropistic response?

9-5 Above an optimum concentration auxin is inhibitory. Why?

9-6 Briefly explain why the cell wall is of such great importance in the process of cell growth. What happens to it during growth and how is auxin thought to produce this effect? (Discussed in Chapter 2 as well as this chapter.)

9-7 Provide one possible explanation for the fact that roots grow downward and stems grow upward.

9-8 Give two pieces of evidence suggesting that starch statoliths are the means by which the force of gravity is detected by plants.

9-9 Discuss briefly the role of cell enlargement in plant growth. How, and by what mechanism, does the water status and the presence of auxin affect this process?

9-10 Auxin affects several cellular phenomena other than cell elongation. What are some of these phenomena, and how are they affected by auxin?

9-11 If coleoptile segments are peeled and placed in an aqueous solution at pH 5 they elongate as if auxin were present. What is the significance of this observation?

9-12. Auxin transport is described as "polar." What does this mean? How might polar auxin transport occur? What aspects of plant morphology and physiology are dependent on polar auxin transport?

9-13. Can you propose an explanation for the fact that auxin inhibits the growth of lateral buds but does not inhibit the growth of the terminal bud that produces it?

9-14. Diagram a longitudinal section of a barley grain to show where gibberellin is produced, where it acts, and what happens to the products of its actions. What basic effect does the gibberellin produce? How does the gibberellin produce this effect? Include the names of tissues essential to your answer.

9-15. If gibberellin is applied to a germinating barley grain which has had its embryo removed (a) What is the result? (b) What effect do inhibitors of RNA and protein synthesis have on the system? and (c) What can we deduce from (b)?

9-16. How might the existence of more than 50 different gibberellins be of significance in controlling the growth and development of plants?

9-17. What is the significance of the observation that dwarf phenotypes will respond to gibberellin whereas tall phenotypes will not?

9-18. It has been suggested that cytokinins induce cell division by acting at the level of protein synthesis on the ribosomes. Briefly summarize the evidence to support this suggestion.

9-19. Suggest two roles of cytokinins in plants. List one part of the plant where they can be found in *high* concentration.

9-20. Using bioassays, how would you distinguish between an auxin, a cytokinin, and a gibberellin?

9-21. Name some effects that can be induced by the external application of auxin, gibberellin, or cytokinin to whole plants or plant parts. Would you expect these effects to differ from those controlled by endogenously produced hormones?

9-22. What hormones influence stem growth? How could one demonstrate their effect? How do they produce their effects?

9-23. Discuss some *positive* evidence favoring the concept that at least part of the mechanism by which plant hormones control development involves the production of new proteins. In the examples you cite, explain how the hormone is believed to act.

10

Hormonal Control of Dormancy, Senescence, and Stress

Ethylene: The Senescence Hormone

The old observation that one rotten apple in a barrel causes the entire lot to spoil can now be explained in simple scientific terms. The one rotten apple produces a volatile agent, *ethylene* (C_2H_4), which causes degradative changes in the healthy fruits nearby. This leads to their spoilage, and they in turn begin to produce ethylene, which affects still other fruits. Thus, there is a chain-reaction, with a small amount of ethylene producing a very widespread result. The common practice of preserving storage apples by enriching the air with carbon dioxide is based on the fact that CO_2 prevents ethylene from exerting its effect.

Although scientists have known these rather simple facts for a very long time, they did not consider the action of ethylene on plant cells and the countereffect of carbon dioxide to be part of normal physiological regulation in the plant. Ethylene production was thought to result from invasion of the plant by some pathogen, or possibly from physiological breakdown of plant cells that results from bruising, unfavorable storage temperatures, or simply aging. Recent experiments, however, have shown that ethylene is a normal plant metabolite, produced by healthy cells and exerting normal regulatory control over morphogenetic phenomena such as fruit ripening and leaf abscission. Since ethylene is produced in minute quantities and may be active in cells other than those at its site of production, it may legitimately be considered a plant hormone.

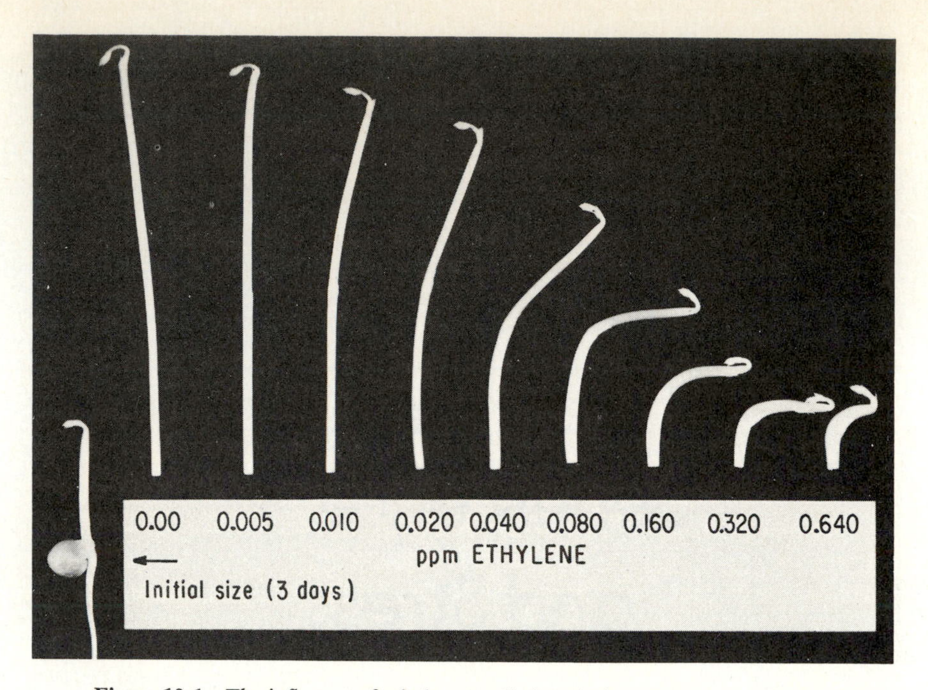

Figure 10-1 The influence of ethylene applied to dark grown pea seedlings during a two day period. Increasing concentrations from left to right. Note how ethylene causes dwarfing, swelling of the stem and a change from the normal response to gravity to *diageotropism*, a positive response to gravity 90° off normal. It is the new growth (during exposure) that changes, while the growth after removal of ethylene is again normal. (Courtesy of H. K. Pratt, University of California, Davis.)

Various morphological responses of intact plants to applied ethylene were known long before ethylene was recognized as a natural regulator in plants. Application of ethylene to etiolated pea seedlings, for example, causes pronounced effects on growth that have been used as a bioassay for this hormone. At the age of seven days, dark-grown pea seedlings consist of an erect, slender, unpigmented stem, a sharply recurved apical hook, a slightly yellow terminal bud, and roots. If a stream of air containing ethylene is blown over these seedlings, their longitudinal extension is inhibited, while their lateral growth is promoted, resulting in a swelling just below the apical bud. The stems also lose their normal sensitivity to gravity, becoming diageotropic or ageotropic (Fig. 10-1). As we have seen, the normal tropistic orientation of stems and roots is at least partly a consequence of the transverse migration of auxin in response to various unilateral stimuli, including gravity and light. Ethylene apparently interferes with this normal lateral transport of auxin, thus preventing the righting reactions that result in the normal geotropic orientation.

The effects of ethylene on light-grown plants are no less striking. In many plants, such as potato and tomato, ethylene causes a pronounced *epinasty* (downward bending) of the leaves, owing to a lateral swelling of the cells on the upper side of the petiolar base and the midvein of the leaf. Sometimes, large numbers of adventitious roots develop on the stem, resulting from promotion of cell division in the region of the cambium. These patches of cells organize, forming a root meristem, which then grows out of the stem as a root. Thus, ethylene not only interferes with cell extension and normal direction-finding through tropistic orientation, but may also cause lateral swelling and influence cell division.

For many years, ethylene was detected and measured quantitatively by a bioassay technique such as the responses of the peas described above. Today, all ethylene measurements are obtained by gas chromatography—a convenient procedure, for ethylene is already a gas and volatile chemical derivatives do not have to be made before assay. A sample of air in or around the tissue is collected in a syringe and then injected into the column which separates the ethylene from the other gases present. At the other end of the column, the gases are detected by a flame ionization detector which is sensitive to as little as one part per billion of ethylene, a concentration well below that

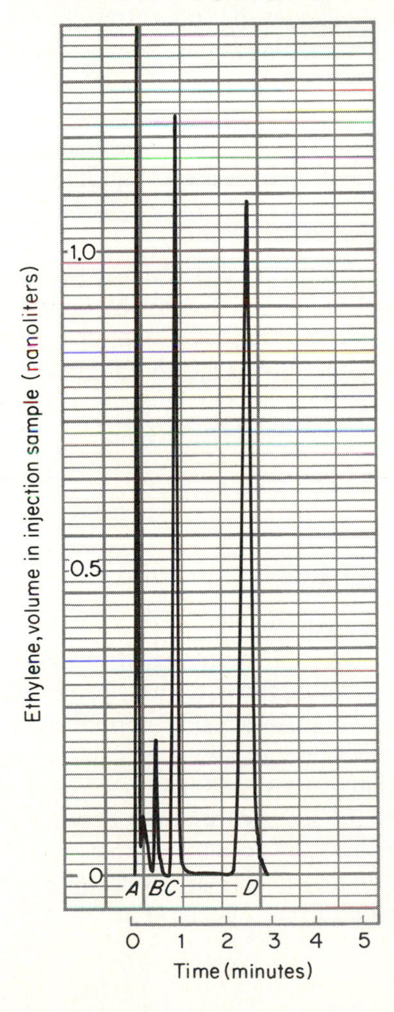

Figure 10-2 Recorder tracing from a gas chromatographic separation of ethylene on an activated alumina column at 45°C, with a flame ionization detector. A, air; B, ethane; C, ethylene; D, propane. In this sample, 1.2 nanoliters of ethylene were detected in a 0.5 ml injection volume = 2.4 ppm. (Courtesy of G. D. Blanpied, Cornell University.)

which causes physiological effects. A typical tracing from a recorder attached to such a gas chromatograph is shown in Fig. 10-2.

The advent of the sensitive gas chromatograph proved that ethylene is indeed produced by healthy plants and contributes to the control of their development. We now know that ethylene is involved in the maintenance of the apical hook in dark-grown seedlings such as the pea. (Such plumular hooks are of importance in protecting the sensitive apical growing point from injury as the apex pushes up through the soil after germination.) If the shoot is given light, ethylene production decreases and the stem straightens up, properly orienting the leaves for photosynthesis; but if ethylene is applied after the etiolated plant is put into the light, the apical hook will not straighten up or may even form again.

Ethylene may also be directly responsible for the auxin-controlled prevention of the growth of lateral buds in a plant showing apical dominance. We noted earlier that auxin, in promoting the growth of excised stem pieces, tends to manifest an optimal range as the concentrations applied to the tissue are increased; now we know that this occurs because auxin concentrations higher than optimum promote the formation of ethylene, which then inhibits rather than promotes growth. Most, though not all, effects of high auxin concentrations can therefore be ascribed to ethylene; both the formation of the apical hook and the maintenance of lateral bud dormancy are examples. Another is the promotion of flowering of bromeliads like pineapple, in which both auxin and ethylene are effective; here again, auxin works by promoting ethylene production. In fact if you have a pet pineapple plant, you can use a well known trick to induce it to flower; this involves putting it into a plastic bag with a ripe banana. The banana produces enough ethylene to trigger the pineapple into flowering! Another way is simply to tip the pineapple plant on its side; the increased auxin level on the lower side caused by geotropic lateral auxin redistribution is frequently sufficient to initiate flowering, via increased ethylene levels.

Plant Senescence, Fruit Ripening, and Leaf Abscission.

Ethylene appears to be a prime controlling agent in many aspects of plant senescence including the fading of flowers, the ripening of fruits, and the abscission of leaves. If an orchid flower goes unpollinated it remains fresh for a long time, but very soon after it is pollinated it starts to fade. The reason for the post-pollination decline is that pollination initiates the production of ethylene, which then causes the senescence of the flower petals. In morning-glory flowers, time, rather than any specific event, is the controlling factor. The flowers open early in the morning, start to droop by mid-morning and by noon are a curled-up mass of soggy petals (Fig. 10-3). The start of senes-

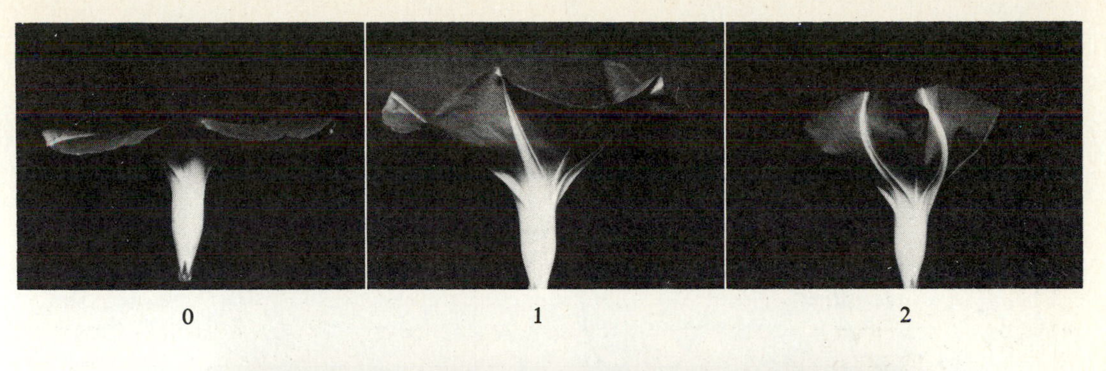

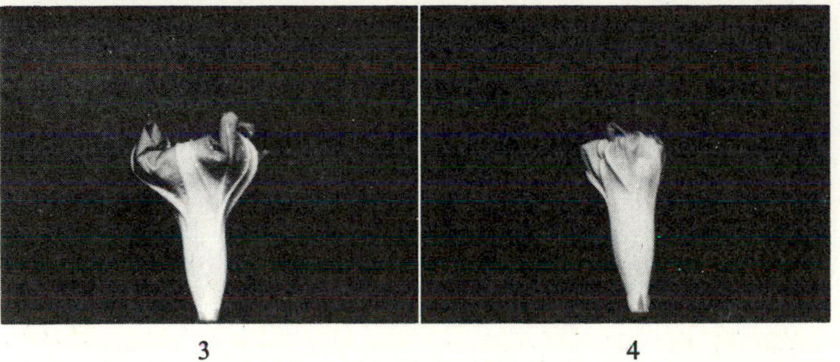

Figure 10-3 Flowers of morning glory (*Ipomoea tricolor*) at various times between 6 am and 1 pm the same day, showing the progress of flower senescence. The senescence is caused by ethylene production by the flower a few hours after it opens. This brings about turgor changes in the cells of the petal midribs, which then cause the curling. (Courtesy of H. Kende, Michigan State University.)

cence, as indicated by the petals beginning to curl, coincides with the onset of ethylene evolution. Premature senescence can also be induced by exposing plants to ethylene. It is difficult to determine whether ethylene is the actual trigger for senescence or whether it simply accelerates the process. In morning glory, the aging of the tissues may cause the tonoplast to become slightly leaky so that methionine, the precursor of ethylene, leaks from the vacuole into the cytoplasm, where it is converted to ethylene. Ethylene in turn might then further degrade the vacuole membrane, accelerating the entire process.

Fruit ripening

It is in the senescence, or ripening, of fruits that ethylene probably plays its most important role in plants. We have already mentioned that ethylene induces the ripening of healthy fruits located near diseased fruits; it is also involved in the natural ripening process. The large fruit develops from the smaller ovary wall under the influence of cell division- and enlargement-promoting hormones such as auxin and gibberellin. Once the fruit has attained its maximum size, subtle chemical changes begin which ultimately cause it to "ripen" and become edible. Many mature fruits, like a full-grown apple, are inedible because of acidity and hardness. The ripening process in

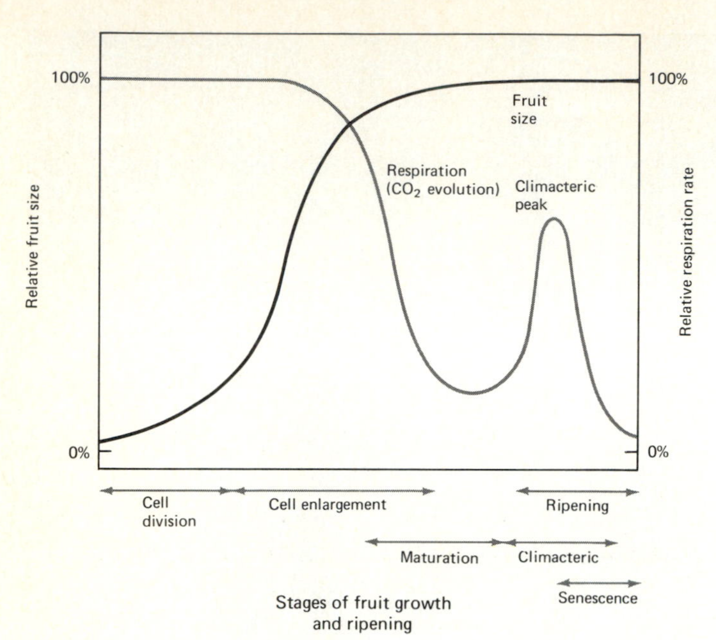

Figure 10-4 The changes that occur in fruit size and in the rate of respiration as a fruit grows and ripens.

the apple consists partly of the disappearance of much of the malic acid which makes the unripe fruit taste acidic. A weakening of the cell walls follows, greatly reducing the mechanical force required to separate one cell from another. Many fruits ripen more rapidly after they are picked, indicating either that the signal for ripening arises in the fruit itself, or that a suppression of ripening is exerted by the other plant parts.

Ripening in certain fruits is closely correlated with an increase in the rate of respiration. If one measures the carbon dioxide output from a fruit or a fruit slice during the course of ripening, one will frequently notice a rather sharp inflection point, denoting a greatly increased output of CO_2 for a short period, followed again by a sharp decline. This period of greatly increased carbon dioxide output is called the *climacteric* (Fig. 10-4). Following onset of the climacteric, the fruit rapidly undergoes those changes that transform it from a fully grown unripe fruit to a ripe fruit ready to be eaten. The climacteric, once thought to be a degenerative process, is really an active developmental change requiring the expenditure of energy. It can be prevented by inhibitors of respiration, by high carbon dioxide or nitrogen concentrations, or by low temperature; application of ethylene, by contrast, promotes both the climacteric rise and ripening in mature fruit. Originally, investigations of the role of endogenous ethylene production in ripening indicated that ethylene production commenced after the climacteric. The more precise measurements available through gas chromatography, however, show that ethylene is consistently produced at or before the onset of the climacteric (Fig. 10-5). A little ethylene is actually present all the time, but this amount is increased by about a hundredfold at the time of the climacteric. When ripening is

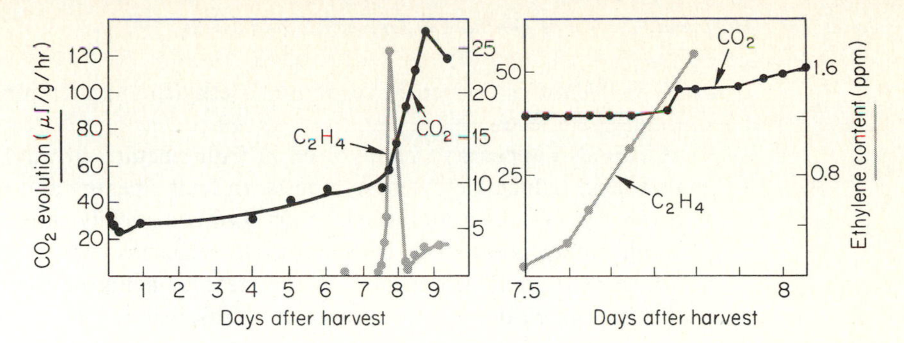

Figure 10-5 Changes in the rate of respiration and the endogenous ethylene content of a banana during a 9 day post-harvest period. Left: gross pattern. Right: a portion of the graph is magnified to reveal small changes that occur at the onset of the climacteric. (From Burg and Burg. 1965. *Bot. Gaz. 126*:200–204.)

prevented by such conditions as low temperature, ethylene production is also inhibited; we may therefore conclude that ethylene is the natural fruit-ripening hormone in plants. Further evidence to support this conclusion has been obtained by removing the ethylene from the fruits as fast as it is formed by subjecting the fruit to reduced pressure, but maintaining the oxygen concentration similar to that of the atmosphere. Under these conditions ripening is delayed.

How does the ethylene induce ripening? Although in some fruits (for example, avocado and mango) ethylene production and the increase in respiration run parallel, in others (such as the banana) ethylene acts only as a trigger, its production declining before the maximum rate of respiration is achieved. This indicates that ethylene must cause the onset of some other process, which then brings about ripening. Ethylene probably has multiple effects. First, it probably increases the permeability of the membranes in the fruit cells in a fashion similar to its action in morning-glory flowers; this would allow enzymes which were previously separated from their substrates by membranes to contact those substrates and start the processes of degradation. Protein synthesis is needed for ripening, and ethylene has been shown to increase the rate at which this process occurs. Paradoxically, the onset of ethylene synthesis itself requires the synthesis of particular proteins, or enzymes. The enzyme that converts S-adenosylmethionine to 1-aminocyclopropane-1-carboxylic acid (ACC), an intermediate in ethylene biosynthesis (Fig. 10-6), generally limits the rate at which ethylene is formed in tissue. This

Figure 10-6 The pathway for the formation of ethylene from methionine. The numerals indicate the fate of the carbon atoms in methionine.

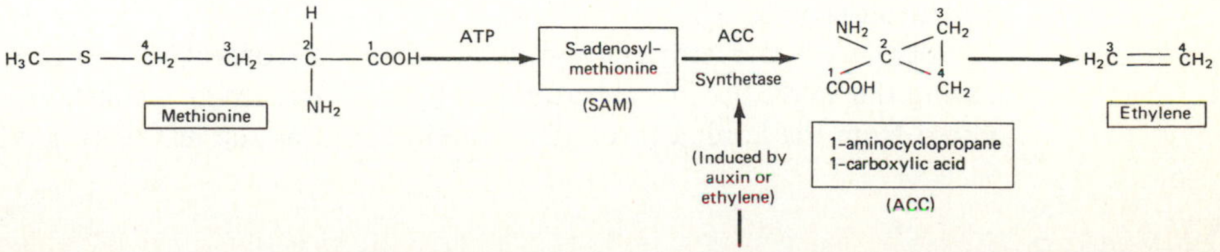

$H_3C—S—\overset{4}{C}H_2—\overset{3}{C}H_2—\overset{2}{C}H(NH_2)—\overset{1}{C}OOH$ (Methionine) —ATP→ S-adenosylmethionine (SAM) —ACC Synthetase (Induced by auxin or ethylene)→ 1-aminocyclopropane 1-carboxylic acid (ACC) → $H_2\overset{3}{C}=\overset{4}{C}H_2$ (Ethylene)

ACC synthetase is formed in response to ethylene application or to supra-optimal auxin levels.

Auxin also appears to be involved in fruit ripening in ways independent of its role in inducing ethylene synthesis. In fruit ripening and in leaf abscission (see below) auxin and ethylene act antagonistically; auxin delays fruit ripening and leaf abscission, whereas ethylene hastens both processes. Which hormone predominates depends on the state of aging of the tissue, as we shall see more clearly in our discussion of abscission.

Leaf abscission

Leaf abscission is one of the most complex processes in the green plant. It may occur progressively as a plant ages, in which case the older leaves being shed are replaced continuously by the formation of new leaves at the stem apex. Alternatively, it may occur completely and all at once as an overwintering mechanism in a deciduous tree. Leaf abscission depends on the formation of specialized cell layers in the abscission zone at, or close to, the base of the petiole (Fig. 10-7). Depending upon the species, this zone may be formed early in the development of the leaf or only after the leaf is fully mature. The parenchyma cells comprising the abscission layer are frequently smaller than the surrounding cells; even the individual vascular elements may be shorter, and fibers may be absent from the bundle in the abscission zone. This combination of anatomical features makes the abscission zone an area of weakness.

Leaf fall is initiated by senescence of the leaf, either because of natural aging or, in deciduous trees, by a signal from the environment. This signal is usually decreasing day length (Chapter 12), which triggers changes within the leaf lamina that result in the entire leaf being shed from the tree. Photoperiodic control of abscission can be inferred from the behavior of trees near street lamps, which frequently retain their leaves longer in the autumn than do other trees; eventually, of course, declining temperatures ensure the abscission of even these leaves.

Prior to leaf fall, numerous changes take place in the abscission zone. Cell divisions frequently occur, forming a layer of brick-shaped cells across the base of the petiole. Active metabolic changes in the abscission zone cells result in the partial dissolution of the cell wall or the middle lamella. The cells thus become separated; the weight of the leaf eventually snaps the vascular connection and the leaf is shed from the tree. A cork layer forms across the petiole stump, protecting the tissues of the tree from microbial invasion and restricting water loss. The xylem vessels become plugged with invaginations of neighboring parenchyma cells called *tyloses*, completing the sealing-off process.

Both auxin and ethylene are involved in the control of leaf abscission. During the active life of the leaf, newly synthesized auxin is constantly transported from the lamina through the petiole, but as the leaf ages, auxin pro-

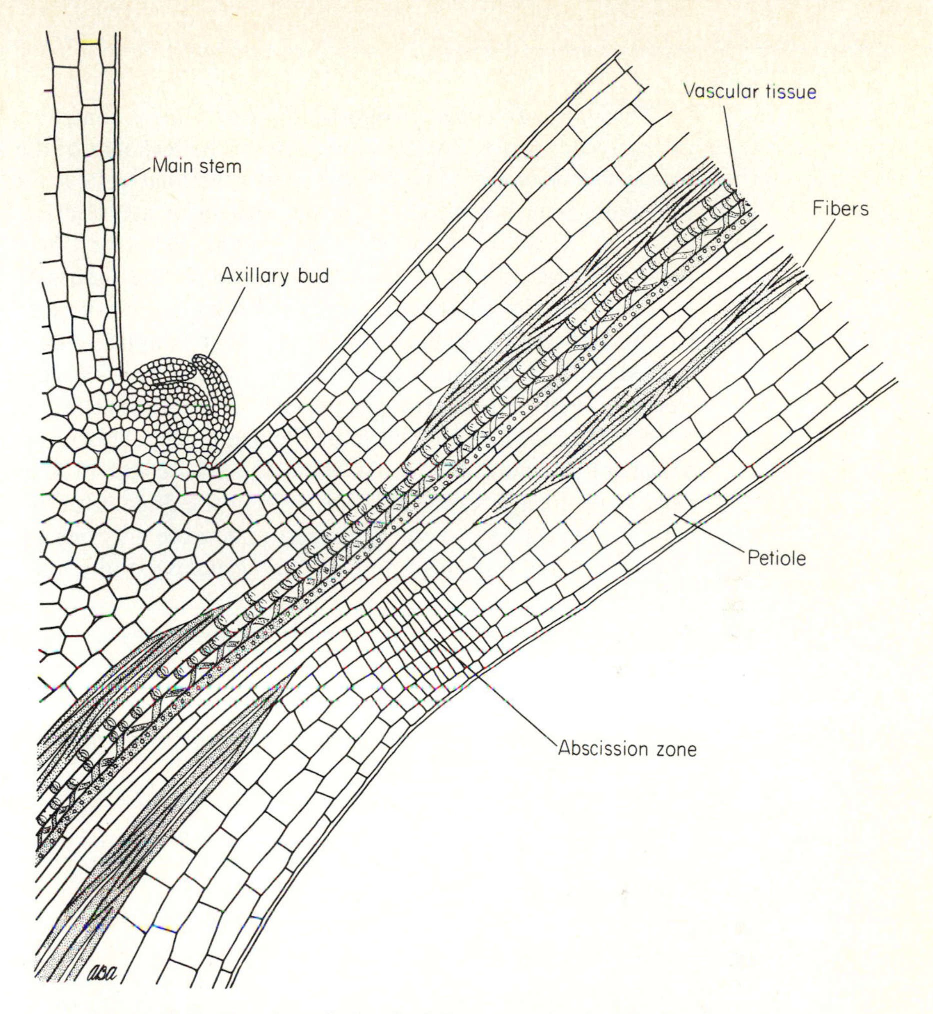

Figure 10-7 Drawing of the abscission zone showing the *smaller* cells across the future line of separation. As the tissue ages, separation of the cells becomes apparent, followed by abscission of the distal portion of the petiole, leaving intact cells at the abscission interface. (Courtesy of F. T. Addicott, University of California, Davis.)

duction and transport decline. This decline is one signal that triggers changes in the abscission zone that lead to leaf fall. Auxin applied to the lamina will delay abscission by maintaining the normal metabolic activity of the cells of the petiole. Decreasing supplies of cytokinins from the roots, as winter approaches, may also contribute to the decline in vigor of the leaf, as cytokinins have been shown to delay leaf senescence (Chapter 9). As the leaf blade begins to age, products from its senescent cells move down the petiole and cause the onset of senescence in the petiole tissues. Despite some speculation that these "senescence factors" include abscisic acid (discussed later), it is now generally considered that abscisic acid, despite its name, has little control over abscission.

The aging of tissues is important not only in its control of the production of senescence factors by leaves but also in the response of the abscission zone tissues to hormones. When senescence in petiolar cells is retarded by auxin, abscission is also retarded, and any treatment accelerating senescence (for example, ethylene) stimulates abscission. Ethylene, however, promotes petiole abscission only if applied after the petiole tissues have aged. Auxin generally inhibits abscission, but can also have promotive effects, because of its stimulation of ethylene production, if it is applied after senescence has started in the tissues of the abscission zone. Thus the age and state of the tissue, as well as the hormones present, determine the fate of the leaf.

Ethylene has been shown to be the main natural hormone controlling leaf abscission. In addition to hastening the senescence of the abscission zone cells, it is directly responsible for causing the dissolution of the walls of the cells of the abscission zone. This occurs because ethylene both promotes the synthesis of the wall-degrading enzyme *cellulase*, and controls the release of

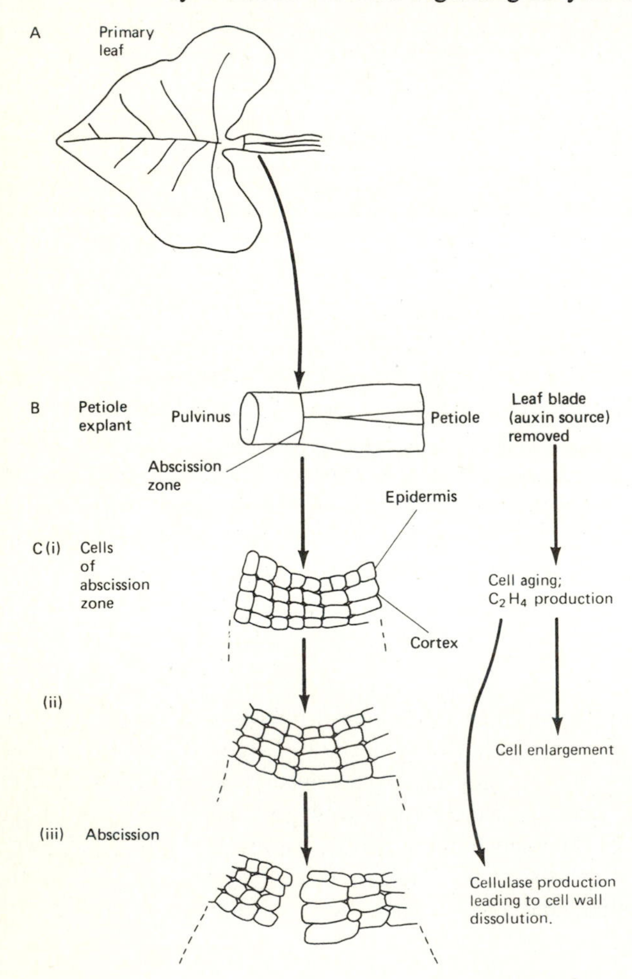

Figure 10-8 Changes in the size of the cells of the abscission zone as abscission occurs in response to ethylene. (From Osborne. 1974. *Cotton Growing Review* *51*: 256–265.)

cellulase from protoplast into the wall. These are active processes which, as with fruit ripening, require active energy metabolism and protein synthesis, and can be inhibited by inhibitors of these processes. Not only does ethylene promote cell wall dissolution through effects on cellulase synthesis and release, but it also cause the cells on the proximal side of the abscission zone (next to the stem) to swell (Fig. 10-8), so that the distal part of the petiole (toward the leaf blade) is physically forced away.

Artificial application of ethylene to petioles will produce the effects just described, but it is clear that ethylene is also the natural agent causing abscission, since ethylene evolution by abscission zone cells increases prior to the abscission process. The abscission zone is in fact a very specialized group of cells, and ethylene exerts its effect mainly on these cells, often on only one layer. These are also the only cells of the petiole, and probably the only cells of the plant, which naturally produce large quantities of cellulase. This again illustrates that there are two interacting systems controlling plant growth and development: the *hormones* which provide the signal, and the *state of the cells*, whose response is determined by their previous developmental history. While it is true that hormones also play a role in determining the state of the cell, still other influences are involved that biologists have yet to unravel completely.

The senescence of whole plants

We have looked so far at the senescence of flowers, of fruits, and of leaves, all organs rather than the whole plant. To date, relatively little experimental attention has been paid to senescence of whole plants, yet the fact that it is under precise physiological control compels closer scrutiny. Consider, for example, that in the wheatfields of Kansas, millions of plants senesce and die at approximately the same time, following the formation of seed. What causes these plants to cease activity and die, leaving only the seeds to carry on in the next season? Why do long-lived perennial species such as the century plant grow for many years, only to die promptly after they complete flowering and fruiting? In plants that flower only once, such as annuals, reproduction seems to trigger a series of events leading to the death of the plant; removal of the flowers or developing fruits frequently delays or entirely prevents senescence. As the developing fruits and seeds require large stores of nutrients, the rest of the plant was thought to decline because of nutrient deficiency resulting from competition among plant organs. This theory was undermined when researchers learned that even the formation of small male flowers in the dioecious* spinach plant led to senescence. Since these male flowers do not

***Dioecious* plants are unisexual; male and female flowers being borne on separate plants. *Monoecious* plants (for example, corn) also have unisexual flowers but both sexes occur on the same plant. *Perfect* flowers contain male and female organs in the same flower (for example, lily.)

develop further, and thus constitute no nutritional drain on the plant, their senescence-inducing effect must involve some other mechanism.

One genetic line of peas produces flowers and fruits in short days, but does not senesce unless transferred to long days; long days alone, without fruits present, will not cause senescence. Since fruits develop fully in short days without any senescence-promoting effect, senescence cannot be nutritionally caused. Rather, some change in the hormonal balance may occur, so that in long days, insufficient senescence-preventing hormone is present to counteract whatever senescence-inducing effects the fruits may have. We know that a graft-transmissible substance is responsible for preventing senescence in these peas in short days, but its nature is still unknown. Since this strain of pea differs by only one gene from peas whose senescence patterns are independent of daylength, the control of senescence may reside in a single gene product, possibly an enzyme controlling the synthesis of some particular hormone.

Whether in the individual organ or the whole plant, senescence involves declining metabolic rates and decreasing rates of RNA and protein synthesis. We have already discussed the changes in respiration rate and membrane permeability that accompany fruit ripening. Most hormones that delay senescence act at least partly through maintenance of RNA and protein synthesis. For example, in fruit tissue such as bean pod, senescence is delayed by either auxin or cytokinin; in some leaves, cytokinin alone is effective while in others, gibberellin alone delays senescence. Many studies indicate that senescence in plants is not a simple running-down and fading-away process but represents rather an active physiological stage of the life cycle, as much controlled by hormones as any that precede it. In plants, death of individual cells or tissues may be a normal, controlled and localized event, helping in the production of the final form of the plant. One example is the death of tracheid and vessel cells, to form the hollow but efficient cells of the water-conducting system.

From the time of germination onward, the plant undergoes a series of developmental changes, each directed through selective control of parts of its genetic apparatus. The induction and repression of particular genes may be controlled by plant hormones, but we still do not understand the master control system that programs the induction and repression of specific genetic activities. Even in mature, fully differentiated cells, the complete genetic complement exists, containing all the information required to form a whole plant. This potential can be elicited when certain mature cells are placed in tissue culture, where they are stimulated to dedifferentiate, to reinitiate cell division, and ultimately to give rise to a whole new plant. This type of experiment indicates that cell maturation and senescence do not result from the loss of genetic material; rather, they appear to result from, or at least to be correlated with, a series of changes in the pattern of protein synthesis in the cell. Cellular senescence may thus result from altered relative activity of different genes, leading to a failure to produce sufficient m-RNA to maintain those functions essential to cellular integrity, while permitting the synthesis of excessive quantities of various degradative enzymes such as nucleases and proteases. Frequently, senescence is also accompanied by an alteration of membrane permeability, leading to leakage between cellular compartments of materials that are normally separated. This is particularly true of the vacuole, which often contains toxic compounds that are released into the cytoplasm if the tonoplast becomes leaky; there, they further promote senescence and death of the cell.

Abscisic Acid—The Stress Hormone

To be able to survive in a world of changing environment and constant challenge, the green plant must be able to stop as well as to start its various physiological processes. Buds must not only be capable of growth when the favorable weather of the spring arrives, they must also be capable of ceasing activity and preparing for the unfavorable conditions of an approaching freezing winter or a long dry spell. Preparation for these unfavorable periods must be made well in advance of their arrival if the plant is to survive them, since some of the adaptations involve the formation of new substances (wax), new tissues (periderm), or new organs (bud scales). The plant must therefore develop a system for sensing and anticipating the approach of the unfavorable period.

Let us consider how a deciduous tree of the North Temperate Zone responds to the challenge of an approaching winter. Examination of a dormant winter twig reveals that the tender terminal growing point is protected from the outside world in several ways (Fig. 10-9). Surrounding it is usually a tuft of cottony insulating material, composed of hairlike organs modified from leaf primordia in the bud. Instead of the usual expanded leaves produced at nodes, there are small, thickened bud scales. The internodes between these bud scales are not at all extended; thus the bud scales overlap in shingle fashion and are closely appressed, forming a tightly fitting protective cap surrounding the insulated growing point. In addition, the scales are covered by a protective varnishlike material that effectively waterproofs the bud and

(a)

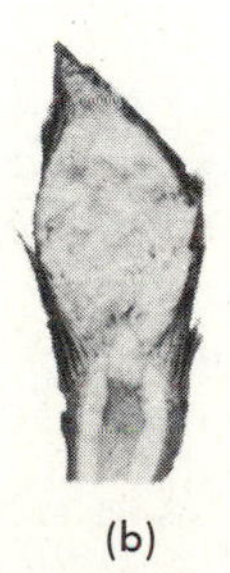
(b)

(c)

Figure 10-9 Winter buds of *Aesculus hippocastanum* (horse chestnut) showing the protective devices to withstand the winter conditions. (a) External view—sticky outer scales. (b) Longitudinal section—overlapping bud scales encase the highly compacted young foliage leaves enveloped in cottony down in the center. (c) Two of the young palmate leaves have been spread out to show the thick protective cottony covering.

furnishes additional insulation protection. Below the terminal bud, all the leaves have neatly abscised and the leaf scar marking the point of their previous attachment to the stem is well protected by a corky layer that effectively seals the broken vascular bundles and the cells around them. Finally, within the trunk, the cambium has become dormant and the parenchyma cells of the xylem and phloem have developed very high osmotic concentrations and have undergone other chemical changes that will enable them to withstand freezing injury. Metabolism, while continuing, does so at a very slow rate. The plant is now *dormant*.

Dormancy is of two types. Many temperate trees grow for only a few weeks after starting growth and then enter a period called *summer dormancy*, in which no further growth occurs even though growth conditions are favorable and photosynthesis is proceeding rapidly. During this period of great metabolic activity, the apical bud produces the leaves and flowers in the bud for the following year's growth. The causes of summer dormancy are uncertain but growth inhibitory hormones may be involved. Summer dormancy can usually be broken by a high level of nutrient or water supply, or by defoliation such as might be produced by insect attack. At the end of the summer, the decreasing day lengths cause many trees to enter a period of true dormancy that can be broken only by low temperatures during the winter. The changing photoperiod is detected by the leaves, where it causes the production of growth inhibitory substances along with a decrease in the growth promotive hormones, so that the metabolism of the tree is switched to a *winter-dormant* state.

Once spring comes, the plant faces further tactical problems. If, in fact, inhibitory substances have shut down the synthetic machinery of the cells of the growing point, how does that machinery start up again? This can be accomplished either by gradually destroying the inhibitor (a kind of chemical hourglass mechanism, in which the rate of disappearance of the inhibitor is nicely linked to the length of the winter frost period), or by producing a substance that competes metabolically with the inhibitor, thus overcoming its action by sheer quantitative preponderance. The plant appears to use both these techniques.

The discovery of abscisic acid

Experiments by various investigators had shown that exposure of leaves to short days leads to the formation of overwintering buds, and it was clear that some hormonal signal was involved. By progressive purification of extracts from birch trees (*Betula*) and their assay in a dormancy bioassay, P.F. Wareing and his colleagues in Aberystwyth, Wales, discovered several dormancy-inducing compounds. That compound producing the greatest dormancy effect is now called *abscisic acid* (ABA) (Fig. 10-10). The name derives from the fact that at about the same time, F.T. Addicott and his

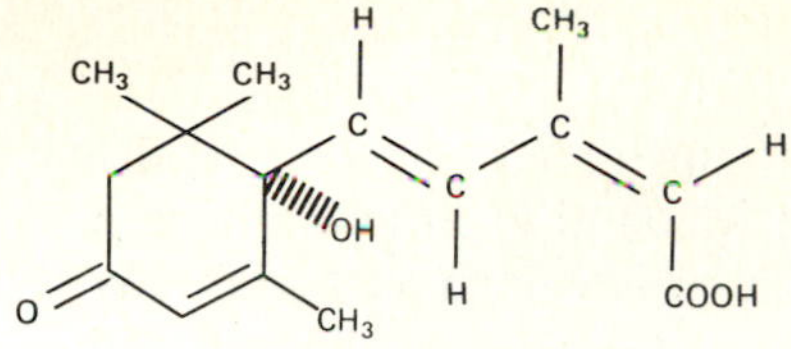

Figure 10-10 The structural formula of abscisic acid.

colleagues in California discovered that the same substance is associated with abscission of cotton bolls. In general ABA appears to play little role in leaf abscission, making the name somewhat inappropriate.

During the last few years the role of abscisic acid in promoting dormancy in trees has been the subject of controversy. ABA applications will delay bud break in twigs when dormancy has been overcome by cold treatment (Fig. 10-11), but when the levels of abscisic acid in trees are assayed throughout a season, or under different growing conditions, high levels of ABA do not always correlate with dormancy, and vice-versa. For example, the level of ABA in actively growing apple shoot tips was found to be higher than in dormant shoot tips, though assays in both peach and grapes did show maximum ABA content coinciding with the maximum depth of winter dormancy. The apparently contradictory nature of these results can probably be rationalized by the concept of hormonal balance; that is how much inhibitory ABA is present relative to growth-promoting hormones such as gibberellins and cytokinins. For example in maple (*Acer*), in which ABA content is highest in young twigs, the level of gibberellin is lowest in short days; the ABA present in twigs can probably induce dormancy during short days, but not during long days, when high gibberellin levels could overcome effects of ABA.

We are not sure how genetic control of summer dormancy may be regulated. Certain trees unfold only the leaves present in the bud and then

Figure 10-11 Abscisic acid treatments have delayed bud break in white ash cuttings. Left to right: control, 0.4 ppm, 2.0 ppm, 10.0 ppm abscisic acid. The photo was taken after 22 days of treatment. (From Little and Eidt. 1968. *Nature 220*: 498–499.)

form another bud, while others continue forming leaves, but stop after a time in midsummer. Environmental conditions such as water stress certainly can encourage this dormancy, but the possibility of hormonal control has been little studied. In willow (*Salix*), ABA rises dramatically in July, just before growth ceases, but in other species no clearcut pattern has been noted. Winter dormancy is usually broken by exposure to several weeks of temperatures just above freezing. This period has been assumed to be related to ABA destruction, but when the total ABA content is measured, often no differences are found between experimental and control plants. The action of the cold period may be explained, rather, by the observation that increasing amounts of ABA become transformed into an ABA-glucoside as the cold period progresses; this may be a way of sequestering or inactivating the ABA. During the breaking of dormancy, the levels of various promoters such as gibberellin and cytokinin also increase, and these probably counteract the effect of any ABA present (Fig. 10-12). The state of the bud is therefore the result of a balance between promotive hormones and ABA, the latter usually working

Figure 10-12 The xylem sap of willow shows a high content of ABA in winter when the shoots are dormant and also in the summer during the period when growth ceases. The breaking of dormancy in spring is associated with a decrease in ABA and a rise in the cytokinin content of the sap. The times of flower bud opening and leaf bud opening (arrows) coincide with the two peaks in cytokinin content. (Redrawn from Alvim, 1976. Plant Physiology *57*, 474–476.)

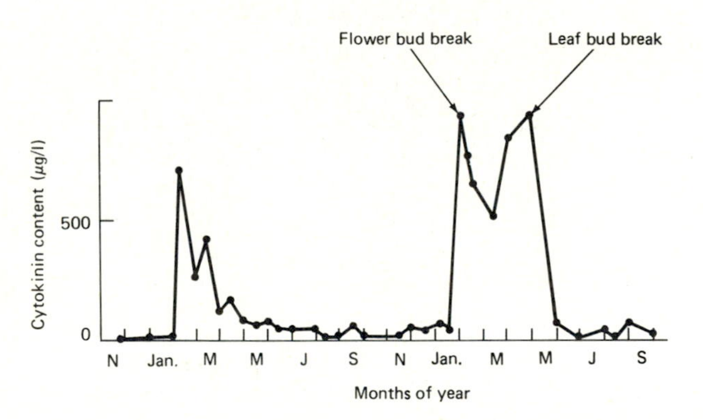

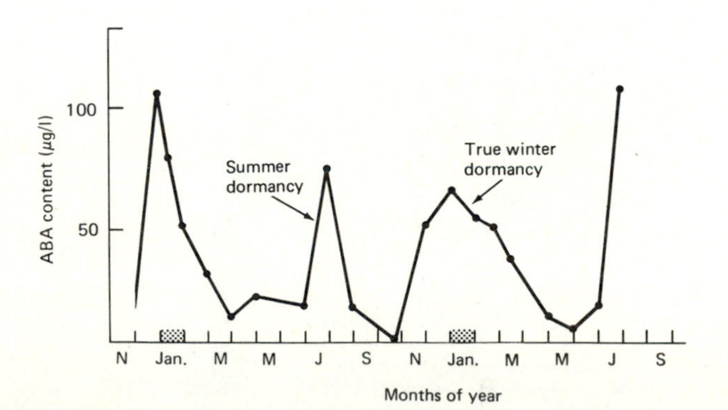

only when the levels of promotive hormones such as gibberellin decline during short days.

The hormones which break dormancy appear to be synthesized, and the ABA degraded, directly in the tissues of the bud itself. Thus, if only certain buds on a tree are given a cold treatment, those buds start growing when transferred to warm conditions, while the other, untreated, buds still remain dormant; this clearly shows that the breaking of dormancy, unlike its onset, is not controlled by hormones moving throughout the plant.

Seed dormancy and germination

Seed dormancy, like the winter dormancy of buds of deciduous trees, has obvious survival value for the plant, for if the seed were to germinate as soon as it matured in the autumn, the tender seedling would be at risk from the first frost. The incorporation into the seed of a critical quantity of an inhibitor such as abscisic acid guarantees a lack of activity during the winter season. In many seeds, germination is prevented by inhibitors found either in the seed coat, in persistent organs of the flower, or in the tissues of the fruit that surround the seed. Frequently, germination can be induced by the simple technique of exhaustively leaching the seed in running water. Although it appears that plants can use many different kinds of inhibitors to achieve such dormancy, abscisic acid seems to be one of the most important. In *Fraxinus americana* (ash), for example, abscisic acid is present in all tissues, but the highest concentrations occur in the seed and pericarp of dormant fruits.

In nature, seed dormancy is normally broken by heavy rainfall, light, or low temperature. In desert plants, inhibitor-induced dormancy is removed when the seeds are washed in the abundant water of a heavy rainstorm. Light control of germination usually involves the phytochrome system (Chapter 11). In seeds requiring exposure to cold to break dormancy, relatively low temperature (2 to 5°C) for a period of about 4 to 6 weeks is needed (Chapter 12). As dormancy is progressively broken by chilling, the level of ABA falls, while the level of growth-promoting hormones such as gibberellins rises either during or immediately after the chilling period.

Once seed dormancy has been broken, germination can take place. Under the favorable spring conditions of warm temperatures and adequate water and oxygen supplies, metabolism quickens. If water is added to dry nondormant seeds, protein synthesis in the embryos starts, often within 15 minutes. Synthesis of messenger RNA to code for some of these proteins may also occur, but in some seeds, such as cotton, protein synthesis precedes the synthesis of new m-RNA. This can occur because dormant messenger RNA, made during the development of the seed in the previous growth season, is stored in the embryo during the long period of seed desiccation, ready to be translated into protein immediately after hydration. These long-lived m-RNAs normally code for enzymes, such as protease and isocitratase in cotton, that are involved in the breakdown of the food reserves. In cotton

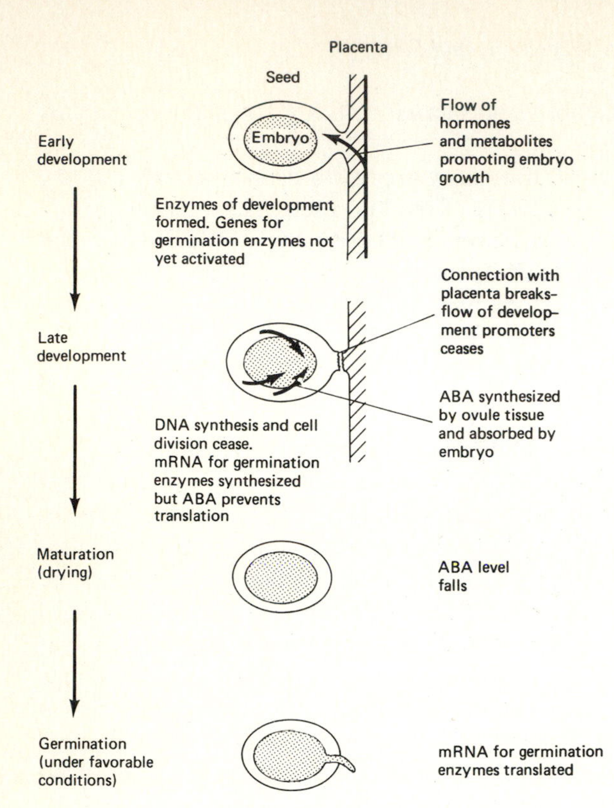

Figure 10-13 Changes in hormones and protein synthesis machinery during the development, maturation and germination of cotton embryos.

the m-RNA that is synthesized during the first 60% of the embryo's development is directly translated into protein needed in development, but in the final 40% of the embryo's growth, much of the m-RNA synthesized is stored and not translated until germination.

We are left with the questions of how the RNA metabolism in the cotton embryo changes as the embryo develops and what prevents immediate translation of the long-lived m-RNAs after they are formed. One clue is that washing the seeds removes the inhibition of translation of the dormant m-RNA, while adding back either the washings or ABA reimposes the block on translation. Immature cotton seeds contain ABA which appears to prevent the translation of m-RNA into the enzymes required during germination. In the cotton ovule, ABA appears only after the ovule reaches about 60% of its final size (Fig. 10-13), an event which coincides with the breaking of the vascular connection to the ovule. Prior to that time, another substance coming from the plant, perhaps a hormone such as cytokinin, maintains cell division in the ovule and activates the transcription of that part of the DNA which forms the m-RNA for the germination enzymes. Once the vascular connection is broken, this supply is cut off. ABA production starts, DNA synthesis and cell division cease, and the m-RNA for the germination enzymes is transcribed but not translated. As the cotton seed matures, ABA production ceases and ultimately its level declines through repeated leaching by rainfall and possible enzymatic degradation. The available m-RNA is therefore ready for translation once the mature seed takes up water in the soil as the first step of germination.

Abscisic acid and water loss

At the beginning of this chapter we referred to abscisic acid as the "stress hormone," which is an appropriate description of its overall role in the plant. Generally it is formed in response to stress or unfavorable conditions, and it in turn changes the plant to accommodate to that stress. The most striking example of such a response is the rapid synthesis of ABA in response to water stress, that is, a shortage of water (Fig. 10-14). When a plant is deficient

Figure 10-14 Changes in the water potentials of leaf cells and in the abscisic acid concentration of *Ambrosia trifida* leaves as the plant loses water to a dry atmosphere. Note that the abscisic acid content does not start to increase until a threshold leaf water potential has been exceeded. (Unpublished data, T. Zabadal, Cornell University.)

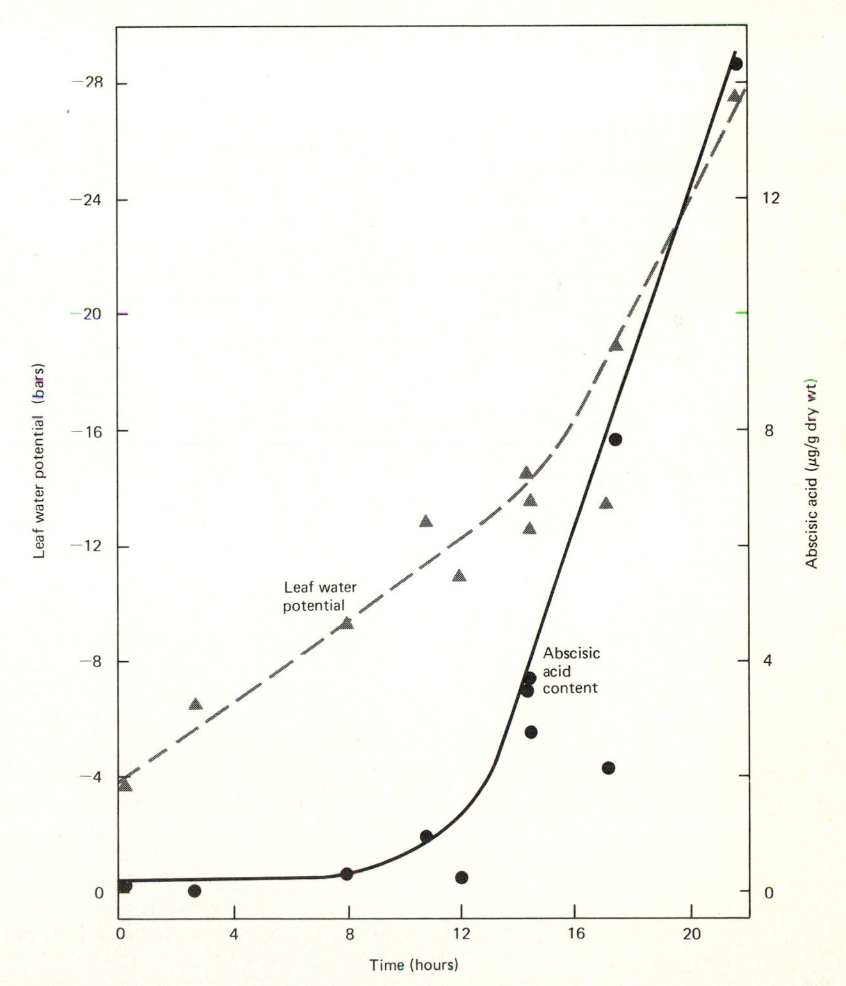

in water, the ABA content of the leaves rises rapidly. This ABA then acts on the guard cells, "deflating" them so as to close the stomata rapidly, long before such a closure would occur from overall water loss by the plant (Fig. 10-15). As we saw in Chapter 6, stomatal movement is caused by the movement of potassium ions into and out of guard cells; the role of ABA in this process is to cause the potassium to leave the guard cells, so that the stomata close. With the stomata closed, the leaf limits its water loss and is better able to withstand the dry period. Other stresses on the leaf such as low temperature can also lead to the synthesis of ABA and the closure of stomata. When water is resupplied to the plant the stomata do not immediately open, as it takes some time for the ABA content to drop. When it does so, potassium is once again transported into the guard cells and the stomata open.

Other examples of ABA-mediated alteration of plant growth are numerous and follow no general pattern, except that the process often occurs in response to unfavorable conditions. For example, in one variety of bush beans (*Phaseolus*), plants flower normally under short days, but abscise their developed buds under long days; this could be correlated with the rise in ABA in the plants under long days, whose unfavorable action would result from promotion of the development of an abscission layer below the bud.

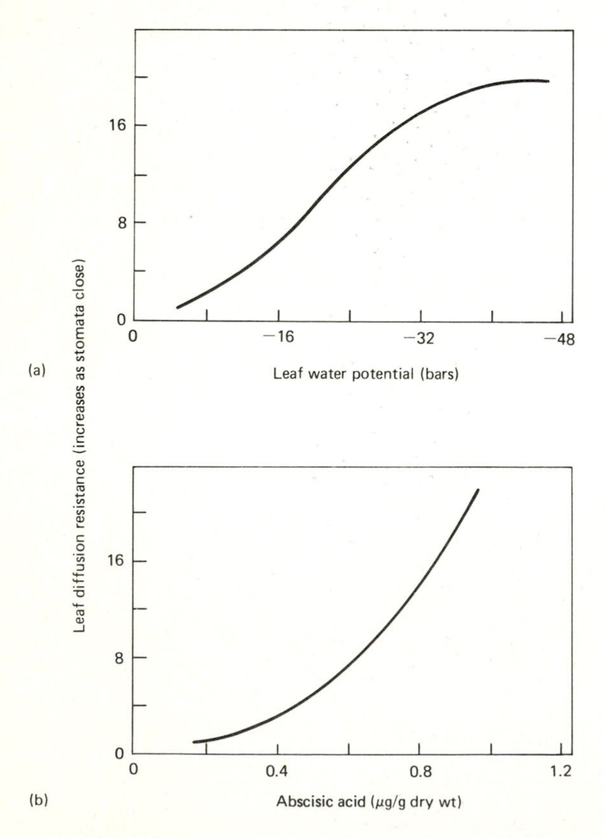

Figure 10-15 These graphs show the relationship between leaf water potential, stomatal aperture, and abscisic acid content of the leaves of *Helianthus decapetalus*. Stomatal aperture is measured as the reciprocal of the resistance to the passage of gas through the stomata; the higher the resistance, the more closed the stomata.
a) As the water potential of leaf cells becomes more negative, the stomata close.
b) Stomatal closure is correlated with the abscisic acid content. (Unpublished data of T. Zabadal, Cornell University.)

How abscisic acid works

One of the main roles of ABA in plants appears to be the control of stomatal closing. This is a rapid reaction, whereas ABA's induction of dormancy involves a slower, generalized change in metabolism. As with auxin, ABA must act through several basic mechanisms. In stomatal closing it affects the movement of potassium out of the guard cells; in inducing dormancy, it appears to act by shutting down synthesis of RNA and proteins. This last action must occur in a selective fashion, since if all cell division and protein synthesis were to stop, even dormant buds would not form. External application of ABA, which presumably mimics an internal rise in ABA level, also rapidly inhibits growth of stems or coleoptiles. This role of endogenous ABA in limiting growth is not yet understood. Finally, one might ask whether abscisic acid is involved in the senescence and death of annual plants referred to earlier. ABA often increases in senescing leaves, but attempts to correlate ABA with the death of the whole plant have so far been unsuccessful. Obviously we still have much to learn about the hormonal control of this aspect of the green plant's life cycle.

SUMMARY

Fruit ripening and other aspects of development and *senescence* in plants are regulated by the production of a gaseous hormone, *ethylene*. The fact that effects of ethylene may be negated by high ambient concentrations of CO_2 has led to the practice of high-CO_2 storage of apples and other fruits. Ethylene causes the formation of the *apical hook* in many etiolated seedlings, and the effect of light in opening the hook is related to its inhibition of ethylene production. Ethylene can also interfere with *geotropic response* and other auxin-mediated responses, such as the inhibition of lateral bud growth. Auxin and ethylene interact in many ways, since high auxin levels trigger ethylene production, while high ethylene levels can cause induction of an enzyme, peroxidase, that inactivates IAA. Ethylene is formed from 1-aminopropane-1-carboxylic acid (ACC), a *methionine metabolite*. Once determined exclusively by bioassay, ethylene levels are now routinely assayed by *gas chromatography*.

The application of ethylene to many unripe fruits results in a marked rise in respiratory CO_2 output called the *climacteric*. Following this change, organic acids decline, intercellular pectins are degraded, and the fruit becomes ripe. Application of ethylene to leaves similarly triggers a new set of metabolic events leading to *abscission*; these include new cell divisions, forming an *abscission layer* of weak-walled cells, whose digestion by newly formed *cellulase* brings about leaf fall. Daylength may regulate leaf fall by initiating the aging process in the abscission zone, which must occur before abscission can take place. In some cases, the application of low concentrations of auxin to leaf blade or fruit can retard senescence by opposing ethylene action; such abscission prevention is important in fruit production. Conversely, high levels of some auxins can accelerate leaf fall, and such planned defoliation has also found some practical applications. In other types of ethylene-induced senescence, stimulating hormones, including auxin, cytokinin, and gibberelline, can oppose the action of ethylene.

The change of seasons in temperate zones poses survival problems for plants which are partially overcome by their induction into and release from *dormancy*. Thus, diminishing daylengths can induce the formation of a hormone, *abscisic acid* (ABA), that reduces the active vegetative growth of a bud or a developing embryo, and sets in motion the events that lead to winter dormancy. Conversely, the resumption of growth in the spring may result from the overwintering breakdown of this dormancy-inducing hormone, or from synthesis of substances, such as cytokinins, that overcome the effects of ABA. Such dormancy breakage in buds is induced by low winter temperatures, while spring bud break follows increasing temperature and daylength. In seeds, water, light, or low temperature may break dormancy. In many kinds of seeds, some messenger RNA coding for proteins needed in germination is produced before the induction of dormancy; it remains stable and nonfunctional in the dry seed, awaiting germination signals.

ABA is generated in high concentrations during water stress of some leaves; this induces stomatal closure through K^+ loss from guard cells, thus preventing desiccation damage. Both ABA and ethylene appear to act in part through effects on differentially permeable membranes and in part through control of protein synthesis.

SELECTED READINGS

Abeles, F. B. *Ethylene in Plant Biology* (New York: Academic Press, 1973).

Phillips, I.D.J. *Introduction to the Biochemistry and Physiology of Plant Growth Hormones* (New York: McGraw-Hill, 1971).

Wilkins, M. B., ed. *The Physiology of Plant Growth and Development* (New York: McGraw-Hill, 1969).

See also Selected Readings *at the end of Chapter 9.*

QUESTIONS

10-1. Auxin and ethylene produce effects that are closely interrelated yet often antagonistic. Provide examples and discuss the interaction.

10-2. Ethylene is now regarded as a naturally occurring plant hormone. How does it affect growth and development?

10-3. What is the climacteric? What does it indicate? What triggers it?

10-4. What changes occur in a fruit during ripening? Are these degradative or synthetic? What evidence supports your choice?

10-5. The rotting of one apple in an enclosed barrel containing many other apples can induce the others to ripen, in a chain reaction. Explain how.

10-6. Describe the various processes in leaf abscission that are controlled or affected by ethylene.

10-7. Flower fading, fruit ripening, and leaf abscission are all examples of the senescence of plant organs. By comparing these processes, show how senescence is controlled by environment, internal hormone changes, and hormone-directed changes in metabolism.

10-8. Abscisic acid appears to have only a minimal role in leaf abscission. If you were to rename the compound what would you call it and why?

10-9. The messenger RNA for some of the enzymes needed in seed germination is formed early in seed development. How does the plant control its translation at a much later time?

10-10. What is meant by "dormancy"?

10-11. What is the possible survival value of seed dormancy?

10-12. Certain "viviparous" strains of corn and barley produce grains that germinate and grow on the ear of the parent plant, undergoing no period of dormancy at all. How might these seeds be treated so as to render them dormant?

10-13. Describe two causes of seed dormancy and the means of breaking dormancy caused by each.

10-14. Freshly harvested seeds of some species require temperatures between 35 and 49°C for prompt germination. As the seeds age, they can germinate at progressively lower temperatures, finally germinating even at 10°C. Present a possible explanation.

10-15. What is the role of abscisic acid in controlling the rate of transpiration?

10-16. The aquatic plant *Potamogeton* produces two kinds of leaves, elongate ones without stomata, under water, and small ones with stomata, floating on the surface. Adding abscisic acid to the water results in the production of stomata-bearing leaves under water. What is the possible ecological significance of this laboratory observation?

10-17. Abscisic acid, normally considered an inhibitory substance, enhances the growth rate of citrus petiole tissue cultures and certain other cells. How might this be explained?

10-18. The abscisic acid level of some leaves rises rapidly after desiccation. How might this occur?

10-19. Compare and contrast the ways in which the various plant hormones are thought to act.

10-20. What chemical substances act as the precursors of the various plant hormones, and by what pathways are they converted into the hormone molecules?

10-21. Explain the role of hormones throughout the life cycle of an annual angiosperm plant, including the function performed by the hormones at each stage. In your answer include seed germination, vegetative growth, fruit ripening, leaf abscission, and dormancy.

10-22. "The growth and development of a plant is controlled largely by a *balance* of hormones." Discuss this statement with examples taken from the different phases of the life cycle of a plant.

10-23. Explain in detail the involvement of the following hormones in the stated processes:

(a) Ethylene, cytokinins, and auxin in apical dominance
(b) Ethylene and auxin in leaf abscission
(c) Abscisic acid and gibberellin in woody plant bud dormancy and seed germination.

11

Control of Growth by Light

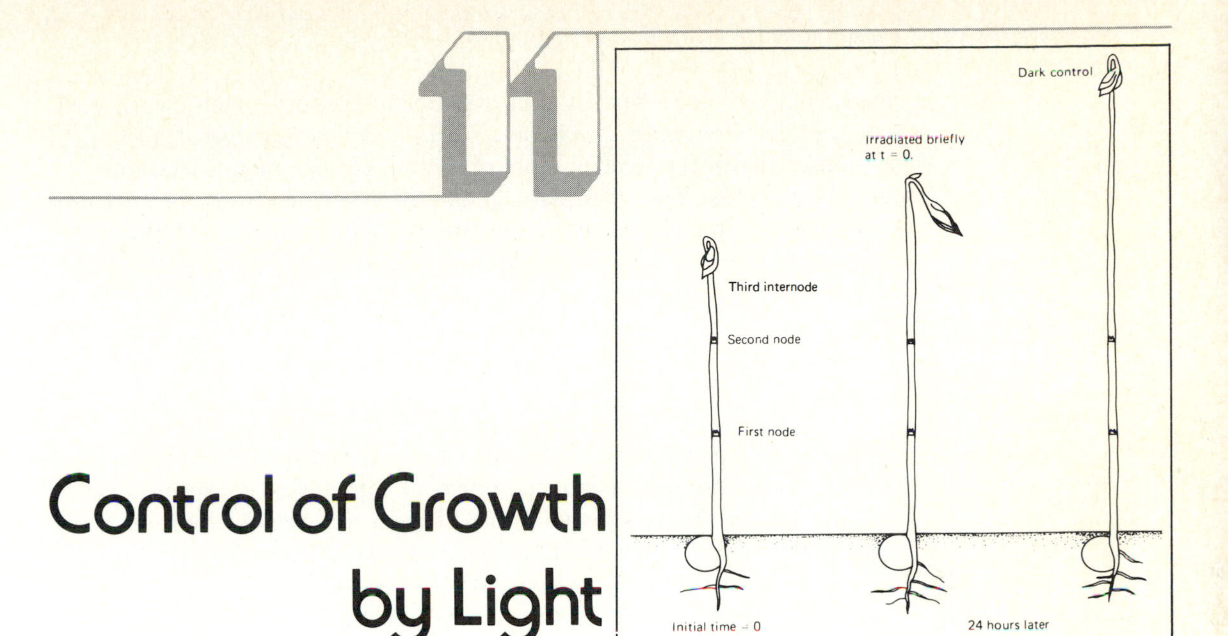

Not only is the green plant a machine that runs on solar energy, but the structure of the machine itself is partially determined by the quantity and quality of the radiant energy it receives. A dark-grown seedling is dramatically altered by the absorption of even a few quanta of light; the rate and direction of leaf and stem growth are modified, the plumular hook is straightened, epidermal hairs are developed, and the pigmentation of leaf and stem and many details of internal seedling anatomy are affected. During later growth, the flowering and fruiting of the mature plant are light dependent; finally, leaf fall, the onset of dormancy in bud and seed, and senescent phenomena are manifestations of light and timing signals previously received.

The various processes through which light influences the form of plants have been collectively called *photomorphogenesis*. Chlorophyll and the various accessory pigments of the photosynthetic apparatus are not the main contributors to photomorphogenic light absorption. Rather it is the phycobilin pigment called *phytochrome* and an as yet uncharacterized yellow pigment, probably a flavoprotein enzyme, that function as light absorbers for these processes. These pigments are present in vanishingly small quantities within the plant, so it is not surprising that the energies required to saturate the photoprocesses involved are several orders of magnitude lower than those involved in photosynthesis. The products of these light reactions do not control plant growth stoichiometrically, as does photosynthetically produced sugar; rather, they seem to affect processes such as membrane permeability, gene function, and enzyme activity, whose perpetuated changed status greatly amplifies the effect of each absorbed quantum. Thus, while 8–10 quanta of light are cooperatively required to liberate one molecule of oxygen in photosynthesis, the same number of quanta per cell might determine the entire reproductive fate of a plant or the direction of growth of an entire stem. In this chapter, we shall inquire into the pigments, the photoprocesses, and the subsequent physiological mechanisms involved in photomorphogenic control.

Discovery of Phytochrome

The history of the discovery of phytochrome in the early 1950s by a group of scientists at the United States Department of Agriculture constitutes one of the most exciting chapters in the history of plant research. The story actually begins about 30 years earlier with the work of two experimenters in the division of tobacco plant investigations. W. W. Garner and H. A. Allard were attempting to propagate a mutant type of large-leaf tobacco, Maryland Mammoth, which had arisen by chance as a single individual in a field of other tobacco plants. As the season progressed, the original type flowered profusely, but the Maryland Mammoth did not. Wishing to obtain seeds of this valuable new type and fearful that the plant might not flower before the autumn frost, the investigators transferred the plant to the greenhouse. Despite every urging, however, the plant steadfastly maintained its vegetative habit until approximately mid-December, when it initiated floral primordia many months after the normal plants had successfully completed seed production.

An analysis of the various factors that might be responsible for this unusual behavior led Garner and Allard to the inevitable conclusion that the plant initiated flowering activity as a result of the very short daylengths characteristic of the Northern Hemisphere at Christmas time. They discovered that flower initiation could be induced at will by transferring Maryland Mammoth plants to special chambers in which the length of day could be artificially shortened (Fig. 11-1). Maryland Mammoth tobacco, which flowers only if the daylength is shorter than a certain critical value, was called a *short-day* plant. Many other plants, including some soybeans, chrysanthemums, and poinsettia fall into this category. In contrast, spinach and certain cereals flower only if the daylength exceeds a certain critical value; they are accordingly called *long-day* plants. Finally, there is a class of plants, such as the tomato, whose flowering is relatively independent of daylength; these are called *day-neutral.* The response of plants to length of day is called *photoperiodism.*

The critical photoperiod varies for different species and even for different varieties of both short- and long-day plants. It is 14 hours for Biloxi, a short-day variety of soybean that grows at a latitude of 35°, while variety Bato-

Figure 11-1 Maryland Mammoth tobacco plants grown under short-day (left) and long-day (right) conditions. (From A. E. Murneek and R. O. Whyte, Vernalization and Photoperiodism. Waltham, Mass.: Chronica Botanica Co., 1948.)

rawka, normally found at latitudes of 45° or higher, will flower even under continuous light. This variation in critical photoperiod is important in regulating the distribution of plants over the earth's surface.

During the years following the discovery of photoperiodism, it became clear that many plants respond to the length of uninterrupted darkness rather than to the length of the light period. Thus, a so-called short-day plant is really a "long-night" plant; it requires a certain minimum duration of uninterrupted darkness for the initiation of its floral primordia. In the same way, a long-day plant may in reality be a "short-night" plant; it will flower only if the night period is not longer than a certain critical maximum.

The effective period of darkness for a short-day plant can be rendered ineffective by the simple expedient of shortening it a bit (even a few minutes will do) or by interposing a brief flash of weak light in the middle of the dark period. This indicates that the plant can "measure" the length of darkness to within a few minutes, and that an extraordinarily sensitive light-receiving system is at work in photoperiodism. In the short-day cocklebur plant, flowering will occur on a regime of 15 hours of light and 9 hours of dark, but not if the dark period is only $8\frac{1}{2}$ hours or if the 9-hour dark period is interrupted by a short light break (Fig. 11-2). But a single appropriate dark period can lead to flowering, even if the subsequent dark periods are unfavorable. This phenomenon is known as *photoperiodic induction.* In various long-day plants, a similar but reverse phenomenon occurs; the interruption of an unfavorably long dark period by a flash of light *leads to* induction and floral initiation. Thus, long-day and short-day plants seem to possess the same kind of photoperiodic mechanism, but they somehow work in reverse fashion.

The short-day Biloxi soybean plant is so sensitive to light that the inductive effect of long dark periods can be cancelled by even one minute of unfiltered incandescent light in the middle of the night. For this reason, H. A. Borthwick, S. B. Hendricks, and their colleagues reasoned that it would be an ideal

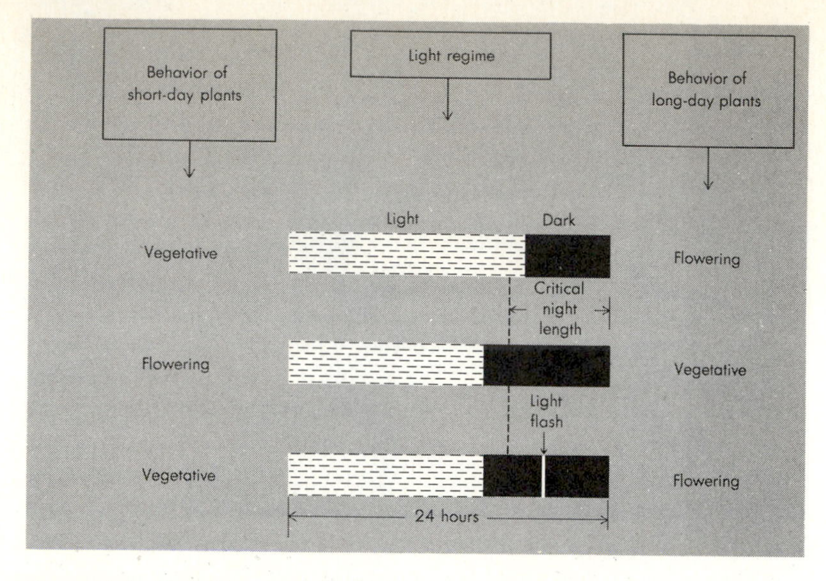

Figure 11-2 The effect of a light-flash interruption of the dark period on flowering in short-day and long-day plants.

Figure 11-3 The method used by Borthwick, Hendricks and colleagues to hold single leaves (these are soybean leaflets) in the image plane of a spectrograph for subsequent irradiation with various wavelengths of light. (From Hendricks and Borthwick, 1954. Proc. First Int. Photobiol. Cong.: pp. 23–35.)

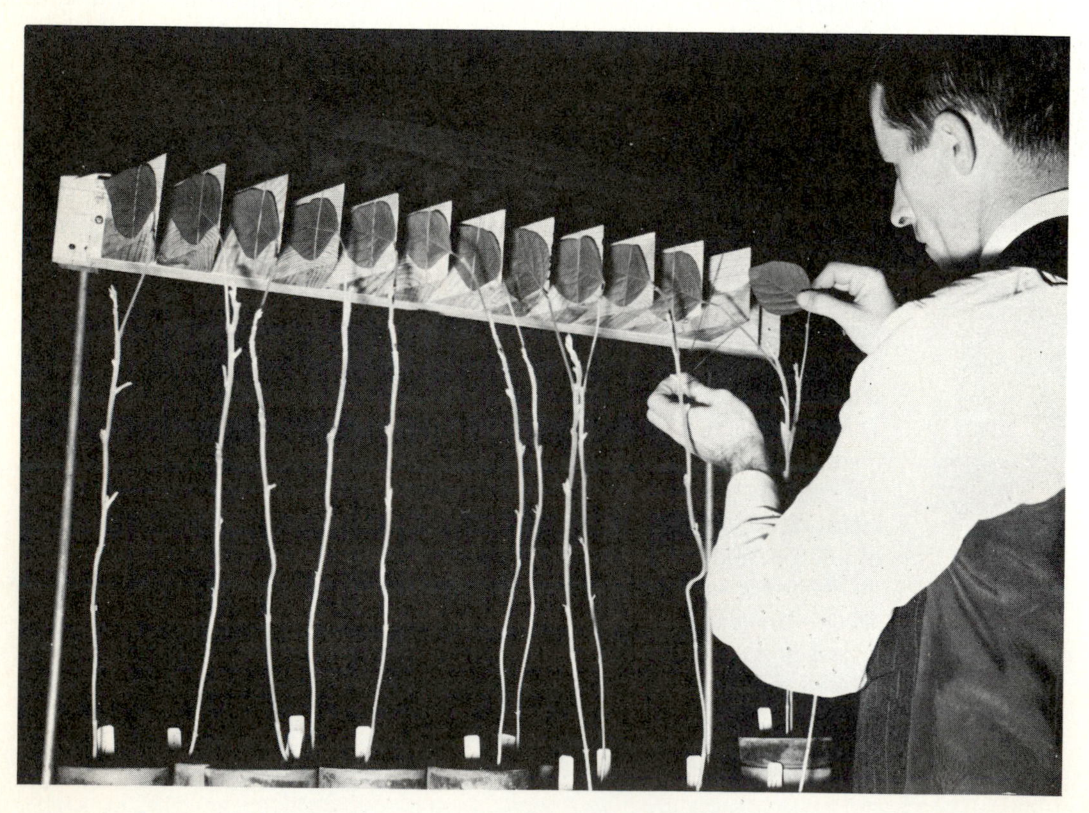

plant for investigating the wavelengths of light most effective in preventing floral initiation; this information, in turn, could be helpful in identifying the photoreceptive pigment involved in controlling flowering. They therefore derived an action spectrum for the process using a large spectrograph (Fig. 11-3) to irradiate groups of plants simultaneously with light of different wavelengths. The resulting action spectra for the inhibition of flowering in the short-day plants soybean and cocklebur, and for the promotion of flowering in the long-day plants *Hordeum* (barley) and *Hyoscyamus* (henbane), were remarkably similar (Fig. 11-4). All showed a maximum of activity in the red region of the spectrum at about 660 nm and very little activity in any other region. The spectra were sufficiently similar to make it likely that the same pigment was controlling the flowering of both long- and short-day plants. Analysis of the action spectrum led to the supposition that the absorbing pigment resembled the algal pigment *phycocyanin*, related to the bile pigments of animals. Unhappily, no such pigment could be detected in their experimental plants, so they continued their investigations in another direction.

It had long been known that the germination of certain seeds is greatly affected by light. For example, moistened seeds of the Grand Rapids variety of lettuce germinate poorly in total darkness but vigorously and promptly if exposed to light for a few minutes. The action spectrum for this effect proved to resemble the action spectra for flowering! Similarly, a pea seedling grown in total darkness has a very long, slender unpigmented stem, a recurved apical hook, and leaves that hardly expand. If such an *etiolated* seedling is exposed briefly to light and then returned to darkness, the leaves expand, the hook

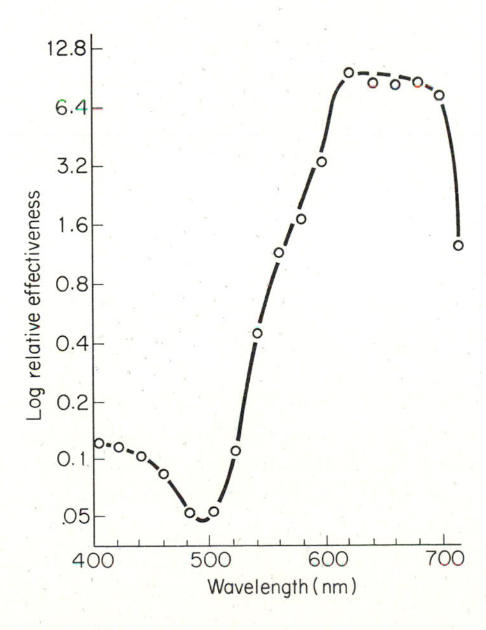

Figure 11-4 A generalized action spectrum for photoperiodism. (Calculated from data of Parker et al., 1949. Amer. J. Bot. *36*: 194–204.)

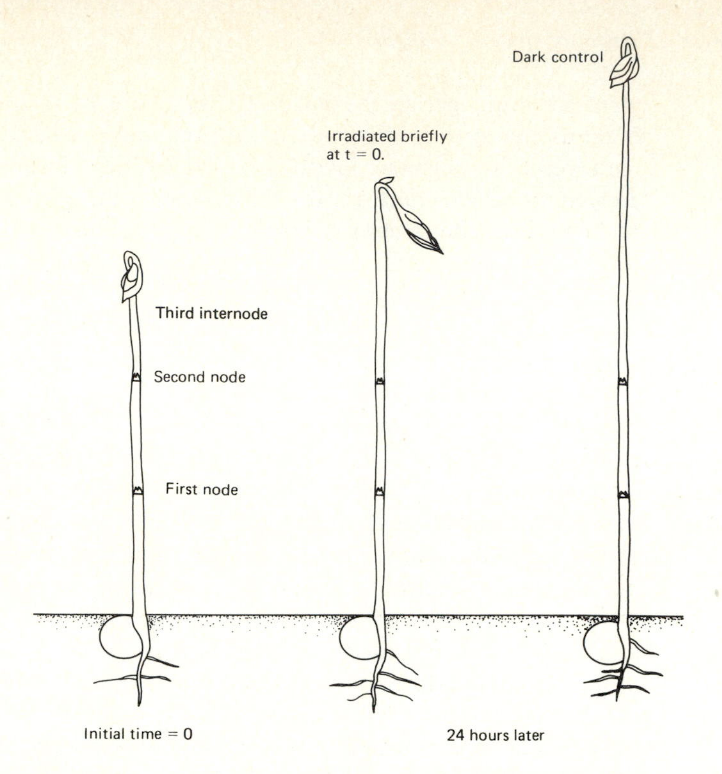

Figure 11-5 A 7 day old etiolated pea seedling before and 24 hours after a 5 minute red irradiation.

opens, and the rate of stem elongation is reduced (Fig. 11-5). Again, the action spectrum was virtually identical with that controlling flowering. The conclusion was inescapable: Reactions as diverse as seed germination, de-etiolation, and floral induction are all regulated by the same receptor pigment.

But what is the pigment? The answer came from a re-evaluation of old experiments on lettuce seed germination, conducted in 1935 by L. H. Flint and E. D. McAlister. These workers had shown that the germination of Grand Rapids lettuce seed is not only *promoted* by red light, but also *inhibited* by light from the far-red region of the spectrum, at about 730 nm (Fig. 11-6). This discovery led to the hypothesis that the effective pigment could be "pushed" in one direction by red light and in the other direction by far-red light. Sequential exposures to red (R) and far-red (F) light confirmed this hypothesis; under such conditions the seed responded only to the last irradiation, as if the form of the pigment fixed by the last exposure was the sole effective control (Fig. 11-7). Even lettuce seed exposed to 100 alternate red and far-red light treatments germinated the same as those exposed to one treatment of each. In all cases, the last treatment before darkness determined the germination response.

Next, Borthwick and Hendricks tested the effect of far-red light on red-light-induced responses in flowering and de-etiolation; again, R–F revers-

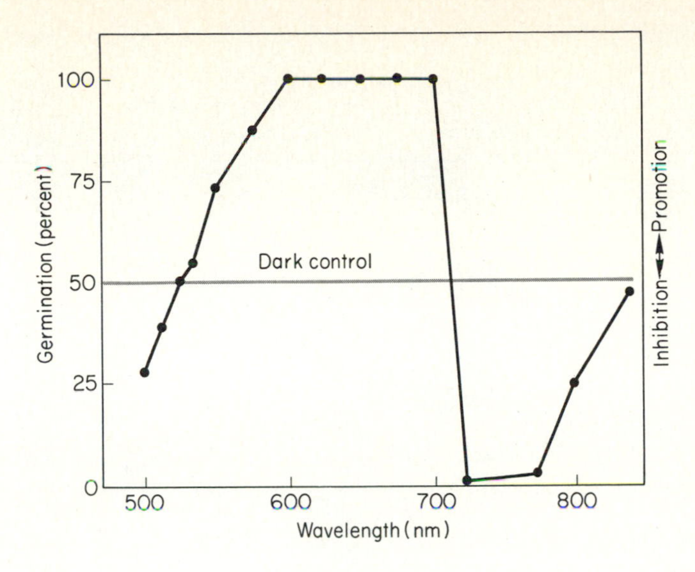

Figure 11-6 Seeds of light-sensitive lettuce induced to 50% germination by weak white light can be further stimulated to germinate by red light (600 to 690 nm) or inhibited by far-red light (720 to 780 nm). (From Flint and McAlister, 1937. Smithsonian Inst. Misc. Collection *96*: 1–8.)

Figure 11-7 Reversibility of the photoresponse for lettuce seed germination. Each lot of seed, after imbibition in darkness, received the indicated succession of irradiations and then was returned to darkness for 2 days.

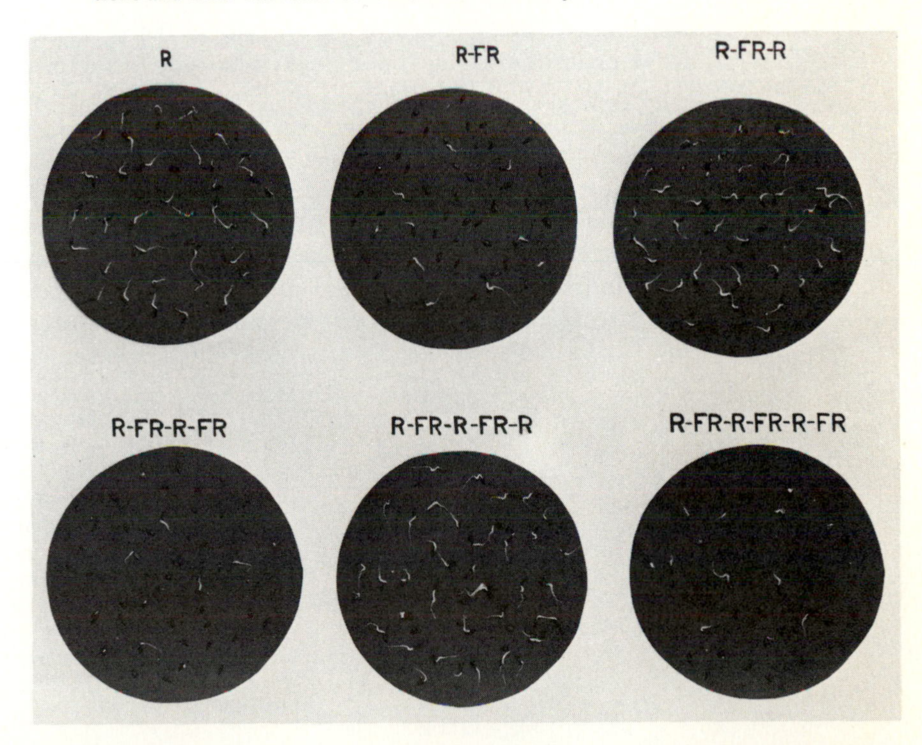

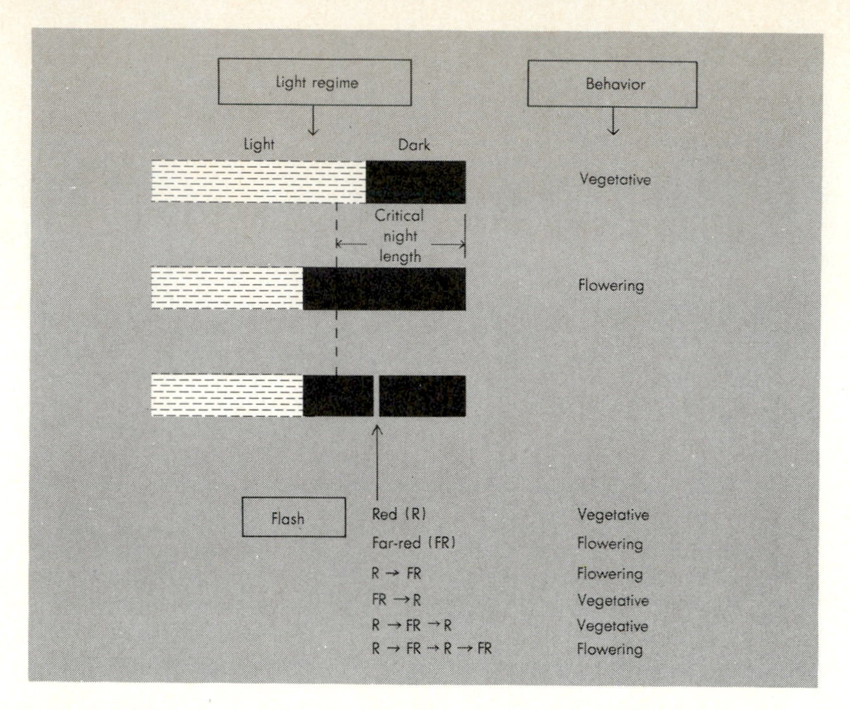

Figure 11-8 Reversible control of flowering in a short-day plant by red and far-red light.

ibility was found (Fig. 11-8). This permitted the prediction that the single active pigment exists in two mutually photoreversible forms, one produced by red light, and one by far-red light (Fig. 11-9). The reversible absorbance changes at 660 and 730 nm following irradiation with red and far-red light, respectively, provided a convenient method for detecting and measuring the quantity of pigment, particularly in etiolated plants where screening by chlorophyll is not a problem (see Box p. 289). Soon the pigment was detected in plant extracts, and it was concentrated, purified, analyzed, and partly characterized. It was named *phytochrome*, from the Greek words meaning *plant* and *pigment*. The two forms are called P_r (red-absorbing phytochrome) and P_{fr} (far-red-absorbing phytochrome). Phytochrome is synthesized in the

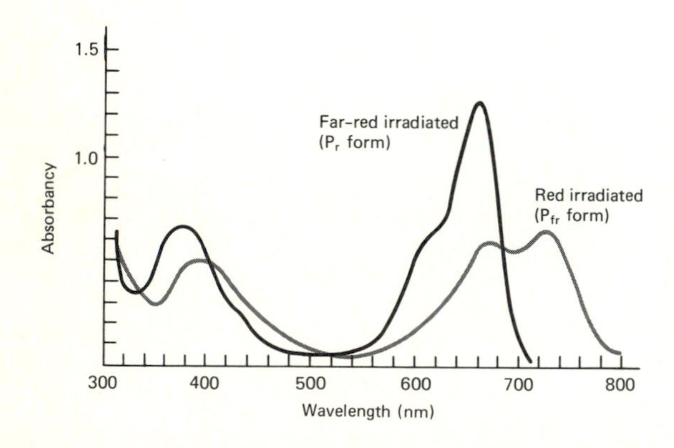

Figure 11-9 The absorption spectra of solutions of phytochrome following irradiation with 660 nm (red light) or 730 nm (far-red light). Note that the absorption spectra of the two forms overlap; 660 nm radiation forms approximately 75% P_{fr} and 25% P_r, while 730 nm radiation forms approximately 2% P_{fr} and 98% P_r. Phytochrome also absorbs in the blue region, producing effects intermediate between those of red and far-red light. (From Siegelman and Butler. 1965. Ann. Rev. Plant Physiol. *16*: 383–392.)

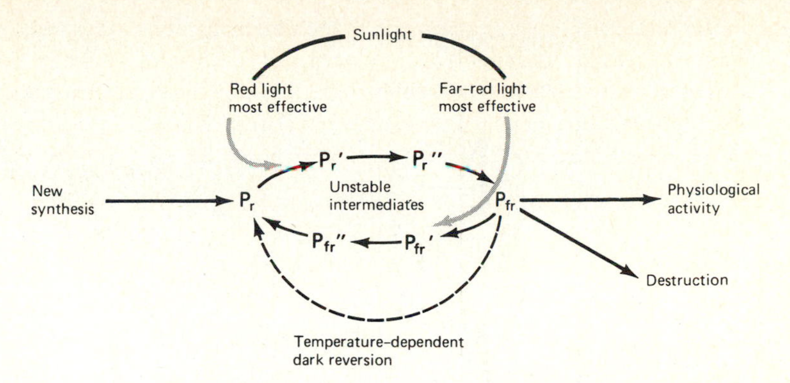

Figure 11-10 The effects of red and far-red light on the state of phytochrome and on physiological activity. Both P_r and P_{fr} pass through short-lived intermediate forms during photoconversion to the more stable forms, as indicated by the series of short arrows. Sunlight establishes ca 50% of the phytochrome as P_{fr}, 30% as P_r, and the remainder as intermediates.

P_r form; therefore etiolated seedlings have P_r but not P_{fr}. Irradiation with red light converts most of the phytochrome to P_{fr}, the physiologically active form, while subsequent irradiation with far-red light converts it back to P_r (Fig. 11-10).

Measurement of Phytochrome by Spectrophotometry

A dual-wavelength spectrophotometer can be used to assay the phytochrome content of plant tissue.

1. Pieces of plant tissue (or tissue extracts) are tightly packed into a cuvette, and irradiated with high-intensity red light (660 nm). The approximately one-minute duration of the actinic irradiation (high-intensity light used to convert the pigment) is sufficient to saturate the conversion of P_r to P_{fr}.

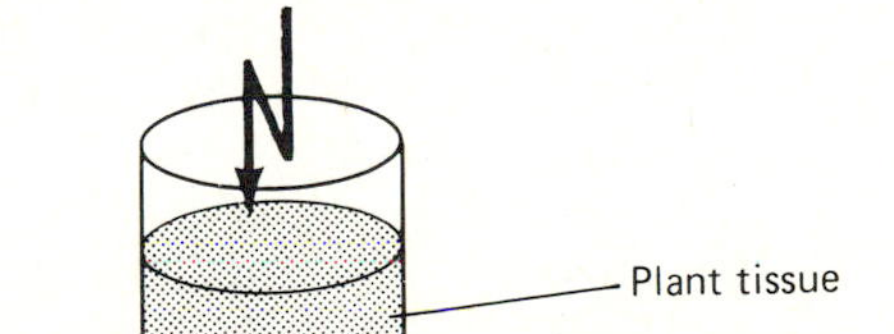

2. Absorbance (A), also called optical density (OD), is measured alternately at 660 nm, 730 nm, 660 nm, etc. The measuring beam is of low intensity and the irradiations are short enough so that they do not have an appreciable effect on pigment conversion.

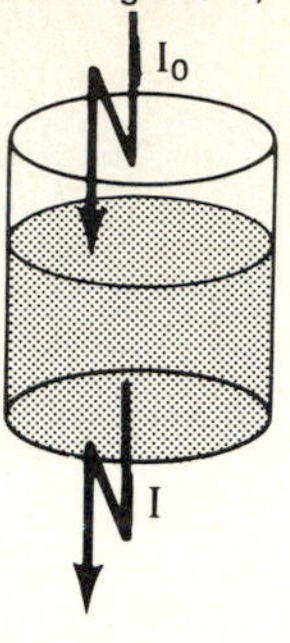

$$\text{O. D.} = \log \frac{I_0}{I}$$

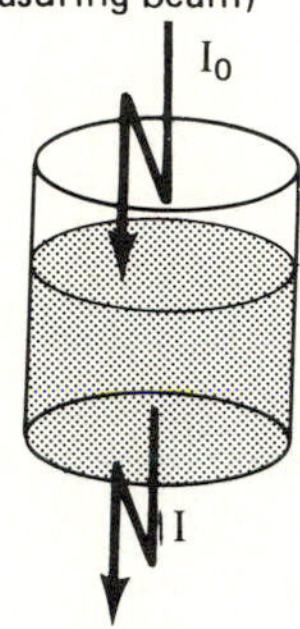

[OD at 660 nm — OD at 730 nm] following 660 nm actinic light is called ΔOD_{660}.

3. Tissue in the cuvette is then irradiated with high-intensity far-red light (730 nm). The duration of the actinic irradiation is sufficient to saturate the conversion of P_{fr} to P_r.

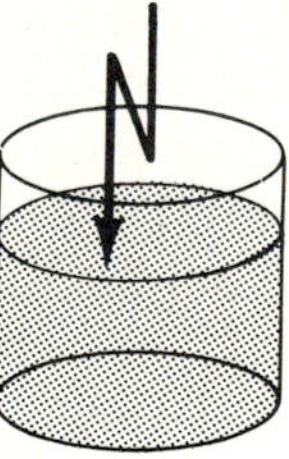

4. Repeat step 2.

[OD at 660 nm — OD at 730 nm] following 730 nm actinic light is called ΔOD_{730}.

5. $\Delta OD_{660nm} - \Delta OD_{730nm} = \Delta(\Delta OD)$, a measure of the phytochrome content of the plant tissue.

As Fig. 11-9 reveals, both P_r and P_{fr} have broad absorption spectra that overlap in the red–far-red region (600–730 nm) and in the blue region (400–460 nm). Therefore any radiation in these spectral regions converts some of

(a)

(b)

(c)

Figure 11-11 Some examples of phytochrome control of morphogenesis. **a)** *Sinningia speciosa* (Gloxinia) seedlings, grown with 8 hours of high intensity white light each day, were exposed to 16 hours low intensity light during each of 15 consecutive nights. The supplementary radiation was: left: far-red light, which establishes a low P_{fr} level; middle: cool white fluorescent light, which establishes a high P_{fr} level; right: dark control. Note the elongated, vertical petioles and the small curled leaf blades of the far-red treated seedling. The plant exposed to white light during the night resembles the dark control except that its leaves are more highly pigmented. (From Satter and Wetherell, 1968. Plant Physiol. *43*: 953–960.)

Figure 11-11b Both *Chenopodium album* seedlings were grown for 21 days under radiation sources providing equal quantum flux densities in photosynthetically active regions (400–700 nm), but different flux densities in the far-red region ($>$700 nm). The plant at the right received more far red; therefore less of its phytochrome is in the Pfr form. (From Morgan and Smith, 1976. Nature *262*: 210–212.)

Figure 11-11c 3 day old seedlings of *Sinapis alba* (mustard) dark grown (left), light grown (right), or dark grown for 42 hours and then exposed to light for 30 hours (middle). (Courtesy of A. M. Kinnersley, Cornell University) the phytochrome to the P_{fr} form. Red light of 660 nm is most effective, producing a 75% P_{fr} level, while far-red light of 730 nm is least effective, producing only 2% P_{fr}. Wavelengths between 600 and 730 nm and blue light all establish intermediary levels. Since all the phytochrome in dark-grown tissue is in the P_r form, virtually any irradiation raises the P_{fr} level. Green light of 500–550 nm is an exception, since neither P_r nor P_{fr} absorbs these wavelengths significantly; for this reason, green light is used as a "safelight" in phytochrome experiments.

Conversions between P_r and P_{fr} act as a metabolic switch that turns reactions on and off. This switching action indirectly regulates a multiplicity of biophysical, biochemical, histological, and morphological processes in plants (Fig. 11-11). Many of these changes occur when etiolated seedlings are first exposed to light and some of their phytochrome is converted to P_{fr}. These changes, collectively called *de-etiolation*, aid the plant in adapting to a lighted environment; they include changes in the activity of many enzymes and in the content of plant hormones; development of chloroplasts from etioplasts; and the synthesis of chlorophyll, carotenoids, and anthocyanin pigments from precursors. Once the etiolated seedling has "greened up," the phytochrome system continues to influence the growth and development of the plant during its entire life. Interconversions of P_r and P_{fr} not only influence floral induction in both long- and short-day plants, but also participate in regulating tuberization, dormancy, abscission, and senescence. However the effects of phytochrome conversion in light-grown plants also depend upon the timing of the light treatment, for the sensitivity of light-grown plants to particular forms of phytochrome shows rhythmic changes. This interesting problem will be discussed in the following chapter.

Characteristics of Phytochrome

The phytochrome molecule consists of two parts, a relatively small, light-absorbing *chromophore* and a much larger, unpigmented protein. The chromophore, like that of phycocyanin and other pigments in algae, is an open-chain tetrapyrrole (Fig. 11-12). When phytochrome absorbs light of particular wavelengths the shape of the chromophore is altered, and this in turn changes the shape of the protein. The protein itself seems to consist of four subunits, arranged in the form of a double dumbbell, with a total molecular weight of about 240,000; this is characteristic of a moderately large protein with about 2000 constituent amino acids.

The conversion of P_r to P_{fr} by a brief red irradiation influences biochemical reactions within the plant during the next several hours of darkness. This remarkable persistence of the effect of a brief light treatment depends upon the fact that P_{fr}, the "active" form of phytochrome, is relatively stable in the dark. Typically, at 25°C, P_{fr} in etiolated tissue has a *half-life* of about

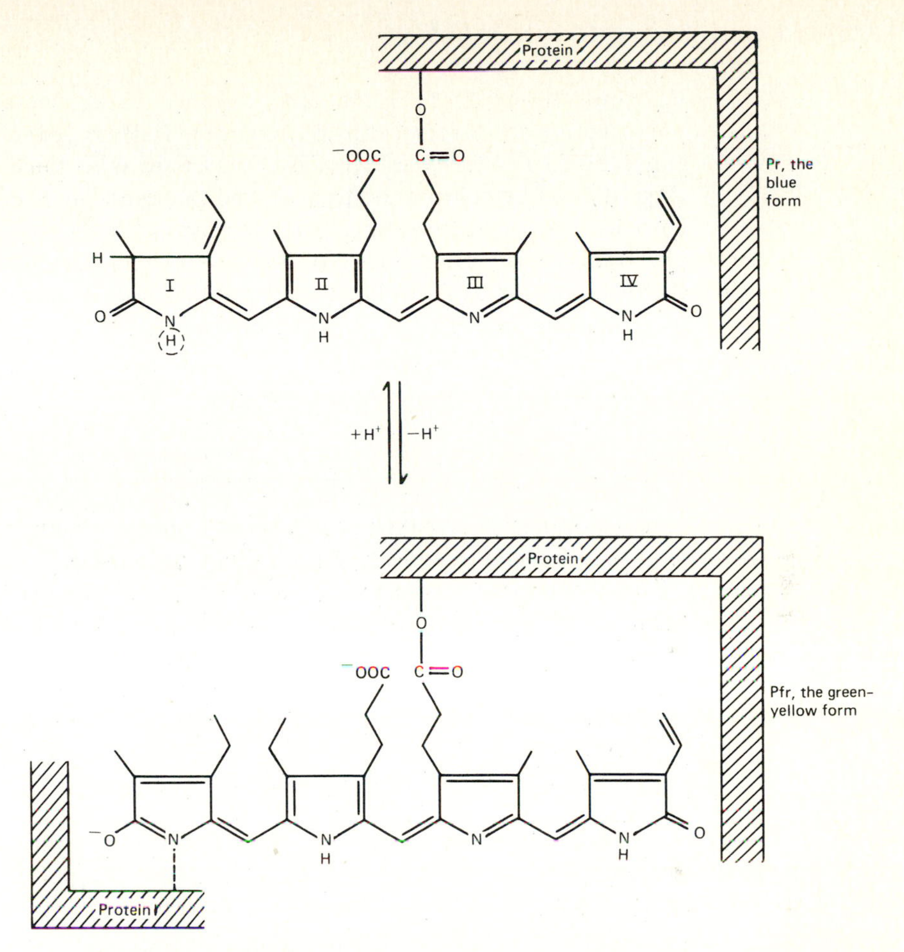

Figure 11-12 Phytochrome consists of a small chromophore (whose structure is altered by the absorption of light) attached to a much larger protein. Note that the chromophore, an open chain tetrapyrrole, contains four individual pyrrole rings joined by carbon bridges, as in chlorophyll (Figure 2-19). According to the model shown above, P_r contains one more proton than P_{fr}. This small difference affects the conformation of the chromophore and the adjoining region of the protein to which it is attached, leading to numerous physiological effects. (Adapted from W. Rüdiger in *Phytochrome.* K. Mitrakos and W. Shropshire, Jr. eds. London and New York: Academic Press 1972.)

two hours (that is, it takes two hours for P_{fr} to decrease to one-half its original level), while P_{fr} in light-grown tissue probably has a half-life of about 8 to 12 hours. In some experiments with light-grown tissues, P_{fr} activity can be detected for more than 72 hours after a red-light treatment. By contrast, the activated forms of other plant pigments are very short lived; activated chlorophyll, for example, has a half-life of only a few milliseconds.

The slow decrease in P_{fr} level in the dark period following red irradiation is due to two processes, known as *reversion* and *destruction.* In some tissues,

P_{fr} slowly reverts to P_r in the dark; in other tissues, the total amount of phytochrome decreases (that is, phytochrome is destroyed) after P_r has been converted to P_{fr}. In some cases, both processes take place simultaneously (Fig. 11-10). Phytochrome destruction and the reversion of P_{fr} to P_r are both hastened by high temperatures, with the precise rates varying for different plants and for the same plant grown under different conditions. In seeds, the thermal reversion of P_{fr} to P_r has important consequences for the regulation of germination patterns, as discussed in Chapter 12.

Effects of Prolonged Irradiation with Broad-Band Light Sources

Plants growing under natural conditions are not normally exposed to brief red or far-red light treatments, but are illuminated instead by the broad-band radiation of sunlight. Under these conditions, approximately half the phytochrome is in the P_{fr} form. Since P_r and P_{fr} have broad and overlapping absorption spectra (Fig. 11-9), they are constantly being transformed in both directions, and the same molecules do not remain as P_{fr} during prolonged irradiation. Any individual molecule probably oscillates continuously between the P_r and P_{fr} states, a phenomenon known as pigment *cycling*. The rate of cycling increases with light intensity and is extremely rapid in bright sunlight.

When phytochrome cycles between P_r and P_{fr}, it is not converted directly from one form to the other, but passes through a series of transient intermediate forms (Fig. 11-10). When P_r absorbs a photon, for example, it changes rapidly to a form with maximum absorption at 692 nm. This form is very short lived; within less than a millisecond, it changes spontaneously to another transient intermediate that absorbs maximally at a slightly longer wavelength. Three or four such intermediates rapidly appear and then disappear before the ultimate formation of the relatively stable P_{fr}. Intermediates also form during conversion of P_{fr} to P_r. Since all these intermediates have short half-lives compared to those of P_r and P_{fr}, they are of minor significance when light treatments are brief. However, during prolonged irradiation with high-intensity light, pigment molecules rapidly cycle between P_r and P_{fr}, intermediates continually form, and those with the longest half-lives tend to accumulate; as much as 20% of a plant's phytochrome may be in intermediate forms when the plant is in bright sunlight. Clearly, the intermediates must be considered when analyzing phytochrome action under natural conditions.

Intermediates between P_r and P_{fr} could also play a role in a series of *high-irradiance reactions* (HIR). HIR include such morphogenetic events as hypocotyl elongation and anthocyanin formation in seedlings, and flowering in certain long-day plants. These processes respond positively to prolonged high-intensity red, far-red, and blue light rather than to the characteristically

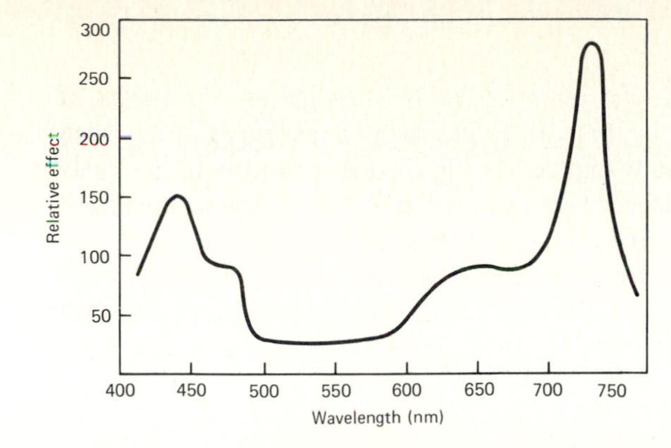

Figure 11-13 The action spectrum of the high-energy reaction in mustard seedlings. (From Mohr, 1962. Ann. Rev. Plant Physiol. *13*: 465–488.)

low intensities affecting phytochrome transformations. They have action spectra with peaks in the regions of 420–480 nm and 710–720 nm (Fig. 11-13). Some of these morphogenetic changes can be elicited by brief irradiation with 660 nm light and prevented by subsequent brief irradiation with 730 nm light, but short light treatments are always less effective than prolonged irradiations. These reactions appear to need a low P_{fr} level for a prolonged period rather than a high P_{fr} level for a short period, and either blue light or 710–720 nm red light could produce such a result. The high-irradiance requirement, however, is poorly understood. Since the rates of pigment cycling and of intermediate formation are intensity dependent, these processes might be important; however, other explanations have also been offered and much more work must be done before HIR are understood. It is not known for example, whether blue light effects are due to absorption by P_r and P_{fr}, or whether a pigment such as the one that regulates phototropism is involved.

The Ecological Role of Phytochrome

Phytochrome or a similar pigment occurs in the green algae, mosses, liverworts, ferns, gymnosperms, and angiosperms, but probably not in the fungi. Its varied roles in these different kinds of organisms demonstrate its remarkable adaptive value. Let us consider a few examples.

Germination of buried seeds

Many seeds need light to germinate, because of a requirement for P_{fr}. Such seeds remain dormant when buried underground, but germinate promptly if brought to the surface of the soil. The common weed mullein (*Verbascum blattaria*) is a plant of this type. In 1973, after 90 years of test burial, 20% of its seed were still viable, and their growth was indistinguishable from that of controls grown from fresh seed. Although such seeds continue to respire during long periods of dormancy, gradually diminishing their stored reserves, such attrition can be very slow if phytochrome is in the P_r form.

Many fresh weed seeds do not require light for germination, but develop a light requirement if buried in the soil, particularly waterlogged soil. These seeds appear to have an initially high level of P_{fr} that reverts to P_r in the dark. Reversion does not take place when seed are stored in a dry atmosphere, since seed phytochrome is dehydrated under these conditions, and dry P_r and P_{fr} are not interconvertible. However, when the soil is sufficiently moist that seed phytochrome becomes partially hydrated, reversion takes place; such reversion in stored seed appears to be an effective survival mechanism.

De-etiolation

Some types of seed germinate deep in the soil, where light cannot penetrate. Since phytochrome is synthesized in the P_r form, the seedlings have P_r but not P_{fr} during their early stages of growth, when they are living on their endosperm reserves. Etiolated seedlings have narrow, rapidly elongating shoots with a recurved hook, and small leaves. This growth habit has survival value, since it permits the seedlings to reach the light quickly and economically; the tender apical growing point is protected by the hook, and energy is not wasted on the growth of leaves that cannot function until the plant reaches the light. Once the seedling breaks through the soil surface and reaches the light, some of the phytochrome is converted to P_{fr}. The hook then opens, the stem thickens, its rate of elongation slows, and the leaves expand and green, increasing their photosynthetic capacity. Thus, many seedling characteristics essential to survival are regulated by the conversion of P_r to P_{fr}.

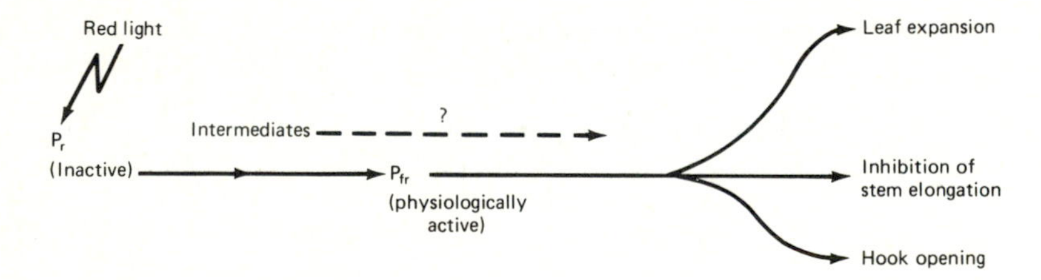

Growth under a foliar canopy

Light passing through a green leaf emerges depleted in the wavelengths absorbed by chlorophyll, and is thus enriched in far-red compared with red (Fig. 11-14). This has the effect of preventing light-sensitive seed from germinating after the leaves of the canopy vegetation have expanded. Since germination in such a light-depleted environment would lead to poor growth, the restraint of germination caused by absence of P_{fr} has positive survival value. The following spring, before the canopy leaves emerge, the seeds can germinate and seedlings become vigorously established.

Shading by green leaves also affects seedling growth. Since the phyto-

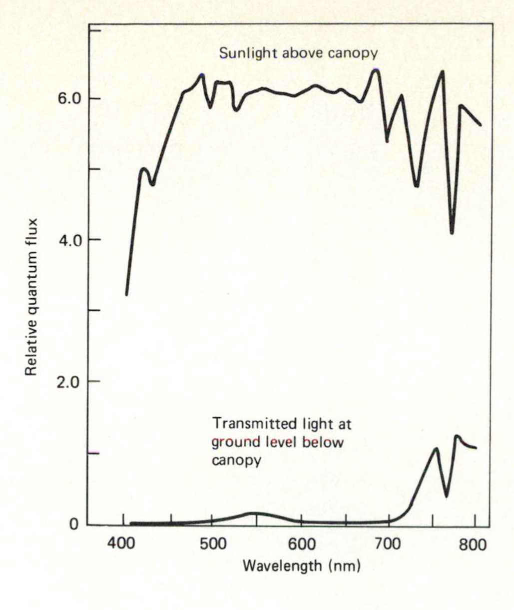

Figure 11-14 The spectral quality of sunlight above a canopy of wheat and below it, at ground level. Note that most of the energy in the 400–500 nm and 600–690 nm spectral regions has been removed by the canopy vegetation. (Adapted from Holmes and Smith. 1977. Photochem. Photobiol. *25*: 539–545.)

chrome in understory vegetation is primarily in the P_r form, the growth of shaded canopy plants resembles to some extent that of etiolated seedlings growing underground: Leaves are small and thin, and stems unusually elongated. Because they utilize an usually large percentage of their energy resources for processes leading to stem elongation, understory plants enhance their ability to reach and penetrate the foliar canopy; this is a distinct competitive advantage, since they receive little energy for photosynthesis when shaded. It seems more than mere coincidence that the absorption peak of P_r coincides with that of chlorophyll *a*, while the absorption peak of P_{fr} falls in the "window" of radiation which escapes the filtering of short wavelengths by photosynthetic pigments, and of long wavelengths by water in plant tissues. Nature has exploited this situation in numerous complex ways!

Location of Phytochrome in the Plant

Knowing where phytochrome is located in the plant would certainly help us to understand its mode of action, and several techniques have been employed to obtain this information. Our most detailed information regarding the distribution of phytochrome at the light-microscope level has been provided by *immunocytochemistry*, a method which uses antibodies synthesized by an animal when a foreign protein is injected into its bloodstream. Rabbits injected with phytochrome isolated from plant tissue synthesize antiphytochrome immunoglobulin. This substance, when purified, binds specifically to phytochrome in plant tissue sections, where its presence can be visualized by coupling its other end to an enzyme, peroxidase, that produces an insoluble,

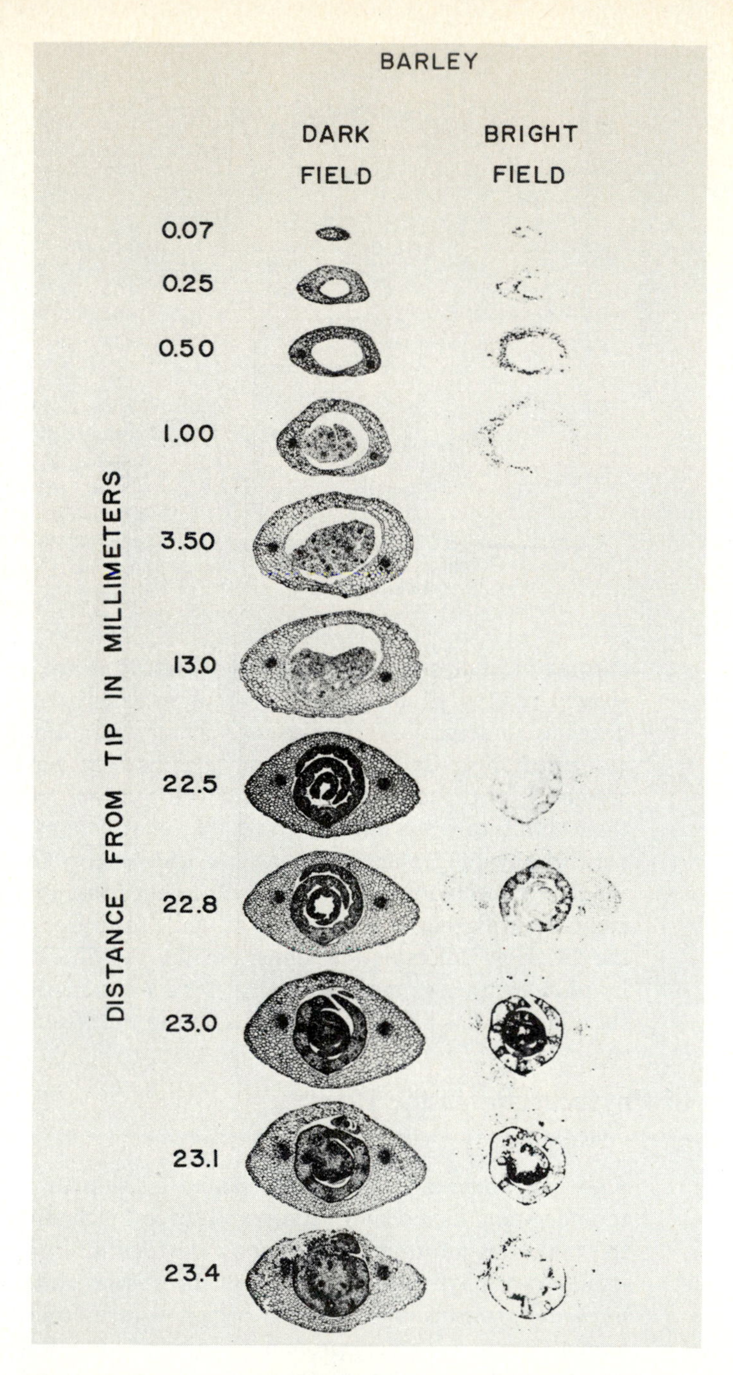

Figure 11-15 Immunochemical pattern of phytochrome in a 3-day-old barley shoot. Cross sections were photographed with both darkfield (left column) and brightfield (right column) illumination. The darkfield micrographs present structural details while dark areas in the brightfield micrographs represent the presence of phytochrome. The numbers indicate distance in mm behind the tip of the coleoptile. ×24. (From Pratt and Coleman, 1974. Amer. J. Botany *61*: 195–202.)

colored product when it acts upon its substrate. The phytochrome distribution in young barley shoots, as revealed by this method, is shown in Fig. 11-15.

More recently, immunochemical techniques have been used to determine the subcellular location of phytochrome in dark-grown oat and rice shoots. Phytochrome is generally distributed throughout the cytoplasm and its membranes prior to a red-light treatment, but becomes associated with discrete regions of the cell after brief irradiation. These regions do not appear to be nuclei, plastids, or mitochondria, but might be endoplasmic reticulum, which is spread throughout the cell. Prolonged exposure to light leads to localization in the nucleus as well. Some studies with isolated cell fractions support these conclusions, but not all workers agree on the interpretation of such results.

Physiological experiments suggest that phytochrome may have many different subcellular locations consistent with its numerous physiological functions. It is apparently located within etioplasts and mitochondria or in their outer envelope membranes, since isolated organelles show physiological responses to red- and far-red-light treatments given after the isolation. Our most detailed information comes from a series of simple yet elegant experiments with the filamentous alga *Mougeotia*. Each *Mougeotia* cell has a single large chloroplast that turns within the cell so that its broad face is parallel to the upper surface if the alga is irradiated with red light, and is perpendicular to the upper surface if the red irradiation is followed by far-red light (Fig. 11-16). It is not necessary to irradiate the chloroplast itself to achieve the effect, and when only part of the cell is irradiated with a microbeam, only a portion of the adjacent chloroplast moves. If plane-polarized light is employed in the irradiating microbeam, the plane of polarization has a

Figure 11-16 The effect of brief red and far-red light treatments on the orientation of the giant chloroplast in *Mougeotia*.
Left: Red light causes the chloroplast to orient face-on toward the light. Far-red reverses the effect of red. Right: The result of irradiation of *Mougeotia* (upper) with a microbeam (3μm diameter) focussed on the region indicated by ● (below) shows that the phytochrome photoreceptor is not in the chloroplast, as beams outside the chloroplast produce motion. One likely location is the plasmalemma. (From Bock and Haupt, 1961. Planta *57*: 518–530.)

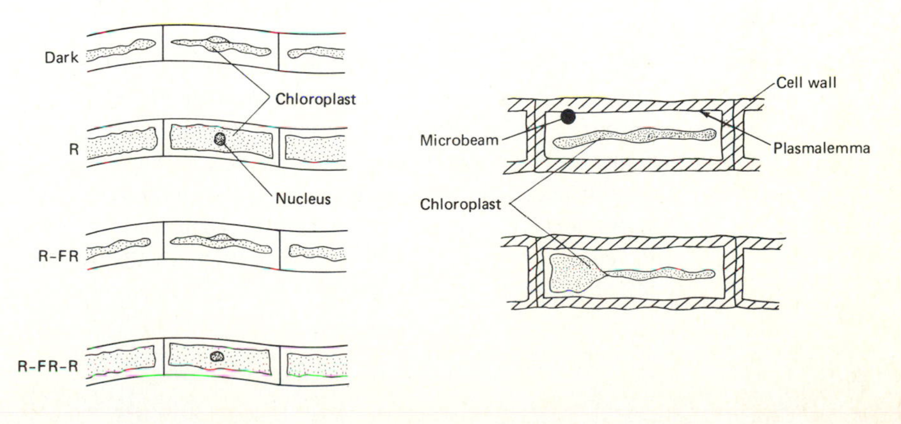

marked influence on its effectiveness. This means that phytochrome is not randomly oriented in the cell, but is fixed in its configuration; extensive probing with microbeams indicates that the phytochrome chromophore has specific orientation in the plasmalemma. Furthermore, since irradiation with polarized red light is most effective when the plane of polarization is parallel to the cell wall, but subsequent far-red is most effective when the plane of polarization is perpendicular to the wall, it appears that phytochrome shifts its orientation in the membrane by 90° during phototransformation.

How Does Phytochrome Act?

Since phytochrome is a quantitatively minute component of plant cells, and since even low energies absorbed by the pigment can trigger large physiological changes, some amplification mechanism must obviously connect the initial photon absorption with ultimate response. Possible amplification mechanisms include control of gene function; control of enzyme activity; control of membrane properties; and control of the level of substances such as hormones, that are themselves effective in minute quantities. All seem to occur in particular situations, but recent evidence indicates that the control of membrane function satisfies requirements for the "master control" that can lead to all the others.

Developmental responses such as floral initiation, seed germination, and de-etiolation clearly involve changes in the basic chemistry, structure, and function of plant cells. These, in turn, depend upon alteration of the activity of many enzymes, and the synthesis of new enzymes as well. Since enzymes are proteins and their synthesis depends upon transcription and translation, the state of phytochrome must influence either or both of these processes. We do not know how phytochrome effects these changes. It could bind to the nuclear chromatin, thereby exerting a direct effect on RNA and protein synthesis, or the effect could be more subtle, perhaps dependent upon phytochrome-controlled alterations in ion compartmentation within the cell, which in turn would affect protein synthesis. The control of protein synthesis is not, however, the sole mode of phytochrome action, since many phytochrome-controlled processes are both independent of and too rapid for protein synthesis.

Plant hormones are known to be rapidly translocated through the plant and to be effective in minute quantities. The effects of phytochrome conversion might be interpreted in terms of hormone synthesis or destruction, or alternatively, of hormone release or binding. Any of these phenomena could probably occur rapidly enough to satisfy even the virtually instantaneous time requirement for phytochrome action. Synthesis or destruction might be effected by direct action of P_{fr}, acting enzymatically, or through phytochrome control of the activity of synthetic or destructive enzymes, existing as "proenzymes" prior to irradiation. Binding or release of hormones, either from an active locus or a storage site, could similarly be effected through modification of the site by P_{fr}.

Levels of four different hormones (gibberellins, cytokinins, ethylene, and auxin) in dark-grown plant tissues change rapidly following a brief red-light treatment, while the level of a fifth (abscisic acid) is altered by prolonged red irradiation. Since the effects on gibberellins, cytokinins, and ethylene are reversible by far-red light, phytochrome is clearly the photoreceptor for those changes; P_{fr} increases the level of gibberellins and cytokinins, and decreases levels of auxin and ethylene. The changes in gibberellins have been investigated most thoroughly, using etioplasts (plastids of dark-grown tissue) isolated from barley leaves. Brief exposure to red light after isolation causes both a rapid gibberellin leakage out of the plastid and a slower increase in hormone level, which probably involves new synthesis. The rapid effects are probably caused by change in the permeability of the etioplast envelope membrane. In barley leaves, either red light or gibberellin causes leaf-unrolling.

Numerous other rapid effects of phytochrome conversion are also most reasonably interpreted as due to changes in membrane structure and function. These effects include changes in the adhesion of barley or mung bean root tips to a negatively charged surface (Fig. 11-17), changes in the electrical potential across cell membranes, and changes in the "sleep movements" of the leaves of certain plants (Fig. 11-18) that are in turn dependent upon changes in ion movements. Since all these processes respond rapidly to brief red- and far-red-light treatments, they could clearly result from changes in membrane properties.

The postulate that phytochrome works in membranes has also been sup-

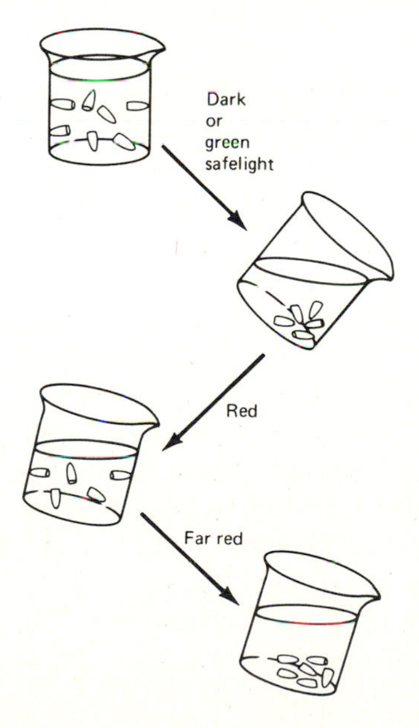

Figure 11-17 Excised tips of dark grown barley roots are submerged in a special solution in a glass beaker whose interior surface has been treated so that it is negatively charged. When the beaker is swirled then tipped, the root tips sink to the bottom if exposed only to green safelight, but adhere to the surface if exposed to red light. Far-red light reverses the effect of red. (Based on data of Tanada. 1968. Proc. Natl. Acad. Sci. *59*: 376–380.)

Figure 11-18 Pinnae of *Mimosa pudica* 30 minutes after transfer to darkness following high-intensity white light terminated with a succession of 2-minute exposures to red (R) and far-red (FR) light. The pinnae remained open if exposure to far-red was last (top row) and closed if red irradiation was last (bottom row). (From Fondéville et al., 1966. Planta *69*: 357–364.)

ported by the direct demonstration of phytochrome incorporation and function in artificial lipid membranes, which were discussed in Chapter 2. Red- and far-red-light treatments given to such membranes causes large changes in the electrical resistance across the membrane (Fig. 11-19). This supports the view that conformational changes in the phytochrome chromophore can instantaneously alter membrane structure. Other more leisurely and distant phytochrome effects, such as gene activation and developmental changes, may possibly derive from this initiating event, or there may be more than one point at which the phytochrome acts. As discussed in the next chapter, not all phytochrome-controlled responses are localized within a single cell; in some cases, irradiation of one part of a plant affects the development of organs some distance away. Some phytochrome-controlled responses of this type occur in etiolated tissues, but the most dramatic examples occur during photoperiodic induction of flowering, tuberization, and dormancy (see Chapter 12). Any explanation of phytochrome action must account for distant as well as localized responses.

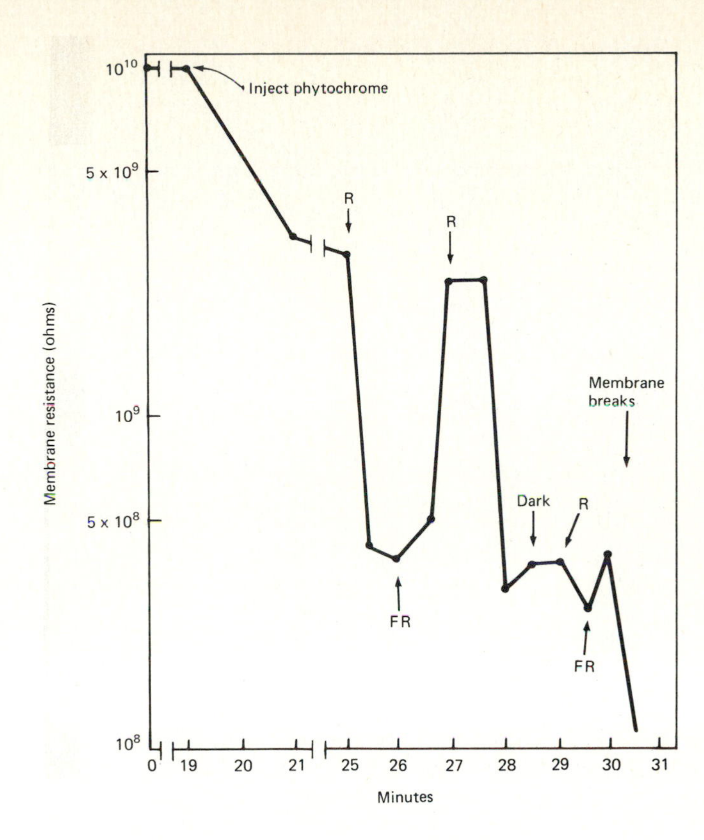

Figure 11-19 Changes in the electrical resistance of an artificial oxidized cholesterol membrane on addition of 6 μg of phytochrome to the bathing solution and illumination with red and far-red light. Note that the resistance decreases after red light and increases after far-red light. (From Roux and Yguerabide, 1973. Proc. Natl Acad. Sci. *70*: 762–764.)

Blue-Light Effects

As already discussed in Chapter 9, blue light, probably absorbed by a flavoprotein, can cause the phototropic curvature of cylindrical plant organs through induction of lateral auxin transport that leads in turn to unequal growth on the two sides of the organ. Blue light also affects a multitude of other processes, such as the opening of stomata and of closed leaves, the rate of cytoplasmic streaming in oat coleoptile cells, the viscosity of cytoplasm in leaf cells of the water plant *Elodea*, the movement of chloroplasts in fronds of duckweed (*Lemna*), and the plane of cell division in fern sporelings (Fig. 11-20). In all these reactions, the Law of Reciprocity is followed; this means that the effect depends on the total energy, and the product of light intensity $\times$ time is a constant ($I \times t = K$). Thus, relative light intensities of 100, 10, and 1 will produce equal effects if given for 0.01, 0.1 and 1.0 seconds, respectively. Since the action spectra for all these processes are remarkably similar, we may deduce that a common pigment produces a common photoproduct that can regulate diverse physiological processes. The nature of this photoproduct is not yet known, although chemical reduction of a particular cyto-

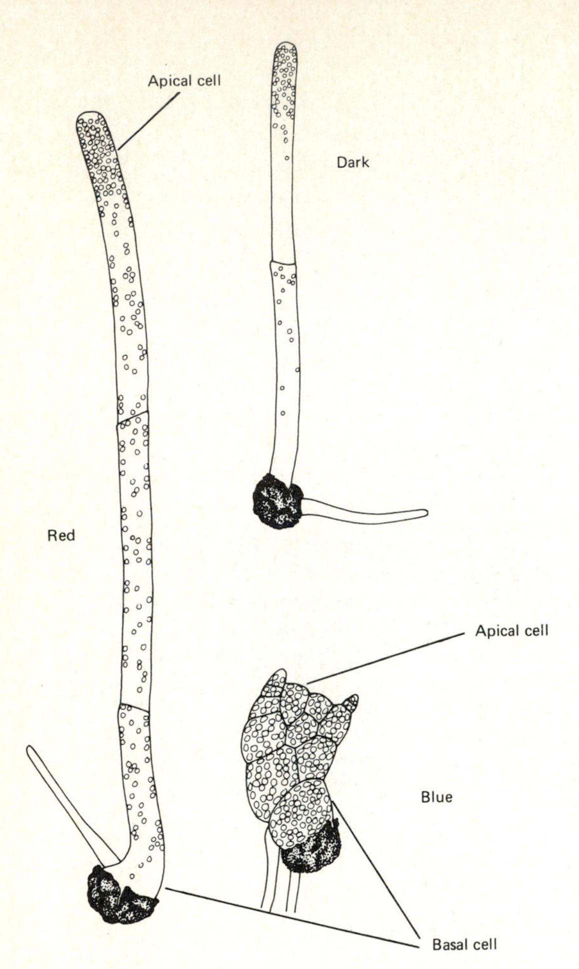

Figure 11-20 Representative sporelings of *Dryopteris filix-mas* grown in darkness, continuous red or blue light. The sporelings were grown for 6 days after germination on inorganic nutrient solution. The blue-grown and the red-grown sporelings have about the same dry weight. (From Mohr and Ohlenroth, 1962. Planta *57*: 656–664.)

chrome has been detected in various plants following flavin photoactivation by blue light. Phytochrome intermediates could also be involved.

Thus the final form of the green plant, stipulated in general terms by its genes, is elicited by light acting through various photoreceptors. Phytochrome can control seed germination, de-etiolation, flowering, abscission, and senescence; protochlorophyll controls the greening process, chlorophyll controls overall autotrophic nutrition, and a yellow pigment, possibly flavoprotein, controls the direction of growth of aerial organs, as well as a host of subtle intracellular processes. The green plant is truly a light-driven machine, living in continuous close dependence on incident quanta.

SUMMARY

Plant form and structure are determined partially by radiant energy impinging on the plant. *Phytochrome*, a biliprotein pigment probably localized in plant membranes, changes its form and absorbancy in response to the light it receives. One form, P_r, absorbs mainly red light centered about 660 nm, and is thereby transformed to another form, P_{fr}, absorbing mainly in the *far-red* region centered about 730 nm. P_{fr} is unstable and slowly decays back to P_r in the dark; it can also be pushed back to the P_r form by far-red radiation. P_{fr} is the "active form" of phytochrome, controlling many reactions and processes, and the $P_r \rightleftarrows P_{fr}$ system forms one component of timing reactions triggered by dark-to-light transitions. Phytochrome is present in minute amounts in the plant cell, and thus low energies suffice to saturate the *photomorphogenic* reactions it controls. These include *de-etiolation* of dark-grown seedlings, promotion of *germination* in certain seeds, the onset of *flowering*, and the onset of *dormancy*. Phytochrome has a molecular weight of about 240,000, and apparently consists of several *subunits*. Several transient intermediates are formed during photoconversion of P_r to P_{fr} as well as of P_{fr} to P_r.

Some photomorphogenic reactions require much higher energies of light, and respond mainly to blue and far-red regions. These *high-irradiance reactions* (*HIR*), which include some aspects of de-etiolation, flowering responses, and anthocyanin formation, may be controlled by phytochrome or a yet uncharacterized blue absorbing pigment. In nature, phytochrome probably controls the form of plants and the germination of seeds found under a canopy, since leaves absorb very well in the 660 nm region, but are relatively transparent in the 730 nm region. The state of phytochrome can control the production and levels of four types of hormones—ethylene, cytokinins, auxin, and gibberellin—in different plant systems. In addition, prolonged irradiation can control the level of ABA. In some instances, hormone application can reverse the effect of phytochrome transformation, as with gibberellin in de-etiolation. Blue light (but not red or far-red) also controls certain photomorphogenic responses; in these instances, the operation of a completely different photoabsorbing pigment, possibly flavoprotein, is indicated.

SELECTED READINGS

KENDRICK, R. E., and B. FRANKLAND. *Phytochrome and Plant Growth* (New York: St. Martins Press, 1976).

MANCINELLI, A. L., and I. RABINO. "The 'high irradiance responses' of plant photomorphogenesis." Botan. Review *44* (2): 129–180. (1978).

MITRAKOS, K., and W. SHROPSHIRE, JR., eds. *Phytochrome* (New York: Academic Press, 1972).

MOHR, H. *Lectures on Photomorphogenesis* (Berlin–Heidelberg–New York: Springer Verlag, 1972).

SMITH, H. *Phytochrome and Photomorphogenesis* (New York: McGraw-Hill, 1975).

SMITH, H., ed. *Light and Plant Development* (London–Boston: Butterworths, 1976).

QUESTIONS

11-1. What is the evidence that phytochrome exists in two interconvertible forms, and that the state of phytochrome following a flash of light is a controlling factor in development?

11-2. How did knowledge of phytochrome photoreversibility aid in efforts to extract and purify the pigment from plant tissue?

11-3. The amount of phytochrome in etiolated plant tissue is routinely assayed in the "Ratiospect" two-wavelength spectrophotometer, but it has not yet been possible to assay phytochrome in green tissue with this instrument. Why?

11-4. The phytochrome molecule consists of a chromophore attached to a protein. What do you think is the function of each?

11-5. List several processes in plants that are under the control of phytochrome. What specific changes does phytochrome bring about in each?

11-6. What is one current theory on how phytochrome may act? What kinds of evidence support this theory? Do you believe that one initial action could lead to such a multiplicity of effects? Justify your answer.

11-7. Many seeds fail to germinate unless phytochrome is in the P_{fr} form. What is the possible adaptive value of this requirement?

11-8. P_r and P_{fr} have broad, overlapping absorption spectra. How does this enable phytochrome to act as sensor for changes in light quality in the environment? Describe some possible functions of pigments that can respond to changes in light quality.

11-9. Exposure of an etiolated plant to red light converts approximately 75% of its phytochrome to the P_{fr} form, yet several hours later, the amount of P_{fr} has decreased. What two processes contribute to this disappearance? How would you determine the contribution of each?

11-10. In an earlier chapter, we discussed the "light" and "dark" reactions of photosynthesis. There are also "light" and "dark" reactions in photomorphogenesis. Describe them.

11-11. P_{fr}, the active form of phytochrome, has a half-life of approximately two hours in etiolated peas maintained at 25°C. Activated chlorophyll, on the other hand, has a half-life of only a few thousandths of a second. This basic difference between phytochrome and chlorophyll profoundly affects the role of each in regulating plant growth. Explain.

11-12. One of the most important environmental factors for a plant is sunlight. What pigments does the plant possess to respond to this light? How does each pigment react or change when irradiated? What is the overall result of such changes? Before you start, make a list of the pigments and their features.

11-13. You will recall that both P_r and P_{fr} absorb blue light as well as longer wavelengths. Describe some experiments that would help in establishing whether phytochrome or some other pigment is the photoreceptor for a growth process influenced by blue light.

11-14. Chlorophyll absorbs heavily in the red region of the spectrum, and less heavily in the far-red. What is the importance of this fact for phytochrome-controlled phenomena in green plants?

Control of Growth by Photoperiod and Temperature

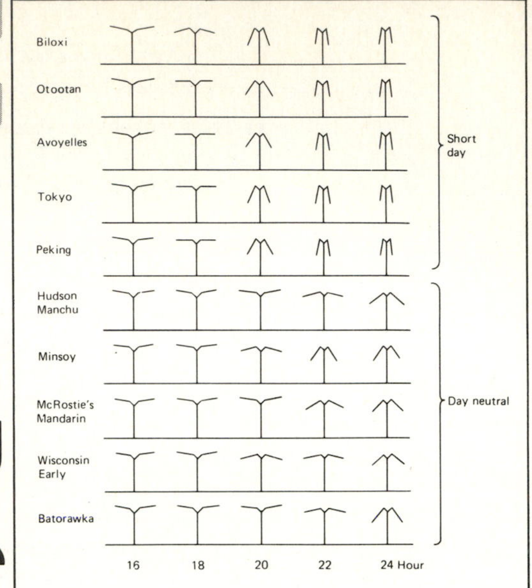

Plants have evolved mechanisms for measuring changes in temperature, daylength and light quality; these permit them to integrate their life cycles with seasonal changes in climate. Length of day is estimated by referring light signals to a *biological clock* whose timekeeping is relatively temperature insensitive, while temperature sensing involves not only chemical reactions that proceed more rapidly as the temperature is increased, but paradoxically, processes that proceed better after a specified period of low temperature. In some cases, low-temperature signals must follow day lengths of appropriate duration; this serves to prevent confusion between days of equal length in spring and fall. In this chapter, we explore the interaction of light, temperature, and the biological clock in regulating temporal aspects of plant growth and development.

Circadian Rhythms

Any event that occurs with a regular periodicity in time or space is said to be *rhythmic*; when the periodicity persists in the absence of external perturbing factors, the rhythm is said to be *endogenous*. The "sleep" movements of certain plants, especially in the family Leguminosae, (the common bean *Phaseolus*, *Mimosa*, *Albizzia*, or *Samanea*) are examples of this phenomenon. The leaves of these *nyctinastic* (night-closing) plants are usually horizontal (open) during the day and vertical (closed) at night (Fig. 12-1). The move-

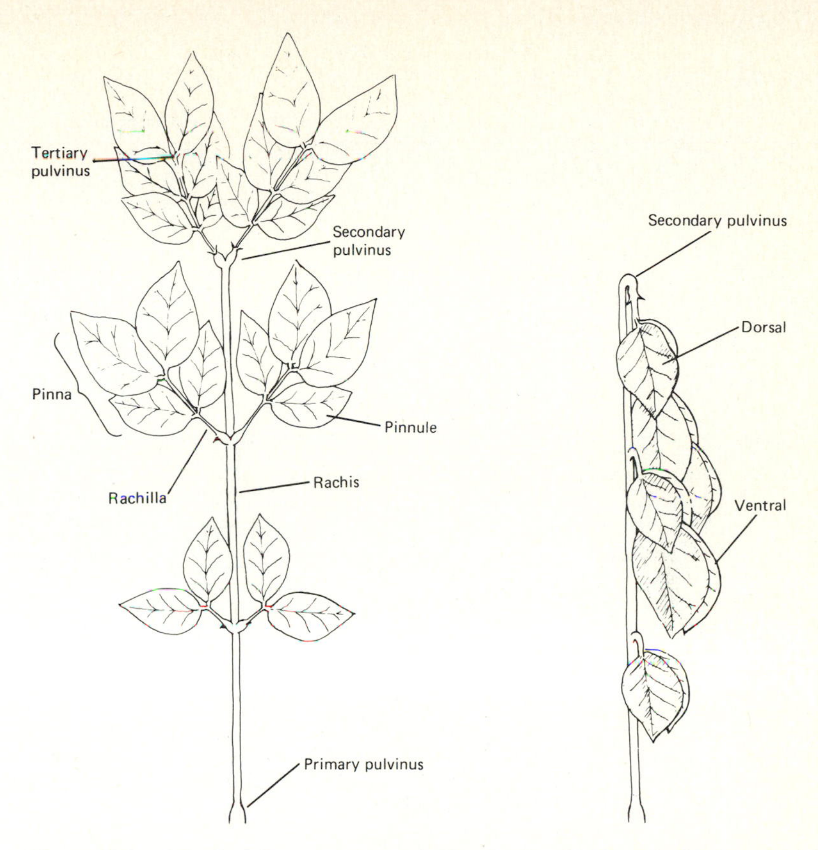

Figure 12-1 A leaf of *Samanea saman* in the daytime (left) and nighttime (right) positions. *Samanea*, a tropical legume also known as the monkey pod tree and the rain tree, has doubly compound leaves divided into paired leaflets (pinnae) in turn sub-divided into paired sub-leaflets (pinnules). Sleep movements are regulated by turgor changes in cells of the primary, secondary and tertiary pulvini located at the base of the leaves, pinnae and pinnules respectively. (From Satter et al., 1974. J. Gen. Physiol. *64*: 413–430.)

ments can easily be visualized by connecting the leaf to the pen of a rotating drum by a fine thread (Fig. 12-2). The tracings show that the leaves begin to open *before* the beginning of the light period, and begin to close *before* the beginning of the dark period. Thus, leaf movement seems to anticipate dawn and dusk, suggesting some other timing control besides light and darkness.

Evidence of the operation of an internal clock comes from experiments in which the leaf is kept in the dark at constant temperature for several days. The opening–closing oscillation persists, with periods of peak opening about 23 hours apart. Such rhythms, with oscillation periods of 20–30 hours (that is, approximately one day's duration) have been named *circadian* rhythms, from the Latin words meaning *approximately* and *day*. The cellular timekeeping machinery that produces the rhythm is usually referred to as

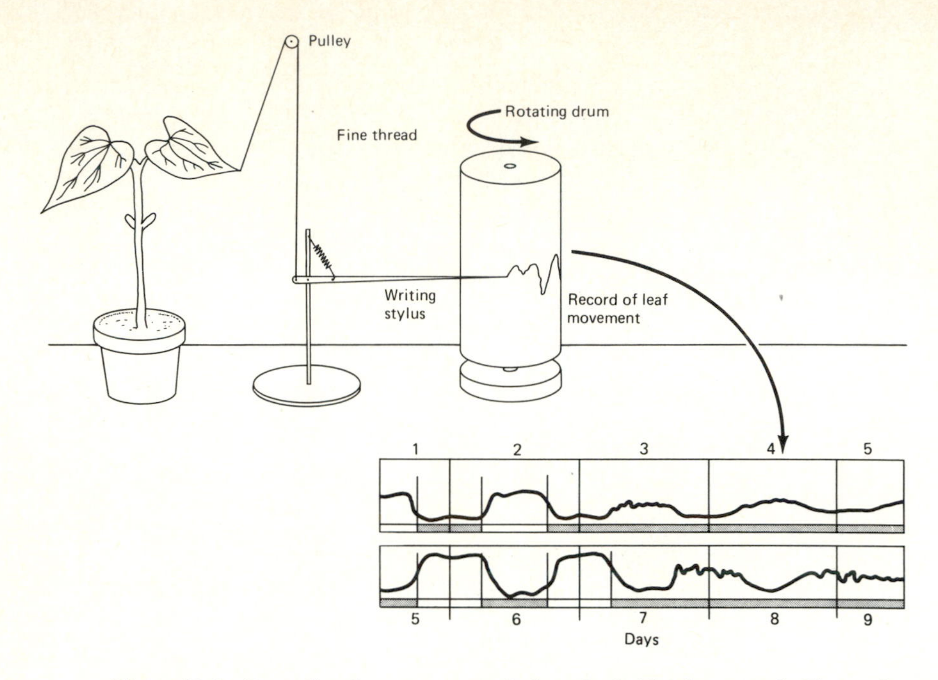

Figure 12-2 Recording the movements of a bean leaf with a kymograph. The tracing indicates movement during 12 hr light (white bars) 12 hr dark (black bars) cycles followed by continuous darkness. Note that the leaves continue to rise and fall during continuous darkness, but reversal of the light-dark schedule (lower tracing) shifts the phase of the rhythm by 12 hours.

Figure 12-3 A diagram representing the circadian clock governing pinna movements in *Samanea.* (a) Secondary pulvini, excised as shown below, will oscillate in the dark for several days if supplied with sucrose. One complete revolution takes 23 hours in continuous darkness, and slightly longer in continuous light. (b) The angle at the secondary pulvinus represented in the figure indicates circadian time.

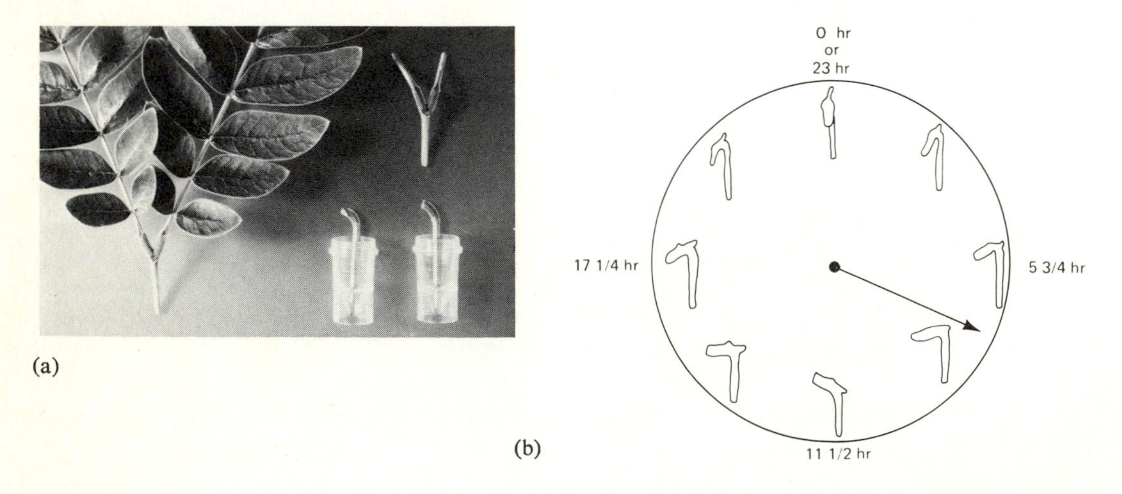

a *biological clock*, and the position of the leaf during any part of the cycle indicates circadian time (time controlled by the internal oscillator). Since leaf movements in nyctinastic plants are such accurate indicators of circadian time, they are sometimes called "the hands of the biological clock" (Fig. 12-3). Circadian oscillations of this type occur in virtually all eukaryotic organisms, including microorganisms, plants, and animals.

Nature of the oscillator

What is the physical basis for these endogenous circadian rhythms? This is difficult to analyze, since each cell is more complex than we yet appreciate, and the organism, as an aggregate of cell units, is even more complicated. Some of the organism's complexity can actually be seen, as with the electron microscope, some can be detected by physical and chemical means, as in the elucidation of the structure of protein and nucleic acid macromolecules, while some can be deduced only from complex behavioral patterns of biological systems. We are faced with the latter situation in regard to endogenous rhythms.

Many investigators have attempted to determine conditions that change the *period* (time required for one complete oscillation) or the *phase* of rhythms, as a means of learning about the oscillator. Any effective stimulus is presumed to interfere with the clock's inner workings. If we were to prevent a *Samanea* leaf from oscillating by clamping it in a fixed position for a few hours, it would, upon release, quickly assume the same angle as an unclamped control. Thus mechanical restraint prevents expression of the rhythm, but has no effect on timekeeping by the clock. In contrast, light and temperature alterations, as well as a few chemical compounds (Li^+ ion, heavy water, ethanol), can change the phase of rhythms. All effective compounds are known to affect properties of cell membranes, leading many researchers to believe that rhythmic changes in membrane properties are an integral part of the clock mechanism. Membrane surfaces surround the cell as well as compartments of the cellular interior. By rhythmic regulation of the flow of metabolites into and out of the cell and its organelles, the clock could produce rhythmic changes in cellular chemistry and behavior.

Rhythmic changes in membrane structure have in fact been observed in the unicellular alga *Gonyaulax*. Some protein particles in the membrane, visualized by electron microscopy after freeze-fracturing, seem to change in frequency in a circadian manner; however, we do not know how these oscillations in membrane structure are generated. This is the key question that must be answered if we are to understand biological clocks.

Temperature independence of time measurement

One intriguing characteristic of biological circadian rhythms is that period length is practically independent of temperature over a wide range

(Fig. 12-4). Functionally, this is not surprising, since any clock would be a poor timekeeper indeed, if the rate at which it ran depended upon external perturbations such as temperature fluctuations! Yet a mechanism for biological temperature compensation is difficult to visualize, and this problem limits the kinds of schemes that might logically be advanced to explain rhythms. Since most metabolic reactions are highly temperature dependent, one theory

Figure 12-4 The rhythm in stimulated luminescence in *Gonyaulax polyedra* in continuous light, 1000 lux, at different temperatures as indicated on each curve. The period length changes only 12% as the temperature is changed from 16.5° to 26.8°C. The rhythm damps at very low and very high temperatures. (From Hastings and Sweeney, 1957. Proc. Natl. Acad. Sci. *43*: 804–811.)

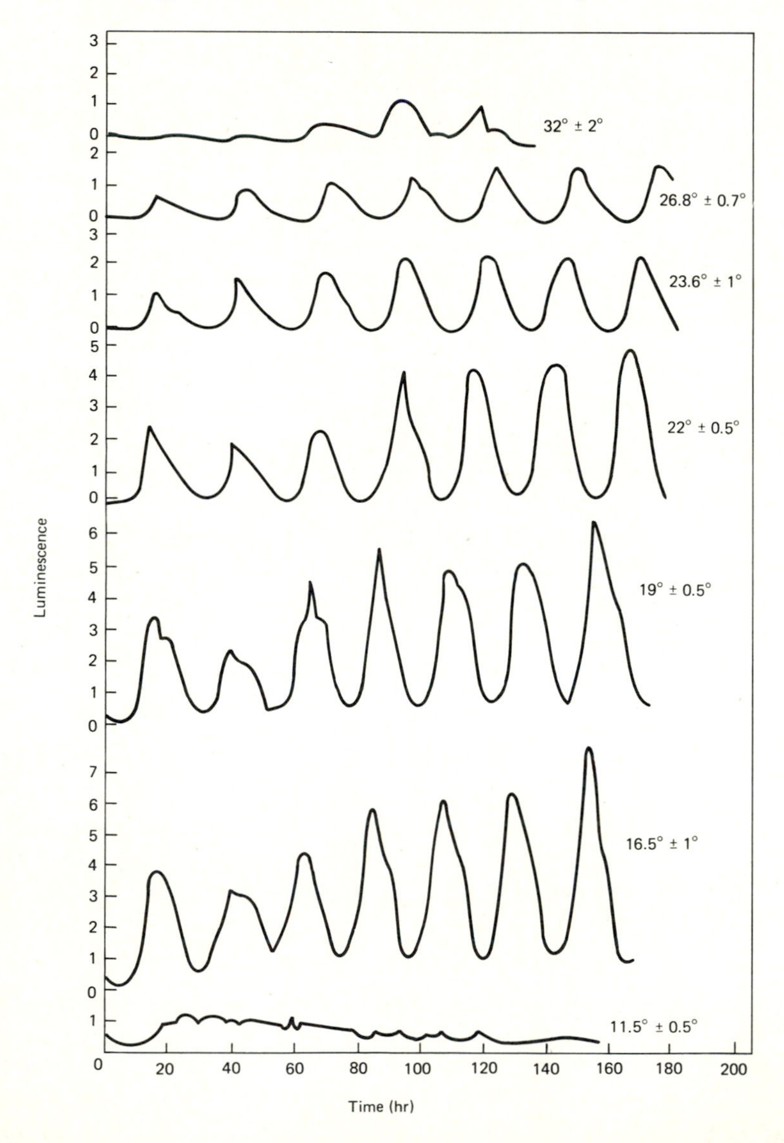

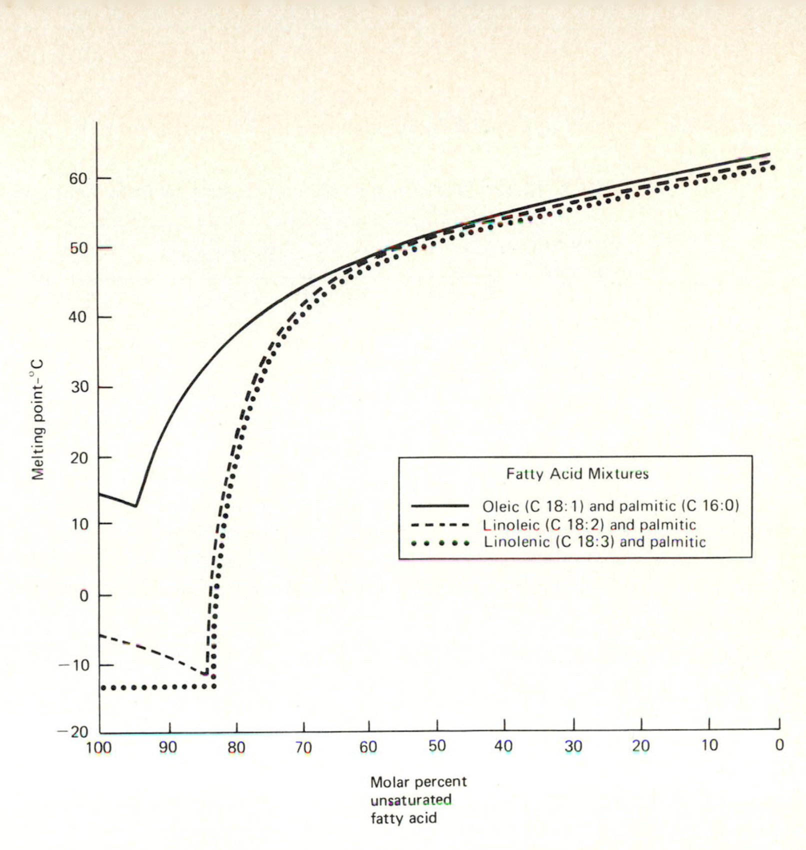

Figure 12-5 The melting point of mixtures of unsaturated and saturated fatty acids. (For each acid the numeral left of the colon indicates the number of carbon atoms in the chain while the numeral right of the colon indicates the number of double bonds. See Fig. 5-12 for structural formulas.). Note that polyunsaturation lowers the melting temperature of the mixture. (Adapted from Lyons and Amundson, 1965. J. Amer. Oil Chemists Soc. *42*: 1056–1058.)

proposes that there are two different temperature-dependent components to the clock, one hastened and the other slowed by the same increase in temperature. As a result of this internal compensation, the rate at which the clock runs can be temperature independent.

Another theory assigns a primary role to lipids in the membrane. These compounds contain long chained fatty acids, which vary in length and in the number and positions of double bonds (Fig. 5-11). The degree of saturation and the length of the fatty acid chains can regulate fluidity, in that shorter chains and unsaturation both lower the temperature at which liquid fats solidify (Fig. 12-5). Changes in membrane fatty acids occur *in situ* in response to changes in temperature, serving to keep membrane fluidity relatively constant over a broad temperature range (Table 12-1). Some investigators believe that changes in chain length and in the degree of saturation of membrane lipids also occur during each daily cycle, and are part of the clock mechanism. If so, we can see how time measurement by such a process could be relatively independent of temperature.

TABLE 12-1. Relationship of temperature to fatty acid content.

GROWTH TEMPERA-TURE	FATTY ACID					RATIO
	C 16	C 18	C 18:1	C 18:2	C 18:3	C 18:2/C 18:3
degrees C				*% by weight*		
30	32.6	3.7	7.5	36.3	13.6	2.67
25	35.8	2.8	5.7	33.4	22.9	1.46
20	32.9	3.3	5.4	31.7	26.8	1.18
15	31.6	4.8	4.2	31.9	27.5	1.16

The fatty acid composition of polar lipids in root tips of cotton seedlings varies with the temperature at which the plants are grown. Note that plants grown at 20°C have twice as much linolenic acid (C 18:3) but only 87% as much linoleic acid (C 18:2) as those grown at 30°. (From St. John and Christiansen, 1976. Plant Physiol. *57*: 257–259.)

Are rhythms really endogenous?

Some people continue to suppose that circadian rhythms are not really controlled internally, but are governed by some as yet undefined cosmic oscillation. The best evidence that circadian rhythms are endogenous, rather than the product of some subtle variable in the environment, is twofold: (1) Circadian rhythms persist in organisms orbiting the earth in satellites, and (2) the period length under "free-running" (constant) conditions is not exactly 24 hours. Under natural conditions, rhythms are *entrained* to run on a 24-hour cycle by periodicities in the environment. *Entrainment* here means that the "natural" period length of a rhythm is altered by some exter-

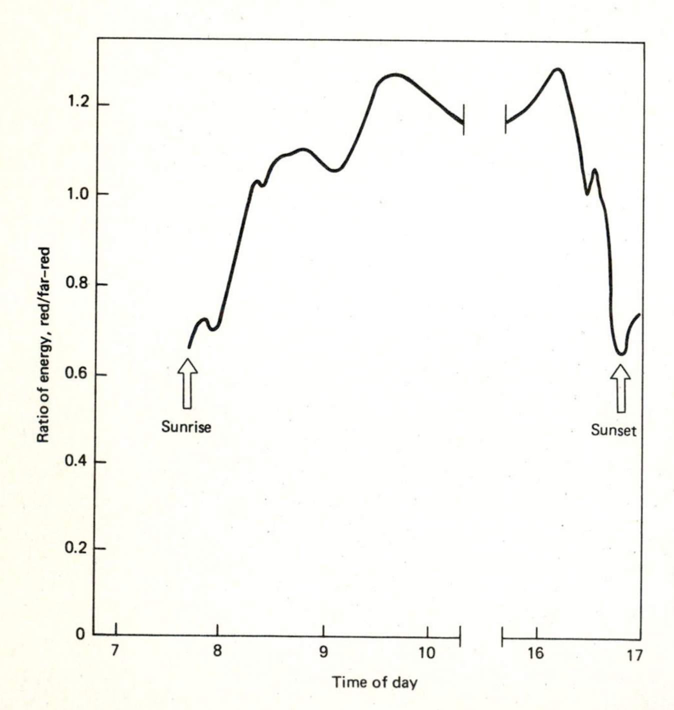

Figure 12-6 Variations in the ratio of red (660 nm) to far-red (730 nm) energy of sunlight from dawn to dusk. Measurements were made at Sutton Bonington, England, latitude 52°50′N. on 22nd January 1974. Sunrise was at 07.53 GMT and sunset at 16.32 GMT. (From Holmes and Smith, 1977. Photochem. and Photobiol. *25*: 533–538.)

nal signal to match that of an external oscillation. Rhythms in most organisms are entrained by daily changes in light (Fig. 12-2) and temperature.

The pigment that acts as phototeceptor for rhythmic entrainment varies in different organisms. The clock is coupled to a blue-absorbing pigment in many animals and in some plants, but phytochrome is implicated as photoreceptor in most plants. Rhythms in these plants are entrained by conversion of P_r to P_{fr}. We would expect interconversions between P_r and P_{fr} to take place in plant tissue shortly after sunrise and before sunset each day. As Fig. 12-6 reveals, the ratio of red to far-red energy from sunlight at the earth's surface is about 1.3 during the day, but drops rapidly to 0.7 at sunset. This causes a small depression in the P_{fr} level, which continues to decrease still further during the night, by dark reversion to P_r. The P_{fr} level rises again in the morning, soon after sunrise, as the ratio of red to far-red energy from sunlight again increases. Thus we can see how an increase in P_{fr} at sunrise could reset the biological clock each day.

Phytochrome-clock interactions

Much of our current knowledge of biological clocks derives from the keen insight and carefully executed experiments of German plant physiologist Erwin Bünning. In observing the behavior of 10 varieties of soybeans, he noted a strong correlation between their photoperiodic behavior and the sleep movements of their leaves (Fig. 12-7); he therefore predicted that both

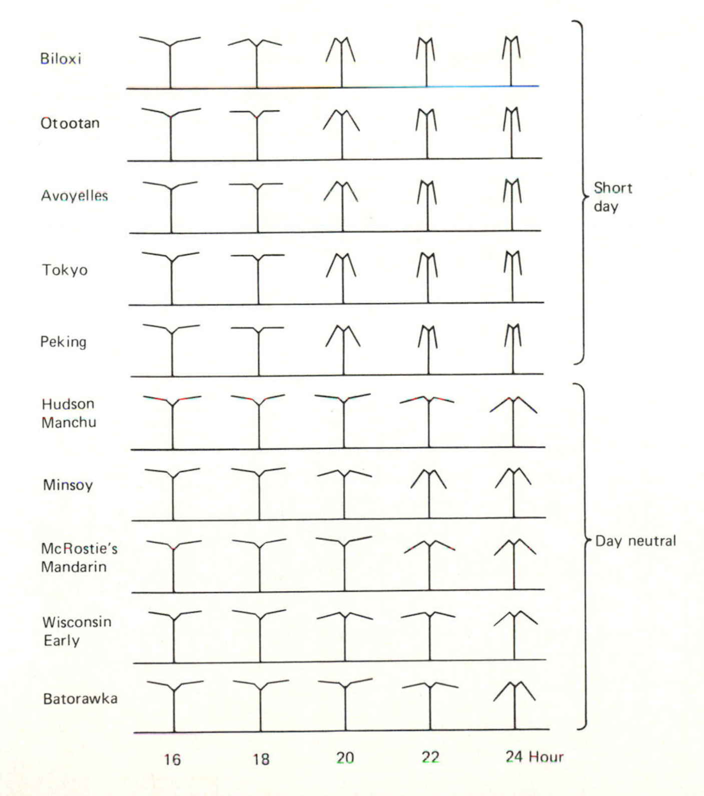

Figure 12-7 Movements of leaves of 10 different varieties of soybeans between 4 PM and midnight. The upper 5 varieties are short day plants, while the lower 5 varieties are day neutral. (From Bünning, 1948. Z. Naturforsch. Teil B. *3b*, 457–464.)

processes were controlled by the same internal clock. Of the many varieties that he studied, those with the most pronounced leaf movements were obligate short-day plants with respect to flowering, while those with less pronounced leaf movements tended to be day-neutral in flowering habit. Presumably, the same phytochrome-rhythmic controlled reactions that regulate leaf movements also regulate the photoperiodic processes that determine what proportions of the plant's resources are committed to vegetative growth, reproduction, storage, and processes leading to dormancy. Since leaf movements are less complex than photoperiodic events, they have been studied to aid in analyzing phytochrome–clock interactions.

Leaf movements in nyctinastic plants result from changes in the volume of motor cells in the *pulvinus*, an organ at the base of the leaf blade (Fig. 12-8). Motor cells on one side of the pulvinus are swollen and those on the opposite side are shrunken when leaflets are open, and the converse is true when they are closed (Fig. 12-9). Changes in the volume of motor cells are in turn regulated by massive fluxes of K^+ and Cl^- into and out of the vacuoles of these cells. Increase in K^+ and Cl^- lowers the cell's water potential, leading to increased water uptake and cellular swelling, whereas decrease in K^+ and Cl^- has the opposite effect. K^+ and Cl^- redistribution is presumed to be regulated by changes in motor cell membranes.

In the nyctinastic plant *Samanea*, whose leaves oscillate with a circadian periodicity during prolonged darkness, the phase of the oscillations can be altered by changes in the P_{fr} level. The amount of phytochrome in the P_{fr} form, initially established by the spectral quality of the light preceding darkness, gradually decreases as P_{fr} reverts to P_r. If the dark period is interrupted by a brief red-light treatment that converts P_r back to P_{fr}, the irradiated leaflets may behave quite differently from nonirradiated controls. For example, if the light treatment is given as the leaflets are opening and have attained about half their maximum angle, they will close prematurely, while controls continue to open. However, if the red exposure occurs 12 hours later, as leaflets are closing, it will have relatively little effect. Thus conversion of P_r to P_{fr} during certain parts of the cycle resets the clock, while similar conversion at other times is without effect.

These interactions can be rationalized by hypothesizing that both phytochrome and the clock change the properties of the same membranes. According to this theory, slow circadian changes in the permeability and transport properties of cell membranes occur during each daily cycle and are part of the clock itself, whereas conversions between P_r and P_{fr} produce more rapid changes in membrane structure and function. Thus we can see how phytochrome conversions could reset the clock by altering the state of cell membranes. This hypothesis would also explain the temporal relationships between brief red irradiations and their ability to alter timekeeping, since the effect of phytochrome conversion would depend upon the condition of

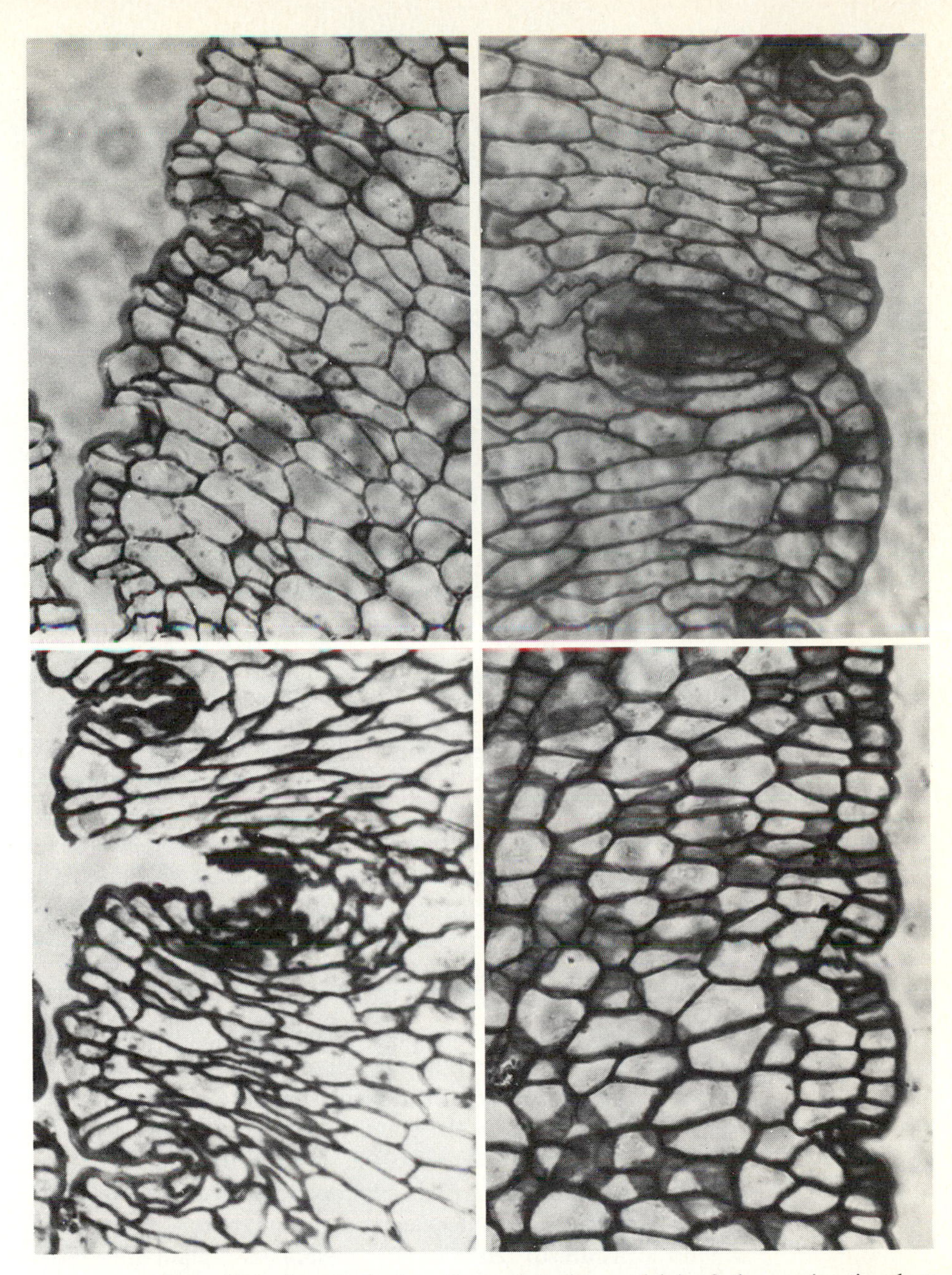

Figure 12-8 Longitudinal sections of tertiary pulvini of the nyctinastic plant *Albizzia julibrissin* during the day (left) and at night (right). *Albizzia* leaflets fold upward at night, controlled by the shrinkage of ventral (upper) cells and swelling of dorsal (lower) cells. Note however that dorsal cells change shape more than size. (Micrographs from Satter et al., 1970. Amer. J. Botany *57*: 374–381.

Figure 12-9 A scanning electron micrograph of the secondary pulvinus of *Samanea*. Note the large number of hairs that originate in cushion-like mounds of tissue separated by creases. The depth of the creases changes as the pulvinus changes its configuration.

the membrane at the time of the light treatment. This explanation seems reasonable and logical, but we need to know much more about phytochrome and the clock to assess its accuracy.

Floral Induction

Participation of phytochrome and the clock

Many developmental processes that require precise seasonal timing also depend upon interaction between phytochrome and the clock. These processes span the entire life cycle of the green plant, including the transition from vegetative to reproductive growth, tuber formation, induction of dormancy, and senescence (Fig. 12-10). Floral induction is the most thoroughly studied of these processes.

It has long been appreciated that most plants, when grown at a particular latitude, flower at roughly the same date each year—we have come to expect violets in the springtime, roses in the summer, and chrysanthemums in the fall. Floral induction in these and many other plants is controlled by the photoperiod, whose measurement involves both phytochrome and rhythms. To determine the role of each, photoperiodically sensitive plants have been exposed to abnormally long dark periods interrupted at varying times by a short exposure to red light. In the short-day plant *Chenopodium rubrum*, the red-light treatment promotes flowering when given during certain parts of the dark period, and inhibits flowering when given at other times (Fig. 12-11). The logical inference is that there are times during a long dark period when P_{fr} favors flowering and other times when it inhibits flowering. The times when P_{fr} promotion is maximal are about 24 hours apart, as are the times when P_{fr} inhibition is maximal, indicating that sensitivity to P_{fr} follows a circadian rhythm.

Figure 12-10(a) A plant of the Japanese morning glory (*Pharbitis nil*, variety Murasaki) induced to early flowering by short photoperiods. In nature, these plants reach much greater heights before flowering, but since *Pharbitis* cotyledons are capable of sensing the length of day, even the first bud to mature can be floral. Normally, only one apical bud is produced; in this experiment that bud has been removed. This permits the normally inhibited cotyledonary buds to sprout and each has given rise to a beautiful violet flower. (Courtesy of S. Imamura and A. Takimoto, Kyoto University.)

12-10(b)(c) The liverwort *Lunularia* grown under 8 hr. and 24 hr. days. In the long day-treated plants, you can clearly see pale edges of the thalli tips where cellular death has occurred. These plants have gone dormant under long day conditions. The small black dots are largely soil particles and are not of any significance. The short day-treated plants show gemmae cups with gemmae. Their production is not confined to short day, but occurs when thalli grow actively. (Courtesy of W. W. Schwabe, Wye College, University of London.)

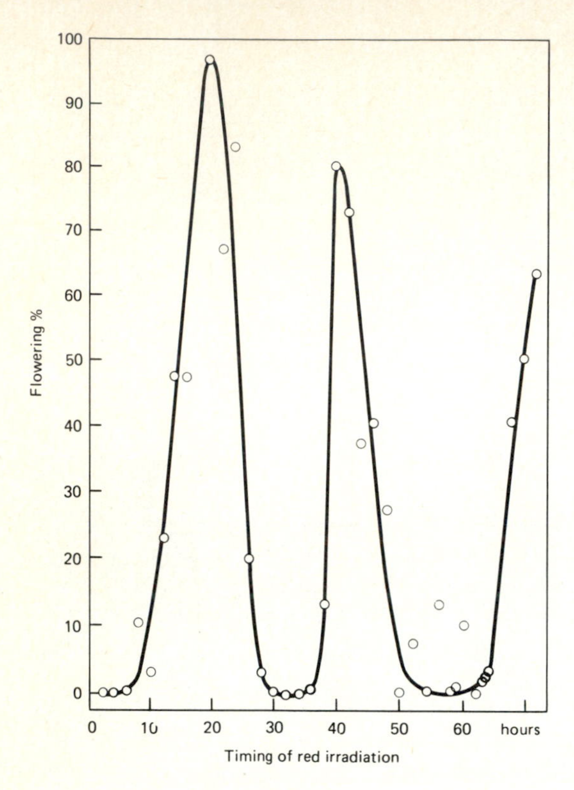

Figure 12-11 Flowering of *Chenopodium rubrum* seedlings exposed to red light for 2 minutes at one of the indicated times during a 72 hour dark period. (Adapted from Cumming et al., 1965. Can J. Bot. *43*: 825–853.)

Floral induction in plants grown on normal 24-hour light–dark cycles also involves some processes that are promoted by P_{fr} and others that are inhibited by it. The temporal sequence and duration of these processes is regulated by the clock. In short-day plants which flower when the dark period exceeds a critical value, the high-P_{fr}-requiring reactions take place during the day and early part of the night, while reactions that proceed best when the P_{fr} level is low take place later in the night (Fig. 12-12). Many plant

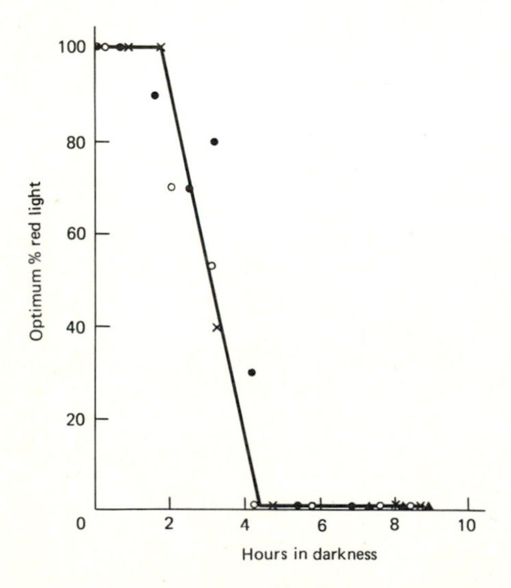

Figure 12-12 Flowering of *Chenopodium rubrum* seedlings irradiated for 5 min. during a 10 hour night with a mixture of red and far-red light. The timing of the light treatment and the proportion of red to far-red energy were varied. The graph shows the percent red in the mixture that gives optimum floral induction. Note that pure red light which establishes a high P_{fr} level is optimal during the early part of the night, while pure far-red light which removes P_{fr} is optimal after the first 4 hours of darkness. (From King and Cumming, 1972. Planta *108*: 39–57.)

physiologists once believed that the critical night length is determined by the rate of P_{fr} reversion to P_r; however, this idea no longer seems to be correct. The critical dark period appears instead to depend on the duration of processes inhibited by P_{fr}, which are in turn regulated by the clock. Thus we can see how photoperiodic time measurement in short-day plants involves neither phytochrome alone nor the clock alone, but an interaction of the two.

The role of the clock in floral induction in long-day plants is less clear. These plants, you will recall, do not initiate flower buds unless the night is shorter than a certain critical length. Some long-day plants also need several hours of high-intensity far-red light (710–730 nm) during each 24-hour cycle. Irradiation during the day with a broad-band light source (such as sunlight) that emits both far-red and shorter wavelengths is effective, as is irradiation with wavelengths shorter than 700 nm during the day followed by long-wavelength irradiation at night. This need for prolonged high-intensity far-red light bears some similarity to HIR in etiolated plants (see Chapter 11), and pigment cycling may be important in both cases. There is however an important distinction: far-red irradiation of etiolated plants, which have only P_r prior to irradiation, raises the P_{fr} level, whereas similar irradiation of light-grown plants usually lowers the P_{fr} level. Thus the high-intensity requirement in long-day plants may be related in part to the need for P_r (or low P_{fr}) during part of the daily cycle.

Morphological changes during flowering

The first evidence of the transition from vegetative to reproductive growth during floral induction is a rise in DNA synthesis and mitotic activity of cells at the apical meristem. The meristem accordingly broadens and elongates,

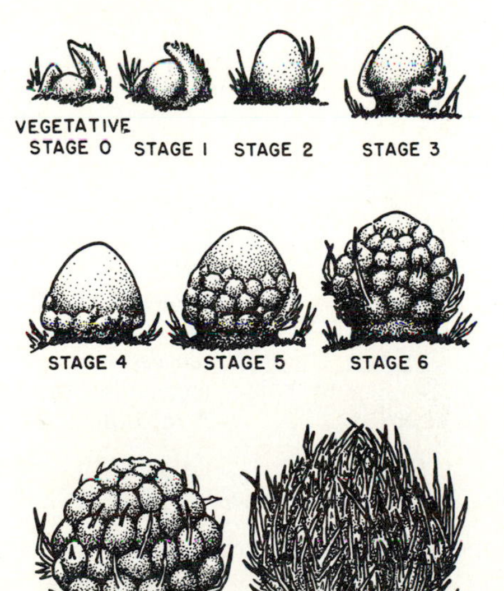

Figure 12-13 Stages of development in the stem apex of *Xanthium*, from vegetative growth to complete inflorescence primordium. (From Salisbury, 1955. Plant Physiol. *30*: 327–334.)

and floral primordia develop. The various anatomical stages involved as *Xanthium* (cocklebur) plants progress from vegetative growth to complete flowering are depicted in Fig. 12-13.

In day-neutral plants such as tomato, flowering is relatively unaffected by the length of day; this does not mean that the plant is insensitive to phytochrome changes, since one can readily demonstrate reversible red–far-red control of some aspects of morphogenesis. Similarly, many varieties of potato can flower on either long or short days but produce tubers only under short-day conditions. Apparently, phytochrome conversion is not always linked to the system that controls reproduction.

Ripeness-to-flower

Some photoperiod-sensitive plants fail to respond to photoperiod at particular times in their life history. For example, some plants must pass from a juvenile phase to a phase designated as "ripeness-to-flower" in order for photoperiod to be effective; in many such plants, the green expanded cotyledons or even first leaves are not sensitive to photoperiod, while later leaves are. In woody plants, the juvenile phase may last years (Chapters 14 and 16). Again, failure to respond is not due to the absence of phytochrome, since such seedlings show the usual morphogenetic responses to red light; rather, some other aspect of the response apparatus appears to be undeveloped. We derive the same sort of conclusion from the fact that photoperiodic requirements for flowering may become less rigid as plants age (Fig. 12-14).

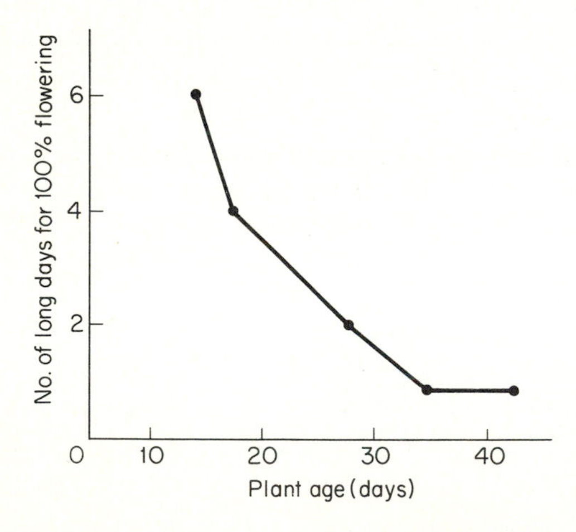

Figure 12-14 The long-day requirement for flower induction in *Lolium temulentum* declines as the plants increase in age. (Data of Evans, 1960. Aust. J. Biol. Sci. *13*: 123–131.)

Quantitative variation in flowering behavior

For some photoperiod-sensitive plants, the response to the length of uninterrupted night is an "all or nothing" phenomenon. Such plants remain vegetative indefinitely in the absence of appropriate inductive dark periods.

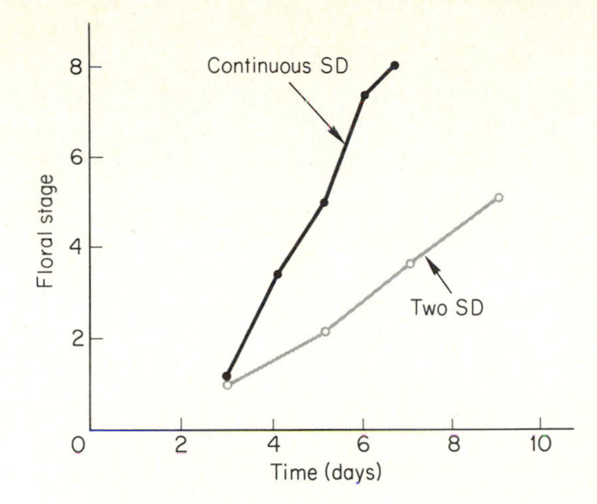

Figure 12-15 If the floral stages (Fig. 12-13) are used as a measure of the influence of light treatments on flowering in *Xanthium*, continuous short days seem to cause more rapid floral development than occurs after only two short days. (From Salisbury, 1955. Plant Physiol. *30*: 327–334.)

In the short-day plant cocklebur, a single long dark period suffices to induce some reproductive activity, although more than one photoinductive cycle can produce a greater flowering response (Fig. 12-15). Other short-day plants, such as soybeans, require approximately four successive photoinductive cycles for floral initiation; other species may require even more.

In other plants, the response to photoperiod may be quantitative rather than qualitative. Plants displaying such behavior flower on long or short days, but can produce more or better developed flowers on appropriate photoperiods. In another variation of quantitative flowering behavior, the flowering of Christmas cactus and some other short-day plants is enhanced by low night temperature, either with or as a substitute for a long night. Some plants require a particular photoperiodic sequence of short days followed by long days (clover) or long days followed by short days (many *Bryophyllum* species). This type of mechanism enables the plant to perceive differences between increasing daylength in spring time and decreasing daylength in fall. Evidently, different responses to complex photoperiodic and and temperature regimes have evolved, enabling plants to adapt to particular ecological niches.

Propagation and perpetuation of the floral stimulus

Plants frequently show a structural separation between the organ perceiving photoperiod and the organ responding to it. An example is the short-day plant *Xanthium*, which will flower if the uninterrupted dark period exceeds 9 hours. Suppose that we put one part of the plant on short days by enclosing it in a light-tight container, while the rest of the plant is exposed to long days. If there were no transmitted effects, one would expect only the portion exposed to short days to flower. However all buds, even those on the portion of the plant exposed to long days, are induced to flower. Clearly, some influence has moved from the portion exposed to short days to the portion exposed to long days.

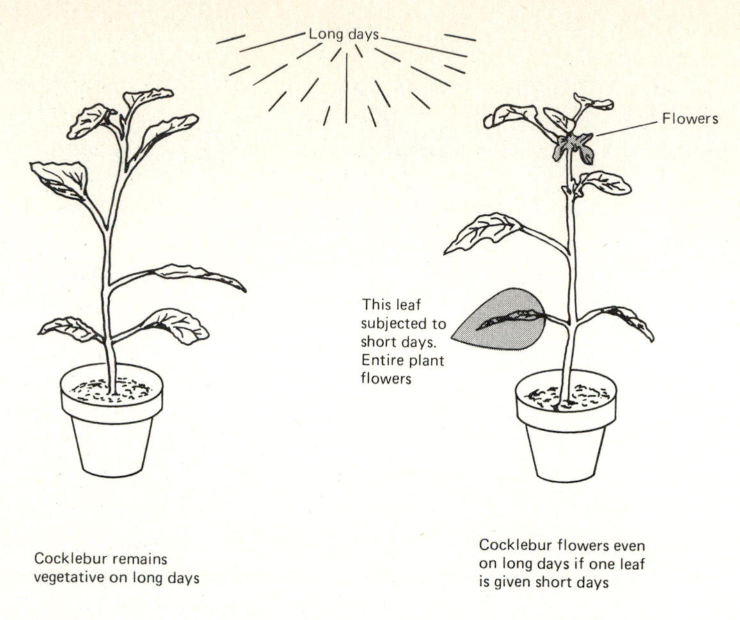

Figure 12-16 In many species, as in the cocklebur, short-day treatment of a single leaf results in flowering of the entire plant. (Adapted from Hamner and Bonner, 1938. Botan. Gaz. *100*: 388–431.)

This experiment can be refined to the point where only a single leaf enclosed in an opaque container is receiving a short day, while the rest of the plant receives a long day; again, buds all over the plant can be induced to flower (Fig. 12-16). This shows that the leaf is the organ of photoperiodic perception and that it influences buds some distance away. The influence is almost certainly due to a chemical substance synthesized in small quantities by the photoinduced leaf. This substance, the hypothetical flowering hormone of plants, has been named *florigen*. Florigen has never been isolated and chemically identified, but its presence is inferred from many different experiments. If plant *A* is induced to flower and plant *B* is not, flowering may be caused in *B* by grafting it to *A* (Fig. 12-17). Even a single leaf from the donor grafted onto the receptor may suffice. This experiment works even if plant *A* is a short-day plant and plant *B* is a long-day plant; thus florigen must be the same or at least functionally equivalent in both types of plants. Through the use of steam-girdling techniques, which kill phloem cells and block florigen transfer, we can deduce that the flow of hormones from donor to receptor occurs in living cells, probably in the phloem, at about the rate of transport of bulk organic substances in plants.

The application of gibberellin to long-day plants under short days generally, though not in all cases, causes *bolting* which is frequently followed by flowering (Fig. 12-18). Similarly we now have evidence that abscisic acid can promote flowering in some short-day plants, while auxin acting through ethylene production can do the same in bromeliads like pineapple. Florigen

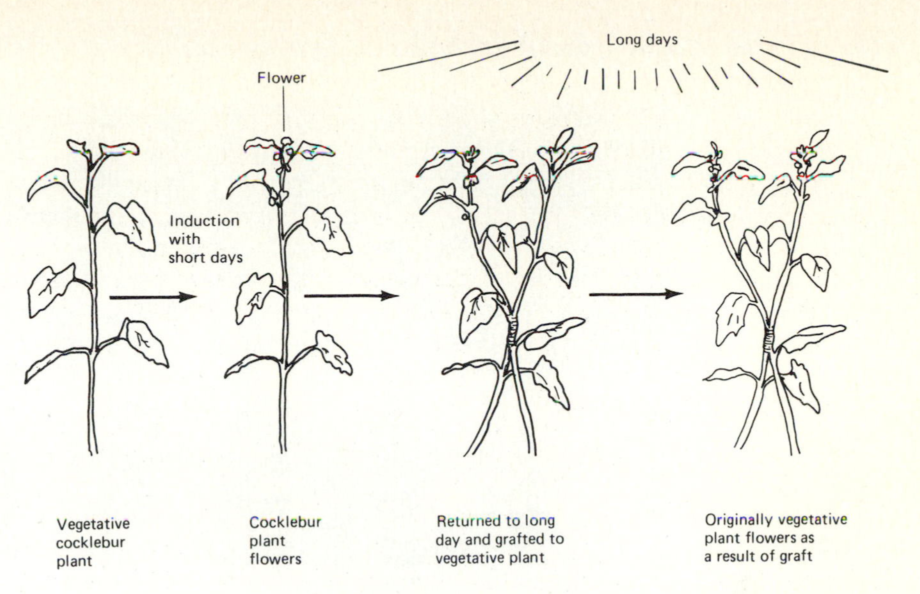

Figure 12-17(a) The flowering stimulus may be transmitted from plant to plant across a graft union. (Adapted from Hamner and Bonner, 1938. Botan. Gaz. *100*: 388–431.)

Figure 12-17(b) This photograph shows plants of 3 different photoperiodic classes: a short day plant (SDP) that normally requires short days for floral induction; a short long day plant (SLDP) that normally requires short days followed by long days, and a long short day plant (LSDP) that normally requires long days followed by short days. All 3 plants were grown continuously under non-inductive long day conditions, and were induced to flower by graft of an induced leaf from the long day plant *Sedum spectabile*. The photographs were taken 84 days (left), 106 days (middle) and 160 days (right) after grafting. (Courtesy of J. A. D. Zeevaart, Michigan State University.)

Figure 12-18 The long-day-requiring plant *Samolus parviflora* can be induced to flower under noninductive short days (9 hours) by applications of gibberellin. From left to right, plants were given 0, 2, 5, 10, 20, or 50 μg GA_3/plant each day. (From Lang, 1957. Proc. Natl. Acad. Sci. *43*: 709-717.)

may thus not be a single substance, but any one or an appropriate combination of different hormones or nutrients which can become critical to flowering in particular plants grown under specific conditions.

One of the remarkable facts about floral induction is that it can be prolonged indefinitely in some plants, even when they are returned to unfavorable photoperiods. Further, induced plants may serve as sources of the floral stimulus, acting as donors in grafting experiments far beyond the time of their initial induction. Their receptors, induced by grafting, can become donors of florigen to still other receptors, even when severed from the donor plants that induced them. Thus it appears that the transition from vegetation to reproduction involves some kind of sustained metabolic transformation, leading to the prolonged and perhaps indefinite production of the floral-inducing stimulus. Some have suggested that the floral stimulus is a kind of latent, self-replicating entity, perhaps like an inactive virus, that is transformed into the active form by appropriate photoperiod. In this context, the initiation of flowering could be compared to catching a disease, which is propagated from one individual to another, until terminated by the spontaneous recovery of the host. As a case in point, in short-day *Perilla*, minimal induction leads to flowering for a time, but this is followed by a gradual return to vegetative growth during an unfavorable photoperiod, with *Perilla* apparently making a spontaneous recovery from the flowering "disease." The problem of the nature of florigen and floral induction remains an intriguing mystery.

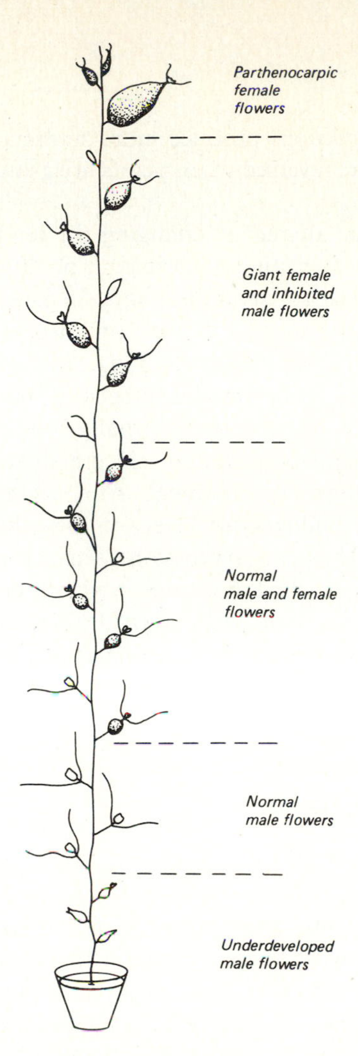

Figure 12-19 Schematic representation of the sequence of flower type in the acorn squash. (Adapted from Nitsch *et al.* 1952. Amer. J. Bot. *39*: 32–43.)

Development of Sex Organs

In some plants, the development of individual sex organs as well as floral initiation itself is under photoperiodic control. This is illustrated most dramatically in plants which produce stamens and pistils on separate flowers, either on separate plants (*dioecious* species) or on the same plant (*monoecious* species). Usually, photoperiodic conditions that accelerate floral initiation promote femaleness and depress maleness. In the short-day dioecious plant *Humulus* (hop), flowers destined to become male develop female characteristics when the plant is exposed to short days: stigmatic apices, a female characteristic, develop on the male anthers, so that stamens resemble pistils. In the short-day *Xanthium*, plants exposed to only one short day before transfer to noninductive long-day conditions produce more staminate than pistillate flowers, but these sex ratios are reversed when plants are grown on continuous short days.

Sex expression can also be altered by changing the temperature or supplying hormones. The most interesting experiments of this type have been performed with cucurbits, such as cucumber, squash, and gherkins. These monoecious plants usually produce male flowers on lower nodes, and female flowers on upper nodes (Fig. 12-19). However, long days and high temperatures can cause some potentially female nodes to become male, whereas short days and low temperatures do the reverse. Externally supplied GA can substitute for long days; this might be expected, since endogenous gibberellin levels usually increase with long-day treatment. Auxin, ethylene, and ABA promote femaleness, and the endogenous levels of these hormones usually increase during short days. Thus, sex expression seems to be controlled by interactions among the several plant hormones, whose levels are in turn daylength dependent. Auxin effects are particularly dramatic, since floral buds destined by their position on the plant to become males can develop as females instead, if excised and cultured on a medium containing IAA. Such extreme plasticity in sex expression is unusual.

Effects of Moonlight and Artificial Lights on Photoperiodic Behavior

Light intensities as low as 0.1 lux (approximately 0.01 ft-candle) during the night can influence photoperiodic time measurement in many plants and animals. Yet the intensity of light from a full moon on a cloudless night may reach 0.3 lux at latitude of 50°, and more than three times this value in tropical regions (Fig. 12-20). This fact led E. Bünning and his colleagues to inquire whether moonlight can disturb time measurement. Surprisingly, their investigations revealed that some plants have adaptive mechanisms that apparently prevent moonlight from interfering with photoperiodism.

You will recall that photoperiodic perception occurs in the leaves. In the leguminous plants soybean, peanut, and clover, sleep movements change the position of the leaves from horizontal during the day to vertical at night. This behavior reduces the intensity of light falling on the leaf surface from an overhead lamp (an "artificial moon") by 85% to 95%, to an intensity below threshold for interference with time measurement. In some nyctinastic plants such as *Albizzia*, *Samanea*, and *Cassia*, leaflets not only orient vertically at night, but also rotate on their axes so that paired leaflets fold together, with the upper surfaces shading each other (Fig. 12-1)—an interesting behavior in view of the fact that the upper surface is more sensitive to light breaks than is the lower surface.

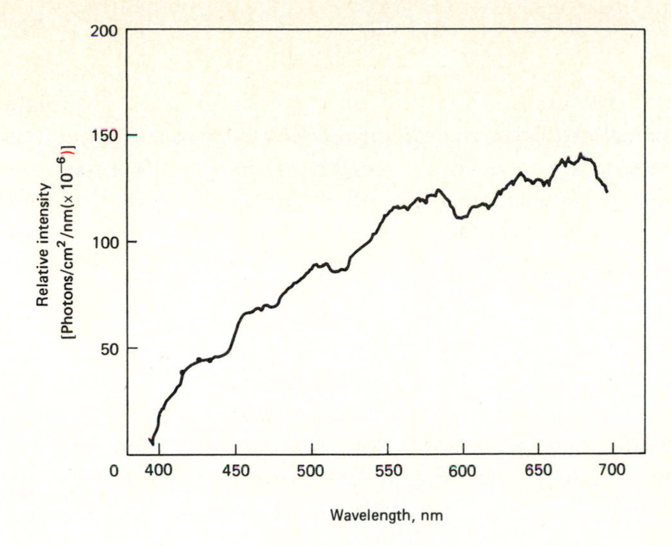

Figure 12-20 The spectrum of moonlight, recorded at Eniwetok, 0200 hr., August 20, 1970. The moon was 15° from zenith. Note that the spectrum peaks near 660 nm. (Adapted from Munz and McFarland, 1973. Vision Res. *13*: 1829–1874.)

Some short-day plants flower most prolifically when grown with low-intensity light (approximately 0.5 lux) rather than complete darkness during the night. In these plants, moonlight probably increases the number of flowers produced by a short-day regime. Although it is not clear why low light intensities promote flowering more than darkness, these examples provide some rational basis for the superstition of planting particular seeds by the light of the full moon. Another full moon one lunar cycle later could have profound promotive effects on flowering.

Street and house lighting, on the other hand, can disrupt important photoperiodic controls in many plants. For example, they can retard the induction of dormancy in rhododendron, dogwood and other trees and shrubs, thereby decreasing the chances for survival during harsh winters. Lamps that maintain a high P_{fr} level because they are rich in red wavelengths are particularly disruptive, a fact which should be considered when installing outdoor lighting facilities.

Effects of Temperature

Plants respond to temperature, as well as to light, in both qualitative and quantitative ways. The rates of almost all chemical reactions in the plant show a graded response to increasing temperature, reaching an optimum and then declining. In contrast, many developmental changes such as seed germination and the breaking of bud dormancy often display "all or nothing" control. In these latter cases, low-temperature induction requires uninterrupted exposure for at least a certain critical period, thus resembling photoperiodic induction, with its requirement for precisely timed dark periods.

Graded responses

With most chemical processes, the rate of reaction increases steadily with increasing temperature. The *temperature coefficient*, or Q_{10}, expresses the ratio of the rate of a chemical reaction after the temperature has been increased by 10°C, to the original rate.

$$Q_{10} = \frac{\text{Rate at } (t + 10)°\text{C}}{\text{Rate at } t°\text{C}}$$

The Q_{10} of most chemically or enzymatically controlled reactions is at least 2.0, while the Q_{10} for physical reactions such as diffusion or photochemical reactions is about 1.1 or 1.2. Plants grown in temperatures over the range of 0°–30°C show a steady increase in elongation with increase in temperature, the Q_{10} approximating 2.0 or more (Fig. 12-21). For reasons that we do not understand, different plants have vastly different temperature optima, indicating that some fundamental biochemical process in their makeup has a different sensitivity to temperature. With all plants, a temperature is eventually reached that is optimal for growth, and if the temperature is raised above this point, the absolute growth rate declines, sometimes very dramatically. For most plants, this optimum is in the region of 28°–32°C.

We do not know why most plants are injured by temperatures above about 30°C, especially since the enzymes or organelles obtained from plants are generally not damaged by this temperature. One possibility is that the delicate membranes of the cell or of its organelles are sensitive to temperature changes because of the melting or solidification of the fatty acids in the phospholipids as the temperature rises or falls. Plants grown at lower temperatures are known to synthesize more unsaturated fats, with appropriately lower melting points; the reverse is true at high temperatures (Table 12-1).

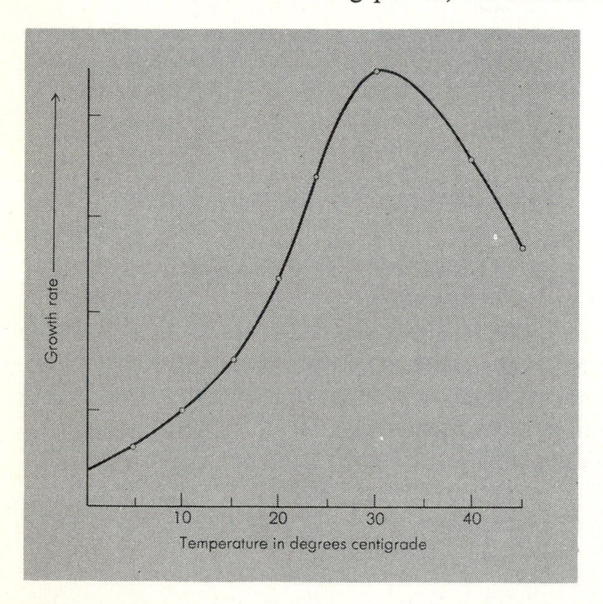

Figure 12-21 A temperature-optimum curve for the short-time growth of pieces of pea stem. (From Galston and Hand, 1949. Amer J. Botany *36*: 85–94.)

Another possibility is that some materials essential for growth are either destroyed very rapidly or not produced in adequate quantity at the elevated temperature. As a case in point, many organisms contain genes that are "temperature sensitive." In the red bread mold, *Neurospora*, the gene responsible for the production of vitamin B_2 or riboflavin works quite well when the organism is grown at low temperatures, but does not function well when the

Figure 12-22 (a) A winter twig of *Aesculus hippocastanum* (horse chestnut) showing the dormant buds and the horseshoe shaped scars left by the abscission of last year's leaves. After dormancy is broken by a cold period in the winter, the buds burst open on the return of spring. Photographs (b–d) show the steady unfolding of the buds over the period of about ten days to reveal the new leaves and flower buds.

organism is grown at higher temperatures. At 35°C, therefore, the organism has an obligate requirement for exogenous riboflavin, whereas at 25°C it is autotrophic for this material. This same situation may pertain to other vitamins, to amino acids, or to hormone synthesis in higher plants. If this is true, and if we knew the chemical basis for the temperature-induced decline in growth rate, we could improve growth at supraoptimal temperatures by application of the required material.

Morphogenetic effects

The very slow rate of chemical reactions in the plant at low temperatures ensures that growth changes are coordinated with changes in climate. In addition, temperature affects many photoperiod-sensitive processes, by changing the critical length of the dark period, although the mechanism whereby it does this is not understood. Since the presence of P_{fr} during the night may interfere with floral induction in short-day plants, we might expect high night temperatures, which enhance phytochrome destruction as well as the reversion of P_{fr} to P_r, to promote floral induction in such plants; in fact, the reverse situation usually occurs. Some plants can be induced to flower either by long dark nights or by low night temperatures. Possibly either treatment can promote some process leading to florigen synthesis.

The most dramatic effects of low temperature involve seed germination, the breaking of bud dormancy (Fig. 12-22), and preparation for initiation of floral primordia. The low-temperature treatment that facilitates seed germination is called *stratification*, while that for floral initiation is called *vernalization*. The former requirement prevents premature germination, while the latter guarantees a biennial habit. Because the optimal temperatures required for germination and vernalization vary among plants, temperature is important both in regulating the geographical distribution of plants, and in regulating the seasonal progress of flowering and fruiting.

Much of our knowledge of vernalization comes from studies with annual and biennial strains of rye (*Secale cereale*). The annual strain completes its reproductive cycle in a single growing season, while the biennial variety requires overwintering in the field if it is to flower. The biennial produces only vegetative organs during the first growing season; no floral primordia can be initiated until the plant has been exposed to prolonged periods of low temperature. Only after such exposure can it respond to a proper photoperiodic stimulus. This requirement for low temperatures can be satisfied at any point in the developmental history of the plant after germination. For example, if a biennial seed is permitted to imbibe some water to start germination, and is then exposed to about six weeks of low temperatures (about 2°–5°C), it will behave just as if it had gone through a cold winter after a year of growth, and will flower following spring planting if exposed to the proper photoperiod (Fig. 12-23). Since the difference between the annual and biennial

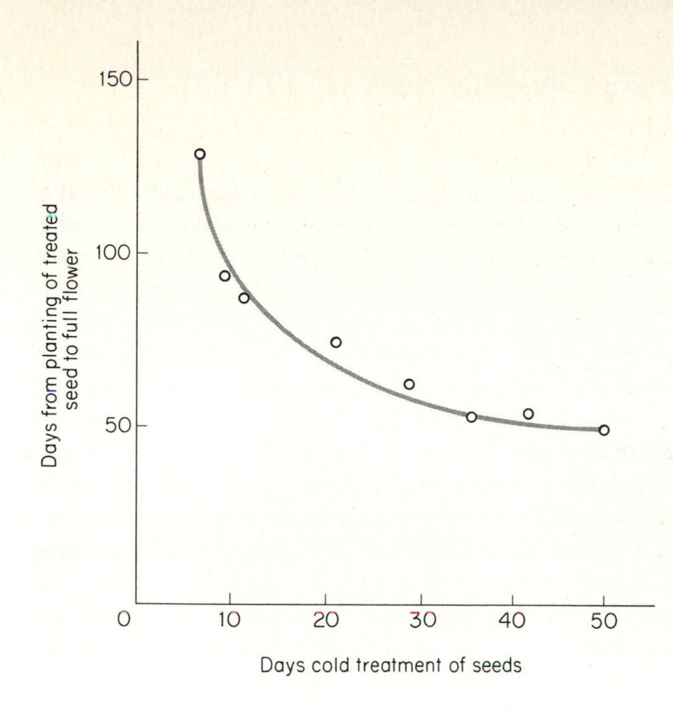

Figure 12-23 Some biennial plants will not flower at all without a cold treatment. Rye gives a quantitative response, the length of time between planting and flowering decreasing as the period of cold treatment is lengthened. (From Purvis and Gregory, 1937. Ann. Bot. N. S. *1*: 569–592.)

races of most plants seems to reside in a single gene, the low-temperature requirement in the biennial form appears to be a substitute for some genetically controlled biochemical event produced without cold treatment in the annual strain.

The cold stimulus is perceived in the shoot tip, the tissue in which floral buds are also initiated. Therefore the effects of the cold-induced stimulus, unlike the florigen produced by photoperiod, need not move through the plant. Most plants that require vernalization are also long-day plants, requiring photoperiodic induction some time after germination. This suggests that the cold treatment produces a localized change in cells which then enables them to respond to florigen. The changes produced by low temperature persist virtually indefinitely after induction in most plants. For example, the vernalized state is perpetuated during the thousands of cell divisions in the apical meristems; thus the vernalization of a bud suffices to transmit the effect of vernalization to cells arising from the bud.

In several biennials, the application of gibberellins to an unvernalized plant seems to substitute for low temperature, causing the prompt initiation of floral primordia when the treated plant is exposed to a favorable photoperiod (Fig. 9-21). But since gibberellins cannot *always* substitute for the cold treatment, it seems unlikely that vernalization works solely by enhancing gibberellin synthesis. An alternate suggestion is that vernalization "derepresses" particular genes in cold-treated cells—that is, it removes a block that had prevented specific transcription and translation sequences. This is an attractive hypothesis, since both stratification and vernalization increase the synthesis of many enzymes.

Seed germination

Temperature requirements for seed germination vary considerably from one species to another; for example, 15°C is optimal for the annual *Delphinium*, whereas 30° to 40°C is best for melon (*Cucumis melo*). However, many seeds remain dormant even when moistened and incubated at a temperature appropriate for germination. Sometimes such dormancy can be broken by light (as discussed in Chapter 11), by preincubation at a specific temperature, by increasing O_2 availability, by leaching out inhibitors, or by rupturing physically restrictive seed or fruit coats. Seeds of temperate plants frequently require low-temperature stratification before incubation at higher temperatures; this delays germination until springtime, circumventing possible frost damage to tender seedlings. On the other hand, some seeds, such as those of white oak, germinate immediately after falling to the ground, and quickly produce a deep root system.

Dry seeds do not respond to stratification treatment, so it is usual to let them imbibe fully before exposure to low temperature. One assumes that some metabolic transformations are proceeding during treatment—such changes are manifested in altered rates of respiration, patterns of translocation, content of enzymes, and even rates of cell division and elongation. One crucial set of changes involves the balance of hormones: ABA levels generally decrease, while gibberellins increase in some seeds and cytokinins in others. ABA is generally associated with dormancy, whereas gibberellins and cytokinins are associated with synthesis and growth. Stratification thus produces an "after-ripening," which prepares seeds biochemically for subsequent germination at higher temperature.

Some seeds, mostly desert types, require stratification at high temperatures prior to germination; in other cases, light and temperature requirements for germination are interrelated. For example, pigweed (*Amaranthus retroflexus*) seeds germinate in the dark if kept at 20°C for several days and then transferred to 35°C, but they require light for germination if the low temperature treatment is eliminated; this apparently indicates a requirement for P_{fr}. You will recall that much of the phytochrome in seeds is in the P_{fr} form, and P_{fr} can revert to P_r after the seed has been hydrated. Since P_{fr} reversion in *Amaranthus* is very slow at 20°C, but four times as rapid at 25°C, P_{fr}-requiring processes in *Amaranthus* that lead to germination can take place during the chilling period. Thus we understand some specific temperature and light requirements for germination in terms of P_{fr}, others in terms of hormones. Whatever the regulatory mechanism, it must ultimately affect, and be affected by, patterns of gene activity.

SUMMARY

The plant's measurement of seasonal changes through *photoperiodism* involves not only the sensing of light and dark signals, but also referral of these signals to an endogenous *biological clock*, whose existence is indicated by overt *circadian rhythms*. Rhythms have been noted in many different processes, including leaf movements, rates of photosynthesis, cell division and bioluminescence in algae, and the activity of some cellular enzymes. While the *amplitude* and phase of rhythmic oscillations are temperature sensitive, the *period* of most rhythmic events is essentially temperature independent. Neither the mechanism of timekeeping nor the nature of the temperature compensator is clear, but some evidence exists that rhythms are manifestations of cyclic changes in *membranes*.

Although rhythms are endogenous, they can be *phased* ("the clock reset") or *entrained* by appropriate temperature and light signals, the latter usually perceived by phytochrome, and possibly by a blue-light photoreceptor. Rhythms can be damped out or changed in period by several chemical agents, including ethanol, heavy water, and Li^+; all are presumably membrane-active substances. The damping out of rhythms at low temperatures may result from solidification of membrane lipids that must be liquid for normal functioning.

Leaf movements in leguminous plants are controlled by the movement of K^+, Cl^-, and possibly other solutes from one side of the *pulvinus* to the other; this in turn controls water movement, turgor pressure of motor cells, and the leaf angle. Movements are initiated by light–dark signals, but are then perpetuated during constant darkness. Leaf movements are sometimes correlated with *flowering* behavior; for example, soybean plants with strong endogenous leaf movements tend to be photoperiod-sensitive, while those without strong *circadian* leaf movements tend to be day-neutral. *Short-day* plants require an *inductive cycle* containing a *dark period* of at least a *minimum critical length*, while *long-day* plants flower only if the unbroken dark period does *not* exceed a maximum critical length. Red light perceived by phytochrome can interrupt the dark period, also resetting the clock; far-red light can reverse the effect of red. If certain short-day plants are exposed to prolonged dark periods of constantly increasing lengths, the intensity of flowering, when plotted against the length of the dark period, describes a circadian oscillation; this implicates the biological clock in the flowering response. Dim light, even that of moonlight, can interrupt a dark period, and *nyctinastic* leaf movements may prevent such interference with the plant's measurement of night length.

Some plants cannot perceive the photoperiodic stimulus until they reach a stage called *ripe-to-flower*; prior to this they are *juvenile*, and exhibit morphological and physiological differences from mature plants. Some plants, when ripe-to-flower, require only a single photoinductive cycle, while others require a succession of several. In some plants, induction leads to the propagation of an as yet uncharacterized floral stimulus (*florigen*) that moves from the leaf in which it is made to the bud. Gibberellin can induce flowering in *bolting*, long-day rosette plants, while auxin is similarly active in pineapples other bromeliads. Photoperiod can control sex expression in flowers of *monoecious* or *dioecious* plants, which have unisexual flowers on a single plant or on two different plants, respectively; temperature and hormone applications can have similar effects.

While most temperature-sensitive processes proceed faster the higher the temperature, roughly doubling in rate with each 10°C rise ($Q_{10} = 2$), some plant processes are promoted by lower temperatures. These processes include promotion of flowering in biennials (*vernalization*) and promotion of germination in dormant seeds (*stratification*). Low temperatures that promote flowering are perceived in the stem tip, and thus require no translocation to be active. Plants requiring vernalization usually then require long days in order to initiate flowers. In seeds, light and temperature signals are also interrelated, in that high temperature can cause some seeds to go dormant, and to require a light signal perceived by phytochrome for the inception of germination. Similarly, low temperature can sometimes remove a light requirement. Gibberellins can mimic the effect of low temperatures in both seeds and buds.

SELECTED READINGS

BÜNNING, E. *The Physiological Clock* (Berlin–Heidelberg–New York: Springer-Verlag, 1973).

HASTINGS, J. W., and H. G. SCHWEIGER. *The Molecular Basis of Circadian Rhythms* (Berlin: Abakon Verlagsgesellschaft, 1976).

HILLMAN, W. S. *The Physiology of Flowering* (New York: Holt, Rinehart and Winston, 1962).

SALISBURY, F. B. *The Flowering Process* (London–New York: Pergamon, 1963).

SWEENEY, B. *Rhythmic Phenomena in Plants* (London–New York: Academic Press, 1969).

VILLIERS, T. A. *Dormancy and Survival in Plants* (New York: St. Martins Press, 1975).

VINCE-PRUE, D. *Photoperiodism in Plants* (New York: McGraw-Hill, 1975).

QUESTIONS

12-1. Although virtually all eukaryotes have internal oscillators that generate circadian rhythms, no such rhythms have been found in any prokaryote. What does this suggest about the nature of the biological clock?

12-2. Since biological clocks are ubiquitous among eukaryotic organisms, it is

reasonable to suppose that they have adaptive value. Speculate on the benefits of oscillatory behavior as compared to homeostasis (steady state maintenance).

12-3. The rate of respiration in many plants varies during the 24-hour day. How would you determine whether these variations were the expression of an endogenous circadian rhythm?

12-4. Seedlings of *Chenopodium* were germinated in continuous light and then transferred to darkness, with the following results.
a) If the plant was given no dark period, it failed to flower.
b) If it was given 13 hours of darkness, it flowered.
c) If it was given 13 hours of darkness interrupted by 15 minutes of white light at the 8th hour, it failed to flower.
d) 28 hours of darkness—no flowering
e) 43 hours of darkness—flowering
f) 58 hours of darkness—no flowering
g) 73 hours of darkness—flowering
On the basis of your knowledge of flowering, explain these results (which are actual experimental results). Comment on anything which may seem unusual and suggest reasons for the plant's behavior.

12-5. "Whether or not flowering occurs depends on synchronization of environmental cues and the endogenous clock." Discuss this statement.

12-6. Describe in some detail how you would proceed to classify a new plant as short-day, long-day, or day-neutral.

12-7. The critical photoperiod of various races of range grasses varies with the latitude at which they grow. What is the ecological significance of this observation?

12-8. What is "florigen"? What is the evidence for its existence? What kinds of characteristics might it have?

12-9. How could you most effectively prevent the flowering of a stand of the short day plant *Xanthium* in an open field?

12-10. Describe several ways in which knowledge concerning photoperiodism has been put to practical use.

12-11. What regulates flowering in those species which are not sensitive to temperature or photoperiod? Is the mechanism of floral initiation in these species necessarily different from that in photoperiod-sensitive plants? Explain.

12-12. How does the dependence of floral initiation on low temperature differ from its dependence on daylength?

12-13. How does the effect of low temperature on flowering differ from its effect on bud dormancy, discussed in Chapter 10?

13

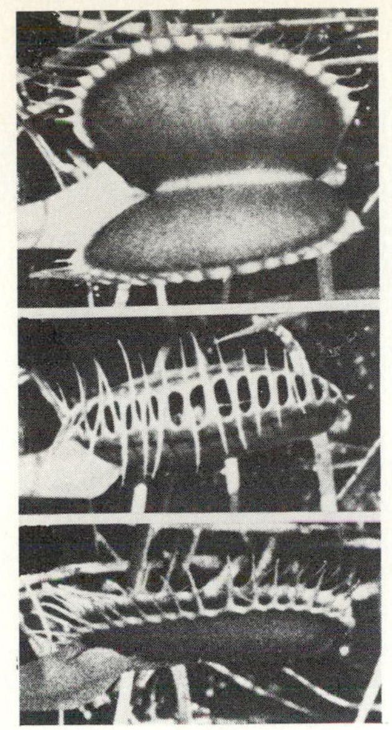

Rapid Plant Movements

Terrestrial plants are firmly anchored in the soil by their root systems, and are thus usually regarded as absolutely fixed in space. This notion is inaccurate in that all plants are able to move slowly by tropistic growth reactions to adjust the orientation of their organs to environmental stimuli such as light and gravity. In addition, some fold their leaves or flower petals together in the somewhat more rapid periodic sleep movements that we considered in Chapter 12. A few unusual plants have a more dramatic and rapid "power of movement" (a phrase used by Charles Darwin) when stimulated by a variety of environmental factors including light, chemicals, touch, wounding, or mechanical vibration. These unusual types of behavior include leaf folding in *Mimosa pudica* (the "sensitive plant"), trap movements in various carnivorous plants, and the curling of tendrils of many vines. The functions of these movements are as varied as the stimuli that evoke them. Rapid movements in *Mimosa* may serve to protect the plant from insects and other predators, whereas trap movements enable carnivorous plants to catch insects which furnish accessory nutrients for otherwise autotrophic plants. The curling of vine tendrils around nearby rigid objects provides mechanical support for the plant body, thus permitting attainment of great heights without a large supporting trunk. Tendrils sometimes also choke out neighboring competitive species, thus improving the plant's ability to survive in a crowded situation.

Rapid plant movements attracted attention more than two millenia ago, and continue to arouse the curiosity of modern scientists. Studies of plant movements have helped to uncover many new aspects of plant behavior. For

example, studies of phototropic curvature led to the first postulates concerning the existence of plant growth hormones. Nyctinastic leaf movements provided the earliest clue that organisms have internal clocks, and current studies are helping us understand how light and clocks interact in biological timekeeping. Studies with *thigmonastic* (touch-sensitive) plants have revealed that electrical signals propagated through plants are, as in animals, an important means of communication between the different cells of the body. In this chapter we shall examine some of the details of these aspects of the green plant's physiology.

Rapid Leaf Movements in the Sensitive Plant (*Mimosa pudica*)

The leaves of *Mimosa*, like those of *Samanea*, show circadian nyctinastic movements, but can also move extremely rapidly in response to mechanical, thermal, chemical, and electrical stimuli. A few seconds after a leaf has been touched, the petiole drops and the pinnules fold together (Fig. 13-1). Such rapid response, demanding a high degree of coordination, is produced by first converting the mechanical stimulus perceived by sensory cells on the petiole to an electrical signal. This signal, probably a propagated depolarization of a membrane, travels rapidly through the tissue until it reaches pulvinar motor cells, which then undergo the volume changes that lead to leaf or leaflet movement.

Figure 13-1 Leaves of *Mimosa pudica* viewed from above. Before tactile stimulation (left); and after stimulation (right).

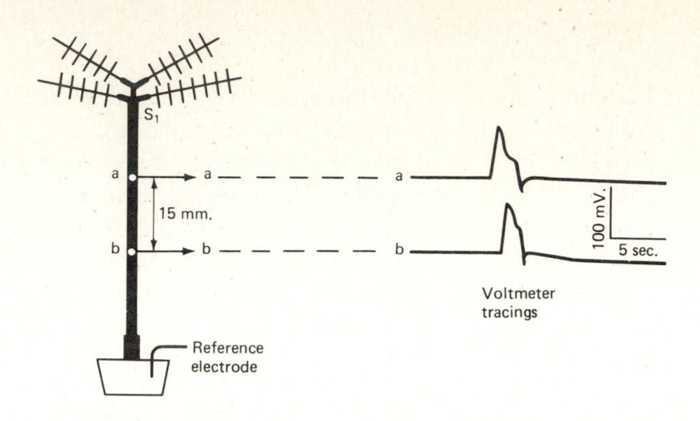

Figure 13-2 Electrical transmission in the petiole of *Mimosa pudica* initiated by a cooling stimulus at S_1. The action potentials (right) are recorded simultaneously at two electrodes a and b, placed as shown in the illustration on the left. Reference electrodes for a and b are in the soil. (Adapted from Sibaoka, 1966. Symp. Soc. Expt. Biol. *20*: 49–73.)

The propagated electrical signal can be measured with a pair of salt-filled glass electrodes, one placed on the petiole, and the other in the soil surrounding the roots, or on the pot containing the plant. The signal consists of a transitory change in voltage between the electrodes, followed by a return to the original value (Fig. 13-2). The stimulus moves at approximately 2 cm/sec through the petiole of *Mimosa*, and about five times more rapidly through the leaf of the carnivorous plant *Dionaea* (Venus' Flytrap). The speed of transmission increases threefold with each 10° rise in temperature, indicating the involvement of chemical reactions. Propagation ceases when the phloem is removed, which implies that the signal moves through phloem cells; the signal cannot cross a barrier of dead tissue, indicating that it probably is not caused by water-soluble diffusible hormones.

External electrodes, while establishing that plants use electrical signals as a means of cell-to-cell communication, provide no clue regarding the cells or mechanisms involved in the transmission. Intracellular microelectrodes (see Fig. 7-8) reveal that certain cells in the phloem and protoxylem of *Mimosa* are "excitable" (Fig. 13-3). Their resting potential, or transmembrane potential prior to excitation, is much more negative than that of surrounding cells. When the voltage across the membrane is sufficiently reduced (that is, when the interior of the cell becomes less negative) by electrical or chemical stimulation to reach a "threshold value," the cells "fire." All stimuli greater than threshold produce a similar all-or-nothing response, consisting of depolarization to zero or even to a positive value, followed by rapid return to the resting potential. Excitable cells in plant tissue are always connected by plasmodesmata, through which the electrical stimulus is presumably transmitted. The propagated signal, called an *action potential*, is similar to action potentials in nerve tissue, except that no chemical neurotransmitter has been identified in plants. Acetylcholine, which transfers the stimulus from one cell to another in nerve tissue, does occur in plants, but no evidence links it firmly to transmission of intercellular signals in plants.

Stimulation of a *Mimosa* petiole by gentle touch or a drop of water initiates an action potential that is limited to the treated leaf, but if a leaf is stimulated by burning or wounding, other leaves are affected as well. These stimuli are said to produce a *variation potential* as well as an action potential. The variation potential is probably transmitted by a chemical substance that moves through the xylem, passes through the pulvinus, and initiates a new action potential on the far side of the pulvinus. The new action potential travels through excitable cells in the phloem and protoxylem until it reaches another pulvinus, which then exhibits the characteristic turgor changes that lead to leaf closure. Depending on the intensity of the original stimulus, the variation potential again moves through the pulvinus, and stimulates still another action potential on the far side. Thus the variation potential coordinates the movements of the different leaves of a plant by initiating action potentials that move through defined regions of the tissue.

Mimosa appears to depend entirely on turgor changes to regulate the movements of its organs. Loss in volume of certain motor cells is accompanied by the secretion of potassium ions and tanninlike compounds from these cells to extracellular space. Such a change, presumably due to rapid increase in membrane leakiness, is accompanied by electrical signals detectable by microelectrodes. The external similarity between this process and muscular contraction in animals suggests that there might be common internal similarities, such as actomyosinlike contractile proteins, but no solid evidence either confirming or contradicting this type of mechanism is yet available.

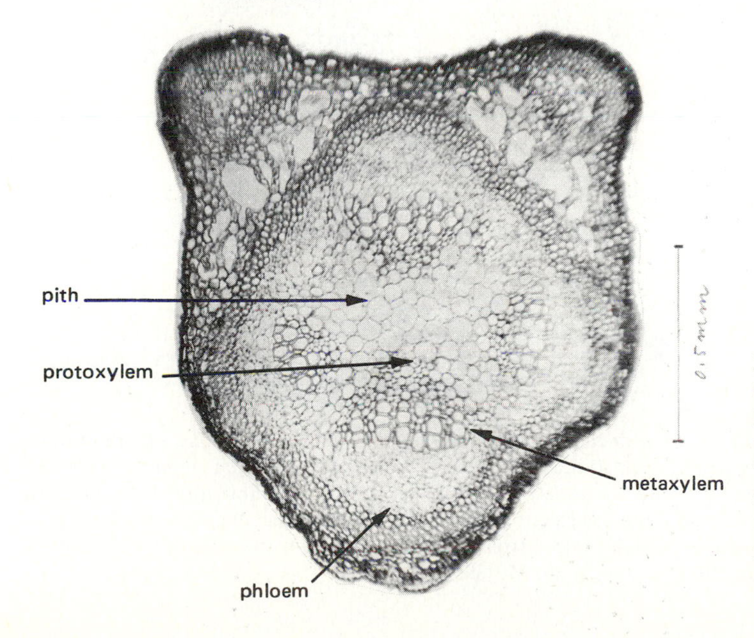

Figure 13-3 (a) Cross-section of the petiole of *Mimosa pudica*. Electrical measurements reveal that all cells in the protoxylem are "excitable." (From Sibaoka, 1966. Symp. Soc. Expt. Biol. *20*: 49–73. Micrograph courtesy of H. Toriyama, Tokyo Woman's Christian University.)

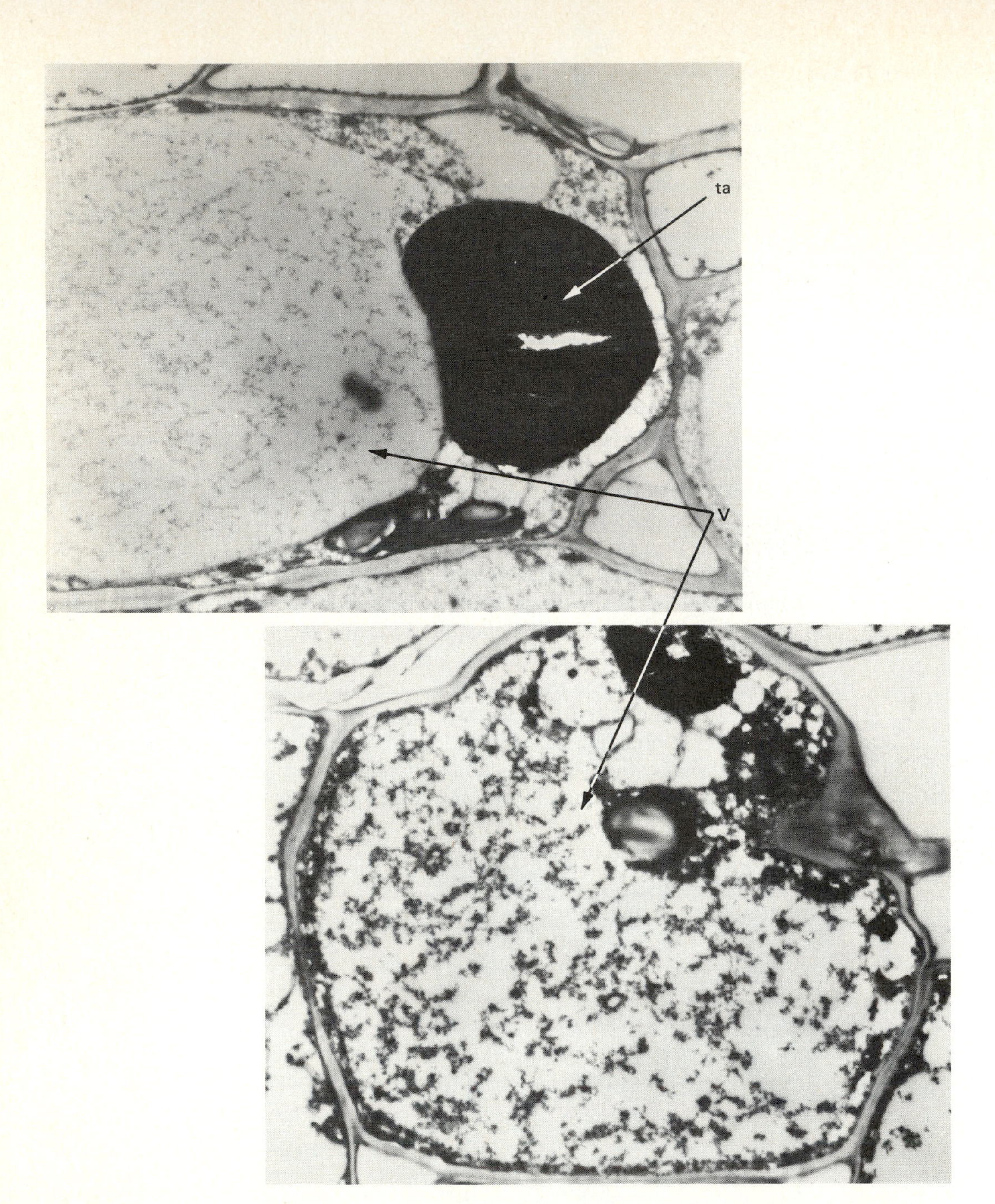

Figure 13-4 A motor cell in the primary pulvinus of *Mimosa pudica* before stimulation (upper) and after stimulation (lower). V: central vacuole; ta: tannins. Note that the large tannin mass breaks up into smaller units after stimulation. Also note the change in finely dispersed electron dense material in the vacuole. × 8,000. Both photomicrographs were produced from cell material fixed with OsO_4. (From Toriyama and Satô, 1971. Cytologia *36*, 359–395.)

Some investigators have emphasized the importance of the tannins in *Mimosa* movements. This heterogeneous group of condensed phenolic compounds can form complexes with both proteins and inorganic ions. Tannins denature proteins and are noxious to the cell, but produce no deleterious effects as long as they are sequestered in membranous sacs within vacuoles. They appear to be particularly numerous in vacuoles of plants with thigmonastic movements (Fig. 13-4). Vacuolar tannin aggregates break up during thigmonastic leaf movements, and some tannins appear to be excreted through motor cell membranes. They then diffuse out into hydathodes (pores) in the epidermis of the pulvinus. Since tannins have an astringent taste, and act as natural insect repellents, the induction by leaf movement of tannin secretion may aid in protecting the plant against predation.

Carnivorous Plants

Many carnivorous plants are native to swampy regions with soils low in nitrogen. These plants satisfy some of their nutritional needs by trapping insects in their leaves and then digesting their prey. This reversal of the more usual plant–animal relationship has intrigued many investigators, including Charles Darwin. His simple experiments revealed how carnivorous plants benefit from a meat diet (Table 13-1). Darwin investigated virtually all aspects of the behavior of these fascinating plants, from methods of insect capture to chemical analysis of the digestive process. His ingenious experiments, carefully collected data, and brilliant deductions should fascinate all students of biology. Some of his books are listed in the bibliography at the end of this chapter.

TABLE 13-1. Differences between a group of starved *Drosera rotundifolia* and a group fed with roast meat (relative values).

	Starved	Fed
Weight (without flower stems)	100	121.5
Number of flower stems	100	164.9
Weight of stems	100	231.9
Number of capsules	100	194.4
Total calculated weight of seed	100	379.7
Total calculated number of seeds	100	241.5

Charles Darwin collected *Drosera rotundifolia* in the field and grew the plants in bowls, each divided into two parts. Plants in one half of each culture were fed roast meat, while those in the other half were controls. After three months of such treatment, the fed plants were superior to the starved plants in all respects, especially those related to reproduction. (From C. Darwin, *Insectivorous Plants*, revised by F. Darwin. London: John Murray, 1908.)

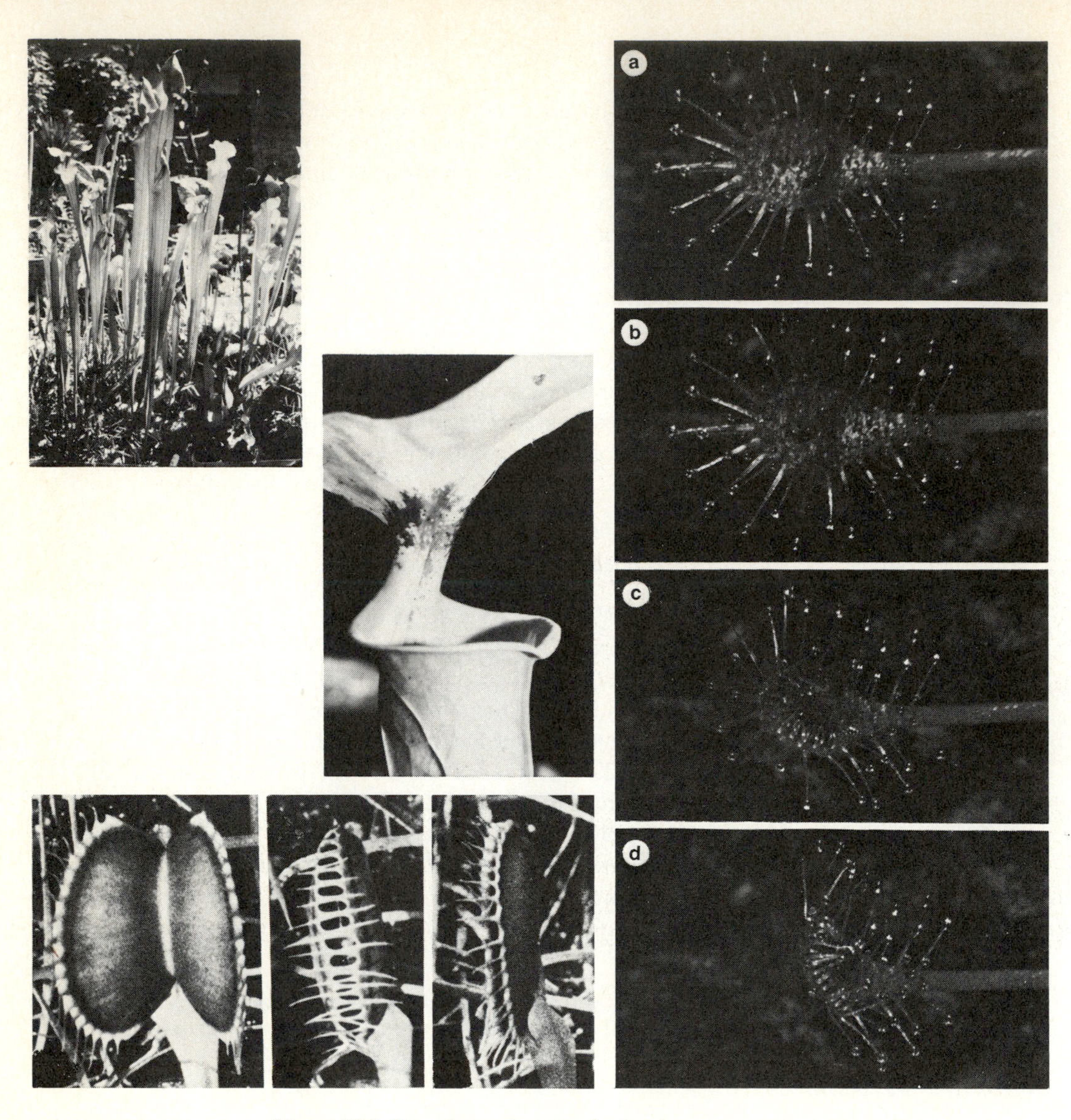

Figure 13-5 Traps in carnivorous plants.
Upper left: *Sarracenia flava,* a pitcher plant, catches insects in its deep cup. The enlargement below shows the beautiful detail of the insect trap. (Courtesy of J. Mazrimas, co-editor Carnivorous Plant Newsletter.)
Right: The behavior of a leaf of *Drosera,* the sundew plant, during insect capture: a. (1 hr. after capture) Insect struggles against some of the tentacles three rows in from the edge of the leaf; b. (1.25 hr. after capture) The stimulated tentacles move and force the insect back from the leaf edge; c. (3 hr. after capture) and d. (4 hr. after capture) Outer tentacles which were not mechanically stimulated move, and the blade curves around the insect.
Lower left: The trap of *Dionaea,* Venus flytrap. a. Before stimulation; b. 10 sec after two stimuli were delivered to the trigger hair; c. narrowed trap one day after the capture of an insect. (*Drosera* and *Dionaea* photos from Williams, 1976. Proc. Amer. Phil. Soc. *120*: 187-204.)

Carnivorous plants, found among the fungi and six families of angiosperms, may use either passive ("flypaper") or active mechanisms for insect capture (Fig. 13-5). Passive traps are generally cup-shaped and deep. An insect that flies or crawls into one of them has difficulty escaping before being injured by the acidity of the secretion and killed by the digestive enzymes it contains. "Flypaper" traps contain mucilaginous substances on the leaf surface that immobilize the insect. These substances, mainly carbohydrate, are synthesized by gland cells on the leaf surface, and are transported to the cellular exterior by secretory vesicles of the dictyosome. Some carnivorous plants, such as the sundew (*Drosera*), employ both flypaper and active trap mechanisms: insects caught in the sticky secretions kick their legs and beat their wings, thereby stimulating sensory cells in nearby tentacles. The tentacles curl around the insect, bringing it into contact with glands that secrete strong acids and degradative enzymes. Trap movements are relatively slow in plants with a combined flypaper–active trap mechanism but are far more rapid in plants such as the Venus' flytrap, which rely solely on trap movements for insect capture.

Dionaea's trap consists of a leaf with two bristled lobes, each containing three triangularly positioned trigger hairs on the upper surface (Fig. 13-6). Mechanical stimulation of one of these hairs elicits a single action potential, which travels through cells in the hair and lobe, finally exciting lobe cells whose volume changes control the movement. Two successive action potentials, or more if the interval between successive potentials is longer than 15 seconds, are necessary for the trap to close. This requirement, like the requirement for successive days of photoperiodic induction, suggests a primitive type of "memory." In *Dionaea*, the plant "remembers" the first action potential if the second one occurs soon enough afterwards, but its "memory" fades with time. Presumably, one action potential leads to an increase in the level of a substance that controls the movement. The level of this substance, which must decay with time, is still subthreshold after a single stimulus, but if another action potential arrives before the level of the substance has declined too far, the threshold concentration for excitation is exceeded and the response occurs.

Trap movement is extremely rapid, as it must be to catch an insect. The total process, including stimulus transduction, transmission, and trap closure, is completed less than 0.1 second after the last required stimulus. Trap closure in *Dionaea*, as in nyctinastic plants, is controlled by changes in the volume of key motor cells; however, it is not clear whether K^+, Cl^-, or other inorganic ions play a significant role in its turgor regulation.

Two stimuli do not close the trap completely; thus very small insects can crawl through the openings and escape between the bristles. However, each time an insect brushes against a sensory hair, another action potential is elicited, the trap closes more snugly, and the secretion of acids and digestive

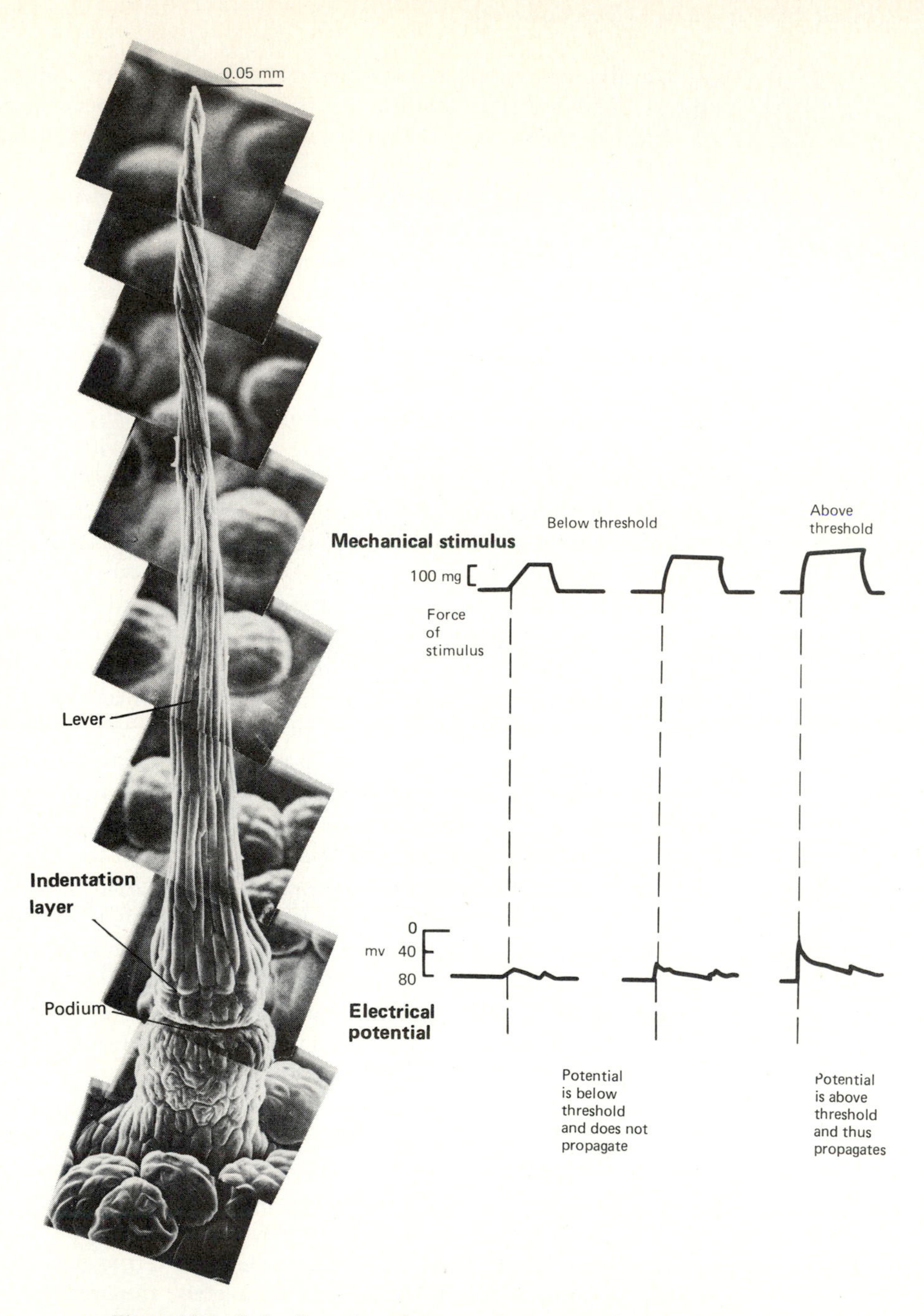

Figure 13-6 Left: Scanning electron micrograph of a trigger hair of *Dionaea*. Material untreated, except for a 10–50 nm gold deposit. (From Mozingo et al., 1970. Amer. J. Botany *57*: 593–598.)
Right: Mechanical stimulation of a trigger hair initiates a propagated action potential, if the stimulus is above a certain threshold value.

enzymes by glands on the upper leaf surface is stimulated. Thus a struggling insect hastens its own death!

The *Dionaea* trap remains closed until the body of the insect, except for its skeleton, has been digested; if the insect is large, this may take a week or longer. If the trap is stimulated to close by a nonnutrient object or a transient mechanical stimulus rather than an insect, it reopens after just a few hours. Thus, as Darwin deduced, the products of digestion themselves keep the trap closed. Reopening after digestion reorients the leaves to the horizontal position which is optimal for photosynthesis; thus in the absence of a meat diet, the plant reverts to its autotrophic, photosynthetic habit.

Tendril Curling

Some plants lacking rigid stems are able to support themselves and rise above neighboring species by the use of thin, whiplike appendages, called *tendrils.* These organs, usually modifications of stems or leaf parts, execute spiral circumnutational movements as they grow, thereby increasing their opportunity for contact with a potential supporting object. Once they meet such an object, they modify their rate and direction of growth so as to grasp the external support, pulling the plant upward.

In some plants, tendrils curl around objects they contact, while in others they form adhesive pads that grow into crevices on the supporting surfaces (Fig. 13-7). In species with curling tendrils, the apical portion of the tendril

Figure 13-7 *Pisum sativum*, the garden pea, supports itself by tendrils (left) while *Parthenocissus tricuspidata* forms adhesive pads that adhere to supporting structures (right) (latter from Reinhold et al., 1972. The Role of Auxin in Thigmotropism. In Plant Growth Substances, D. J. Carr, editor, Springer-Verlag, New York).

forms a coil around the supporting object; soon afterwards, the more basal portion forms a free coil in space, thereby pulling the plant closer to the supporting structure. The free coils may become lignified, which adds to their tensile strength. A coiled tendril of the passion flower (*Passiflora coerulea*), which forms both contact and free coils, may develop a force equivalent to 350–750 grams, the weight of several feet of vine.

Some tendrils are *thigmonastic* (the directionality of curling is predetermined by asymmetry in the structure of the tendril), while others are *thigmotropic* (the directionality depends upon the direction of the stimulus). Whether nastic or tropic, tendrils that have started to curl usually straighten if the contact stimulus is removed prematurely. Thus transient contact with an animal, a windblown branch, or other mobile object does not ordinarily produce a permanent change in tendril orientation.

Curling is promoted by light, but unlike leaf movements in nyctinastic plants, phytochrome does not seem to be involved. The light effect appears to be directly related to the need for ATP supplied by photosynthesis, since the light requirement can be replaced, in part, by supplying ATP to the incubation medium on which tendrils are floating. The breakdown of large quantities of ATP to yield ADP + inorganic phosphate (P_i), has been measured during curling.

Pisum sativum (garden pea) is a favorite experimental organism for studies of tendril curling. Its curling response is thigmonastic, and can be stimulated by stroking the lower (abaxial) side of the tendril with a glass rod. Curling, which is initiated within two minutes (Fig. 13-8) and continues for more than 48 hours, consists of two distinct phases: Initially, abaxial cells contract and upper (adaxial) cells expand; following this, cells on both sides of the tissue expand, the adaxial more rapidly than the abaxial.

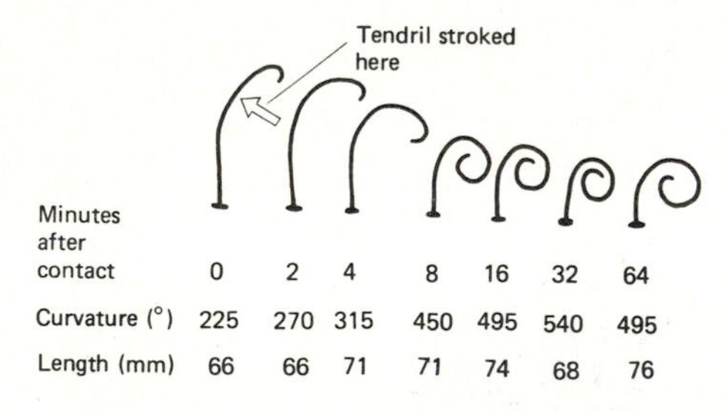

Figure 13-8 The movements of a pea tendril that had been stroked 5 times with a glass rod. (From Jaffe and Galston, 1966. Plant Physiol. *41*, 1014–1025.)

The hormones auxin and ethylene both seem to be involved in regulating the changes in cell size that lead to curling. Excised tendrils of peas and all other species tested will coil in the absence of tactile stimulation if treated with auxin (Fig. 13-9)—this is not surprising, since auxin is a growth-stimulating hormone and coiling involves a large increase in the rate of cell elongation. However, coiling depends upon asymmetric growth of abaxial and adaxial cells, yet occurs in response to symmetrical auxin treatment; adaxial and

Figure 13-9 The effect of IAA on the curling of tendrils of *Marah fabaceus* (upper) and on the formation of adhesive pads in tendrils of *Parthenocissus tricuspidata* (lower). The first and third tendrils of *Marah* were immersed in H_2O while the second and fourth were immersed in IAA (150 mg/liter). (From Reinhold, 1967. Science *158*: 791–793); the left tendrils of *Parthenocissus* were treated with lanolin (control), the middle ones with lanolin + IAA (0.05%) and those at the right were stimulated by contact. (From Reinhold et al., 1972. The Role of Auxin in Thigmotropism. In Plant Growth Substances, D. J. Carr, editor, Springer-Verlag, New York.)

abaxial cells must therefore respond differently to auxin. The explanation depends in part upon the hormone ethylene. Auxin stimulates ethylene synthesis in tendrils, as in many other plant tissues (see Chapter 10), but such synthesis appears to be several times higher on the abaxial than on the adaxial side of the tendril. Ethylene makes membranes leaky, and increase in the ethylene content of the abaxial cells is accompanied by the loss of solutes from the vacuole, causing these cells to shrink during the early phase of curling. Differences in the ethylene content of abaxial and adaxial cells might also explain differences in their elongation during the second phase.

The tip of the tendril is most sensitive to tactile stimulation, but such stimulation causes basal regions as well as the tip to curl, indicating that a stimulus is translocated through the tissue. You will recall that auxin is translocated through plant tissue in a basipetal direction; nevertheless, auxin cannot be the primary transmitter of the stimulus, since the tendril stimulus moves far more rapidly than auxin. It appears that some other event, possibly involving ion movements, must precede auxin intervention. We do not yet know whether tendrils contain excitable cells that could transmit an action potential, as in *Mimosa* and some of the carnivorous plants, or whether some other form of electrical signal is involved.

Thigmonasty: Its General Significance

The three types of movement described in this chapter, that is, rapid movements in *Mimosa*, trap movements in carnivorous plants, and tendril curling, all depend upon specialized organs or systems not found in all plants; yet they may have more general significance than has been appreciated in the past. All plants are subjected to mechanical stimulation from many sources during their usual growing conditions: they are buffeted by the wind; brushed by insects, animals, and neighboring plants; and their roots are continually subjected to frictional forces as they push through the soil. These forces are known to alter the morphology and development of the plant (Fig. 13-10). This complex of morphological responses, called *thigmomorphogenesis*, seems to be mediated by ethylene, as is tendril curling. There are some functional similarities between these two responses, and studies of tendril curling might therefore aid in understanding other, more widespread, phenomena.

Some plant physiologists believe that variation potentials, first described in *Mimosa*, may also occur in virtually all plants as a generalized response to injury. We are just beginning to appreciate the role of electrical signals in coordinating the activities of cells in different parts of the plant body, and we should expect progress in this area in the near future. Studies of rapid plant movements which depend upon specialized, easily identifiable cells should aid in this endeavor.

Figure 13-10 Mechanical stimulation has a marked effect on the growth of *Bryonia dioica*. The plant on the right was rubbed gently with thumb and forefinger 30x, twice daily, each day for 7 days. The control plant on the left was untreated. (From Jaffe, 1973. Planta *114*: 143–157.)

SUMMARY

Although plants are anchored by their roots to a single location, they are capable of a variety of *movements*, ranging from the slow curvatures of tropistically stimulated stems, leaf petioles, and roots to the sudden closing of the traps of insectivorous plants. *Mimosa pudica*, the sensitive plant, opens and closes its compound leaves slowly in response to light–dark and rhythmic signals, but very rapidly in response to mechanical, thermal, chemical, and electrical stimuli. All movements depend on turgor changes in motor cells of the pulvinus. When the leaf is touched an electrical signal, in the form of a transitory voltage drop, moves along the petiole at about 2 cm/sec. The rate of transmission rises with rise in temperature, and is dependent on living cells of the phloem. Certain *excitable cells* "fire", that is, electrically depolarize, when the strength of the stimulus exceeds a *threshold value*; this firing causes a reaction in neighboring cells, leading to a propagated *action potential*. It is not known whether plants, like animal nervous systems, depend on chemical transmitters to connect the cells that respond electrically.

Carnivorous plants trap and digest insects by a variety of devices, both passive and active, and use the digested material as a source of nitrogen. *Passive* traps (for example, pitcher plant) lure insects into situations from which they cannot emerge because of sticky, slippery, or blocked surfaces, whereas *active* traps (for example, Venus' flytrap) close rapidly on an insect after it has stimulated sensitive triggers. In the Venus' flytrap, electric signals and sudden turgor changes in motor cells precede movement; two stimuli, administered within about 15 seconds, are required for firing, which then causes closing within about 0.1 sec. The products of digestion keep the trap closed, so that when the insect is fully digested, the trap opens and is ready for action again.

Tendrils frequently help nonrigid plants attain great heights by curling around nearby rigid objects. They "find" the object by chance contact during *circumnutation*, then curl around it in response to the mechanical stimulation. Subsequently, *free coils* form, tightening the connection, and *lignification* may then result in strong attachments. In peas, whose tendrils start curling within two minutes after stimulation, light may favor curling through ATP formation. In the curling reaction, abaxial cells contract and adaxial cells expand. Symmetrical auxin treatment can lead to the asymmetric reactions leading to curling; this may be related to asymmetric ethylene formation and response on the two sides of the tendril. Cells sensitive to touch are located near the tip of the tendril, whereas responding cells are both at the tip and further down.

SELECTED READINGS

DARWIN, C. *Insectivorous Plants* (New York: D. C. Appleton and Co., 1896). (Reprinted in New York: AMS Press, 1972).

———. *The Movements and Habits of Climbing Plants* (New York: D. C. Appleton and Co., 1893). (Reprinted in New York: AMS Press, 1972).

———. *The Power of Movement in Plants* (New York: D. C. Appleton and Co., 1896). (Reprinted in New York: AMS Press, 1972).

HAUPT, W., and M. E. FEINLEIB, eds. "Plant Movements," *Encyclopedia of Plant Physiology*, New Series, Volume 7 (Berlin–Heidelberg–New York: Springer-Verlag, 1979).

LLOYD, F. E. *The Carnivorous Plants* (Waltham, Mass.: Chronica Botanica Co., 1942).

TRONCHET, A. *La Sensibilité des Plantes* (Paris: Masson and Co., 1977).

QUESTIONS

13-1. Leaflets of the sensitive plant *Mimosa pudica* fold together in response to any one of three stimuli: touch, transfer from light to darkness, and rhythmic signals during a dark period. What similarities would you expect among these different types of movement? What differences? Discuss (a) rate of movement, (b) sensory receptors, (c) turgor changes, (d) possible control mechanisms.

13-2. How do tropic and nastic movements differ? In what ways are they similar?

13-3. The directionality of nastic movements is controlled by structural or physiological asymmetry in the organ of curvature. Describe some likely asymmetries.

13-4. Several types of plant movement involve permanent changes in cell size and shape, while others involve completely reversible changes. What differences in control mechanisms would you expect for irreversible and reversible movements? What are some features common to both types of movement?

13-5. When the tip of a tendril of some plants is dipped into a solution of IAA, the tendril curls along its entire length. Explain how this could occur, using information on auxin transport and action discussed in chapters 2 and 9.

13-6. Experimenters have demonstrated that the hormones auxin and ethylene can participate in regulating tendril curling. What other hormones might be involved? Describe experiments to test these possibilities.

13-7. Cells in the trap of *Dionaea* perform several specialized functions: (a) They are capable of rapid changes in size and shape following stimulation of sensory receptors 1–2 cm away; (b) they secrete digestive enzymes and other molecules that degrade their prey; and (c) they absorb large quantities of inorganic and organic compounds from the captured insects. Speculate on specialized ultrastructure and biochemical features which these cells might possess.

13-8. Venus' flytrap (*Dionaea*) is said to possess a primitive type of memory. On what evidence is this statement based? What does this indicate regarding the mechanism controlling movement?

13-9. How many tendril-bearing plants can you list? What advantage does the tendril habit confer on each of these plants?

13-10. Of what possible adaptive value is the rapid movement of *Mimosa*? Can you name other rapid plant movements? Are they of any value to the plant?

14

Some Physiological Bases of Agricultural and Horticultural Practice

Up to now, we have considered processes occurring in individual organelles, cells, tissues, or entire plants. In nature, however, plants grow in communities, in which one individual may influence another, either competitively or cooperatively. The physiological conditions in communities are not identical with those encountered by an isolated plant, in that the environment experienced by each individual plant is greatly affected by the other plants surrounding it. In fact, other plants are a part of the environment. Modern man manipulates plant communities by growing horticultural and agricultural crops in single-species aggregations (*monoculture*), then using some product of the crop either as food, animal fodder, fiber, or other special product. If the farmer understands the growing conditions in a plant community, he can manipulate them and the plants to produce an optimum yield. These manipulations usually take the form of controlling the nature and density of the crop, the supply of water and fertilizers, and aid in warding off competitors, predators, and pests.

Most plants used by man are grown as mass agricultural communities, but some are grown under special conditions referred to as *horticulture*. Horticulture is an intensive form of crop cultivation in which the value of each individual plant or plant product (such as a fruit) is sufficiently great that it is important to produce high-quality individual specimens rather than a large number or a great mass of product. In horticultural practice, plants may be propagated and treated individually to obtain the desired commercial product. In this chapter we will consider those features peculiar to crop plants growing in communities and the physiological manipulations used by farmers and growers to increase the value of the harvested product.

Nutrients

The nutrient relationships of crops are similar to those of individual plants. The more of a particular nutrient applied, the greater is the yield, provided that: the concentration employed is not toxic; does not interfere with the uptake of other nutrients, is not so great as to produce deleterious osmotic effects because of salt buildup in the soil; and other nutrients are not rate limiting. As the supply of mineral nutrients is increased, however, the increase in growth achieved by each additional increment of nutrient diminishes. Since fertilizer costs remain constant, eventually the increased yield produced by the extra increment does not justify the further addition of fertilizer; at this point, optimal yield has been obtained in terms of the cost per unit of crop produced.

To achieve optimal productivity and economy of production it is more important to provide a steady supply of fertilizer over the entire growing season than to furnish the whole amount at one time. Under the latter regime, some elements that are not immediately absorbed by the plant may be adsorbed onto the clay colloids and organic matter in the soil and may be taken up by the plant during the growing season; most soils, however, have a limited adsorption capacity, and any excess fertilizer applied may be leached to the water table and subsequently into rivers and lakes. This causes *eutrophication*, the development of dense algal and other plant growths in lakes previously clear because of nutrient limitation. The development of fertilizers of limited solubility, such as nitrogen-supplying urea–formaldehyde, that slowly release their contained nutrients, has decreased such nutrient loss. This innovation represents, in fact, a return to one of the desirable attributes of some of the original organic fertilizers such as bonemeal, but the modern successors have the advantage of providing the required nutrients in a controllable amount, and are more available commercially.

The main nutrients used in agriculture are nitrogen, phosphorus, and potassium (N, P, and K.). Most fertilizers contain one or more of these nutrients, and the actual nutrient percentages, must, by law, be recorded on the container. Thus, a 5: 10: 5 fertilizer contains 5% of the total weight as nitrogen, 10% as oxides of phosphorus, and 5% as oxides of potassium. The best value is that fertilizer which provides the highest nutrient content, not total weight of fertilizer, per unit price, with nutrients in a slowly released form. The fertilizer required will depend on the kind of crop and on the natural nutrient content of the soil. A high content of N compared to P and

K promotes stem and leaf growth at the expense of reproductive activity; therefore, crops in which the leaves are desired products typically require fertilizers with a higher nitrogen content than do other crops.

Plant nutrients other than N, P, and K are not normally included in fertilizers, as sufficient quantities generally occur in the soil. If deficiencies in the micronutrients are detected, however, the required mineral can be added to the soil. If sufficient nutrient is present in the soil, but is unavailable because soil conditions cause it to become insoluble, the soil conditions may be changed or a dilute solution of the salt sprayed on the leaves. The leaves can take up enough salts to satisfy the plant's needs, if the mineral is required in only low quantities.

One of the principal reasons for crop failure or poor growth is improper soil pH (Fig. 14-1). In most conditions in which a pH imbalance occurs, acid soil causes the problem. Crop plants commonly start to suffer when the pH drops below about pH 5.5–6.0 and suffer more seriously when the soil pH falls below 5.0. The principal cause of plant damage in low-pH soils is fre-

Figure 14-1 The availability of a nutrient is directly related to its solubility which, in turn, depends upon soil pH. The thickness of the horizontal band in the graph represents the solubility of each mineral at the various pH values. Some minerals (e.g. iron) become toxic at pHs at which they are very soluble. (From C. J. Pratt: *Plant Agriculture*, W. H. Freeman and Company, San Francisco, 1970.)

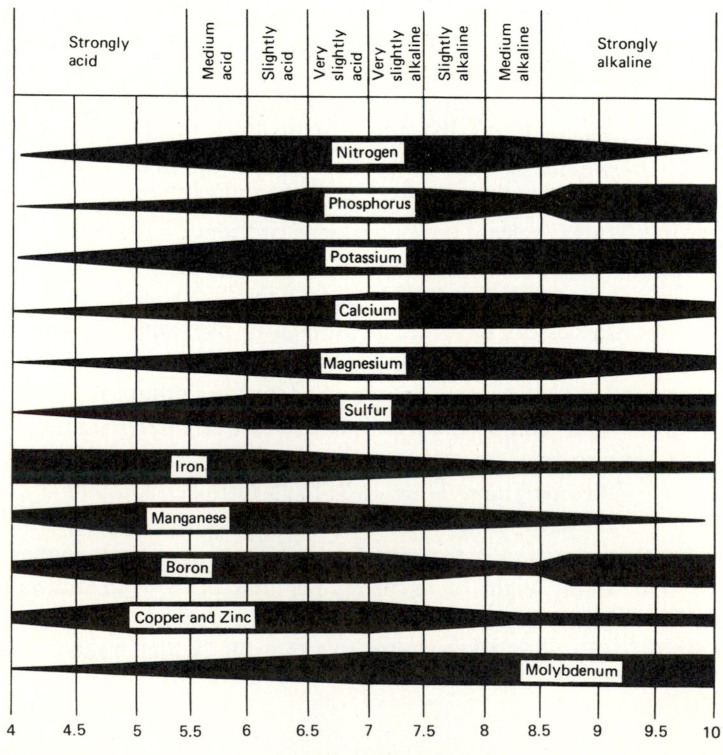

quently not the high H^+ ion content itself, but rather the excess availability of iron and, in tropical soils, also of aluminum. These elements become toxic at the high levels caused by acidity-induced solubilization of soil minerals (Fig. 14-1). At the same time, low pH limits the supply of some elements, such as phosphorus. Soil acidity can usually be cured by the addition of lime in the form of calcium carbonate or calcium hydroxide. Alkaline soils are less common and less of a problem in agriculture than acid soils, but extreme alkalinity (above pH 8) may render both iron and phosphorus relatively insoluble and thus limiting to growth. Correct nutritional management involves an adjustment of the soil pH to about 6–8, followed by the addition of any deficient nutrients to the extent to which they are removed annually by the crop. The nutrient elements required can be estimated by soil analysis, foliar analysis, or visual estimation from plant deficiency symptoms.

Hydroponics and the nutrient film technique

At one time, *hydroponics* (the growing of plants with their roots in a nutrient solution) was proposed as an answer to agricultural and horticultural problems in areas where good soil is lacking; yet over the years little has come of these suggestions. Although hydroponics can work well in the laboratory or research greenhouse, it is more difficult to operate on a commercial scale, principally because the roots need vigorous aeration for the plants to grow well, and the plants also need support in the solution. A technique has been developed, however, which has proved highly successful in intensive horticulture, producing very rapid plant growth and large yields. Termed the *nutrient film technique*, it will probably see extensive commercial use in the near future.

The nutrient film technique consists essentially of growing plants with their roots in shallow films of flowing nutrient solution (Fig. 14-2). Long troughs are made by folding the sides of a long strip of black polyethylene sheet, leaving a slit along the top. These troughs are arranged with a slight slope so that the nutrient solution delivered at one end runs down the trough into a collecting tank at the lower end. The nutrients are then pumped from the collecting tank back to the top end of the trough, so that the solution is kept cycling. Plants are placed in the slit in the trough, planted for support either in pots of soilless medium (such as peat plus perlite) with holes in the base of the pots, or in blocks of material such as polyurethane foam, through which the roots can penetrate. The roots grow out of their pot or support block into the film of nutrient solution. As the liquid film is very shallow and continually flowing, it contains adequate oxygen for the roots to remain aerobic. Under these conditions very rapid growth occurs, and since the plants never suffer from water stress, growth is continuous as well. The nutrient solution pH (6–7 is best) and the level of water and nutrients can be measured and corrected. In large-scale commercial systems this is all accomplished by continuous automatic monitoring, with the monitors connected

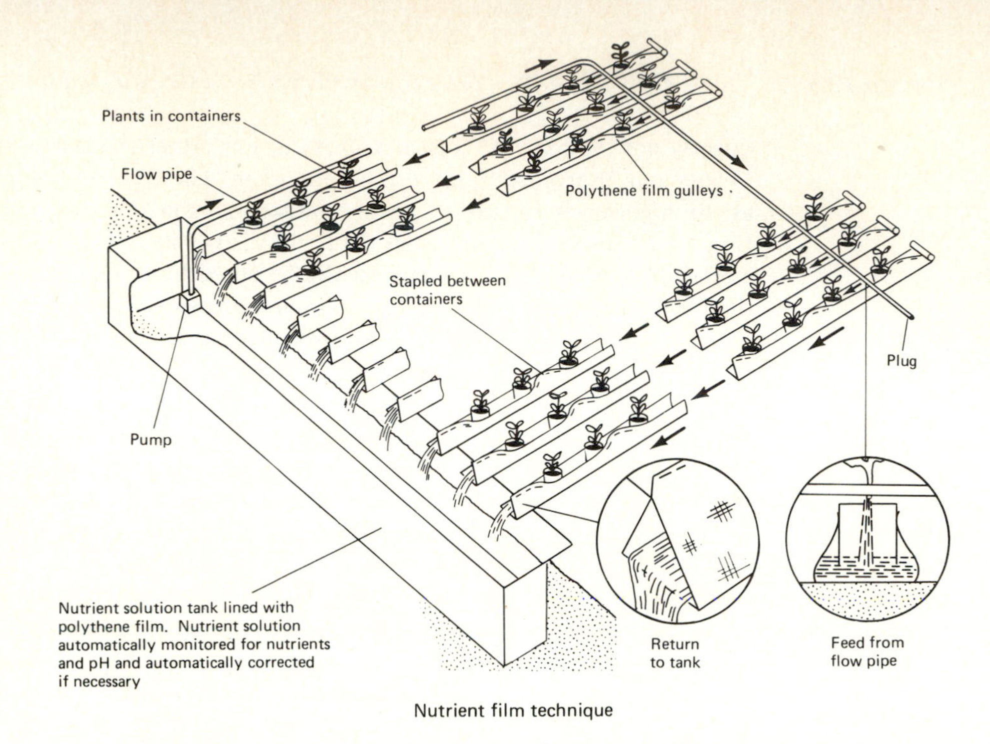

Figure 14-2 Diagram of the nutrient film technique (from A. Cooper, Nutrient Film Technique of Growing Crops, London: Grower Books, 1976).

to supplemental water, nutrient solution, or phosphoric acid (to reduce the pH), all of which are added automatically as needed.

Because the nutrients run in enclosed plastic trenches, water loss from the soil is eliminated; the system therefore can be used in arid areas. Its main use at present is with greenhouse or nursery stock; the technique is valuable for the latter use because the plants can easily be removed for sale with no root damage. In the future, as transport costs increase, this system may be also used in combination with lighting systems to grow crops in winter in high latitudes. Such a system makes it possible, under artificial conditions, to grow plants in a completely pest- and disease-free environment, thus eliminating the need for costly (and sometimes hazardous) pest control. The cost of light and heat, however, would represent a considerable offsetting disadvantage.

Water loss by Crops

All the leaves of an isolated plant will transpire because they are surrounded by relatively dry air. In a crop growing in the field, however, the lower leaves are mostly surrounded by air made moist by the transpiration of other leaves, and only the upper leaves are in contact with the drier atmosphere itself. Furthermore, in the crop canopy below the top leaf layer, the leaves restrict the movement of air to such an extent that the air in the crop space becomes water saturated, and the lower leaves therefore transpire at a much lower rate. The molecules of water vapor diffuse from the surface of the saturated air layer in the crop into the dry air above the crop, and provided the crop surface is approximately even, as in a field of grass, the effective transpiring surface becomes the surface of the crop rather than the surface of the individual leaves (Fig. 14-3). It makes no difference how many plants are in the crop or how many leaves are on each plant; once the crop covers the ground, the water loss will be determined by the area of the crop surface. The water loss by such a crop is relatively high; during the daytime it is almost identical to the loss of water from an open water surface such as a lake. Because most plants close their stomata at night, thereby limiting transpiration, the daily water loss from a crop is somewhat lower than from an open water surface over a similar period. Since the loss by each crop is usually a definite fraction of the loss from an open water surface, it can be calculated either by measuring the water loss from an open water surface or from measurements of the heat energy of the sun available for evaporation. Such data are now usually available from meteorological stations.

The growth of plants depends basically on the water needed to produce turgor in growing cells. Crop yield is thus directly related to the amount of water available to the plant, most of which is transpired. Since over the long term, water input must equal water loss, a knowledge of the amount of water transpired by the crop permits an estimate of the amount of irrigation needed (transpiration loss + drainage − rainfall = irrigation need). This enables the farmer to supply irrigation water only when needed and thus to avoid wasting water and energy. Ideally, water should be provided at the same rate

Figure 14-3 Differences in the transpiration of a single plant and of many plants in a community. The isolated plant loses water over its entire surface but the crop loses water only at the exposed surface, since the air surrounding the lower leaves becomes saturated with water vapor. Net loss of water molecules occurs where this saturated air meets the dry air above the crop.

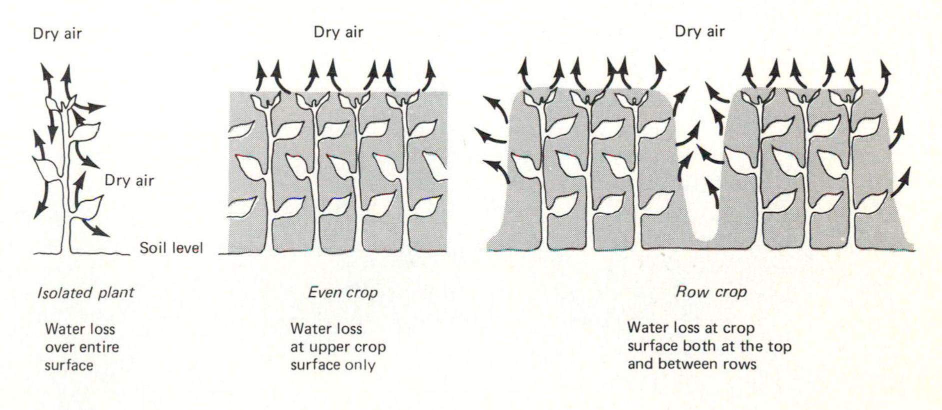

(a)

Figure 14-4 In arid areas it is almost impossible to grow crops unless additional water is supplied for transpiration and growth. The effect is dramatic when seen from an airplane. Giant irrigators rotating around a central pivot (a) produce circles of green up to a mile in diameter in an otherwise arid landscape (b). The photos show corn (*Zea mays*) growing in northeast Colorado, an area with an annual rainfall of only 35 cm. (Photos courtesy H. R. Duke and D. F. Heerman, Agricultural Research Service, USDA Fort Collins, Colorado.)

(b)

as it is transpired, but this is often impractical. Crops differ in their ability to withstand a water deficit before growth is curtailed, but to avoid a loss in yield, irrigation must make up the transpiration loss before the deficit has reached the growth limiting level (Fig. 14-4).

Depending on the nature of the crop, water loss may range from very little to an amount greater than that from an open water surface. The surface of a bare soil soon dries out, thereby limiting the loss of water from layers below. This is why bare soil is left fallow to conserve water in areas of limited rainfall and no irrigation, since most of a year's rainfall can be stored in the soil in this manner. When plants start to grow, the roots can draw water from the lower layers of the soil. Water loss then increases until the crop completely covers the ground and forms the transpiring surface. Transpiration from a row crop, or trees in orchards, may exceed the loss of water from a smooth, even crop because they have a larger transpiring surface area than the area of ground on which the crop is growing. In addition, wind may blow between the rows, replacing the moist air with dry air. Irrigation of row crops may therefore have to be greater than that for an even-surfaced crop, depending largely on the total surface area of the crop.

Salinity

Throughout the irrigated areas of the world, the soil is becoming more saline. High concentrations of sodium chloride and other salts in irrigation water damage plants both because the osmotic effects hinder water uptake and because sodium has an inhibitory effect on enzymes in the cytoplasm. Thus when salt accumulates, plants become stunted and eventually fail to grow. The salinity of the soil increases even though irrigation water usually contains only dilute concentrations of salt, because in dry climates virtually all the irrigation water is evaporated; the salts remain behind in the soil, slowly building up in quantity until their concentration starts damaging the plants. At this point the soil must either be abandoned for agricultural use or leached by adding excess water to remove salt from the soil. The large quantities of water required for leaching, however, are often not available in areas requiring irrigation.

Two approaches can be used to circumvent this problem. The first, *drip irrigation*, provides adequate water to individual plants, but none to the soil in between. It is accomplished by laying a pipe along each row of plants, with holes so placed as to supply water for each plant—the plant thus receives an adequate water supply while none is wasted on the bare, unplanted soil areas. The irrigation water, even if mildly saline, tends to wash the salts to the edge of the root zone while maintaining fresh water in the root zone, so that plant growth is less affected by salt buildup. Drip irrigation is used for orchard trees or individual plants such as lettuce, though it is not applicable to a dispersed crop such as wheat. Clearly, however, this technique can only delay the day of reckoning in a completely arid climate. Some arid regions, however, do experience a rainy season, which can leach the soil; in such cases the technique not only conserves water during the dry season, but may permit the use of mildly saline water which would be damaging if provided by sprinkler or flooding irrigation.

Another approach is the selection and breeding of salt-tolerant plants. Such *halophytes* do occur in nature. They cope with the problem of salinity by absorbing considerable amounts of salt, so that their cell water potential permits water uptake even from a saline soil. The salt is, however, accumulated in the cell vacuoles, so that cytoplasmic enzymes are unaffected by the high sodium content. Trials have shown that special strains of some plants not normally considered halophytic may in fact demonstrate considerable capacity for salt tolerance. Thus, a salt-tolerant barley strain developed by Emanuel Epstein at the University of California at Davis can produce a crop under conditions so saline that normal barley plants die. Further breeding from this strain may produce a variety for agricultural use in saline soils.

Sunlight for Photosynthesis

The amount of photosynthetically active sunlight received by the leaf of a crop plant depends on its position in the leaf canopy. As the light intensity on a single horizontal leaf increases, so does the rate of photosynthesis, up to a maximum which occurs at a light intensity considerably below that of full sunlight (Fig. 14-5). Light above this saturating light intensity is wasted, at least for that leaf, and since leaves at the top of the canopy receive full sunlight, they have light to spare. Leaves at the top of the canopy are usually

Figure 14-5 The rate of photosynthesis of a horizontal leaf as a function of light intensity. Note that the leaf is light saturated by intensities much lower than those existing on a bright summer day; consequently, much of the light energy is wasted.

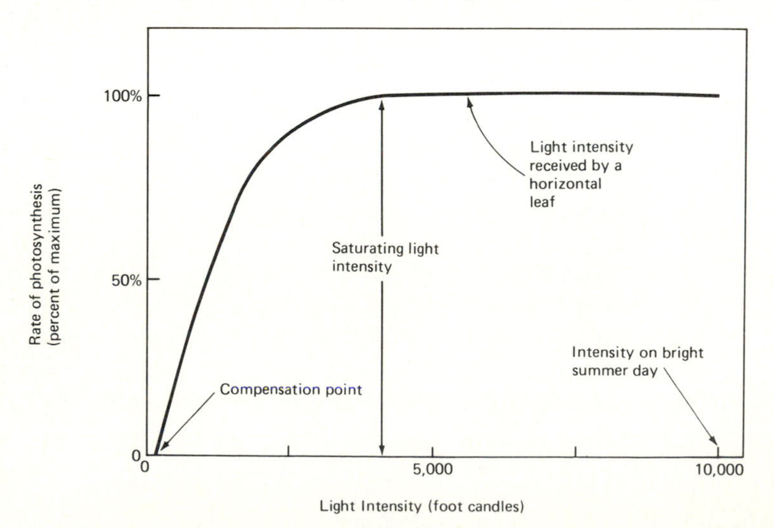

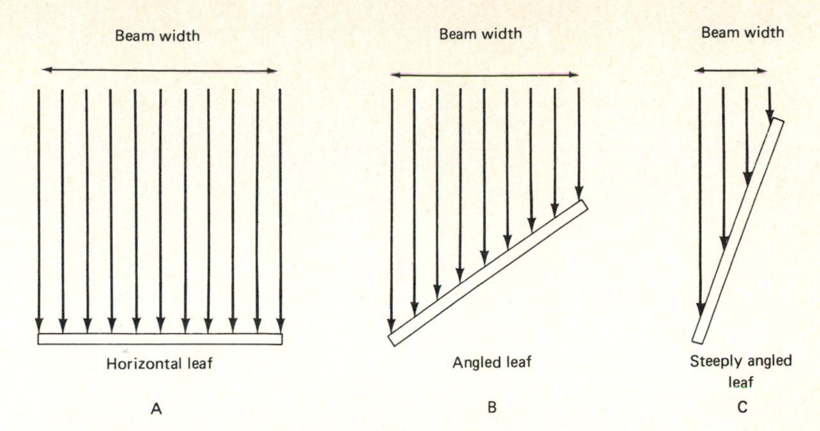

Figure 14-6 In a given quantum density in sunlight striking a leaf, a horizontal leaf (A) intercepts a greater total number of quanta than an obliquely inclined leaf (B, C). The upper leaves of plants that use sunlight efficiently are usually slanted while the lower ones are horizontal. This maximizes quantum absorption by the entire canopy.

oriented not at right angles to the incoming sunlight, but at an oblique angle. At this oblique leaf angle, a given amount of light is distributed over a greater total leaf area than if the leaves were at right angles to the direction of the light (Fig. 14-6). The light intensity per unit leaf area is therefore reduced, but this does not reduce the rate of photosynthesis, since the intensity is already supraoptimal. When leaves are slanted, more leaves can be accommodated in the uppermost layer of the canopy and more leaf area operates at optimal efficiency. Photosynthetic production is thus greater than if all the upper leaves were horizontal.

Lower down in the canopy, the light intensity decreases rapidly. The leaves do not usually overlap completely, so that flecks of full sunlight still may penetrate through the small gaps in the upper canopy to reach the lower layers. Outside the sunflecks, the sunlight will have passed through one or more leaves. The amount of light absorbed by a leaf varies with its chlorophyll content, but is usually about 90% of the incident light. Thus, the second layer of leaves receives 10% of full sunlight and the third layer 10% of 10%, or only 1%. Although the uppermost leaves make best use of the full sunlight when at an acute angle to the sun's rays, the lower leaves operate best in the low light intensity if the leaf surface is at right angles to the light, capturing the greatest light intensity per leaf surface (Fig. 14-7). The ideal plant thus has its lower leaves horizontal, with the leaf angle increasing at each successive upward layer, to almost vertical in the uppermost leaves. Such plants are often selected and bred by plant geneticists (Fig. 14-8).

Even if they absorb all the incident energy efficiently, the lowermost leaves in the canopy are likely to be close to their compensation point. If a leaf receives insufficient light to attain even the compensation point (see Fig. 14-5), it will be respiring more than it is photosynthesizing, and will

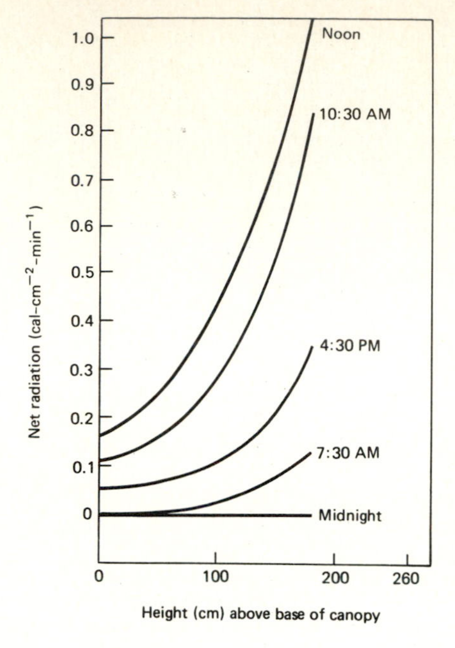

Figure 14-7 The light intensity is high at the top of a crop canopy, but decreases lower down in the canopy due to shading by the upper vegetation. The leaves of an ideal plant are almost vertical at the top of the plant to minimize shading of one leaf by another, while leaves at the base of the plant are horizontal, to make maximum use of the reduced light intensity. The amount of light penetrating the canopy also varies with the time of day, as more light can reach the lower leaves at midday when the sun is overhead than early or late in the day when the sun is at a lower angle.

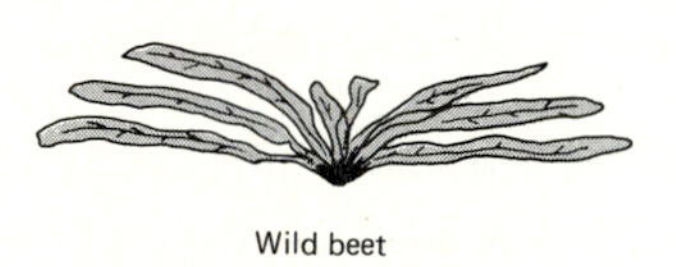

Figure 14-8 A cultivated beet can produce dry matter more rapidly than a wild beet as it has a more efficient leaf arrangement. The reasons can be found in Figs. 14-6 and 14-7.

thus be a drain on the plant. Such leaves usually senesce, turn yellow, and are shed from the plant.

In agricultural analysis, the quantity of leaves carried by a crop is expressed as the *Leaf Area Index* (LAI). This is the ratio of total leaf area to the area of ground they cover. Thus, when LAI = 4, there are four square meters of leaves over every square meter of ground. Using such a measurement, one can examine the efficiency of crops in producing dry matter, the ultimate result of photosynthetic activity. For a crop with horizontal leaves, such as clover, the optimal LAI (that quantity of leaves at which the lowest leaves are just at the compensation point) is lower than for a crop with largely vertical leaves, such as wheat, because the vertical leaves shade the lower leaves to a lesser extent than would horizontal leaves. Thus the optimal LAI for clover might be 4, while for wheat it would be about 7. Using such information, one can assess the management of a field of grass. If it is allowed to produce too many leaves, the yield will be reduced, as the lowest leaves will be below the compensation point. On the other hand, heavy grazing will allow the removal of so many leaves that light will reach the ground without being fully utilized in photosynthesis. The best management practice is therefore to keep near the optimal LAI, which can be achieved by frequent grazing or mowing as soon as the LAI becomes supraoptimal, and reducing the leaf area to just below optimal LAI. In this way the production of dry matter can be kept at the optimal rate.

The objective of leaf area management is to maintain an optimal leaf area for the incoming sunlight. Clearly this varies with the intensity of the sunlight, and therefore with the season. A plant community can have a greater leaf area for optimal growth rate at midsummer than it can earlier or later in the season when the sunlight is less intense, a fact which becomes important in timing the growth of crops. In temperate zones, the cold weather in spring limits how early a crop can be sown. Frequently the crop develops very little

Figure 14-9 Crops vary in their photosynthetic efficiency, due not only to changes in the amount of sunlight (maximal at the end of June) which alters the efficiency of each leaf (top graph), but due also to the number of leaves they carry (lower graph). For optimal photosynthesis a crop should have its maximum leaf area in late June, as does barley. Since sugar beet and potatoes have very low leaf area in midsummer, the total production is low even though each leaf photosynthesizes rapidly. Later, in September, they have a high leaf area, but then with the decreasing amounts of sunlight they often have too many layers of leaves so that the lower leaves are heavily shaded and respire more than they photosynthesize. (From Watson. 1947. Ann. Bot. *11*: 41–76.)

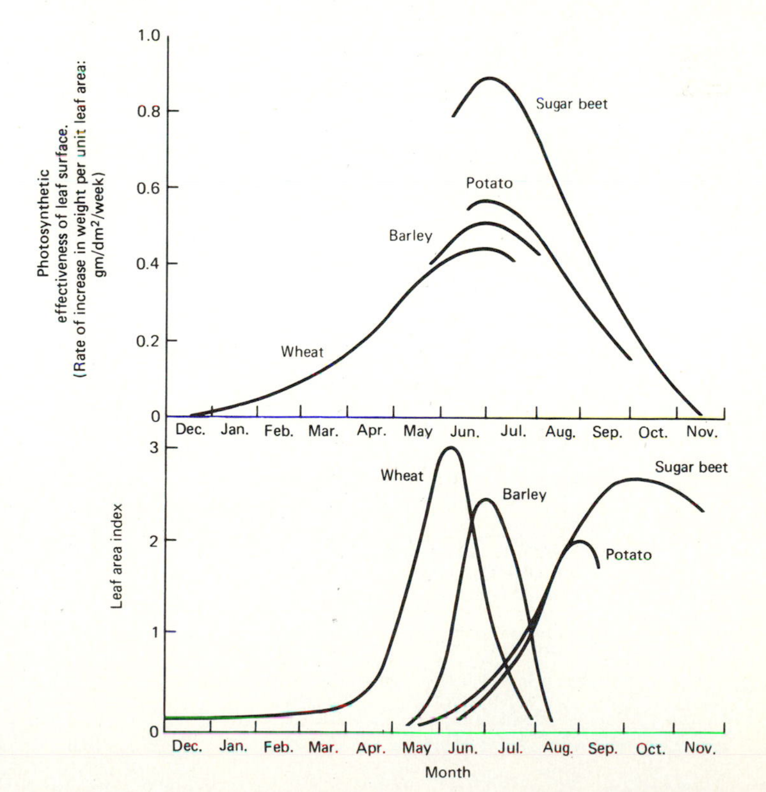

leaf area by midsummer, so that sunlight is wasted, yet it continues to grow so that the optimal leaf area is exceeded later in the season (Fig. 14-9). An ideal crop is one that can be sown early, rapidly developing leaf area so that maximal sunlight is intercepted in midsummer, at the time when it is most available. The breeding of cold-tolerant varieties of crops for early spring sowing holds promise for increasing yields.

Carbon Dioxide in Plant Communities

In a dense plant community, when CO_2 is removed more rapidly than it can be replaced by diffusion from the air above the crops, carbon dioxide as well as light may be limiting for photosynthesis (see Fig. 4-5). The action of wind will speed up the mixing of the air in the crop with that above it, thus replenishing the supply of CO_2 in the canopy (Fig. 14-10). Whether light or CO_2 is limiting will depend on light intensity, plant density, and wind speed. In enclosures such as greenhouses, CO_2 fertilization (usually achieved by burning gas or oil) can be practical, but it is obviously out of the question in the open field. In addition to enhancing photosynthesis by making CO_2 more available, the addition of CO_2 will result in lower photorespiration because of an increased CO_2/O_2 ratio. This occurs because the same enzyme in the Calvin–Benson cycle may act both as RuBP carboxylase and RuBP oxygenase, and how it acts will depend on the CO_2/O_2 ratio (see Chapter 4).

Figure 14-10 Many conditions including carbon dioxide concentration vary with depth in a dense crop. This chart shows how certain features change within a corn crop. (From Lemon *et al.*, 1978. *Science* 174, 371–378.)

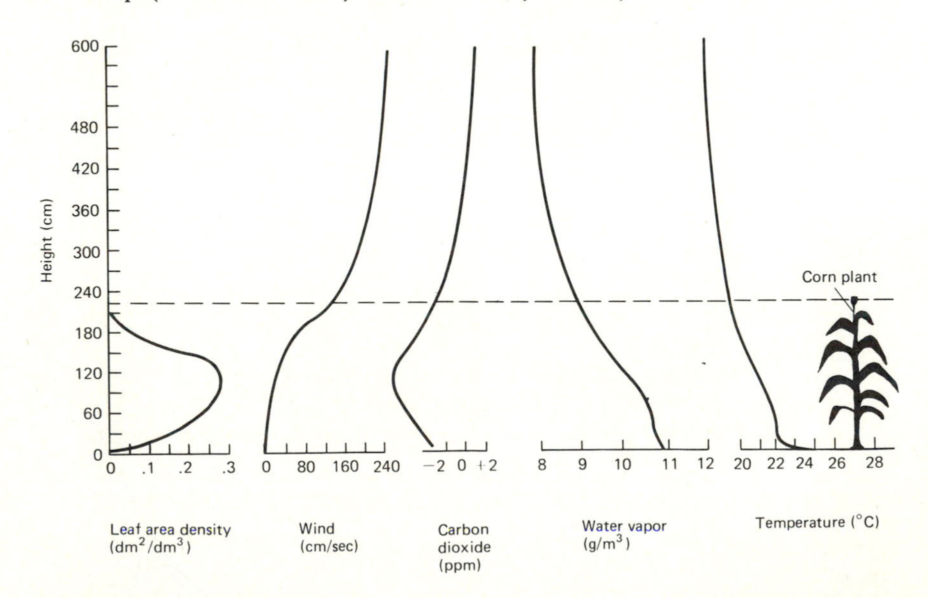

Light and Temperature in the Control of Growth and Development

Photoperiodic control of flowering

As we saw in Chapter 12, the flowering of many plants is controlled by daylength, through the action of phytochrome and endogenous rhythms. Horticulturists have made use of this property to control the flowering habit of several valuable species of plants so that they can be brought to market at a favorable time. *Chrysanthemums*, for example, are short-day plants, naturally flowering in the declining daylength of autumn. By placing artificial lights above the plants, one can prevent their flowering. Six weeks before flowers are required for market, the artificial lights are turned off and, provided that the natural days are short, flowers will be initiated; in this way flowers can be produced throughout the winter months. *Chrysanthemum* flowers can also be produced in summer by covering the plants to exclude light for part of the day.

Although it is not generally practical to use light to control flowering of agricultural crops in the open field, this practice has proved of value in sugarcane cultivation. Sugarcane stores sugar in its stems only as long as it does not flower; when flowers and fruits are formed, some of the sugar is converted to starch in the seeds. Flowering in some strains is induced by short days, and in these instances flowering can be delayed by interrupting the long night with a short flash of light (see Fig. 11-2). Commercially, this can be achieved by shining searchlights onto the crop for a short period each night. Unhappily, the expense of such a practice limits its application.

Formation and growth of storage organs

Several crops and ornamental species are propagated through the use of storage organs, which may also constitute the harvest itself. One familiar example is the tuber of potato, a swollen form of stem. Tuberization is promoted by cool nights and short days, though varieties differ in their response to, or requirement for, these factors. The daylength signal is perceived by the leaves; then a hormonal influence, not yet fully elucidated, causes the tip of the horizontal stem (*stolon*) to cease elongation growth and swell into a tuber. There is some indication that an enhanced supply of abscisic acid and cytokinins may promote tuber formation.

Bulbs, which form after flowering from swollen leaf bases surrounding an unelongated stem tip, go through a period of dormancy prior to sprouting. Proper management of bulb-producing crops such as tulips has furnished the basis for an important horticultural industry in the Netherlands. Temperature is the key to this management. Cool summer nights promote the translocation of carbohydrates synthesized in the green leaves (the source) to the developing bud (the sink) (see Chapter 8). When the leaves die, bud primordia for next year's flowers form. This step is promoted by warm temperatures in late summer or during storage. A period of low temperature (vernalization, discussed in Chapter 12) is then required before flower stalk elongation occurs—this normally takes place during winter, but bulbs can be "forced" for indoor flowering in winter by exposing them to 4–6 weeks of warm temperatures (20°C) followed by about 6 weeks at 5°–9°C. Rapid growth and flowering will then occur when warm temperatures are restored.

In biennial root crops, such as sugar beet, formation of the storage root occurs during the first growing season; the plants then require cold vernalization prior to flowering the following summer, but if sugar production is the aim, the roots are harvested the first autumn. The longer the growing season in the first summer, and the earlier a full leaf cover is formed, the higher is the yield. However, if the seed is sown too early in the year, the seedlings may become vernalized—if this occurs, flowers are formed the first season and no root crop is produced. The timing of the sowing is therefore crucial. The situation has been eased somewhat by the breeding and selection of varieties (of sugar beets in particular) which require more extended cold periods so that the seeds can be sown earlier in the spring without becoming vernalized.

Growing plants under artificial lights

In dwellings, experimental laboratories, and in some kinds of horticulture, plants are increasingly grown under artificial lighting. Some house plants originally derived from tropical vegetation growing in the deep shade of jungles can survive and grow in amazingly low light intensities, while others require supplemental light if they are to grow in places other than in the open or in greenhouses. The primary purpose of the supplementary light is, of course, to provide energy for photosynthesis. We should not overlook, however, that light also has an effect on plant form (Chapter 11), and that the correct quality of light is needed to grow healthy, well formed plants.

Many different sources of illumination are available for plant growth. A prime consideration is that the lighting should provide a high intensity of useful radiation without excess production of heat. Fluorescent lights most closely fit this requirement. Mercury lamps can be used in greenhouses, but they are large and produce a considerable amount of heat as well as much light in the near ultraviolet region that is not usefully absorbed by plants, and can even cause damage. Normal "daylight" or "cool white" fluorescent lights do not emit peak energy at the main photosynthetic wavelengths, giving principally a band near 560 nm with a minor peak of output at 475 nm (Fig. 14-11). Special fluorescent lamps, termed "Gro-Lux" lamps, emit a spectrum of light which more closely fits the absorption of chlorophyll, having high peaks of output at 440 nm and 660 nm. These lamps appear pink to the naked eye. African violets, gloxinias, and other plants grow luxuriantly under such lamps. However, both standard fluorescent and "Gro-Lux" lamps still produce virtually no light at wavelengths above 700 nm, and, as both photosyn-

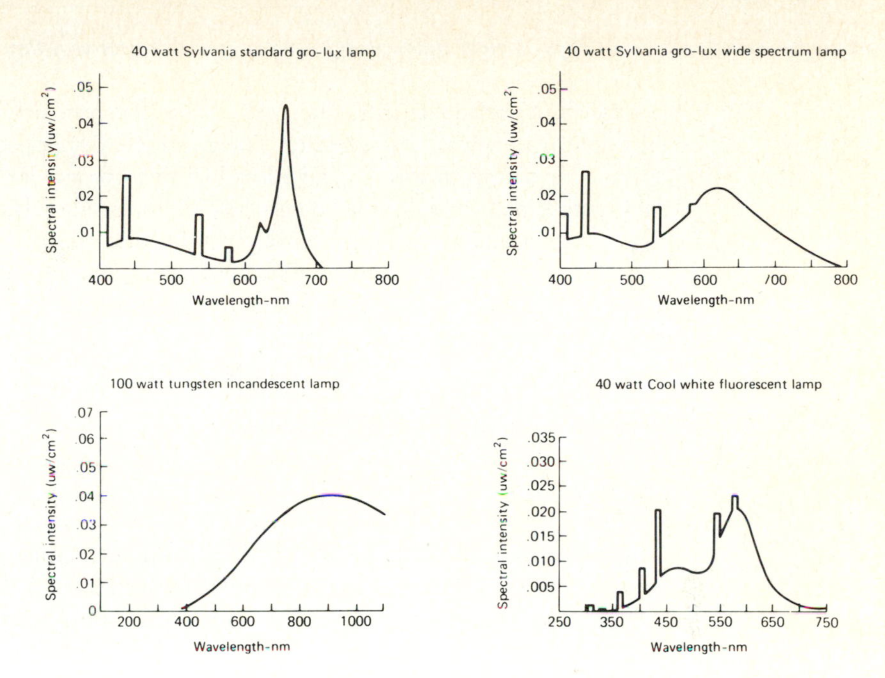

Figure 14-11 Emission spectra of some lamps used in horticulture.

thesis and more especially morphogenesis are influenced by light in this region, many plants given these fluorescent sources alone do not grow optimally. Two solutions are available: the use of different light sources, or the addition of supplemental lamps to the fluorescent tubes. Wide-spectrum "Gro-Lux" fluorescent tubes which produce a more balanced illumination, with a considerable output above 700 nm, can grow healthier plants. These lamps are very expensive, however, and similar results can be obtained much more cheaply simply by adding a few incandescent lamps to the standard fluorescent lights. Special mercury lamps are available which emit light at 660 nm, and some that emit above 700 nm are available, but they still suffer from size problems, and act as point sources of light, rather than as providers of broad, uniform light fields. Not only is light at the prime photomorphogenic wavelength of 720 nm important, but there is considerable evidence that light in the near-infrared (720–800 nm) may be also beneficial to plant growth, as the form of plants grown with supplemental illumination in this range more closely approximates that of plants grown in sunlight. As incandescent bulbs put out much of their energy in this range (even more is in the infrared, which is wasted both as visible light and photomorphogenic light) they make a suitable supplement to fluorescent tubes. It is important not to illuminate plants with incandescent bulbs alone, as they have insufficient light in the blue and red compared to far- and infrared; these sources are detected by the plants primarily as far-red, which produces unnatural, etiolated-looking plants.

Considerable effort is currently being expended by industrial concerns to produce lamps suitable for mass use in the horticultural industry. Such lamps must produce adequate intensities and a suitable spectrum, but must also be economical to install and operate. The products developed so for are sub-optimal in several respects; "Gro-Lux" lamps are not as efficient in total light output as "daylight" or "cool white" fluorescent tubes and are expensive, while incandescent bulbs waste most of their energy by emitting in the infrared. At the present time regular fluorescent plus incandescent bulbs are the most economical sources to purchase and operate, but there is certainly still scope for much improvement.

Horticultural Practices: The Role of Physiology

Pruning

Pruning is one of the oldest horticultural techniques used by man to improve the aesthetic qualities or optimize the productivity of a woody shrub or tree. The technique involves the selective cutting away of some of the branches in order to alter the pattern of growth. The excision of dead, diseased, or damaged wood helps to maintain the health of the plant by removing probable pathways to generalized infection. Where a bush or tree of a certain shape is desired for landscape purposes, that shape can be obtained by pruning away branches that do not conform to the desired pattern. Pruning can also be used to improve the quality, and sometimes quantity, of flower or fruit production. A general thinning enables light and air to penetrate into the leaf canopy, maximizing the number of leaves photosynthesizing at an optimal rate, while selective thinning removes weak branches and increases the vigor of those remaining by increasing the availability of water and nutrients for each remaining branch. Even though pruning reduces total growth by reducing the total number of leaves, the remaining growth can be more vigorous and productive. Roses furnish a common example of the benefits of pruning; when all weak side shoots are removed and each strong branch cut back to one or two healthy buds, these buds grow into vigorous branches with larger flowers.

If the end of a branch is removed, the number of branches is increased, mainly because removal of the source of auxin supply from the excised apical bud breaks apical dominance in that branch, and lateral buds sprout (see Chapter 9). This practice is used to create a denser appearing specimen and may assist fruit production in trees such as apple, where the fruit is borne on short lateral spurs. Pruning thus redirects the tree's resources to the desired form or product rather than into random, weaker vegetative growth.

Juvenility

Plants that are too young to bear flowers and fruit are called *juvenile*. Extended juvenility is frequently an economic problem in tree crops; it deter-

(a) (b)

Figure 14-12 Numerous plants have different juvenile and adult forms. The familiar English Ivy (*Hedera helix*) leaves are in fact the juvenile form (a). The adult is a large bush with simple leaves (b).

mines how long a fruit tree must be grown before a crop is obtained, and is important to foresters in that juvenility prevents the early production of seeds on vigorous conifer trees. Sometimes the leaves on juvenile trees (for example, *Eucalyptus*) differ in shape from those on the adult plant; other juvenile plants (for example, citrus) may bear thorns, while adult plants do not; cuttings of still others root readily, while adult forms do not. In some cases juvenility is considered an advantage horticulturally; the juvenile foliage of junipers, for example, is considered more desirable for ornamental purposes.

The causes of juvenility are not understood, but a hormonal balance is certainly part of the story. In English ivy (*Hedera helix*) (Fig. 14-12), gibberellin sprays given to the adult causes some reversion to the juvenile form. Gibberellins also appear to be involved in conifer juvenility, and intensive research is being carried out to elucidate the situation in these trees more fully.

Plant propagation

Plants can be reproduced either sexually through seeds, or asexually (vegetatively) by using some part of the plant body to yield an entire plant. Almost all plants can be vegetatively propagated, differing in this way from animals, where only the simpler members can increase their numbers without passage through a sex cell stage. Each method of reproduction has its advantages and disadvantages. The production of seeds enables large numbers of potential new individuals to be produced, then distributed over wide areas through the agencies of wind, water, and animals. In addition, as seed production is usually the result of genetic recombination, new variants are produced which may be better able to adapt to a new and changing environment. Vegetative reproduction, on the other hand, while producing fewer individuals, usually enables continued "parental" support or provides a nutrient supply for early growth; by contrast, seedlings generally have a minimal nutrient supply in the endosperm or cotyledon and must rapidly become dependent on their own photosynthesis. Vegetative reproduction usually does not allow a wide dissemination of the species and, as all individuals are genetically alike, less adaptation to changing environmental situations can occur. It is important commercially as a means of propagating desirable varieties.

Many plants have natural methods of vegetative propagation, exploited by man since prehistoric times. A frequent technique involves storage organs (bulbs, rhizomes, tubers) that may produce new plants (Fig. 14-13). Trailing stems, stolons, and runners may root at intervals, each rooting point capable of becoming an independent plant (Fig. 14-14). In addition, horticulturists have developed numerous artificial techniques of vegetative propagation. Based on cell and tissue culture for the commercial propagation of plants, these will probably play an increasing role in the future. In this section we will look at some of the physiological principles on which plant propagation is based.

Seed Germination Many crop species germinate without special treatment under favorable conditions of temperature and moisture. This results from generations of selection by man against seed dormancy or any unusual requirements for germination. Seeds of some ornamentals, however, require light or cold to break dormancy by reducing their content of inhibitors, or increasing that of germination promoters (see Chapters 10, 11, and 12). Other seeds possess impermeable seed coats that prevent the entry of water and oxygen. In nature such hard seed coats would be slowly degraded by bacteria.

Figure 14-13 Storage organs act as agents of vegetative propagation because each plant produces several such storage organs each of which can give rise to a new plant. (a) Section of bulb of onion (swollen leaf bases); (b) sprouting gladiolus corm (swollen stem base); (c) sprouting rhizome of iris (swollen horizontal underground stem); (d) potato plant (*Solanum tuberosum*) with tubers (swollen underground stem) and (e) a potato tuber sprouting. (From W. H. Muller: Botany-A Functional Approach, 3rd Edition, 1974. Macmillan, New York.)

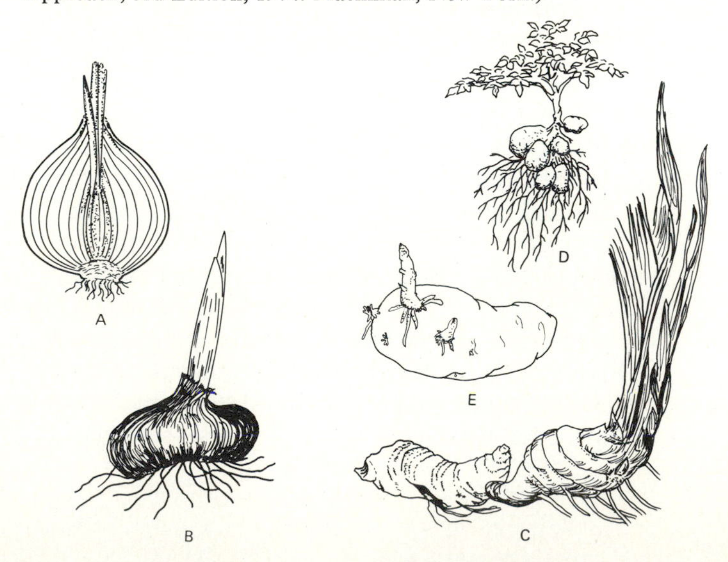

Figure 14-14 Some plants reproduce vegetatively by trailing stems. This strawberry plant produces runners which root at intervals, producing new plants. (From W. H. Muller: Botany-A Functional Approach, 3rd Edition, 1974. Macmillan, New York.)

As this would produce a very uneven crop stand, such seeds may be mechanically or chemically *scarified* (abraded) to break open the seed coat. If the pretreatment permits water and oxygen to penetrate, rapid germination will ensue. Soaking in water or leaching of the seeds may also be used to soften the seed coat and remove inhibitors. Light-requiring seeds must be sown on the surface of the soil, while cold-requiring seeds are moistened and placed in a refrigerator or outdoors for 1–4 months at 2°–7°C prior to sowing. It is important that oxygen and moisture be available during the chilling period; this is generally accomplished by *stratification*, putting the moist seed between layers of a mixture of moistened sand and peat. Hormone treatment to induce seed germination is not extensively used, although gibberellin may be applied to enhance germination of light- or cold-requiring seeds.

Propagation by Cuttings and Grafting In propagation by cuttings, a stem tip is removed and its base placed in a well-aerated, moist rooting medium of soil or sand. To prevent these rootless cuttings from wilting, they are frequently kept moist by enclosure in plastic bags or by water mist from fine overhead sprays. Callus formation, xylem and cambial proliferation, and root initiation occur in response to elevated auxin concentrations resulting from the accumulation of auxin transported basipetally from the stem tip. Rooting may be increased by slightly warming the soil and treating the cut surface with synthetic auxins such as indolebutyric acid (IBA) or α-naphthalene acetic acid (NAA) (Fig. 14-15). Callus generally develops at the cut surface and root initials start growing just exterior to the vascular bundles or, in woody plants, in the young secondary phloem at the base of the cutting. Some cuttings will root better without applied auxins, probably because they possess a sufficiently high endogenous level to do so. Those that do not produce enough of their own auxin will not root naturally and need applied auxin as a rooting stimulus.

In grafting, a stem tip or bud of one plant (termed the *scion*) is inserted into the stem of another plant (the *stock*), and the two permitted to join

Figure 14-15 The effect of auxins in the induction of roots on cuttings. Here camellia cuttings were untreated (left) or treated with a dilute solution of IBA and NAA (right). (Photograph courtesy of the Boyce Thompson Institute for Plant Research.)

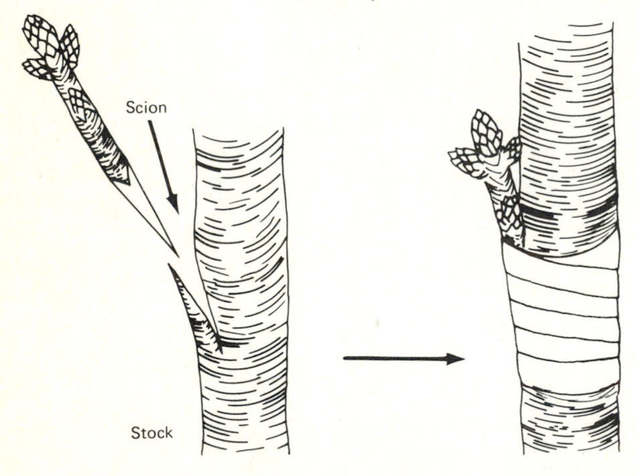

Figure 14-16 In a graft, a bud on a scion is inserted into a stem (stock) and bound tightly until a tissue union occurs. Auxin from the bud stimulates a vascular connection.

organically to make a single individual (Fig. 14-16). This technique is used when the roots of the stocks are more vigorous or disease resistant than those belonging to the plant whose shoot is desired. The graft union is bound and treated with grafting wax or other protectant to prevent desiccation. When the graft "takes," a vascular connection is formed through differentiation of the tissues of the rootstock into vascular elements joining up with those of the scion. This vascular differentiation is caused in part by auxin formed by the stem tip of the scion diffusing from the base of the grafted stem. Various types of grafting have found favor with horticulturists; which type is used depends on the species being propagated.

TISSUE CULTURE IN PLANT PROPAGATION The success of plant tissue culture in the laboratory has naturally led to its use in horticulture for plant propagation. A piece of sterilized plant tissue is placed on agar containing proper nutrients and an experimentally determined balance of auxin and cytokinins, where it proliferates to form a callus. This can be subdivided many times, producing much tissue from an original small explant. By changing the hormone balance, one can stimulate the callus to produce roots and shoots (Fig. 14-17); the plantlets can then be removed from the parent callus and cultured individually on agar until they are large enough to be planted into pots of soil.

Figure 14-17 The formation of roots and shoots from undifferentiated *Pelargonium* (geranium) stem pith.
(1) Pith tissue cultured on basal medium with IAA (5 μM) and benzyladenine, a cytokinin (1 μM) a = day 1, b = day 15, c = day 30, d = day 60. By 60 days, organized nodules have appeared. (2) Plantlets bearing roots and shoots formed from isolated nodular bodies of 1d. (3) Details of buds formed on callus. Note glandular hairs on leaves. (4) Larger buds and leaves. (5) Root (arrow) developed from base of a growing bud. (From Chen and Galston, 1967. Physiol. Plant. *20*: 533–539.)

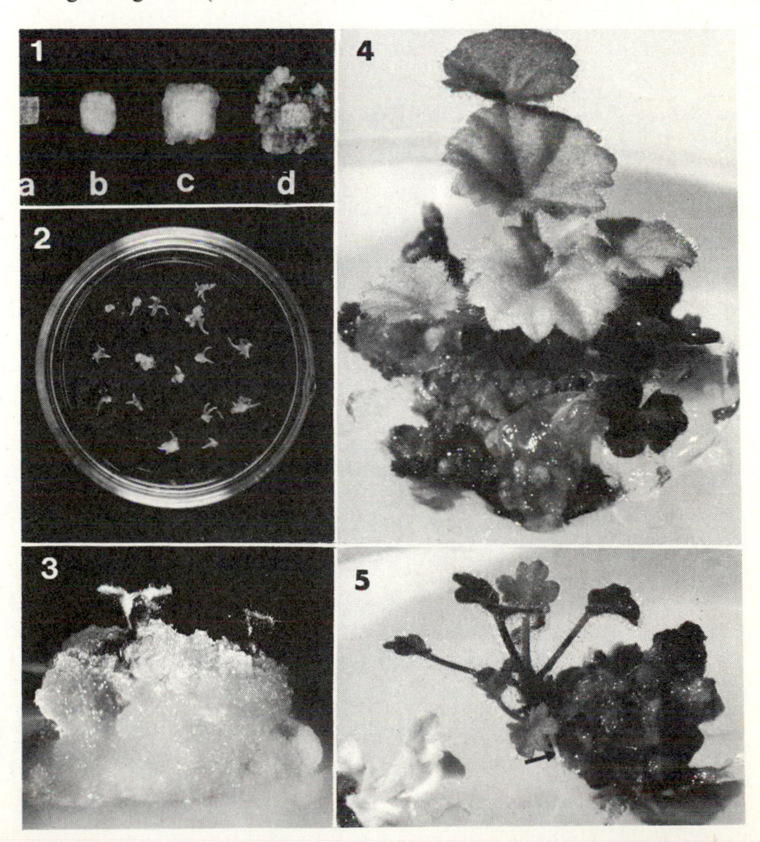

Figure 14-18 The use of tissue culture for the commercial propagation of horticultural plants. An apex of chrysanthemum is excised and placed in a sterile culture solution containing 0.02 mg/liter naphthalene acetic acid and 2 mg/liter kinetin in a flask on slowly rotating wheels. The apex grows into a callus (A) on which plants may start to develop. This callus breaks up into many pieces, all of which continue growing. When pieces are removed from the solution after about 3 months and placed on a solid agar medium of similar composition, many small plants develop (B, at 4 months). These can be separated (C), individually planted on agar and, when large enough, planted in pots (6 months). A very large number of plants can be produced for commercial production in this manner, all identical to the original parent. (Courtesy of R. W. Langhans, Cornell University.)

A faster method of vegetative propagation through tissue culture involves direct excision and culture of stem apices on agar. Under these conditions, the apex produces a callus and ultimately many small shoots which can be separately cultured and potted (Fig. 14-18). The main advantage of such a system is that it permits rapid clonal propagation of plants which do not breed true from seed, and which can be propagated only very slowly by normal methods of vegetative propagation. It has also been used to produce virus-free strains of plants, since many systemic viruses, being transported in the phloem, do not penetrate into the apical growing point, which is not connected with the phloem.

These methods, now common, have been used for the production of orchids for decades. The shoot apex of an orchid is cultured to form a mass of tissue which can then be divided into many pieces. These pieces are rotated in nutrient solutions to prevent development of polarity and differentiation (Fig. 14-19). Pieces of the tissue are then transplanted onto nutrient agar, where they produce shoots and roots and develop into plantlets, which can then be transplanted into pots. Similar methods are now in use for carnation, chrysanthemum, and some other plants. It is likely that many large greenhouses and plant nurseries growers will soon include a tissue culture propagation facility.

Tissue culture will almost certainly prove useful in the propagation of tree species which are slow to produce seed, or which do not transmit desired characteristics through the seed. Selection has recently produced some rapidly growing conifer tree lines, but as they do not produce seed for many years, propagation continues to be a problem. Intensive research programs are

Figure 14-19 Parts of a tissue culture propagation facility at a large commercial grower. The right photo shows flasks containing the tissue and culture medium, rotating on a clinostat to prevent gravitational stimulation of polarity and differentiation. The left photo shows the thermostatically controlled incubation room containing many cultures. (Courtesy of R. W. Langhans, Cornell University.)

under way to try to induce conifer tissue formation and proliferation in culture, followed by regeneration of the whole plant. This last phase is often the stumbling block, and without it tissue culture has limited practical value. Certain excised organs, such as roots, may synthesize desired chemicals, such as alkaloids; such organ cultures can be useful economically.

CELL AND TISSUE CULTURE AS TOOLS IN PLANT BREEDING In addition to its role in vegetative propagation, tissue culture may serve as an unusual path to sexual reproduction. Suppose we have a high-yielding crop plant marred by susceptibility to a certain disease, and a related though sexually incompatible species possessing resistance to that disease. If normal crossing cannot be achieved through pollination, there is no sexual way to introduce the resistance character into the crop; however, *parasexual fusion* is now possible through the use of cultured somatic cells. Ordinary plant cells cannot fuse in culture because their cell walls prevent the protoplasts from uniting. It is possible, however, to dissolve away the cell wall by using a mixture of enzymes which break it down. Pectinase is first used to separate cell from cell, and cellulase can then be used to digest the walls of the individual cells. The living cell contents, or protoplasts, can then be harvested in bulk by gentle centrifugation and treated as free-living naked microorganisms (Fig. 14-20). If wall digestion is carried out in a hypertonic medium to prevent the protoplasts from bursting, the naked protoplasts are viable, and under appropriate conditions can reform a wall, divide, then regenerate an entire plant. If protoplasts from two different species are mixed in the presence of fusion-inducing agents such as polyethylene glycol, a small fraction of the protoplasts fuse, producing heterocaryons (Fig. 14-21), or cells containing multiple nuclei from different sources (Fig. 14-22). If nuclear fusion occurs, true *parasexual hybrids* may result.

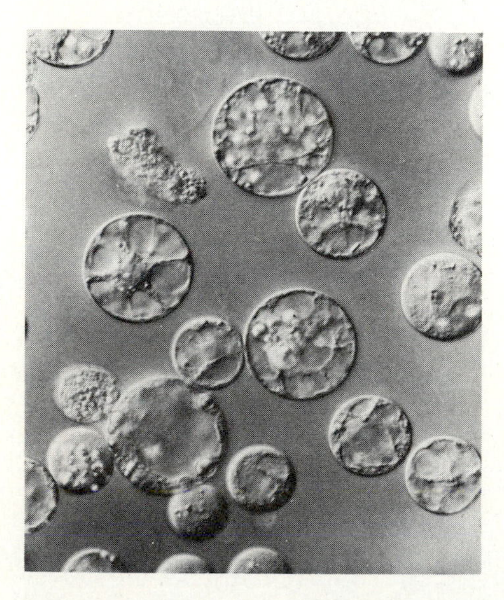

Figure 14-20 Protoplasts derived from a suspension culture of tobacco (*Nicotiana tabacum*) cells. Magnification ×500, taken under Nomarski differential interference contrast optics. Note the spherical shape and the clearly defined membrane, cytoplasmic strands, vacuoles, chloroplasts, and nucleus.
(Courtesy of L.C. Fowke, University of Saskatchewan.)

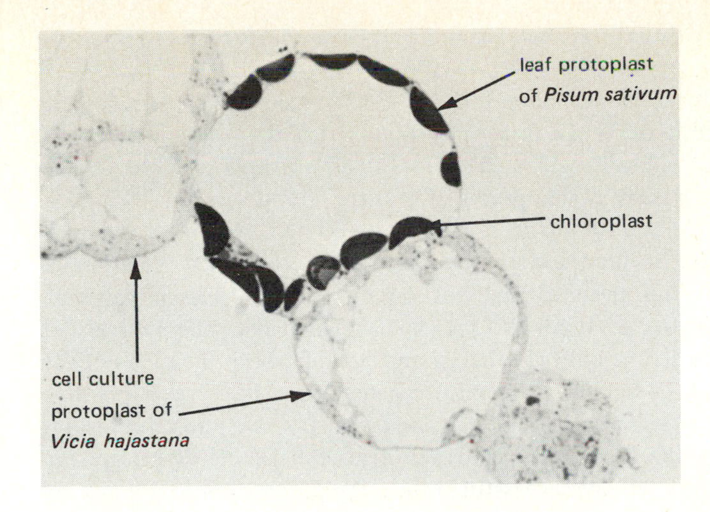

Figure 14-21 A pea leaf protoplast agglutinated with several vetch protoplasts derived from a tissue culture. Surface agglutination has been accomplished by the use of high concentrations of polyethylene glycol (PEG). This 1 μm thick section was stained with toluidine blue. (Courtesy of L. C. Fowke, University of Saskatchewan.)

Once fusion products have been obtained, a selection technique must be available to permit recovery of the hybrids from the self-fusion products of each original cell type. If one protoplast species is resistant to chemical *A* but sensitive to chemical *B*, the latter drug may be used to kill it or its intraspecific fusion products. A reverse pattern of drug resistance and sensitivity can be used to eliminate cells of the other species with chemical *A*. Provided that resistance to the different chemicals is carried into any fused hybrid protoplast, it alone should be resistant in the presence of both *A* and *B*. Thus, only hybrid (fused) protoplasts will be able to grow on a medium containing both drugs. Alternatively, if protoplast species *A* is autotrophic for auxin but not cytokinin, and protoplast species *B* autotrophic for cytokinin but not auxin, only their fusion product can grow in a medium lacking both hormones.

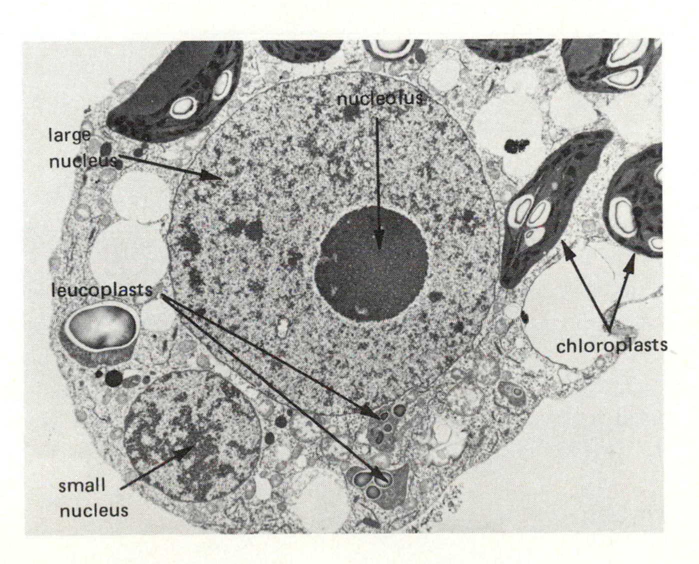

Figure 14-22 An electron micrograph of a thin section cut through a heterocaryon produced by PEG-induced fusion of protoplasts from cultured *Vicia hajastana* cells and *Vicia norbonensis* leaf cells. The large nucleus (with dark central nucleolus) and the leucoplasts belong to *V. hajastana*, while the small nucleus and chloroplasts belong to *V. norbonensis*. Nuclei and cytoplasmic organelles from both parent protoplasts may continue to function in such a heterocaryon. (Courtesy of L. C. Fowke, University of Saskatchewan.)

Parasexual hybrid plants started from somatic cells will probably be tetraploid. Diploid hybrids can in turn be obtained if the starting materials are haploid; such haploid cells can be obtained through culture of the microspores, or immature pollen grains of anthers.

This area of research is now in a great state of flux, with new methods evolving frequently. So far the complete cycle from protoplast isolation to plant regeneration has been carried out on only a few plants (tobacco, petunia, carrot, asparagus, rape) and parasexual fusion with the recovery of hybrids has been achieved only in tobacco and petunia, both members of the *Solanaceae* family. As yet, no successful regenerations have been possible with members of the *Gramineae* and *Leguminosae*, the two most important families of food plants. It is likely, however, that somatic cell hybridization and plant regeneration will be successful in these families as well, and will form an important part of future crop improvement technology (See Chapter 16).

Post-harvest storage

Because of the seasonal nature of fruit and vegetable production, some form of storage is necessary to maintain a year-round supply. In bygone times, houses had a cool root-cellar, where potatoes, carrots, and other vegetables were stored over the winter. Considerable losses ensued under such conditions because of shrinkage and, later in the season, sprouting. Fruits such as apples could be kept for about two months, but by the end of that time were barely edible. Modern storage techniques have changed all that. It is now possible to purchase apples in June that were harvested the previous September, the fruit being almost indistinguishable from those on sale the previous autumn. Vegetables that were previously unstorable can now be purchased year-round in the frozen state.

With frozen vegetables, such as peas, the essence of successful storage practice is to freeze as quickly as possible following harvesting. The peas are given a dip in hot water to rapidly inactivate degradative enzymes, and are then immediately frozen. Provided that sub-freezing temperatures are subsequently maintained, such produce in sealed packets will keep almost indefinitely.

With most fruits, however, freezing destroys the texture of the product, and is therefore unacceptable. As a substitute, apples can be kept by cooling to about 4°C in a controlled atmosphere to keep respiration at a minimum and prevent desiccation. In general, the environment includes a low oxygen level (5%) to decrease respiration rate, but not so low as to cause the onset of anaerobic respiration which would result in a greater loss of dry weight. The nitrogen level of the atmosphere is correspondingly increased and extra carbon dioxide introduced to counteract the effects of any ethylene that the fruits might produce. In this manner apples can be kept for many months.

Changes do occur in storage, however, and even though apples are fresh-looking and good-tasting upon removal from such storage conditions, they must be eaten within a few days, since they quickly soften and decay.

Onions, potatoes, and root crops are also stored at a reduced temperature. For potatoes this temperature must not be too low, for starch is enzymatically changed to sugar by temperatures slightly above freezing, making the potatoes taste unacceptably sweet. The storage life of potatoes is also shortened when this occurs. Sprouting, and its accompanying softening and shrinking of the bulb, tuber, or root, is inhibited by treating the plants with a growth inhibitor such as *maleic hydrazide* a few weeks prior to harvest.

Controlling Plant Growth with Chemicals

The discovery of plant hormones as agents controlling growth and development led naturally to attempts to modify the growth of crop plants by applying these compounds in agriculture and horticulture. In this section we will discuss the use of natural and artificial chemicals to control the growth pattern of a crop plant; the application of pesticides and herbicides will be considered in the next chapter. In many cases, early hopes of vastly increased yields were not fulfilled, because hormones do not generally increase photosynthesis, which is the ultimate basis of crop yield. Also, applications of natural hormones are frequently followed by destruction of the excess hormone through adaptive metabolism of the plant. Over the years, however, many chemicals useful for the control of plant growth have emerged. Most of these are structurally related to but not identical with the natural hormones, so that they are not immediately degraded by the plant. The changes induced by these substances are often subtle; they may alter characteristics of the harvest to maximize the quantity of some desired component or may improve quality rather than quantity of yield. Recently a long-chain alchohol named *triacontanol* [$CH_3(CH_2)_{28}CH_2OH$], isolated from alfalfa (*Medicago sativa*), has been shown experimentally to increase seedling growth and crop yield when applied to several different plants. How triacontanol produces its effect is unknown and whether the substance will have any practical value in increasing crop yield in agriculture or horticulture remains to be seen. At present no known hormonal application can increase the yield of a cereal crop or pasture grass, but antigibberellin growth inhibitors can do so indirectly by preventing the elongation that predisposes the crop to *lodging* (being blown down by the wind). Optimizing the fertilizer application and improving the genetic makeup of the variety are the main ways to achieve yield increase in cereals. Hormones can, however, increase fruit yields, and growth regulators may be used to prevent post-harvest losses.

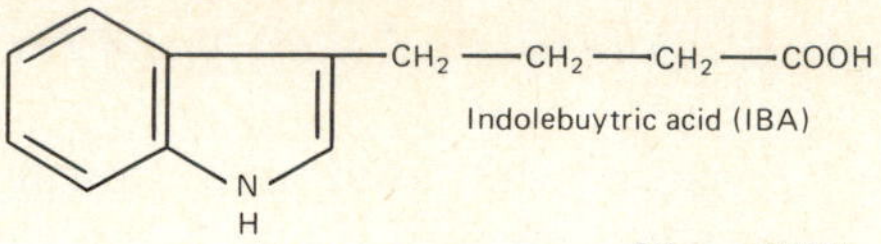

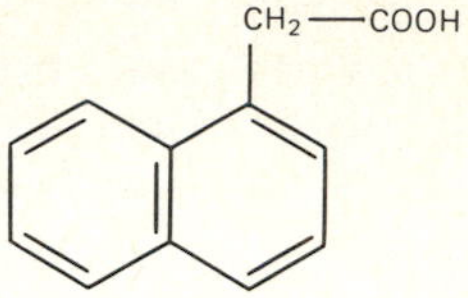

$CH_2—CH_2—CH_2—COOH$

Indolebuytric acid (IBA)

$CH_2—COOH$

α-naphthaleneacetic acid (NAA)

Figure 14-23 Synthetic auxins used in horticultural applications.

Auxins

The uses of synthetic auxins in horticulture can be traced directly to the natural roles of IAA in the plant. In general, compounds such as α-naphthalene acetic acid (NAA) are used because they resemble IAA in action but are resistant to degradation by plant enzymes (Fig. 14-23). Auxins are used for a variety of agricultural purposes, including:

1. Promotion of rooting of cuttings. The base of the cutting is dipped in a powder containing NAA or indolebutyric acid (IBA) prior to planting.

2. Induction of flowering in pineapple (actually caused by the auxin-induced production of ethylene). NAA is generally employed as the auxin.

3. Increased fruit set and induction of seedless tomatoes. Both result from an auxin-induced growth of the pericarp in the absence of fertilization.

4. Thinning of fruit during early development, as well as

5. Prevention of preharvest fruit drop. These two effects seem paradoxical and opposite, but both are under auxin control. In apples, for example, whether auxins enhance or delay fruit drop depends on the timing and level of application. The enhancement of fruit drop probably occurs because of auxin-induced ethylene formation. Thinning is required early in fruitset to prevent too many small fruits from forming, while later on, the developing fruits need to be kept on the tree until they ripen.

6. Auxin type herbicides—see next chapter.

Gibberellins

Commercially, gibberellins are produced by fungal cultures, and it is the purified natural products that are applied to plants. Generally, gibberellic acid (GA_3) is used, because this is the only gibberellin obtainable in commercial quantities, although an expensive mixture of GA_4 and GA_7 is now commercially available for specific purposes. Gibberellins are used for the following purposes:

1. Enhanced production of seedless grapes. Bigger, more uniform bunches with larger fruit are produced. Among other effects, the gibberellin causes lengthening of the peduncle (stalk) attaching each grape to the cluster, thus permitting larger grapes to form. Virtually all the grapes that go to market are now treated with gibberellin (Fig. 14-24).

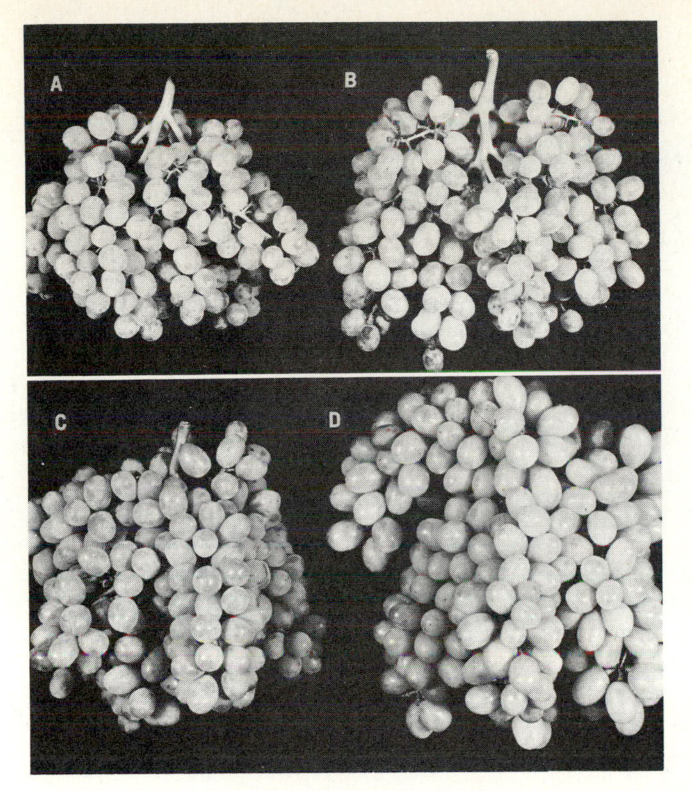

Figure 14-24 The effect of gibberellin on the growth of Thompson Seedless Grapes. (A) Control grapes. (B, C, D) Grapes sprayed with 5, 20 and 50 parts per million, respectively, of gibberellin. (Courtesy of R. J. Weaver and S. B. McCune, University of California, Davis.)

2. Treatment of oranges to prevent rind senescence, to permit longer storage on the tree, and thus to extend the marketing period.

3. Enhancement of flower bud formation and improvement of fruit quality in cherries.

4. Improvement of fruit setting in apples and pears, particularly under weather conditions poor for setting.

5. To substitute for a chilling requirement in instances such as:

a) Flower induction for seed production (radish)
b) Increased elongation (celery, rhubarb)
c) Earlier flower production (artichokes)

6. The production of hybrid cucumber seed. Most high-producing cucumbers are F_1 hybrids. GA sprays induce the production of male flowers on cucumber plants that normally produce only female flowers. The seed from neighboring all-female plants of a different strain is then exclusively hybrid.

7. Increased malt production. The addition of GA to germinating barley during beer production enhances α-amylase production so that more malt is produced more quickly. As the malt is the raw material for fermentation, a greater production of beer is made possible by this technique.

8. Increased sugarcane yield: GA promotes the elongation of sugarcane stalks with no change in the sugar concentration, so that the net yield of sugar is increased.

Figure 14-25 The two celery stalks on the right were treated with 10 ppm of benzyladenine, while those on the left were not treated. (Courtesy of S. H. Wittwer, Michigan State University Agricultural Experiment Station.)

Cytokinins

The major use for cytokinins derives from their ability to delay senescence and maintain greenness. The artificial, highly active cytokinin, *benzyladenine*, is the main compound used. The treatment of holly for festive decorations enables its harvest many weeks prior to use. Post-harvest sprays or dips are now available to prolong the storage life of green vegetables (Fig. 14-25) such as asparagus, broccoli, and celery, but such uses are not yet licensed. Because cytokinins resemble nucleic acid components, they may possibly have cytotoxic properties, and until proper toxicological tests are made, they will probably not be cleared for use.

Ethylene

Ethylene enjoys a wide variety of uses, but its gaseous nature precludes its use in nonenclosed spaces. Ethylene itself can be used to enhance the ripening of fruits such as bananas in storage following their shipment in an unripe condition; this is of great benefit, since the green bananas are rugged and do not bruise or spoil easily. The tender ripe bananas can then be carried safely to market from the nearby warehouse. Recently, an ethylene-producing liquid chemical, 2-chloroethylphosphonic acid (commercially called *Ethrel* or *Ethephon*) has been introduced into commerce. This compound is sprayed onto the plant at a slightly acid pH. When it enters the cells and encounters the cytoplasm at about neutral pH, it breaks down to release gaseous ethylene. Numerous commercial applications for this compound have appeared, mostly in relation to the natural effects of ethylene:

1. The most important commercial use involves enhancing latex flow in rubber trees in Southeast Asia. When a rubber tree is "tapped," the latex flows for a certain period before the cut seals and the flow stops. Ethephon delays the healing of the cut so that the latex flow continues for a longer period, thus yielding more latex with less tapping.

Figure 14-26 Ethrel can be used to produce uniform ripening prior to mechanical harvesting. Here untreated tomatoes (left) are still largely green while those sprayed with 250 parts per million ethrel have ripened uniformly and turned red. (Courtesy of W. L. Sims, University of California, Davis.)

2. Enhancement of uniform fruit ripening and coloration. This has been shown to be of particular value in field tomatoes picked at a single time by machine (Fig. 14-26).

3. Acceleration of fruit abscission for mechanical harvesting. This provides a potential area of use in a wide variety of fruits such as grapes, cherries, and citrus.

4. Promotion of female flower production in cucurbits (cucumber, squash, and melon) so as to increase the number of fruits produced per plant.

5. Promotion of flower initiation and controlled ripening in pineapples.

Synthetic growth inhibitors

At present, the natural inhibitor abscisic acid has not found any commercial use, partly because it is so costly to produce. Various synthetic growth inhibitors have, however, found extensive use in post-harvest storage and in horticulture. The first synthetic growth inhibitor to be widely used was maleic hydrazide, which prevents sprouting during storage in onions, potatoes, and certain root crops (Fig. 14-27), and also prevents the sprouting of lateral "sucker shoots" in detopped tobacco plants.

Figure 14-27 Pre-treatment of onions with the synthetic growth inhibitor maleic hydrazide (below) prevents them from sprouting during subsequent storage (above—untreated). (Courtesy of F. M. R. Isenberg, Cornell University.)

A range of growth retardant chemicals presently available can perform essentially similar tasks; the most suitable in any particular instance depends on the species and the required effect. Most of the growth-retardant chemicals act by inhibiting gibberellin synthesis in the plants so that dwarfing results. In addition, any other change produced, or inhibited, by gibberellin will often be counteracted. These inhibitors include "CCC," "Alar" (or "B9"), "AMO-1618," "Ancymidol," and "Phosphon." Their uses include:

1. Preventing of lodging in wheat. Under high fertility levels in wet climates such as that of western Europe, wheat frequently becomes "top-heavy" just before harvest and falls over so that it cannot be harvested mechanically. Application of CCC causes the production of a short stiff straw so that lodging is prevented.

2. Reduction of height of ornamental plants such as poinsettias, chrysanthemums, and lilies. Commercially, the optimal height of a flowering ornamental plant is about 30–45 cm rather than the natural height of a meter or more. Growth retardants applied to the leaves or soil when the plant is young have no effect on flower size but result in very short stems (Fig. 14-28).

3. Control of shrub growth. This is particularly important along highways, even if the plants are cultivated ornamentals, or below power lines. Spraying after bud break but before stem elongation leads to the same number of leaves, but the stems are dwarfed so that the need for pruning is reduced.

Figure 14-28 The application of a growth retardant decreases the stem height of many ornamental plants, making them more commercially acceptable. Here, ancymidol [α-cyclopropyl-α-(4-methoxyphenyl)-5-pyrimidinemethanol] known commercially as A-rest has been used to dwarf Easter lillies. The compound was applied when the plants were about 15 cm high. Left to right: no treatment, 0.25 mg soil drench; 0.5 mg soil drench, 1.0 mg soil drench, 50 ppm leaf spray; 100 ppm leaf spray. (Courtesy of J. G. Seeley, Cornell University.)

4. Specific effects on fruit quality or ripening. A wide range of miscellaneous effects of growth retardants have been noted. One such example is the effect of "Alar" on tree fruits. In cherries it advances maturity, enhances fruit color, produces a firmer fruit, and thus facilitates harvesting.

It should be clear that plant physiological research has uncovered many phenomena that have resulted in improved or more economical horticultural practice, in increased yield, or in increased quality of plant product. Considering the rapidity with which new science is now making its way into technological innovation, we should expect the next few decades to witness further remarkable transformations in the way man regulates his crop plants. These will be exciting times for the experimental botanist!

SUMMARY

When plants are cultivated under conditions of intensive agriculture designed to optimize yield, special problems arise whose solutions are based on knowledge of plant physiology. Mineral nutrients, supplied as fertilizer, must not be used to excess, or harm to plant and environment may occur. *Eutrophica-*

tion, the development of dense plant growths in previously clear lakes, may result from raised levels of nutrients leached from soils. Fertilizers of limited solubility can help correct this problem.

Soil pH may cause poor plant growth; usually the problem is low pH, which leads to toxicity from iron and aluminum and a deficiency of phosphorus. *Lime* applications can prevent such damage. Alkaline soils, low in available iron and phosphorus, must be corrected by the addition of acid or acid-producing systems. Optimal nutrient and water supply can be provided by growing plants in solution culture using the nutrient-film technique.

Optimal water supply, as in *irrigation*, must be based on knowledge of transpiration rates and water deficiency levels leading to growth inhibition. Optimization of sunlight received by a leaf canopy depends on precise knowledge of *leaf angles*, mutual shading, and light intensity limiting photosynthesis. The *leaf area index*, the ratio of leaf area to ground area covered, varies with each crop, and must be managed to secure optimal yield. Addition of carbon dioxide to the air can improve plant growth in enclosed chambers and greenhouses, but is probably impractical in large-scale, open agriculture. Control of *photoperiod* can improve yield through regulation of flowering; fruiting; tuber, bulb, and storage root formation; abscission; dormancy; and senescence. *Light quality*, as controlled by supplemental fluorescent or incandescent sources, can also be important in enclosures, but not in open fields. *Juvenility*, a phase of plant life in which flowering and fruiting cannot occur, can sometimes be controlled by hormonal applications. *Vegetative propagation* can be used to produce many genetically identical individuals.

Seed germination can be controlled by temperature, light, and genetic manipulation. *Pruning*, the deliberate shaping of a woody plant by surgery, can control productivity and sturdiness, as well as form for aesthetic purposes. *Grafting* of *scion* onto *stock* can aid in the production of sturdier, more productive crops. Cells, tissues, and even protoplasts can be used for plant propagation on artificial media. Typically, a rapidly growing *callus* is produced, from which differentiated organs and ultimately entire plants can be produced in quantity. *Protoplasts*, cells whose walls have been removed enzymatically, can be used in *somatic hybridization*; the resulting fusion product can reform walls, divide, and then produce new types of individuals. Culture of excised stem apices can also be used to "cure" plants of systemic viruses. The conditions of post-harvest storage of fruits and storage organs can greatly affect their longevity and quality; control of temperature and the gas phase are frequently important.

Chemicals can be used to control growth and productivity of crops. *Auxins* are used to promote rooting, induce flowering and fruit set, control number of fruits, and differentially kill plants. *Gibberellins* enhance production of seedless grapes; improve flower and fruit quality in citrus, cherries, apples, and pears; substitute for chilling requirements; aid in hybrid seed production by effects on monoecious flower sexuality; increase sugar production in sugarcane; and increase beer production through control of *amylase* content. *Cytokinins* can control senescence in excised plant organs, while ethylene controls fruit ripening, latex rubber flow, abscission, flower initiation, and sex expression. Synthetic chemicals can aid production in various ways, by preventing axillary bud growth, excess stem growth and *lodging*, and improving the quality of fruits produced on important crops.

SELECTED READINGS

BLEASDALE, J. K. A. *Plant Physiology in Relation to Horticulture* (London: Macmillan, 1973).

HARTMAN, H. T., and D. E. KESTER. *Plant Propagation—Principles and Practices*, 3rd ed. (Englewood Cliffs, N. J.: Prentice-Hall, 1975).

JANICK, J., R. W. SCHERY, F. W. WOODS, and V. W. RUTTAN. *Plant Science*, 2nd ed. (San Francisco: Freeman, 1974).

MILTHORPE, F. L., and J. MORBY. *An Introduction to Crop Physiology* (New York: Cambridge University Press, 1976).

NICKELL, L. G. Plant growth regulators, Chemical and Engineering News *56* (41): 18–34 (1978).

WEAVER, R. J. *Plant Growth Substances in Agriculture* (San Francisco: Freeman, 1972).

WITTWER, S. H. "Growth Regulants in Agriculture" *Outlook on Agriculture* 6: 205–217 (1971).

QUESTIONS

14-1. What is meant by the "law of diminishing returns"? How does it affect agricultural practice?

14-2. If a micronutrient is unavailable in the soil, how may a crop be treated for deficiency of the element?

14-3. Why is soil pH important to plant growth?

14-4. You are an agronomist managing a potato (or sugar beet, or wheat) field. If it were possible, which metabolic process(es) would you like to eliminate or diminish in your plant and which would you like to augment, to obtain increased yields on a nutrient-poor soil? Give reasons for your answer.

14-5. Why do plants often grow better in nutrient film culture than in solution culture? How might one improve the growth of a plant whose roots are immersed in an optimal nutrient solution?

14-6. Which will transpire more: (a) an isolated plant or (b) a similar plant in a dense crop of the same species? Briefly give reasons for your answer.

14-7. Explain why crop yield is often observed to have been greater when the crop has transpired more water.

14-8. Why is it advantageous for a field crop to grow rapidly early in the season?

14-9. A leaf one layer down in a crop canopy of four layers experiences conditions which affect photosynthesis differently, as compared with a leaf in full sunlight at the top of the canopy. What environmental conditions would differ and how could both leaves be operating near their optimal efficiency?

14-10. Why is the leaf area index (LAI) important in crop photosynthesis and how does the optimum LAI vary with light intensity?

14-11. How do you account for the fact that "shade" plants are much more efficient at using low light energy than are "sun" plants?

14-12. Compare and contrast the physiological processes occurring in a plant in a dense plant community with those in an individually grown plant.

14-13. Discuss some physiological principles involved in (a) bulb storage, (b) vegetative propagation, (c) induction of seed germination.

14-14. What qualities are important in lamps used as the sole source of light for plant growth?

14-15. Give some reasons why cell and tissue culture of plants has a place in horticulture and agriculture.

14-16. Some sugar beet seedlings experience a period of several weeks at about 5°C because of a cold spell shortly after planting very early in spring. Explain what will probably happen and why.

14-17. Name one natural and one artificial (a) auxin, and (b) cytokinin. Why are the artificial rather than the natural compounds often used in agricultural and horticultural applications?

14-18. Give the scientific basis for (a) the use of lights to control flowering in chrysanthemums; (b) spraying grapes with gibberellin; (c) putting fruit in an atmosphere of ethylene or high CO_2.

14-19. Growth substances have many practical uses in agriculture and horticulture. Briefly describe one different use of each major group of a natural or artificial growth-regulating substance. As far as possible give the physiological property of the growth-regulating substance on which the use is based.

14-20. Synthetic compounds that retard stem growth frequently produce desirable agricultural effects. Explain.

14-21. What is meant by somatic hybridization? How can it be accomplished?

15

Plant Protection

Like all living things, plants live in a potentially hostile world. From the moment of germination on, they must compete with other plants for space, light, water, and mineral nutrients. As young, tender seedlings, they are prone to attack by bacteria, fungi, insects, nematodes, and a host of other predators. By the time they have grown, they face not only their old enemies, but also herbivorous animals of various kinds. Survival to completion of the life cycle requires the evolution of defense mechanisms to avoid or ward off pests and predators. Wild plants, by definition, have succeeded in that task, but our modern agricultural crops, having been bred to satisfy man's needs, frequently lack their own defenses and must be bolstered artificially. Man has responded to that need by inventing a host of agents, mainly chemical, to help the plant attain maturity.

Plants possess both structural and physiological mechanisms that help them withstand the extremes of environmental stress, such as very high or low temperatures or prolonged drought. In coping with potential predators and diseases, the nonmobile green plant must possess multiple deterrents, whether structural, physical, or chemical. Thorns and stinging hairs are obvious and effective systems of defense against larger animals, but do not occur in all plants and may not be effective against minute insects. The most important device used by almost all plants against a variety of enemies is a chemical defense system, involving thousands of different compounds, only a few of which are essential to the life processes of the plant. The remainder form the armory with which the plant defends itself against all potential pests and predators. Let us view some of the plant's defense mechanisms in detail.

Temperature and Water Stress

On a temperate, moist, summer day, an herbaceous plant has little trouble surviving and growing at an appreciable rate. Yet if that plant were transferred to a hot desert in summer or to a high mountain in winter, it would not survive for long without special protection. In both these stressful environments, it would have to face an extreme temperature, in which water is either absent or unavailable through being frozen. Yet plants survive and even grow in both situations. How do they do it? The answers reveal the "ingenuity" and versatility of mechanisms developed under the challenge of natural selection.

There are three principal ways by which plants can survive in harsh climates: (1) by adaptations which permit them to avoid the stress; (2) by structural modifications; and (3) by physiological features permitting them to overcome the stress. Plants may avoid unfavorable conditions by passing the stressful season as a resistant seed or in a state of dormancy. Thus, deserts may seem barren in the hot, dry season, but after spring rains, herbaceous plants that had previously existed as dormant seeds lying in the sand rapidly appear. Once germinated, these plants grow rapidly, bloom, and, after assuring the continuity of the species by producing a new crop of resistant seeds, die as the searing dry summer starts. Similarly, summer annuals in the temperate zones complete their life cycle in the favorable days of summer and overwinter as resistant seeds. Herbaceous perennials may survive as underground food storage organs such as bulbs or rhizomes, with the soil and perhaps the overlying snow protecting them from the chilling winds. Trees in temperate regions cannot, of course, disappear underground; they nevertheless avoid some of the effects of winter by dropping their tender foliage, leaving only the more resistant branches with their buds encased in hard, scalelike leaves to face the harsh climate. This same mechanism may be also used to avoid stresses produced by heat. Thus some desert shrubs, such as the Ocotillo (*Fouquieria splendens*), carry leaves only during the rainy season, losing them during the dry season when water loss from their expanded surfaces would become a burden on the limited supply in the soil.

Structural adaptations

Structural features of special significance for survival under stress may be present throughout the life cycle, or may arise at a particular stage of development, enabling the plant to withstand the unfavorable part of the life cycle. The main structural defenses usually act by restricting water loss, because it is water that is usually in shortest supply. The leaves may be covered with a thick, waxy cuticle as a waterproof barrier, and abundant hairs and sunken stomata may trap a layer of moist air, thus decreasing the transpiration rate (see Chapter 6). Leaves may be absent or reduced in size, thereby cutting down the surface area from which water loss takes place, and the leaves or stems may often be swollen from the storage of water. The latter feature can be found not only in desert plants but also in arctic and alpine perennials living under conditions in which water is often frozen in the soil at a time when the bright sun is promoting evaporation from the leaves. In addition, such plants often display a low growth habit, lying close to the ground and thereby minimizing the drying and chilling effects of the wind. Those who live in cold areas will be familiar with the "wind chill factor" referred to in weather forecasts; by evaporating water, wind cools the evaporating surface to below the temperature of the surrounding environment, and thus may cause cold damage to the organism even though the air temperature itself may be above the damaging level. By adopting a prostrate habit, the plant minimizes these adverse effects.

Man can make modifications to assist plants during periods of natural stress. The most obvious are greenhouses, which trap the heat of the sun and whose enclosed space can also be heated in winter. In summer, however, solar heat trapping may result in excessive heating, so that greenhouses are frequently painted with whitewash to cut down on such energy absorption. Cooling may also be achieved by a steady flow of air, precooled by passage over wet pads before entering the greenhouse. In dry climates, where evaporation occurs readily, the energy absorbed in converting water from the liquid to the gaseous state cools the air appreciably. In cold regions, a blanket of snow provides effective protection. Gardeners may partly protect their cold-sensitive shrubs with a covering of burlap; while this cannot protect the plant against excessively low temperatures, it shelters the shrub by protecting it against the chilling effects of wind, so providing an extra margin of safety. In cold-sensitive orange or peach orchards, prevention of damaging frost may be achieved by wind machines, which look like giant aircraft propellers on towers. These machines force the warmer air of upper levels down to the ground, thus preventing the leaf surfaces from cooling to below freezing by radiative heat loss. Growers may also light small oil or smudge fires in orchards to warm the air by several degrees and to minimize radiation loss to the clear open sky. In larger-scale agriculture, such operations become impractical, but the farmer can improve his chances for a successful crop by planting the seed after the last probable killing frost has passed. Breeders have also produced many cold-tolerant varieties (discussed later), thereby permitting earlier sowing and a longer growing season.

Physiological adaptations

In addition to structural protective devices, plants also possess physiological adaptations to hot, cold, and dry conditions. Many desert succulents have a unique mechanism of photosynthesis which reduces water loss. By the use of Crassulacean acid metabolism (CAM), described in Chapter 4, these plants are able to separate the time of stomatal opening from the time of maximum water loss, thus "saving their transpirational water while eating their photosynthetic cake." They open their stomata to admit and fix carbon dioxide during darkness, when transpiration is minimal, and keep their stomata closed in the light, when excessive water loss might desiccate the plant. In the light, the internal shuttle of CO_2 from the C_4 to the C_3 system makes the operation of efficient photosynthesis possible. This unique adaptation is important for survival in the desert.

Similarly, physiological adaptations help plants avoid freezing injury. If a greenhouse-grown plant is suddenly taken outside to temperatures slightly below freezing, it will probably be severely damaged or killed, even though that same plant in nature could easily withstand subzero temperatures in winter. The development of *cold-hardiness*, or *acclimation*, is a process which begins during the decreasing daylength and falling temperatures in autumn. During acclimation, numerous physiological changes take place; we are not yet certain which of these produces cold-hardiness, but most likely a combination is required for the plant to resist subfreezing temperatures. One process can be compared to putting antifreeze in the radiator of an automobile, where it acts to prevent ice formation and the resultant expansion and cracking open of the radiator. Like the automobile, the plant contains freezable water whose expansion during ice formation could split cells open. Early in acclimation, various solutes accumulate in the cell, lowering the osmotic potential and decreasing the likelihood of freezing by lowering the freezing point. The major damage arising in freezing plant cells comes from the development of ice crystals inside the cells; as these crystals grow, they rupture the various cell membranes and kill the cells. Raising the solute concentration protects the plant by diminishing the likelihood of large ice crystal formation. During acclimation, there are also changes in the membranes, making them less brittle at cold temperatures. This can result from an increased degree of unsaturation of the membrane lipids, lowering their melting points and permitting them to remain semiliquid at lower temperatures.

Another kind of protection against freezing injury involves the synthesis of greater quantities and new types of proteins which have the property of becoming highly hydrated. Like the water that hydrates the flour of baker's dough, this water is virtually nonfreezable; it is held in the vicinity of the protein molecule by forces that prevent ice crystal formation. Clearly, the more such proteins there are, the more resistant to freezing the cell will be. In many plants, there is a regular cycle of synthesis of such proteins in the fall and their utilization as metabolites in the spring. According to some workers, the proteins associated with cold hardiness have an especially high content of sulfhydryl (—SH) groups, characteristic of the amino acid cysteine. If this is true, cold hardiness can be assayed by chemical determination of protein-bound sulfhydryls in plant cells. While arguments continue over the significance of "bound water," it seems to provide one mechanism for protection against freezing damage.

Insects and Plants

Insects and flowering plants, while appearing independently in the course of biological evolution, have interacted and evolved together in many ways. The evolution of each of these groups has not been the result of an independent development through time; rather *coevolution* seems to have occurred, in which gradual changes persisting in one group become beneficial to the other group, and *vice versa*. This is most strikingly displayed in the relationship between flowers and pollinating insects. The flowers provide the insects with food in the form of pollen and nectar; in turn, the insects cross-pollinate the flowers, thus facilitating genetic recombination and the continuance of the plant species with the variation required to survive the challenges of environmental change. Just as the plant does not "purposely" produce nectar to attract insects, so it is probable that the insect is unaware that it is carrying

Figure 15-1 To us, various flowers simply appear yellow (left) and it is sometimes difficult to distinguish between two species with similar flowers. However, bees which pollinate the flowers always collect pollen from only one species at a time. They are able to do this because they can see ultraviolet light which we cannot see. If we photograph yellow flowers with ultraviolet-sensitive film (right), the flowers change sharply in appearance, indicating that a bee can distinguish differences we cannot perceive. At the top is Black-Eyed Susan (*Rudbeckia serotina*); at the bottom Marsh Marigold (*Caltha palustris*). (Courtesy of T. Eisner, Cornell University.)

pollen from flower to flower and so expediting seed formation; yet, these randomly evolved patterns fit together biologically, benefiting both organisms. Some plants have evolved very specific forms of flowers that assist pollination by insects. Bright colors and strong scents abound, but, in addition to the colors we see as visible light, bees can see patterns of ultraviolet light absorption that are invisible to us. If we look at flowers with an ultraviolet camera, we get a "bee's-eye view," in which striking new patterns emerge (Fig. 15-1). Glowing lines converging on the nectaries at the base of the petals can be seen, apparently guiding the bee right to the food source. But these ultraviolet markings also help the bee distinguish among numerous types of flowers which appear very similar in visible light; thus, the insect can collect nectar of the same type while accomplishing the intraspecific transfer of pollen.

Insect deterrents in plants

Having attracted an array of insects for pollination, the plant is then faced with the problem that many insects, especially in their larval or caterpillar stage, like to eat plant parts, especially leaves. Fortunately, this applies only peripherally to bees, which mainly collect nectar and pollen, but which occasionally destroy wood. This problem of insect deterrence is usually solved by the synthesis of a wide variety of "secondary metabolites" that make the plant unpalatable to insects. *Secondary metabolites* is the name given to the array of plant compounds that have no role in the primary metabolic processes such as respiration or construction of cell parts. We infer the absence of any role for these compounds in basic metabolism because neither all nor even most plants contain any specific compound. Secondary metabolite chemicals are often restricted to a single family of plants and, on occasion, to a single species or subspecies. Secondary metabolites number in the thousands; there are so many that we still do not know them all, and many organic chemists spend their professional lives characterizing "new" secondary metabolites. For a long time, biologists wondered about the role of such metabolites, but it now seems clear that they form the plants' defensive system against insects, disease, and in some cases, other plants as well.

How did such a defense system develop? It appears to have occurred by chance. Let us assume a plant mutation in a metabolic pathway appears, giving rise to a new compound. If that compound is poisonous or distasteful to predators, the plant stands a better chance of surviving than its better tasting relatives, which would be eaten by preference; thus, the gene-regulating synthesis of the compound stands a better chance of being passed on to the next generation. If all plants changed in this way, the insects would have nothing to live on—but insects also have an amazing ability to change. A chance insect mutation might appear, making the insect able to tolerate the poison, and so able to eat what the others cannot; it therefore survives and prospers where the others in its species do not do as well. Thus, the mutated insect will pass the resistant gene to a larger number of individuals in the next generation. This results in parallel evolution, not involving single individuals, as mentioned above, but rather populations of plants and insects changing together.

The types of compounds involved in the plant's defense against insects are varied. These substances may actually be poisonous, and kill any animal that eats the plant; therefore, unless the animal can detoxify the compound, it senses and tends to avoid the plant, which is then not eaten. Alternatively, compounds may make the plant distasteful and therefore deter an animal from feeding on it, perhaps completely. Numerous *isoprenoid* compounds, such as terpenes* and essential oils, probably fill this role. Alkaloids*, cyanogenetic glycosides*, and some "unusual" amino acids* in plants are compounds that may be toxic to any animal feeding on them. Cyanogenetic glycosides release hydrogen cyanide when the tissue is injured, thus forming an extremely effective deterrent. Toxic amino acids, prevalent in numerous tropical legume seeds, are lethal because they become incorporated into protein, substituting for the correct amino acid. The insects which can feed on plants containing a poisonous compound can do so because they have developed a means of breaking down the compound or chemically binding it in an inactive form. A bruchid beetle resistant to the toxic amino acid canavanine has another method of avoiding the poison. It feeds on seeds of *Diodea megacarpa* which contain 8% of canavanine, of which as little as 0.25% is usually lethal. It is immune to the poison because its arginyl–tRNA synthetase can discriminate between arginine and canavanine.

A further change may occur in which the substance, poisonous to other insects, is not only detoxified by the species adapted to feed on it, but becomes an attractant to that insect species; the insect then tends to be found only on that particular plant species and nowhere else (Fig. 15-2). Caterpillars of *Papilio ajax*, for example, feed only on the leaves of certain umbelliferous (wild carrot family) plants which contain essential oils of a definite composition, mainly methylchavicol* or the terpene carvone*. The caterpillars will take food only in the presence of these compounds; they will even eat filter paper treated with pure carvone!

One might wonder if poisonous plants gain anything once an insect becomes adapted to eat them. Occasionally they appear to gain more than they lose, since the insects feeding on them in a controlled, nonlethal way may accumulate the plant poison, thus producing a powerful smell that deters other herbivores as well as potential predators of the insects. This makes for a tidy symbiosis.

Insects are not the only animals feeding on plants; many mammals, for example, are herbivores and if they ate poisonous plants they would not

*Structural formula provided, pps. 400–401.

Figure 15-2 These two caterpillars are adapted to feeding on milkweed (*Asclepias syriaca*), which contains cardiac glycosides poisonous to vertebrates and non-adapted insects. While the milkweed tussock moth (*Euchaetias egle*) (a) can either tolerate or degrade these poisonous plant compounds, the monarch butterfly (*Danaus plexippus*) (b) actually stores the cyanogenic glycosides and uses them for its own protection against avian predators. (Courtesy of P. Feeny, Cornell University.)

survive for long. They avoid such plants because they receive a taste signal: that is, many poisonous plants do not taste good and are therefore rejected. While we can specify only what does not taste good to us, and cannot assume that a cow would make a similar judgment, plants that taste bitter to humans seem to correlate well with the plants rejected by many animals including cattle. Bitterness, in fact, seems to have a universally repellent character. *Astringency*, a drying sensation in the mouth caused by tannins*, also seems to be an important determinant of the food habits of wild animals, with the animals showing a preference for plants with little or no tannin. In fact, most bitter or astringent compounds are toxic, though certain animals may possess detoxification mechanisms for them. The diet of gorillas consists of only a few species, all of which tasted bitter on sampling by one human observer. The gorillas had become adapted to and then desired these tastes, much as we initially tolerate the mild bitterness in tea, or beer, or some of our culinary herbs and spices. Sweetness of taste or of odor, on the other hand, is attractive to animals. Compounds with such properties may play a role in various plant–animal interactions, such as seed dissemination by animals that eat succulent fruits.

Insecticides—universal poisons or selective assistants?

A wild plant population consists of a mixture of many species, each filling its own niche in the ecological fabric of the community. Each plant species has its diseases and predators but in a stable wild community, the hunted and the hunter are in balance. The limited food supply puts a limit on the population of the herbivorous insect or organism causing plant disease. In addition, the distance between communities of the plant species may be large, so that transfer of insects or other predators from one community to the next is limited. At the same time, each herbivorous insect has its own predators to keep its numbers in check. Monocultural agriculture, involving vast unbroken acreages devoted to a single crop, changes all that. Instead of only a few plants isolated within similar communities, an insect "pest" is faced with an essentially unlimited supply of food and a means of spreading over vast distances, because food is available all the way. The result is frequently an explosion of the insect population which cannot be kept down by natural means, because the predators of the herbivorous insect are unable to breed or feed at the same speed as the pest itself. The quickest and most convenient way to control such an infestation is through the use of chemical insecticides.

One may ask why modern agriculture has allowed this to happen. The answer lies in the rationalized, massive approach to agricultural productivity now in practice almost all over the world. In the past, low levels of agricultural pests were always present but seldom reached plague proportions because a mixture of crop species was being grown, there was no unbroken area devoted to one crop, and great distances were maintained between fields of the same crop. This has all changed with modern mechanization and transport—crops are now grown over vast acreages, to be harvested by massive combines, in areas best suited to that crop. Modern economics has forced monoculture agriculture upon us.

The development of organochlorine compounds such as DDT* during World War II provided an insecticide of great potency for use in agriculture as well as in public health. This compound acts on the nervous system of insects, destroying nervous coordination. Originally, it did not seem to have undesirable short-term side effects on man or domestic animals; however, certain environmental problems have arisen in the long term as a consequence of its use, and DDT is now banned in the United States. The difficulties arose on several fronts. First, DDT is nonselective, destroying useful insects, such as those involved in pollination and the predators of the pest insects, as well as the pests themselves. As we shall see below, such effects can cause the pesticide to "backfire" and actually increase the insect pest population. Second, resistant strains of pest insects appeared which were essentially unaffected by DDT; since their predators had already been destroyed by DDT, they were able to increase rapidly, so that the final problem was worse than before DDT use. The most important problem arose from the extreme persistence of DDT in the environment, since it is only moderately susceptible to microbial degradation. Its great stability, together with its higher solubility in lipids than in water, causes DDT to accumulate in membranes and fatty

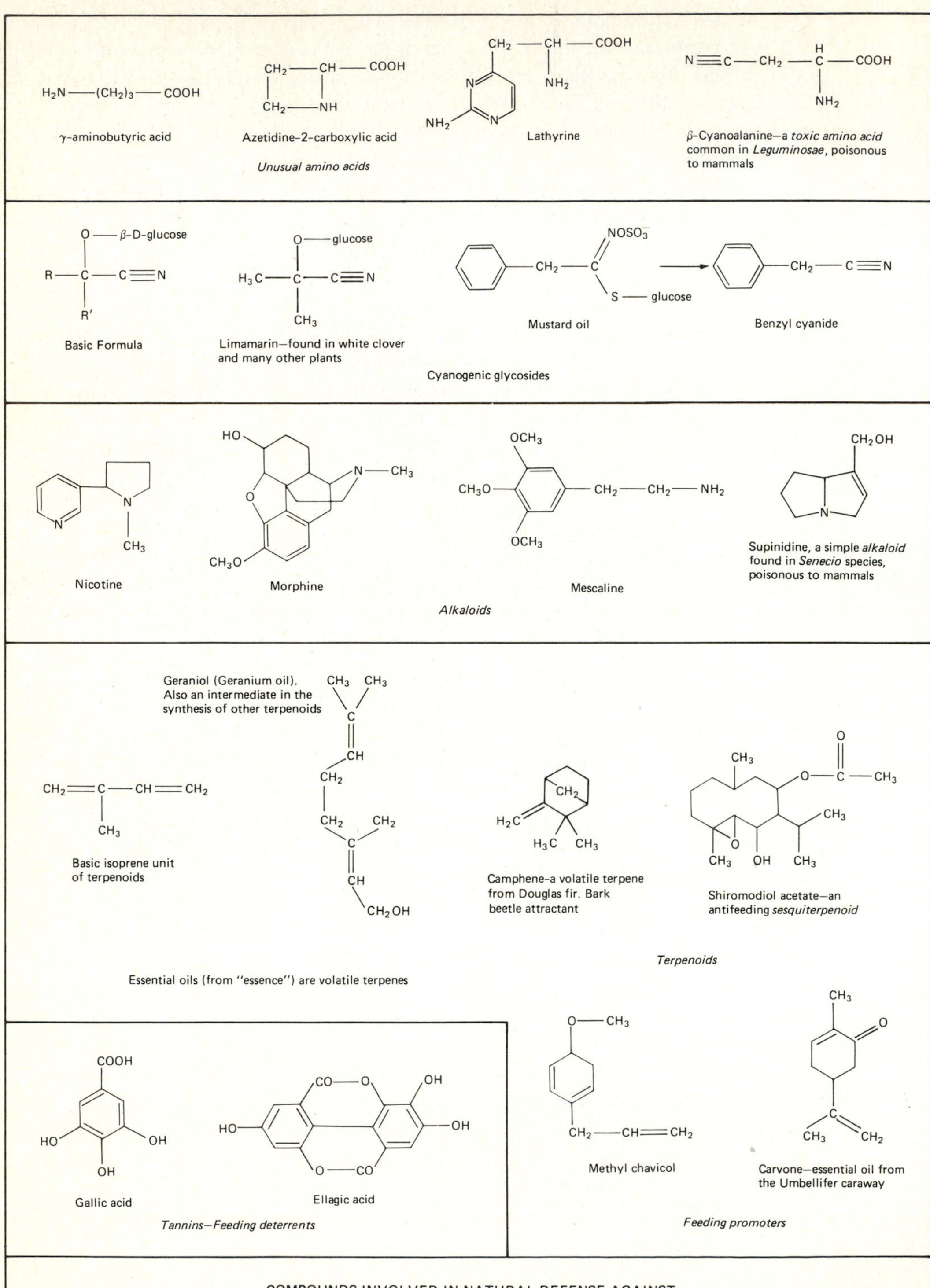

COMPOUNDS INVOLVED IN NATURAL DEFENSE AGAINST, AND ATTRACTION OF, INSECTS AND OTHER ANIMALS

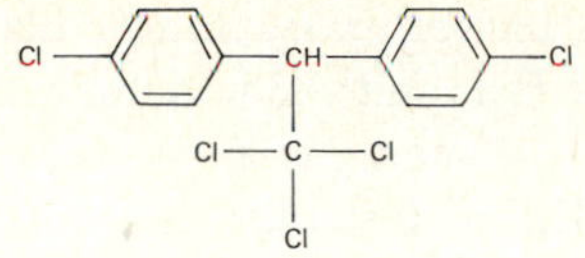

The synthetic insecticide *DDT*

INSECTICIDES DERIVED FROM PLANTS

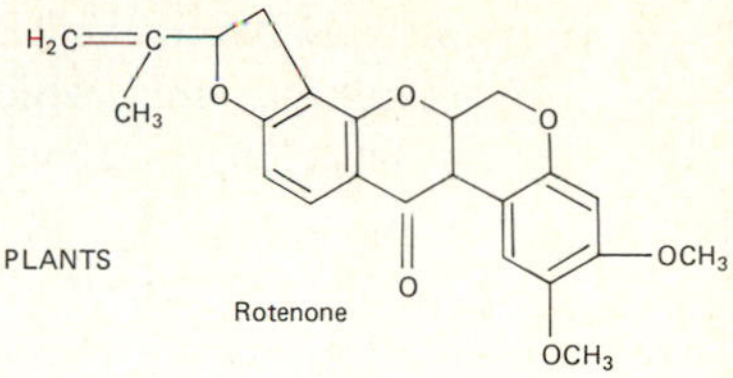

Rotenone

Pyrethrin I

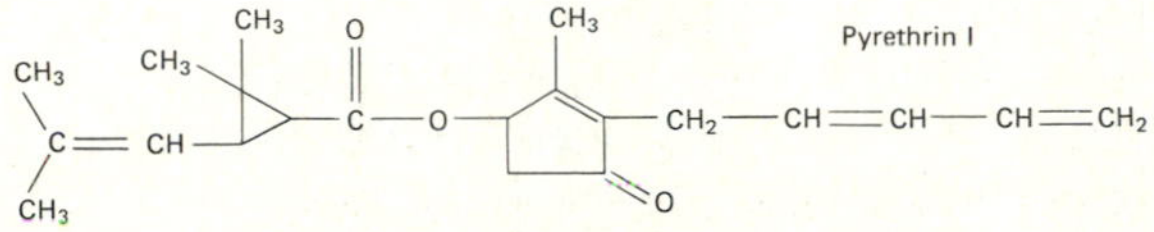

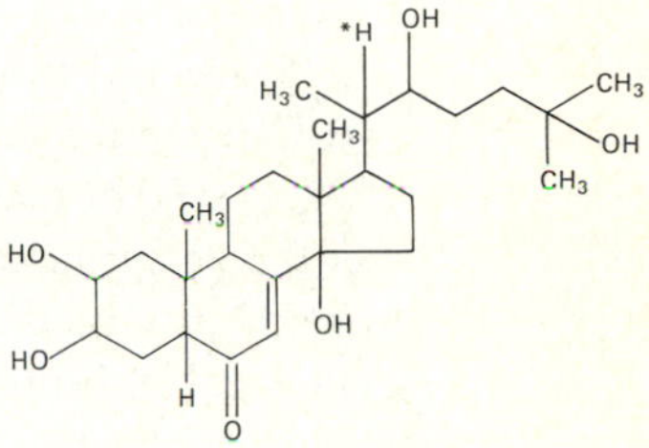

Ecdysone—the insect moulting hormone—also found in some ferns. With an —OH in place of —H at * the compound is hydroxyecdysone which is found in a wide range of plants

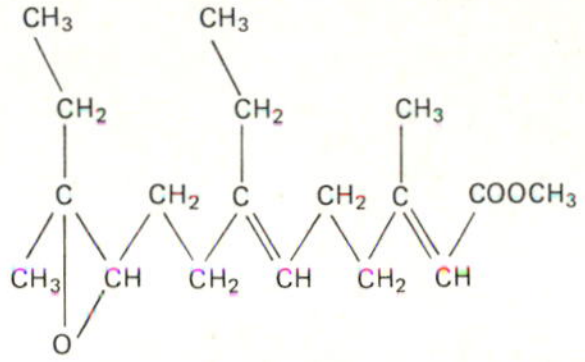

An insect juvenile hormone

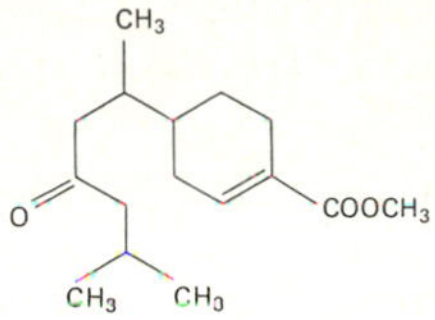

Juvabione—a *juvenile hormone* of insects found in conifers

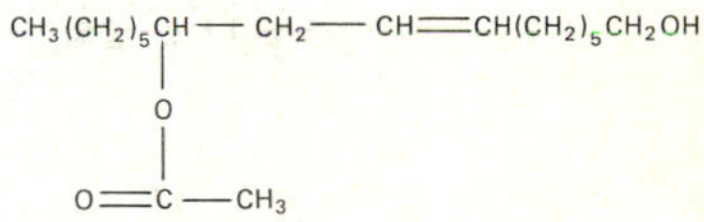

$CH_3(CH_2)_5CH(OCOCH_3)—CH_2—CH{=}CH(CH_2)_5CH_2OH$

The terpene *sex attractant* of the gypsy moth

NATURAL COMPOUNDS THAT WILL PROVIDE THE BASIS FOR NEW SELECTIVE INSECTICIDES

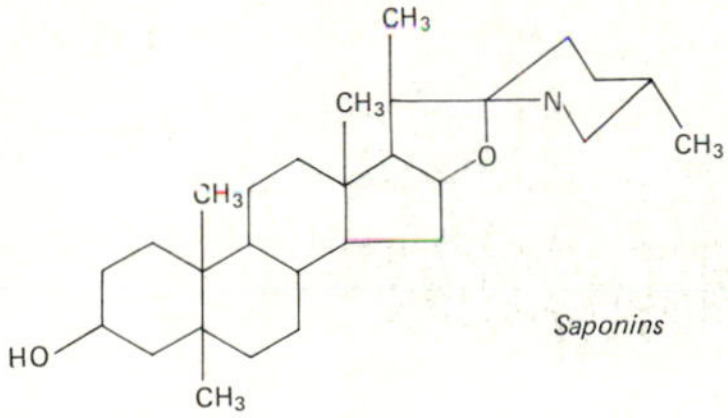

Saponins

The alkaloid α–solanine

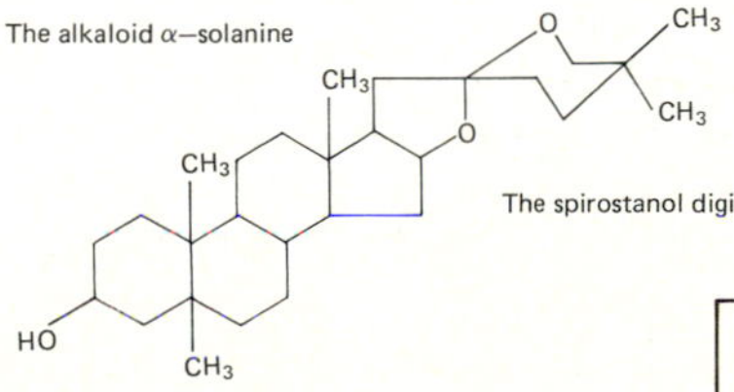

The spirostanol digitonin

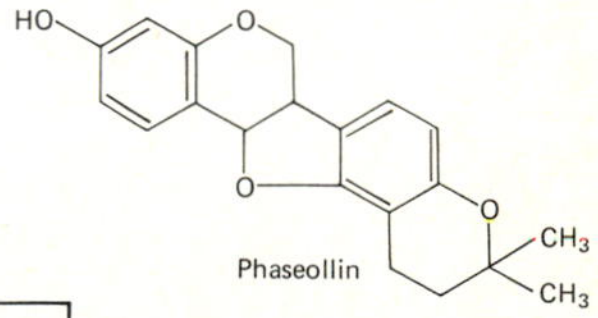

NATURAL ANTIFUNGAL COMPOUNDS

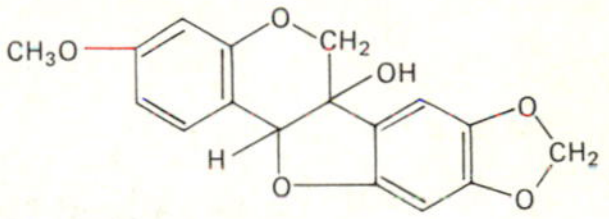

Phytoalexins

Pisatin

Phaseollin

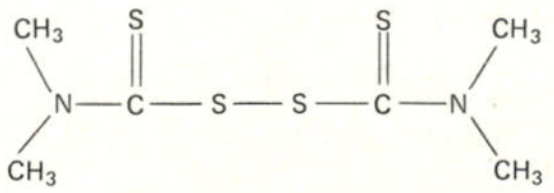

Thiram
A sulfur *fungicide*

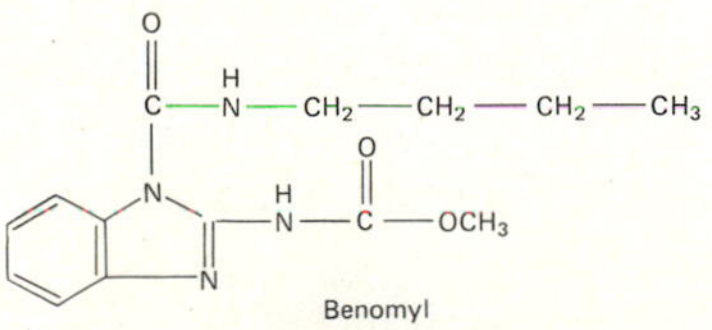

Benomyl
Systemic fungicide

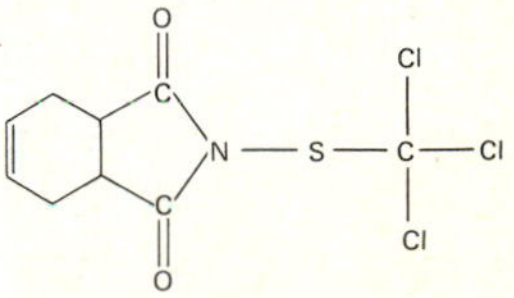

SYNTHETIC FUNGICIDES

Captan

deposits of living creatures, thereby becoming concentrated in the food chain. Thus, DDT washed into our rivers and lakes is picked up by minute planktonic organisms, which are then eaten by small fish, and these in their turn by larger fish, other predators, and ultimately by man. At each stage the larger organism eats many times its weight of the smaller ones as it grows, and at each step in the food chain, DDT is effectively retained and concentrated in the fat of the larger animal. So what may start as a part per billion of DDT in a planktonic organism may end up as several hundred parts per million in a predator high up the food chain.

Some fish are sufficiently sensitive to these organochlorine compounds that massive fish kills have occurred, particularly in the Mississippi river. The creatures that have suffered most are probably the predatory birds such as the peregrine falcon, eagle, and osprey. As they are at the top of the food chain, they pick up vast quantities of DDT. Among the side effects of the chemical in birds is an altered calcium metabolism that makes the eggshells thinner. The eggs of affected birds break during incubation in the nest and no offspring are produced. The sharp decline in the populations of predatory birds in America during the 1950s and 1960s can almost certainly be blamed on DDT and related compounds. Fortunately, since the banning of DDT in the United States, some of these bird populations have started to recover. DDT toxicity also has implications for plant populations, since birds and plants do interact ecologically in various ways, and DDT has also been reported to inhibit important plant processes, such as algal photosynthesis.

But the problem is more insidious than this, and is not limited to areas where DDT has been sprayed. DDT is a global contaminant because it is both stable and volatile. Much of the sprayed compound evaporates into the atmosphere and drifts around the world, so that DDT can now be detected even in the fat of penguins in the Antarctic, thousands of miles from the nearest point of DDT application. Although this is disquieting, some have proposed that we continue to use DDT because in producing its beneficial effects, it is so inexpensive and powerful. But in answer, one notes that man is simply one creature in the community of nature, and an extensive disruption of the earth's fragile ecological balance will probably hurt man more in the long run than it will help him. One especially alarming possibility is that DDT may not be harmless for man; the fact is that we do not yet know its long-term effects on humans. What will be the effects of 50 years of exposure, which have not yet elapsed? There is no telling what lies in the Pandora's box opened by such wide distribution of a chemical—will it turn out to be a harmless molecule or potent carcinogen, mutagen, or teratogen? On the general thesis that it is better to be safe than sorry, and since we have only one global ecosystem to play with, we must probably adopt additional measures to eradicate the use of this chemical-Godsend turned demon. A worldwide ban would probably be desirable, but many "Third World" nations are more interested in eradicating tsetse flies and malaria mosquitoes than in worrying about possible distant deleterious effects on environment and world health.

Natural deterrents and toxins

The production of sufficient food for the modern world of four billion people growing at a rate of about 1.8% per year necessitates intensive agricultural techniques, in which insecticides and herbicides play a vital role. But, once again, we can turn to nature. Plants have certain natural defense mechanisms, and insects have certain points of natural vulnerability; both can be exploited. Most of the work in this field is still in the research or developmental stage and as yet without practical application, but better new products will probably be in widespread use in the near future. In the meantime, other insecticides are being used which meet more closely, though still imperfectly, the properties of an ideal pesticide, which should (a) kill only the target organism; (b) be cheap, easily synthesized, and safely handled; (c) have no toxic effects on man, domestic animals, or environment; and (d) break down rapidly to nontoxic compounds. Many synthetic pesticides are not broken down because their chemical structure is not found in nature and thus cannot be degraded by microorganisms. The answer is therefore to use natural compounds as a model; both plants and insects provide a range of compounds that can be used to reduce the population of insect pests.

Antifeeding substances found in many plants have not yet proven to be of value because of the incomplete and transitory protection provided by spraying such substances onto the surfaces of plants. They are effective only if the insect meets the chemicals as endogenous constituents of the plant. Despite these problems, two compounds from plants, pyrethrin* (a monoterpene) and rotenone* (a flavonoid) have been utilized as insecticides. These substances actually kill the insects onto which they are sprayed, and can be extracted from plants in sufficient quantities to be marketed commercially as insecticides. Pyrethrin is extracted from the daisy, *Pyrethrum*, grown as a crop for the insecticide in East Africa. As both pyrethrin and rotenone are natural compounds, they meet one of the requirements for an ideal insecticide, in that they can be broken down by bacteria and are not persistent in the environment. Pyrethrin, unfortunately, is also unstable to ultraviolet light, so that it may be broken down before it has had its effect; one notable recent advance is the development of synthetic derivatives of pyrethrin which are stable to ultraviolet light and thus more potent as insecticides. Luckily, bacteria can still recognize and degrade these pyrethrin derivatives, so we have partly achieved our goal of the ideal insecticide. These compounds are still nonselective among different insect species, and thus leave some room for further development.

The best approach to fabrication of a completely successful insecticide probably lies in the life cycle of the insect itself. A chemical poisonous to an insect is probably poisonous to man, because both have a similar basic metabolism. However, the developmental control systems of insects vary widely from man's, so these form a point of attack. Do particular compounds control stages of the life cycle specific to insects, and are they specific to only one insect species? Three lines of attack are currently being developed in the approach to this problem. Insects have hormones that control their developmental stages, the larval stage being maintained by a juvenile hormone*. The adult cannot develop until the level of juvenile hormone drops. The weak link in the insect cycle of metamorphosis is that it cannot reproduce, no matter what its age, unless it has developed into an adult. Some plants have compounds that mimic insect juvenile hormones, a fact which first came to light when it was found that paper towels from almost any paper of American origin could prevent the maturation of the European bug *Pyrrhocoris apterus* reared on them. This led to the discovery of the effective compound juvabione* in conifers, particularly in the balsam fir, *Abies balsamea*. Such a compound may well prevent a range of insects from infesting such plants. If we could spray juvenile hormone onto insects, their reproduction might be prevented. No succeeding generations could appear in the absence of adults; thus with the death of the sprayed generation of larvae, the pest would have been wiped out or at least severely reduced if 100% contact had not been achieved.

Another point of attack represents the opposite of the first. Two chemicals have been found in the common bedding plant *Ageratum* which stop the secretion of the juvenile hormone, causing the premature development of immature larvae into tiny sterile adults. The extraction or synthesis of these substances would provide a safe insecticide, as they act on the juvenile hormone system specific to insects. In addition they have an advantage over the use of the juvenile hormone mentioned above in that they transform voracious larvae into nonfeeding adults, thus saving the crop. While we cannot expect an absolutely unique selectivity of such insecticides, some degree of selectivity to a particular group of insects is quite likely to be found. Another distinct advantage of such compounds is that just as an insect would not be expected to develop a resistance to its own hormones, neither would it be likely to do so to such compounds sprayed to control the insect population. A further effect of the *Ageratum* substances is that they also induce a hibernationlike condition in the insects. If a crop is sprayed with these chemicals, by the time the effect wears off the insect, the crop would be harvested and the insect killed by winter. While such insect control systems represent a new venture for man, they almost certainly represent an old strategy for plants, which have always had to deal with insect depredations.

Other insect-modifying chemicals have been found among the secondary metabolites of plants. A whole range of "*phytoecdysones*" which mimic the ecdysone molting hormones of insects occur in a wide variety of plants. These steroid derivatives* were discovered only recently in quite high concentration in plants—up to 2% of the dry weight in rhizomes of the fern *Polypodium*, for example. One species often contains several different related compounds. The effectiveness of this group of compounds as a defense against predators is questionable, since when fed to insects they produced no discernible effect. It has been suggested that in the presence of certain other natural compounds they might be able to penetrate the insect cuticle, but a natural role for phytoecdysones in plant defense remains to be demonstrated.

Nutritionally interactive defense mechanisms

The wide range of similar compounds found in plants and insects indicates that many insects rely on plants to provide the chemical skeletons of many molecules which they cannot themselves synthesize. Insects depend on the isoprenoid skeleton from plants and use these plant substances, after modification, in producing their own hormones and in fabricating substances for their defense. An interesting case is that of the relationship between the Douglas-fir tree (*Pseudotsuga taxifolia*) and the Douglas-fir beetle (*Dendroctonus pseudotsugae*). Damaged Douglas-firs release volatile terpenes which attract a "pioneer female," which in turn secretes an attractant that brings other males and females from afar. While that does not act as a feeding deterrent with respect to the Douglas-fir, it does indicate a potentially most useful tool in man's arsenal of weapons against insect pests. The females of many insects secrete *pheromones*, or sex attractants*, a few molecules of which can attract males from many miles distant. Each species has its own different pheromone, of which more than 50 have now been identified and synthesized in the laboratory. Such compounds can be used to lure all the males from a large area into a trap where they are subsequently destroyed. Alternatively, a general spray could produce mass confusion in the males so that they could not locate the females, thus preventing reproduction of the population.

Another type of chemical signal emitted by insects is a warning signal. An attacked aphid, for example, releases an alarm substance which causes neighboring aphids to drop to the ground. Such a compound, or simple analogs, could be used as a permanent repellant. Only minute amounts of pheromones are needed; in some cases spraying is not required and, since insect pheromones are species specific, only the targeted species is affected. There would also be no effect on predators, which are then able to take care of any remaining members of the pest species. This kind of biological control is another weapon in the war against insect pests, and while currently only of significance in isolated cases it is likely to come into widespread use in the future. The most important aspect of selective insecticides is that they permit natural biological control to continue without interruption. (For an informative account of the dangers of chlorinated hydrocarbon pesticides and alternative control methods, the reader is referred to Rachael Carson's excellent account, *Silent Spring*, and to Frank Graham's sequel, *Since Silent Spring*.)

Fungi, Plant Disease, and Disease Resistance

Most plant diseases are caused by parasitic fungi. To start the infection, a spore of one of these minute, filamentous, unpigmented plantlike organisms must land and germinate on the plant surface (Fig. 15-3). The resulting *hypha* penetrates into the plant cells from which the heterotrophic fungus obtains its nourishment. At the same time it produces the symptoms of the disease, because of either the resulting cellular damage or the production of fungal toxins. Plants possess various defenses against such invasion; the fact that most fungi can infect only one species or a group of related species indicates that other types of plants must somehow be resistant to their attack. Even among susceptible species, there is often some degree of resistance in certain individuals, a fact used by plant breeders in the selection and breeding techniques used to improve the disease resistance of crop species.

The resistance of plants to disease takes numerous forms, of which structural resistance is the most common. If the plant has a thick cuticle through which a germinating fungal spore cannot penetrate, the plant is well on its way to preventing the disease from starting. Included in the plant's second line of defense are various chemical compounds, again belonging to the group of secondary metabolites; these also provide resistance to insects and other animals. For convenience, researchers have divided these chemicals

Figure 15-3 Scanning electron micrograph (X5000) of a germinated spore (conidium) of mildew (*Erisyphe graminis hordei*) on the epidermis of a leaf of barley (*Hordeum vulgare*). The furrows are the outlines of the epidermal cells. The upper germ tube attaches itself firmly to the surface in the trough formed by the wall between two epidermal cells. The fungus then penetrates into the epidermal cell located behind the tip of the germ tube. The plant cell responds by secreting materials on the inside of the cell wall that act to oppose the infection. If the infection succeeds, the fungal hyphae ramify over the surface of the leaf, drawing nutrition from the leaf cells and producing the typical fungus-covered leaves of a mildew infection. (Courtesy of H. Kunoh, Cornell University.)

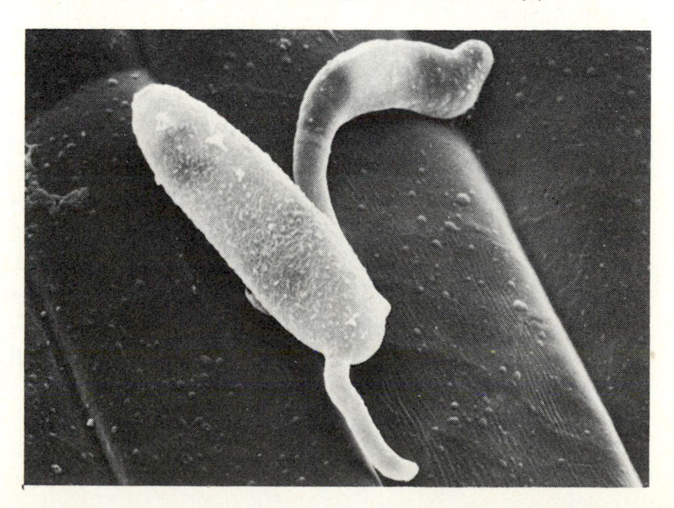

into two groups: those that are present in the plant prior to the infection and those that are produced only in response to the fungal infection. There is some degree of overlap in these categories, but compounds that are present in very low concentrations prior to infection, increasing markedly on infection, are usually included in the second group.

Preformed compounds

Many preformed substances in the plant seem to provide some resistance to fungi. Probably the most important single group are chemicals called *saponins**, including alkaloids and triterpenes with a sugar chain attached in such a way as to make them water soluble. Many defensive substances are in fact *glycosides* of this type; if the sugar is removed, water solubility is lost. These compounds possess pronounced antifungal properties and can also produce marked changes in the permeability and specific transport property of cell membranes. These two properties seem to be connected, since saponins exist in the vacuole, and on invasion of the cell by a fungus the cell membranes are damaged, releasing the saponin. Often the saponin may be stored in an inactive form, and is activated only when it comes into contact with an enzyme in the cytoplasm. If the membrane of the fungus contains the enzyme β-glycosidase, the saponin loses the sugar part of the molecule, thereby losing water solubility and dissolving in the lipid part of the cell membrane. There it attacks certain sterols of the membrane, causing them to change from a liquid to a more solid state—this damages the fungal cell membrane and kills the fungal cell. Not all fungi are susceptible, however; some lack sensitive sterols in their membrane while others can inactivate the saponin and are therefore resistant to this defense.

Since plants possess defenses such as the saponins, how is it that they may still be susceptible to disease? We must emphasize that, by definition, these mechanisms are not effective against actual pathogens, but they can and do prevent yet other organisms from becoming pathogens. Thus, each plant is attacked by some specific pathogenic organisms, but is resistant to others that may be found on other species. Preformed defensive substances affect mainly the fungi which severely damage the host cells and release or activate their defensive substances. Successful parasites often manage to invade the cell with *haustoria*, small hyphal processes that penetrate the cell membrane and absorb food without causing much cellular disruption, thus avoiding the triggering of defensive reactions. Alternatively, as mentioned above, they are either resistant to or can degrade any antifungal compounds, so that the infection progresses unchecked.

Phytoalexins

Phytoalexins are antifungal compounds formed in plants in response to infection by a fungus. The name *phytoalexin*, which is Greek for "a warding-off compound," was coined when it was incorrectly thought that such compounds represented a plant immunity response equivalent to that of animals. These compounds do not, however, represent a specific response to a fungal invasion, but rather a general response to any form of injury.

Several compounds of this type have now been isolated from a wide range of plants and appear to differ greatly in chemical structure. Each appears to be specific for the plant in which it is produced, and the pathway of production depends on the phytoalexin under consideration. Pisatin*, produced in peas, requires little change in the metabolism of the tissue since closely related isoflavonoids of similar structure are widespread in the *Leguminosae*, the family to which peas belong. Pisatin appears to be formed from the amino acid phenylalanine, which is initially deaminated by the enzyme phenylalanine ammonia lyase (PAL) to cinnamic acid, a pathway common to all flavonoids (Fig. 15-4).

Whether a phytoalexin confers resistance depends on many factors. The speed with which the plant recognizes the infective agent and responds is all important. Often phytoalexin production in an infection occurs so late that the fungus has already spread to other parts of the tissue; alternatively, a fungus may be able to degrade the phytoalexin and so spread. On broad beans (*Vicia faba*), the fungus *Botrytis cinerea* is susceptible to the phytoalexin and is limited to the infection site, producing small brown spots on the leaf, giving to the disease its name—"chocolate spot"; but the related fungus *Botrytis fabae* is able to destroy the phytoalexin and so spreads throughout the leaf.

The interrelationship between plant host, fungal disease, and phytoalexin production is well illustrated by the anthracnose disease of bush beans (*Phaseolus vulgaris*) caused by *Colletotrichum lindemuthianum*. In this disease, no relationship can be found between resistance or susceptibility of different bean varieties and their production of high levels of the phytoalexin *phaseollin** within 24 hours following artificial injury. In resistant beans, the immediate response to fungal penetration is death of the host cell and failure of the fungal germ tube to develop further beneath the spherical *appressorium* on the surface of the cell. In susceptible cells, the germ tube grows through the cell wall and then continues between the wall and the protoplast without injuring the cell. The hyphal growth continues for several days, passing from

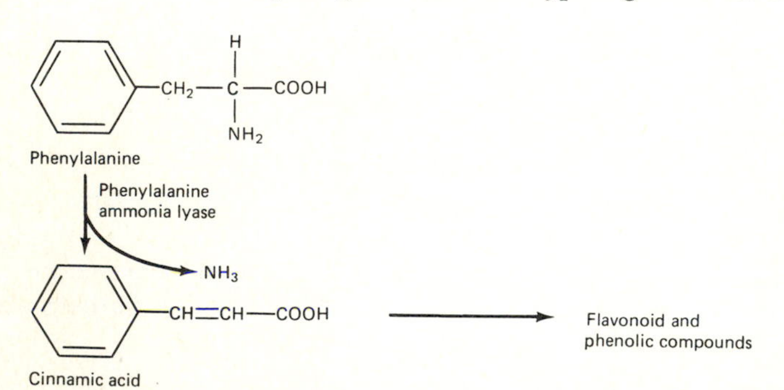

Figure 15-4 Phenylalanine is the main precursor in the formation of flavonoids and phenolic compounds, many of which are associated with disease resistance.

cell to cell with no signs of damage. Browning, necrosis, and tissue collapse then occur suddenly, producing the typical symptoms of anthracnose disease. In the resistant plant, rapid phaseollin production occurs on the second day, coincident with hypersensitive browning. In the susceptible plant, however, no phaseollin is formed until the anthracnose lesions are produced, by which time it is too late! We see here that while the phytoalexin probably has a role in preventing the disease, the essential difference between resistant and susceptible varieties lies in their ability to recognize the presence of the fungus and to react to it. Conversely, the specificity of a disease may depend on the ability of a particular race of pathogen to avoid provoking, or to suppress processes leading to, cellular damage, and thus the production of a phytoalexin. *Hypersensitivity* may therefore be a major response through which a plant can resist a fungal invasion. The response involves a rapid death of the plant tissue infected, and often also surrounding tissue, so that the fungus is isolated among dead cells and unable to spread the infection to other cells of the plant. Research in this exciting and interesting area is lively at present.

Protection of plants from disease

One of the accomplishments of successful agriculture is the minimizing or prevention of plant diseases. As with animal pests, the treatment can take the form of avoidance or counteraction. Avoidance, if it can be achieved, is usually cheaper and more successful. Various cultural practices can also help prevent disease; obvious ones include the destruction of any infected material and the mixing of crop species, in both time and space, to minimize the survival and spread of a disease organism through the crop. Some diseases are spread by particular organisms (*vectors*), in which case control of the vector is also desirable.

FUNGICIDES Fungicides can be applied either as external protective coatings or as internal, systemic treatments. Fungi such as mildew that grow on the outside of plants are vulnerable to treatment with external fungicides and are therefore somewhat easier to treat than fungi growing inside the plant tissue. Since fungi are biologically further removed from humans than are insects, fungitoxic compounds are less likely to be poisonous to man than are insecticides, but care must still be exercised. Fungicides can be broken down into three basic classes: First, there are the old inorganic compounds, which have been largely superseded, but are still useful. One example, sulfur dust, is quite effective against some surface fungi; Bordeaux mixture, a mixture of calcium hydroxide and copper sulfate, is also widely used. On mixing, a reaction takes place between these two compounds so that the copper, the active ingredient, becomes sufficiently "available" to prevent the growth of the fungus but not of the plant. Devised and originally used to prevent the stealing of wine grapes, Bordeaux mixture was found to prevent mildew and so has been used as a fungicide ever since. The second group of fungicides is composed of organic compounds that act at the surface of the plant but are more effective than the inorganic compounds. One of the most important, versatile, and widely used groups are organic compounds containing sulfur*. Another very effective compound is captan*, widely used as a seed dressing to prevent infection of seedlings with seed-borne diseases.

A most significant advance in the field of plant protection is the development of the third type of protectant, the *systemic* fungicide or insecticide. These compounds enter and are transported throughout the plant via the xylem or the phloem, so that they are quickly able to combat a fungal or insect invasion in various tissues. Systemics must have a chemical structure that allows both penetration into the plant and movement within the transpiration or food-transport systems; basis for both lipid and water solubility is therefore required. In addition they must be relatively stable in the plant, otherwise no protection would be obtained.

*Benomyl**, one of the most successful of the new systemic fungicides, has a wide spectrum of fungitoxic activity. Applications are best made to the leaves; soil applications provide a continuous supply to the roots, but require much larger quantities; bark injections have had some success in controlling Dutch elm disease, though such treatment is not a certain cure. Benomyl is transported in the xylem, so there is not much movement from sprayed mature leaves to new leaves. This is a distinct disadvantage, as it necessitates respraying new growth. It also means, however, that no movement occurs to the fruit, so that fungicide-free fruit can be harvested from fruit trees sprayed prior to fruit development. Benomyl acts by interfering with fungal cell replication, through the prevention of mitotic spindle formation. Though Benomyl would seem to be an almost ideal fungicide, trouble has been appearing in the form of resistance in the pathogenic fungi. The fungicide acts at a point in the cell cycle where a single gene mutation is sufficient to confer resistance, and this has been happening in an increasing number of cases. Thus the practical role of Benomyl will most probably be short lived, and the continuing development of new compounds to protect crops from disease will be needed.

BACTERIA AND VIRUSES Some plant diseases are caused by bacteria and viruses. The defenses of the plant against bacteria are quite similar to those against fungi, but the resistance of plants to viruses (which are not cellular organisms) relies on the resistance of the cell to the take-over of its nucleic acid replication machinery by the virus. In agriculture, protection against viral diseases is achieved in part by isolation from the source of infection and in part by spraying aginst insect vectors.

In horticulture, cures from plant viral infection can be achieved in two ways. Heat treatments of the plant tissue, ranging from prolonged periods at 35°C for growing plants to 30 minutes at 55°C for dormant tissues, can achieve destruction of the virus without drastically harming the plant tissue, though often the margin of safety is rather slim. Meristem cultures (see Chapter 14) are gentler, but effective only with certain species. Shoot tips of plants such as chrysanthemum can be excised and grown rapidly in culture. In this situation, the plant tissue grows more rapidly than the virus can replicate so that the extreme apex eventually becomes free of the virus. The apex is then excised and grown into a whole plant free of virus.

Chemotherapy against plant viruses is in its infancy, although one notable success has been achieved. Western yellows virus of lettuce causes the lettuce leaves to yellow and decay; this can decimate the crop. The systemic chemical *carbendazym*, applied to the soil and taken up by the roots, prevents the virus from breaking down the chlorophyll; thus, even though the lettuce has the virus in its cells, the leaves appear perfectly healthy and the lettuce crop is fully marketable. Perhaps we can look forward to similar successes with other viruses in the future.

HIGHER PLANT PARASITES One does not normally think of higher plant parasites as being very important, but serious infestations can tap the plant's water and food conducting systems to the point where serious depletion of nutrients occurs. Mistletoe, the common winter festival plant (*Phoradendron flavescens*), is in fact a serious parasite of oak and conifer trees and can considerably reduce the rate of tree growth. Dodder, a vinelike parasite (*Cuscuta gronovii*), can be a serious pest, especially in warmer climates (Fig. 15-5). Two genera of parasitic plants attack the roots of plants:
Striga is found on semitropical cereal crops such as maize and sorghum, while *Orobanche* infects many kinds of plants, including legumes, tomato, and tobacco. In parts of the world *Striga* is a serious agricultural problem. Low yields of maize in northern Africa, usually blamed on poor soil, are often due to *Striga* infection, and in certain parts of the United States, especially the mid-Atlantic region, shipments of corn are quarantined because of this root parasite. *Striga* is not completely parasitic; it has small green leaves usually visible around the base of infected maize plants, and can perform some photosynthesis, but is still a heavy drain on its host plant. Herbicides have not yet proved effective in controlling *Striga*. The main potential attack

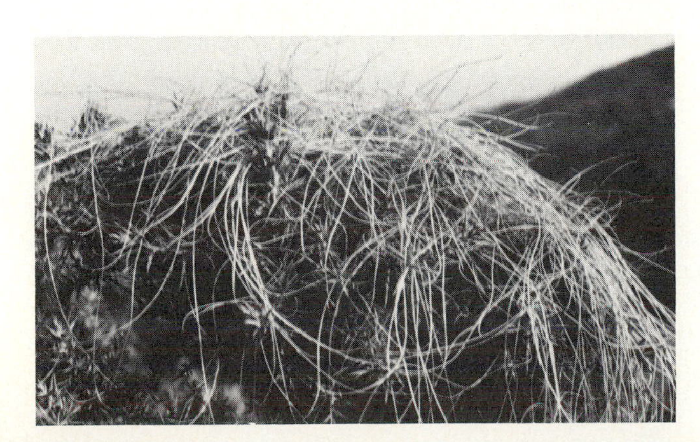

Figure 15-5 Dodder, (*Cuscuta gronovii*) a higher plant parasite, is seen here enveloping a gorse bush, (*Ulex europaeus*).

point in its life history is seed germination: *Striga* seeds lie dormant for long periods in the soil, and are stimulated to germinate only by certain compounds produced by the roots of host species plants. The chemical nature of the compounds stimulating germination has recently been elucidated. The effective compound, *strigol*, bears a faint structural resemblance to the gibberellins. Synthesis and artificial dispersal of this substance might control the disease by promoting *Striga* germination in the absence of corn plants; the young parasitic seedlings could not survive without additional nutrition from a host, and would thus probably die. Application of such germination promoters would probably be necessary over several years to clear the soil of *Striga* seeds, since it is unlikely that all the *Striga* seed would germinate at the same time.

Competition with Other Plants in the Community

Most plants grow in communities consisting of many individuals of the same or different species. If the basic resources available to each community—space, nutrients, water, and light—are limited, each plant will compete with its neighbors for these resources. Those plants that can grow rapidly in the early stages usually capture a major share of the resources and outgrow their neighbors, becoming dominant. In agriculture we almost always plan to grow crops in monoculture communities—only one species in the community is wanted, the others being weeds. (A *weed* is simply defined as a plant unwanted in a particular situation.) Unhappily, the naturally adapted weeds can often grow faster than the crop plants, because the latter have been selected for their high-yielding capacity, not for their competitive ability. Thus crop plants frequently need help against their competitor weeds. In previous times, and at present in small-scale gardens, weed removal is accomplished by hand or by hoeing. With vast acreages devoted to monoculture, and with limited and expensive work forces, such manual methods are not practical in Western agriculture; agronomists therefore resort to herbicides or weedkillers. Even if the weeds are removed, competition between the individual plants of the crop species still exists. This is relatively unimportant, since we are concerned with yield per acre, not per plant. If each plant must reach a minimum size for it to produce, as with pineapple or sugar beet, then some form of controlled transplanting or thinning is still required, though this can often be done easily by machine.

Phytotoxins

As we have discovered before, nature has frequently preceded man in the matter of inventions, and herbicides are no exception. Many plants produce their own "herbicides," termed *phytotoxins*, which inhibit the growth of their competitors. The negative effect exerted by one plant on another through

(a)

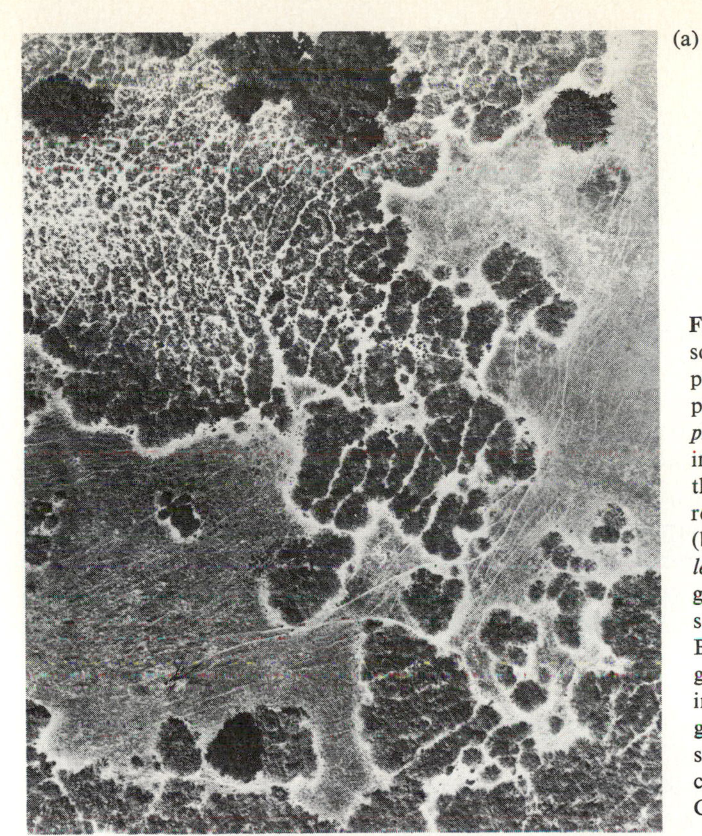

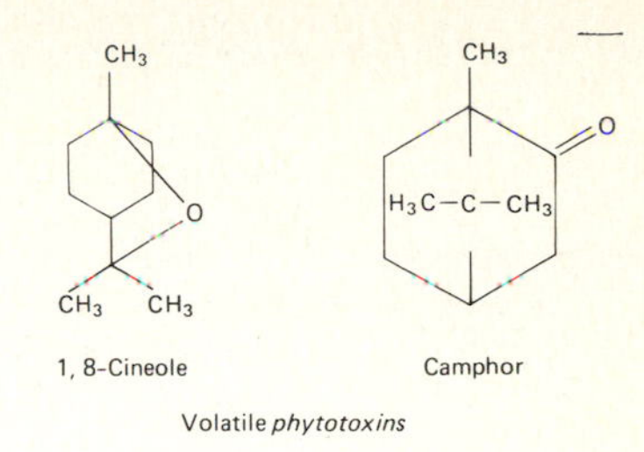

Figure 15-6 (a) The inhibitory effect of some shrubs on the growth of other plants can be clearly seen in this aerial photograph of intermixed *Salvia leucophylla* and *Artemisia californica* invading annual grassland in California. Note the light shaded areas of bare soil surrounding the darker areas of the shrubs. (b) The phytotoxic effect of *Salvia leucophylla* on the surrounding annual grasses can be seen clearly by the bare soil surrounding the shrubs (A to B). Beyond the bare zone (B to C) the growth of the grasses is still clearly inhibited and normal growth of the grasses (right of C) is not seen until several feet from the *Salvia*. (a and b courtesy of C. H. Muller, University of California, Santa Barbara.)

(b)

chemicals released into the environment is termed *allelopathy*. Most examples of allelopathy occur in semiarid climates, since abundant rainfall washes away water-soluble phytotoxic compounds and also facilitates their microbial degradation. The evidence for the presence of an allelopathic condition is the absence of competitive plants in the vicinity of one type of plant. Frequently, the plant producing the phytotoxin is surrounded by bare soil, even though water, light, and nutrients are available in the bare area, and other species can be found growing at some distance from the isolated species (Fig. 15-6).

Phytotoxins may be either volatile or water-borne. We will use an example of each to illustrate how they work. *Salvia leucophylla* (sage) is a small shrub that invades California grassland. The center of the shrub thicket is devoid of herbaceous weeds, as is a zone 1–2 meters beyond the shrubs. For several meters beyond this, only selected herbs show stunted growth. Studies showed that the volatile terpenes *cineole** and *camphor** are found in the atmosphere adjacent to the branches and in the surrounding soil; these compounds are most toxic to the germination of seeds of herbaceous species. The terpenes are released by the *Salvia* late in spring when the rains have virtually ceased. They are adsorbed onto soil particles and remain in the soil over the long dry summer. On the resumption of the winter rains, when the herbaceous seeds start to germinate, the terpenes enter the seedling roots by dissolving in the waxy surface cutin. The phytotoxins then suppress growth sufficiently so that the seedlings cannot survive under the periodic drought stress typical of the region. After several weeks of rains the terpenes can no longer be detected in the soil, but further production at the end of the rainy season again restores the level of the phytotoxins, thereby ensuring the continuing dominance of the *Salvia*. *Salvia* roots also form a dense mass near the soil surface, effectively absorbing all available water, and making the establishment of other plants difficult.

A wide variety of chemical toxins make up the water-borne phytotoxins. The California "chaparral" consists of shrub vegetation in the dry mountain foothills, with *Adenostoma fasciculatum* dominant and an almost complete lack of herbs between the shrubs. Here phytotoxins accumulate in soil neither from the leaf litter on the ground, nor from the roots, but by rain drip from the leaves. The toxins accumulate on the leaves in dry periods and are rapidly washed off by very little rain or even mist. Seed germination and growth of herbaceous plants is, as a result, inhibited in the area surrounding the plants. When the leached material was examined it was found to contain 9 phenolic compounds with phytotoxic properties. The toxicity in the soil is of short duration if the shrubs are removed, so the shrub stands of *Adenostoma* must depend on a periodic washing of their leaves to renew their allelopathic dominance. The toxins of some plants are more persistent, however. *Arctostaphylos glandulosa* produces a persistent phytotoxin that accumulates both in its roots and in the leaf litter under the plants, while *juglone*, a quinone-type material produced by certain oaks, severely limits competing vegetation

*Formulae p. 413.

around large trees. Though natural phytotoxins have received little attention as potential herbicides in the past, such compounds might well prove valuable in providing us with some natural safe herbicides in the future.

Herbicides

A herbicide is a chemical that kills plants. Although some general plant poisons find use in clearing paths, railroad tracks, or areas under power lines, some degree of selective herbicidal activity is usually needed in agriculture. The herbicidal value of a compound depends on its toxicity to the target weeds, its degree of selectivity, the absence of unwanted ecological or public-health side effects, and ultimate biological degradability, if it is an organic molecule. Sulfuric acid was among the earliest of herbicides; it could certainly kill plants by scorching their leaves and even had a degree of selectivity, because it bounced off cereal leaves but would kill broadleaved weeds. It suffered, however, from obvious disadvantages related to its corrosive properties. During the late 1930's research workers began searching for chemicals that would selectively regulate plant growth. Their labors, many carried on in secret during the World War II, led to the discovery of the chlorinated phenoxyacetic acids, 2,4-D* and MCPA. These compounds, for reasons still largely unexplained, have the desirable property of toxicity to broad-leaved plants only (Fig. 15-7); thus, following the war, their production for use as

Figure 15-7 Auxin-type herbicides have the ability to kill dicotyledonous plants while having no effect on cereals. The photograph shows the effect of 2,4-D on (left to right) wheat, charlock (*Sinapis arvensis*) and groundsel (*Senecio vulgaris*). (Photo courtesy of the Agricultural Research Council, Weed Research Organization, Oxford, England.)

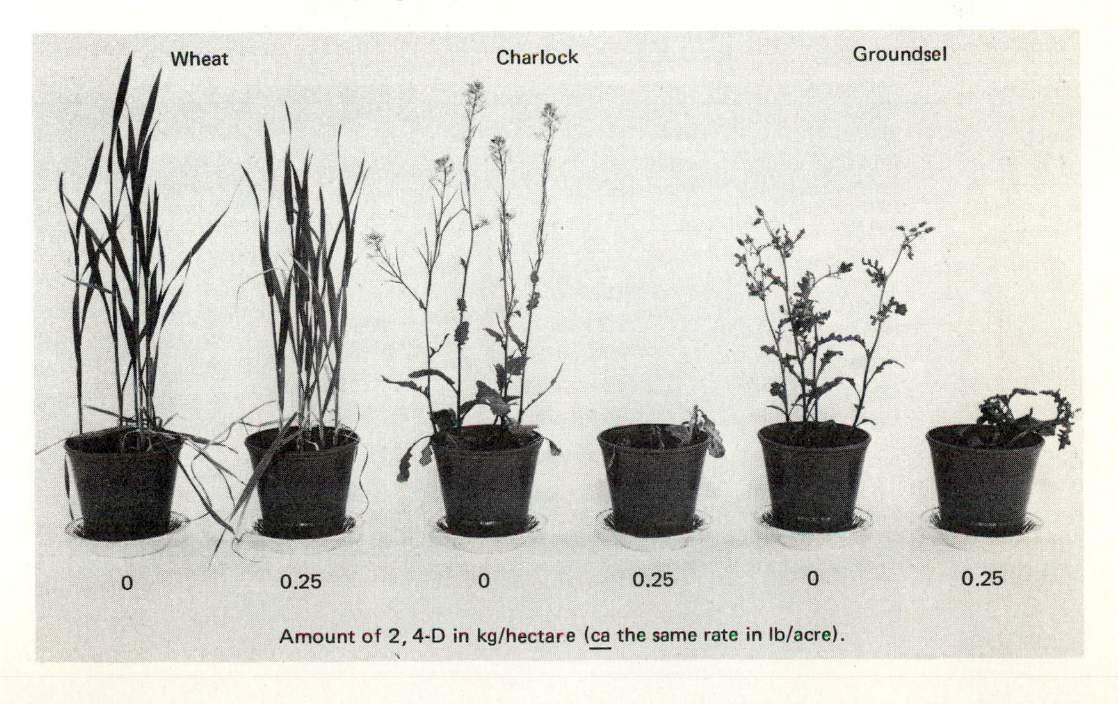

selective herbicides commenced. Since that time a vast collection of chemical compounds has been developed for use in agriculture. Many of these compounds are for specific uses, especially for controlling particular weeds in food crops. So versatile have the chemicals become that one can now even manage to control various species of annual and perennial grass weeds in cereal crops, which are themselves grasses (Fig. 15-8). The wide array of compounds is too vast to cover here; we will examine only some of the principles on which the use of a herbicide relies.

Herbicides must manifest their plant toxicity at low concentrations, at which they can affect numerous reactions in plant cells. Some herbicides such as *diuron* are inhibitors of photosynthetic electron transport, and kill plant cells quickly because they favor the accumulation, in the light, of highly oxidizing substances resulting from the photolysis of water. These products in turn promote the destruction of the integrity of the cell membrane. Growth-regulatory compounds such as 2,4-D are structurally related to auxin and at low concentrations they can indeed function as auxins. At higher concentrations they cause distortions and imbalance in the growth of the plant, frequently resulting in its death. Even though 2,4-D has been known for over 35 years, we do not yet know how it works. At the very least, it affects membrane-localized proton pumps, as well as nucleic acid and protein synthesis

Figure 15-8 While the selectivity of the older auxin-type herbicides was only between widely dissimilar plants, some modern herbicides have the ability to remove a grass weed from a cereal, itself a grass. Here barban [4-chlorobut-2ynyl N-(3-chlorophenyl) carbamate] sprayed at 0.35 kg/ha [hectare (ha) $\approx$ 2.47 acres] onto the soil on the left of the photograph has removed wild oat (*Avena fatua*) from a crop of barley. The wild oat can be seen growing vigorously in the unsprayed area in the right of the photo. (Photo courtesy of the Agricultural Research Council, Weed Research Organization, Oxford, England.)

that are also promoted by the natural auxin, IAA. But normal controls do not operate, possibly because 2,4-D cannot be destroyed by IAA oxidase, and also forms conjugates inefficiently. Thus, 2,4-D levels rise until they are toxic, cellular order is lost, and disruption ensues, followed by death. Several other herbicides, such as the grass killer *dalapon*, probably act by causing conformational changes in both enzymatic and membrane proteins leading to metabolic and cellular disruptions.

To be effective, a herbicide must be able to get into the plant; as with systemic fungicides, a balance between lipid and water solubility is needed. The compounds are usually sprayed as a water solution but need some lipid solubility to pass through the cuticle and cell membrane. Sometimes, penetration through the cuticle is aided by formulating sprays of the herbicide with *adjuvants*, agents that depress the surface tension of water and facilitate entry into the waxy layer. The most effective herbicides are translocated via the xylem or phloem to other parts of the plant. To kill a perennial rhizomatous plant such as bindweed or poison ivy, the compound must move in the phloem to the food storage organs and young developing shoot apices at the tips of the rhizomes. A new herbicide, glyphosate, does this.

Herbicide selectivity depends on a wide range of properties, any one of which is in itself probably unable to provide selectivity for one species over another. The nature of these properties changes with the herbicide and more specifically with the mode of application, whether to the soil prior to the emergence of both crops and weed seedlings (pre-emergence) or to the leaves of the crop and weeds (post-emergence). If the herbicides are applied to the soil, the position of the chemical in the soil is of importance in determining its selectivity. For example, if the herbicide remains only in the upper soil layer, any weed seed germinating in the surface layers of the soil will tend to be killed, while the more deeply-drilled crop roots may not pick up the herbicide. Even though seeds of wild oat (*Avena fatua*) germinate below such a surface herbicide layer, they are affected while wheat and other cultivated cereals are unaffected, because the elongating stem of wild oat pushes the meristematic region into the herbicide-containing zone, whereas the cultivated cereals do not elongate in this way and remain below the herbicide zone. Other selective herbicides are applied to the soil after the crop has appeared but before weeds start their growth. Many crabgrass killers in lawns depend upon their inhibition of germination of the annual crabgrass seeds while having no effect on the perennial lawn grass plant.

After a post-emergence herbicide is sprayed onto the plants, its effectiveness depends on its reaching the target cells of the plant. This in turn hinges on a variety of considerations. How much spray hits the leaves, for example, is determined by the arrangement of the leaves on the plant, since the vertical leaves of cereals intercept much less spray than plants with broad horizontal leaves. Whether the spray droplets that hit the leaves are retained or lost is affected by the nature and angle of the leaf surface: Drops bounce off an inclined waxy surface easily, while a horizontal nonwaxy surface will tend to retain them. Peas, with their very waxy leaves, are often completely resistant to sprays that become highly toxic if artifically caused to remain on the leaf. Once the spray droplet is retained on the leaf, the effective compound has to get into the cells. Here again, a waxy cuticle functions in retarding the entry of water soluble compounds, as compared with a nonwaxy surface.

If a herbicide is not significantly translocated from its point of entry, the morphology of the plant can play an important role in toxicity. For example, grasses have a short stem and an apical bud protected by the sheathing bases of the leaves, whereas the apical bud of a dicotyledonous plant, as well as its stem, is exposed to the spray. In these cases, the sprayed grass may suffer a little leaf "scorch" from which it easily recovers, while the injury to the bud and stem of the broad-leaved weed results in its death. Translocatability may also provide a basis for selective herbicidal action. A compound translocated from leaf to bud may kill one plant type; a related species may be resistant because the compound does not move from its point of application.

Once the molecule is widely distributed within the plant, further selectivity may occur. The clearest instances occur when the resistant species is able to break down the herbicide to a nontoxic compound, whereas the susceptible species is not able to do so. The most outstanding example of this is the detoxification of the herbicide *simazine* by maize. From the opposite standpoint, the sprayed compound may not be phytotoxic but can be converted to a phytotoxic compound by plants, which are then killed; any plant unable to achieve this conversion remains immune to the chemical. If the side chain of a chlorophenoxy alkylcarboxylic acid is increased in length, some plants are able to remove the carbons in the chain in pairs by a process called β-oxidation. If the chain contains an odd number of carbon atoms in addition to the carboxyl group, the resulting compound after β-oxidation is always the toxic acetic acid derivative, if an even number, the compound is broken down to the relatively nontoxic phenol derivative (Fig. 15-9). Not all plants can perform β-oxidation of these compounds, however; members of the *Leguminosae* are conspicuously unable to do so. Thus, if leguminous plants and nonleguminous weeds are sprayed with a chlorophenoxybutyric acid, such as 2,4-DB, which has three intermediate carbon atoms, the weeds degrade the compound to 2,4-D and are thus killed, while the chemical remains 2,4-DB in the legume, with no resultant effect. This has led to the use of 2,4-DB as a herbicide in cereals undersown with a legume forage crop for use after the crop has been harvested. Finally, selectivity may rest on the action in an essential biochemical process which exists in the susceptible species, but is absent, or in some way protected or less sensitive, in the resistant species. Although no specific cases have yet been proved it seems that the resistance of some species to certain herbicides (for example, that of grasses to 2,4-D) must depend to some extent on such a property. Only when

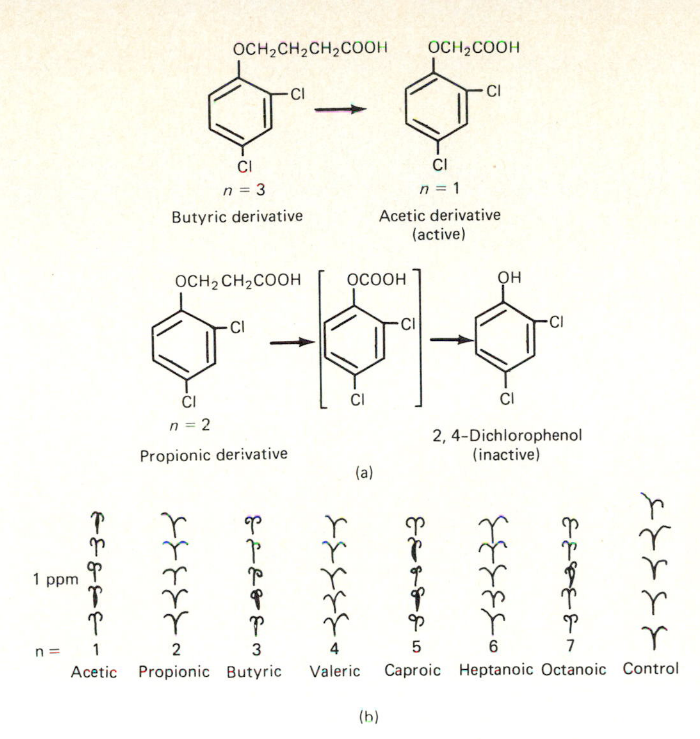

Figure 15-9 (a) The side chain of the ω-(4-chlorophenoxy) alkylcarboxylic acids is cleaved 2 C's at a time, by β oxidation. If the final product possesses one methylene carbon, the product is active, whereas if this is lost so is the activity. (b) Thus, compounds with 2 (acetic), 4 (butyric), 6 (caproic), and 8 (octanoic) carbons in the side chain are active, whereas those with 3 (propionic), 5 (valeric), and 7 (heptanoic) carbons are inactive. The test pictured here uses split pea stem, in which the activity is measured by the angle of inward bending of the 2 halves of the split stem. With the active compounds at 1 ppm the half stems have completely crossed and coiled. [From Wain, 1964. Pp. 465–481 in Audus (ed.), *The Physiology and Biochemistry of Herbicides.* Academic Press, New York.]

the actual mode of action of the herbicide is fully known can one hope to determine such a specific reason for the differential phytotoxicity of a herbicide such as 2,4-D.

Most herbicides appear reasonably safe to use when the prescribed dosages and mode of application are followed. Occasionally, as newer and untried types of chemicals make their way into agriculture, unexpected problems arise. A particularly vexing problem has arisen from the use of the herbicide *2,4,5-T* (Fig. 15-10). Developed as a woody-plant controller, this compound was used extensively as a defoliant in Vietnamese forests and mangrove swamps during the recent war. Although 2,4,5-T was originally assumed to be harmless to man and animals, commercial preparations were found in later laboratory tests to produce many developmental malformations in the embryos of pregnant mice and rats. These *teratogenic* (abnormality-inducing) effects were later traced to a contaminant, 2,3,7,8-tetrachlorodibenzodioxin (TCDD) (Fig. 15-11), an unwanted byproduct of the synthetic reaction by which 2,4,5-T is produced. Although the original 2,4,5-T had about 25 parts per million (ppm) of TCDD, the best modern samples have less than 0.1 ppm, and 2,4,5-T is accordingly being used for forest thinning operations, even near human habitation, on the assumption that this lower level is safe. Recently, several additional disturbing discoveries have led to serious questioning of the continued advisability of the use of 2,4,5-T. First, in laboratory experiments with rodents and primates, it has been found that even 5 parts per *trillion* (!) of TCDD in the diet can significantly increase the

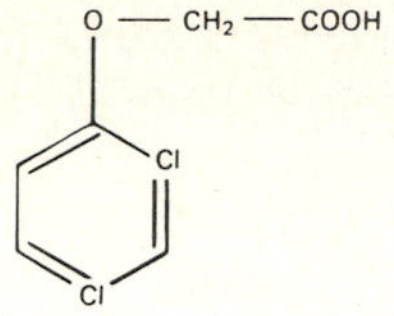

O — CH_2 — COOH

Cl

Cl

2, 4-D

Cl Cl Cl — OH + I — CH_2COOH → Cl Cl Cl — O — CH_2COOH + HI

2, 4, 5-Trichlorophenol | Iodoacetic acid | 2, 4, 5-Trichlorophenoxyacetic acid (2, 4, 5-T) | Hydriodic acid

Synthesis of 2, 4, 5-trichlorophenoxyacetic acid (2, 4, 5-T)

Figure 15-10 The selective herbicides 2,4-dichlorophenoxyacetic acid (2,4-D) and 2,4,5-trichlorophenoxyacetic acid (2,4,5-T). The pathway of synthesis of 2,4,5-T is also shown.

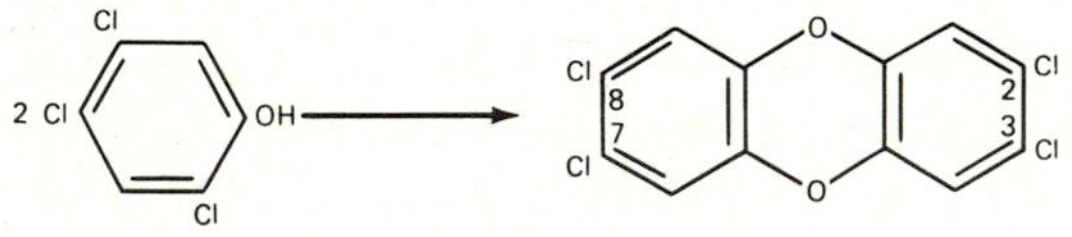

2 Cl Cl Cl OH → Cl Cl O O Cl Cl

2, 3, 7, 8-Tetrachloro-p-dibenzodioxin (TCDD)

2, 4, 5-Trichlorophenol

Formation of 2, 3, 7, 8,-tetrachloro-p-dibenzodioxin (TCDD) through a side reaction.

Figure 15-11 The structure of TCDD, the undesirable toxic contaminant of 2,4,5-T, and its formation from the reaction mixture producing 2,4,5-T.

occurrence of cancers of various kinds. Second, minute quantities of TCDD have been found to induce mitotic malformations in plant cells, an indication that the compound could be mutagenic, as well as carcinogenic and teratogenic. Finally, in regions where 2,4,5-T has been sprayed for forest thinning or on rangelands, significant quantities of TCDD have been found in the beef fat of cattle and even human mothers' milk. The danger of TCDD to the human population has not yet been adequately estimated, since it is not known whether it is concentrated in the food chain, or whether it can be degraded or excreted.

These facts have led the Environmental Protection Administration to ban 2,4,5-T. Proponents of continued use argue that the dangers produced by its use are much less than their probable benefits. Those who support the 2,4,5-T ban emphasize that even one human stillbirth, cancer, or malformation is too high a price to pay for the use of what is essentially a labor-saving device. They point out that we could cut down unwanted forest tree seedlings by hand, and that at the very least, we should defer the use of 2,4,5-T until we find some other herbicide with less toxic side effects.

In the same way, atrazine, the most popular maize herbicide, has recently been shown to be metabolized within the plant to derivatives which are mutagenic in microbial and plant test systems. Thus, although the original chemical had been tested and found nonmutagenic, the plant metabolites of this compound are harmful. To discontinue the use of this valuable herbicide will certainly produce dislocations in maize production, but the step may have to be taken in the interest of public health. Herbicides are one of modern agriculture's greatest boons, but constant vigilance is required to keep their use within safe bounds.

Biological control of weeds

The control of weeds through the use of another organism is not a new idea, and insects have been specifically used to reduce the population of specific weeds in rangelands and similar areas. Such methods, however, simply keep a weed in check rather than achieving its complete eradication, because the spread of the insect is dependent on an adequate number of plants for its food supply. The use of insects to control weeds in a crop has therefore proved of little value.

The need for increased selectivity and lower toxicity of herbicides, as with insecticides, has caused a re-examination of the potential use of biological weed control. The agents of promise are, however, not insects, but fungal diseases. If a unique disease can be introduced at a sufficiently high rate, it has the potential for complete and rapid eradication of a weed in a crop. One instance in which success has been achieved is the control of water vetch (*Aeschynomene*) in rice fields in the southern USA. This weed reduces yield, and its seeds in the harvest make the rice commercially unmarketable. Although the weed can be controlled by 2,4,5-T, the control is not complete, and 2,4,5-T may damage the rice and neighboring crops, particularly soybean and cotton, in addition to producing the toxic effects discussed above. A highly specific race of the fungus *Colletotrichum gleosporioides* has been found to naturally infect the vetch but no other plants. Thus, to achieve control of the weeds, fungal spores are produced by the hundreds of kilograms in an industrial fermenter; if sprayed by air at a concentration of 10 million spores per milliliter at 10 gallons per acre, these achieve a complete kill of the weed. Promising results have also been obtained in the control of water hyacinth (Fig. 15-12), a rapidly growing weed of waterways in the tropics and subtropics, by the use of the fungus *Cercospora rodmanii*. Such a control method may prove feasible for more and more weeds, providing that a

Figure 15-12 The beautiful water hyacinth is one of the most noxious weeds of the tropics and sub-tropics because of its rapid rate of spread over water channels and lakes. Various methods of control are being examined, including the use of fungal diseases.

specific fungus can be found which does not affect the crop or surrounding natural vegetations.

Use of information and methods suggested by natural systems in the control of our agricultural pests could result in maintenance of high food productivity through selective control, without harm either to man or to the other components of the ecosystem.

SUMMARY

In addition to the dangers they face from exposure to adverse temperature and water stress, plants are subject to attack by myriad microorganisms, animals, and other plants. Wild plants have evolved defense mechanisms against these stresses and predators, but cultivated plants, often devoid of natural defenses, must frequently be helped to survive, sometimes by physical means, but also by chemicals that ward off attackers or lessen the stress.

Plants cope with temperature and water stress by *avoidance*, as through dormancy, or by structural and chemical modifications that help them *overcome* the physical challenge, for example, a waxy cuticle that minimizes water loss and hairs that reflect light and prevent overheating of leaves. *Cold-hardiness* involves a series of adaptations induced by environmental conditions such as short photoperiods; the adaptations include higher solute concentration and an increase in highly hydratable proteins that bind water and prevent its freezing, and membrane lipids with lowered freezing points.

Insects and plants have frequently evolved together, achieving both beneficial and harmful interactions. Plants synthesize many *secondary metabolites* like phenolics, alkaloids, isoprenoids, flavonoids, and tannins that taste bad, and thus repel insects and animals. Some insects mutate and overcome the repellent, and the plant in turn frequently evolves a new deterrent. Unusual amino acids and cyanogenetic compounds can actually kill the insect feeding on the plant. Synthetic *insecticides* such as DDT can protect plants, but the insects can mutate to produce strains capable of metabolizing DDT to innocuous compounds. While necessary and useful, many synthetic insecticides produce undesirable side effects such as the broad-spectrum killing of useful as well as harmful insects, and deleterious effects on soil microorganisms and nontarget animals, including man. During passage from plants to primary, secondary, and tertiary animal consumers, these compounds are frequently accumulated, thus amplifying their biological toxicity. The use of naturally occurring defense compounds in plants, such as insecticidal *pyrethrins*, or natural insect hormones interfering with completion of the life cycle, may minimize these problems in future years. Some of the latter compounds exist in plants as well. *Pheromones*, compounds that provide attractant or warning signals to insects, may also be used to lure them into traps or make them avoid crop plants.

Plant defenses against fungal invasion are both physical (tough epidermis) and chemical (toxins present or developed in response to the invasion). *Phytoalexins* are chemicals inhibitory to fungal growth, developed in the plant tissue during fungal invasion. They are varied chemically, and only partially specific. Preformed chemicals may include glycosides of compounds toxic only in the free form; the pathogen may cleave the sugar off, thus generating the toxin that poisons it. As with insects, plants can be protected against fungal invasion by chemical sprays and dusts that act locally, or by chemicals that enter the plant and become *systemic*. Viruses, frequently transmitted by insects, can sometimes be "killed" by incubating plants at higher temperatures, while higher plant parasites, frequently dependent on specific exudations from their host plants, can be diverted by synthetic analogs of these compounds.

Plants excrete organic materials that deter other plants from growing nearby. Such *allelopathic* substances may be important in ecological relationships, especially in harsh environments such as deserts. Synthetic herbicides, frequently analogs of auxins or other plant hormones, can be generally toxic to all plants, or may show partially delimited patterns of toxicity. 2,4-D, one of the most common herbicides, is much less toxic to monocotyledonous than to dicotyledonous plants, and can thus be used as a selective herbicide. Effective herbicides must be cheap and easily made, must enter the plant readily, kill effectively, and be biologically degradable. In some instances, herbicides, their metabolites, or their impurities may cause unwanted, sometimes serious health problems for man, a danger which requires some kind of societal scrutiny and regulation.

SELECTED READINGS

Stress

LEVITT, J. *Responses of Plants to Environmental Stresses* (New York: Academic Press, 1972).

Natural defenses

DEVERALL, B. J. *Defense Mechanisms of Plants* (New York: Cambridge University Press, 1977).

HARBORNE, J. B. *Introduction to Ecological Biochemistry* (New York: Academic Press, 1977).

HARBORNE, J. B., ed. *Phytochemical Ecology* (London: Academic Press, 1972).

HEROUT, V. "Plants, insects and isoprenoids," *Progress in Phytochemistry* 2, 143–202 (1970).

RICE, E. L. *Allelopathy* (New York: Academic Press, 1974).

SWAIN, T. "Secondary compounds as protective agents," *Annual Review of Plant Physiology* 28, 479–501 (1977).

WALLACE, J. W., and R. L. MANSELL. "Biochemical interaction between plants and insects," *Recent Advances in Phytochemistry* Vol. 10 (New York: Plenum, 1976).

Plant pathology and fungicides

AGRIOS, G. N. *Plant Pathology* (New York: Academic Press, 1969).

ERWIN, D. C. "Systemic fungicides," *Annual Review of Phytopathology* 11, 389–422 (1973).

HEITEFUSS, R., and P. H. WILLIAMS. "Physiological Plant Pathology," *Encyclopedia of Plant Physiology*, New Series, Vol. 4 (Berlin–Heidelberg–New York: Springer-Verlag, 1976).

ROBERTS, D. A., and C. W. BOOTHROYD. *Fundamentals of Plant Pathology* (San Francisco: Freeman, 1972).

SIEGEL, M. R., and H. D. SISLER. *Antifungal Compounds* (New York: M. Dekker, 1977).

Weed control

ANDERSON, W. P. *Weed Science: Principles* (St. Paul: West Publishing Co., 1977).

ASHTON, F. M., and A. S. CRAFTS. *Mode of Action of Herbicides* (New York: Wiley, 1973).

AUDUS, L. J. *Herbicides: Physiology, Biochemistry, Ecology* (New York: Academic Press, 1976).

KLINGMAN, G. C., and F. M. ASHTON. *Weed Science: Principles and Practices* (New York: Wiley, 1975).

General

CARSON, R. *Silent Spring* (Boston: Houghton Mifflin, 1962).

GRAHAM, A. *Since Silent Spring* (Boston: Houghton Mifflin, 1970).

QUESTIONS

15-1. How do plants cope with environmental stress imposed by (a) low temperature, (b) water shortage?

15-2. Which of the following are important in determining the geographical range of a plant species? Why? (a) average annual temperature, (b) lowest night temperature, (c) highest daily temperature, (d) highest night temperature, (e) total number of days with temperatures below freezing.

15-3. Warm-blooded animals maintain constant internal temperature despite large fluctuations in external temperatures. Plants do not have this ability, but have other adaptive mechanisms for coping with changes in external temperature. What are some of these mechanisms?

15-4. What is the primary cause of frost damage to plants? How may the plant be made more resistant to low temperature? From what is this increased resistance probably derived?

15-5. Describe those attributes a pesticide should possess if it is to be effective in protecting plants, yet safe ecologically.

15-6. What natural defense mechanisms do plants possess against (a) competition from other plants, (b) fungal invasion, (c) predatory insects?

15-7. Plants contain many toxic compounds that help to protect the plant from animal predators. How does the plant prevent these poisons from interfering with its own metabolism?

15-8. How might an insect that becomes adapted to a particular plant species utilize those plant products that deter other insects?

15-9. What do you understand by the term *secondary metabolite?* How did this name originate? What is the probable function of these substances in plants?

15-10. What types of insecticides give promise of a precise, safe control of specific insect pests?

15-11. What is a phytoalexin, and how is it involved in disease resistance in plants?

15-12. It is said that "hypersensitivity" plays an important role in the defense of plants against disease. What is hypersensitivity, and how can it be a defense mechanism?

15-13. What agricultural and horticultural practices are important in maintaining a disease- and pest-free crop?

15-14. Why are weeds detrimental to crop growth?

15-15. Some plants avoid competition through *allelopathy*. Define this term and explain how the process works.

15-16. Describe some ways in which herbicides are able to kill one plant species without harming a neighboring species.

15-17. As we find with many discoveries, "nature invented it first." Discuss how this statement applies to chemicals manufactured by man to protect plants from pests and disease.

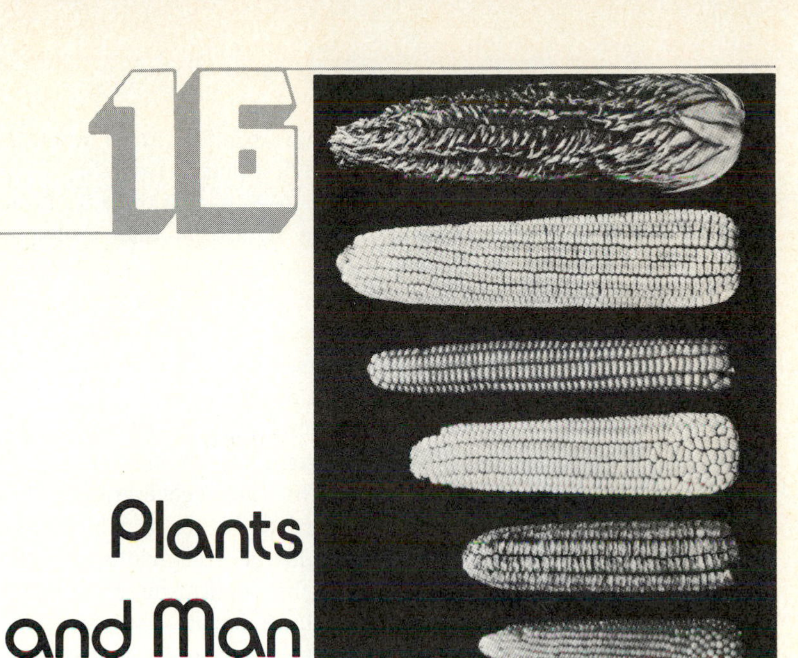

16 Plants and Man

A Look Back and a Look Ahead

Early man was a hunter and gatherer of foods provided by the unaided bounty of nature. Small nomadic human populations could feed themselves by ranging their travels over a sufficiently broad expanse of territory to harvest the fruits, nuts, grains and wild animals supported by the balance of the ecosystem. It is estimated that approximately 10 square miles of territory were needed to provide sufficient food for one nomadic hunter-gatherer per year. Under such a system, man's contribution to the ecological framework could be minor; he ate what he wanted, disseminated seeds in the process, and his wastes neither fouled the ecosystem nor significantly altered the kinds of creatures growing in his environment.

Many of man's other needs were also satisfied by the wild vegetation of the environment. For his fires, he harvested wood, a product of plant cell-wall growth. To protect himself against the elements, be used leaves, straw, branches, and tree trunks. Through chance and random observations, he came to know that certain wild plants had medicinal value, and that others produced odoriferous or tasty components that appealed to him. Still others produced curious products like rubber, resin, and fibers which could be extracted and put to good use. Most of the economically important plants used today were probably "discovered" accidentally by what we would now call "primitive people"; modern man has made very few additions to the basic list of cultivated useful plants.

We do not know exactly how or when man began to cultivate plants for his needs, but the first person to realize that seeds could be collected, stored and planted *en masse* to guarantee a stable food supply brought about a revolution no less important than that caused by the control of fire, the industrialization of manufacturing, or the introduction of atomic energy. The most important effect of the domestication of plants and animals, i.e., the beginnings of agriculture, was the change from the nomadic habit to fixed population centers. This reduced the territorial needs per person to less than 0.1 square mile, a decline of two orders of magnitude from the requirements of the nomad. Domestication of plants began a process of constant improvement. Likely progenitors of better yielding crops were selected and cultivated, their seeds saved for future generations. This *selection* and *breeding* of desirable organisms led to higher and higher yields of both plant and animal crops. But the higher genetic yield potentials of the new plant crops could be realized only by the application of *fertilizer* to the soil—manure, decaying vegetable and animal matter, bone meal, bird guano—any available source that could be readily obtained was utilized. At the beginning, the soil additions must have been made empirically, without understanding the reasons for the action. Certainly the eastern American Indians who placed a herring or several fish heads in each hill of corn planted with several seeds must have been innocent of any chemical understanding of the consequences of their action, yet without this source of organic matter and minerals, their yields would have been vanishingly small in the rocky soils they tilled. Much modern human technology is based on such empirically validated procedures, introduced before a theoretical understanding could have been possible, given the state of science at the time of the innovation. This pattern continues to this day, when, for example, medicinal products, many of plant origin, are used efficaciously without an understanding of their mode of action.

Genetically superior plants adequately supplied with fertilizer can grow lushly and produce bountiful harvests, but as we have seen, they may also be attractive food sources to a vast assortment of potential predators, ranging from bacteria, fungi, and insects to birds, rodents, and larger mammals. Thus, to guarantee that the fruits of his labors came to him and not the predators, the farmer was ultimately compelled to employ an arsenal of physical and chemical agents to ward off, repel, or kill the offending organisms. The usual tools ranged from barbed wire, nets, scarecrows, and electrically charged fences to *pesticides*. Pesticides, by far the most important of these tools, have been generally effective, and have become absolutely essential to the progress of modern, intensive agriculture. Yet they are not without their drawbacks; we have considered the fact that most pesticides are not specific enough in their target organisms, some are not biodegraded rapidly enough, others contain noxious impurities, and in many instances, target organisms mutate or select out variants that are resistant to the previously effective pesticide, making it no longer useful. While alternatives to pesticide usage such as crop rotation, catch crops to attract pests, and biological agents to combat pests have been described and are being developed, pesticides cannot now be abandoned or even significantly phased out without serious immediate crop loss that would lead to hunger and famine.

Even genetically superior, well fertilized, and pesticide-protected crops can run into difficulties. A highly productive wheat field ready for mechanical harvest can be ruined by a windstorm or hailstorm that flattens the plants against the ground. *Growth-modifying chemicals*, like the antigibberellin substance CCC (chlorocholine chloride, or Cycocel) can help protect against such an adverse possibility by producing a stockier and sturdier plant with shorter internodes. Genetic improvement may lead to the same effect without the use of chemicals. Similarly, a heavy crop of fruit on a tree may be ruined by premature abscission; this catastrophe can frequently be avoided by the well-regulated use of auxin-type sprays such as NAA. The chemicals that affect germination and dormancy, flowering, fruiting, root initiation, leaf senescence, and organ abscission are of growing importance in the field of agriculture, and we may anticipate further developments along these lines.

After the crop is harvested, there are still dangers. Many fruits and vegetables, once off the plant, have relatively short useful lives. Not only must they be protected against predators; they must also be subjected to *storage conditions* that will minimize the effects of their own self-destructive metabolism. Thus, ethylene-producing fruits such as apples are frequently stored at relatively low temperatures to retard ethylene formation and in the presence of high ambient carbon dioxide levels to retard ethylene action. Modern fruit transport vehicles also employ hypobaric (reduced atmospheric pressure) conditions, under which ethylene produced by the tissue tends to be removed and dissipated before it can do damage to the stored food material. Certainly, further improvements in food storage facilities and technology can be anticipated in the future; for example, research in new methods of freeze-drying and reconstitution, radiation sterilization of plastic-enclosed meats and vegetables, and chemical retardation of spoilage is under way in many laboratories.

The final step in the long process of food production for mankind is the *distribution* of the product to the consumer. We are well aware that some portions of the earth, like North America and Australia, are areas of food surplus and export, while other areas, like Africa, Asia, and South America must import considerable food. What makes a country a food exporter? Many components, to be sure, but they certainly include good soil and climate, an efficient agriculture backed up by technical capability in biology, chemistry and engineering, a vigorous economy, and a small enough rate of population increase to permit a food surplus to be realized. Technically advanced countries can support research programs that yield better genetic strains, more and better fertilizers, pesticides and growth regulators, irrigation procedures, and storage and transport facilities. The less developed countries may be blocked at one or more of those steps. Unfortunately, economic and political systems that are inefficient, inequitable, repressive, or unstable are frequently the biggest barriers to ultimate consumption of produced foodstuffs on a mass basis. Thus, while politics and economics are not generally considered as part of plant biology, they are certainly relevant to agricultural practice, and thus, ultimately to patterns of food productivity and consumption. Even in prosperous countries like the United States, there are pockets of poverty in urban and rural areas that lead to inadequate nutrition, especially among children and the aged. Our society has attempted to ameliorate these inequities by various social devices such as unemployment insurance, food stamp programs, and hot lunches served at schools, but these practices have not yet guaranteed that "no American goes to bed hungry." Much remains to be done, and future policy will depend on the wisdom and competence of the present generation of students, including, we hope, some readers of this book.

The "Green Revolution"

In the 1960's some visionary forecasters predicted an entirely mechanized agriculture by the 1980's, with vastly increased yield in all major crops. Corn yields of 500 bushels per acre (a sixfold increase) were discussed in this context. Rapid advances in technology stimulated by the space race led to the first prediction, while the prodigious increases in yield that had occurred since the advent of scientific plant breeding, together with the intensive use of fertilizers and pesticides, led to the feeling that similar improvement could go on forever. The "Green Revolution," based on varieties bred for high yield when provided with adequate levels of fertilizer, water, and pesticides was believed to provide the basis for an agricultural Utopia. But today widespread hunger continues and much of the vast grain surplus previously in storage has been dissipated. The Green Revolution, as originally conceived, has fizzled. What went wrong?

First and foremost, the world in the sixties did not anticipate the energy crisis that has now become a reality. It seemed then that oil was cheap and limitless, but now, because of too extravagant use and certain political developments, the price has increased severalfold. We now realize that the world's oil supply will run out within a century or less, and until it does, the price will continue to rise. Yet, farm machines, fertilizers, pesticides and irrigation systems all depend on fossil fuels for their manufacture and use. The vast majority of farmers in the world cannot now afford to grow the demanding genotypes produced by the Green Revolution and are returning to the old varieties which, although lower yielding, produce reliable harvests with little or no fertilizer or irrigation, and show some resistance to pests and disease organisms without pesticides.

At the same time that the "Green Revolution" was providing varieties that thrived in energy intensive agriculture, the agricultural frontier was being expanded into areas where crop production had previously been uneconomical or impossible. The rainforests of the world, such as those of the Amazon basin, seemed lush, highly productive, and suitable for cultivation. Cleared for agriculture through high-energy technology, they have yielded only temporary benefits. The lateritic soils of these regions are actually unstable and poor in nutrient elements, most of which are in the vegetation of the rainforest, through which they constantly recycle. The forest prevents erosion and maintains soil structure partly through the excretion and deposition of organic matter that feeds the microorganisms involved in the maintenance of soil stability. When the trees are removed, the nutrients leach out, the surface of lateritic soils often erodes, and the remainder can change into a hard mass in which no vegetation will grow. The valuable stabilizing influence of the balanced rainforest community, once lost, is difficult or impossible to reconstruct. Hopefully, we will learn our lesson before all the tropical rainforest is destroyed. At the very least, a good part of the tropical rainforest must be maintained because of its stabilizing effect on the world's CO_2-O_2 exchange.

Another disappointment has involved the attempt to cultivate some previously arid or semi-arid environments. Certain crops planted in semiarid areas can thrive, once established, as their roots penetrate down to a water reserve not tapped by more shallow-rooted native vegetation. Such subterranean water, however, may have accumulated over many years, and once exhausted is not replenished rapidly enough for continued crop growth. In addition, as the native vegetation is removed, the dry soil is exposed to the wind, which frequently blows away the fertile topsoil, leaving only relatively infertile subsoil, and creating a "dust bowl." With adequate irrigation, the situation changes and the desert blooms-but again often only temporarily. Almost all irrigation water contains considerable dissolved salt, which remains in the soil after the water is transpired or evaporated. This salt buildup can be negated only by extensive leaching, but it is rare that enough water can be applied to wash the accumulated salt out of the soil. Thus, fertile soils slowly become more saline and eventually have to be abandoned for agriculture. The land of the world available for continuous agriculture is indeed limited, and almost all of it is now being used.

Even in the highest technological societies, crop yield has not continued to increase at the same rate as in the last fifty years. All the "easy" improvements, such as introduction of hybrids and heavy application of fertilizer, irrigation, and pesticides have already been made. New gains in yield will now be harder to obtain. Thus, maximum yields of corn appear to have stabilized at just over 100 bushels per acre. Future gains in crop production will probably come by modifying plants to increase the quality of the product, as in high-lysine corn, to prevent losses where they now occur, and by wiser use of harsher environments. Social critics have also noted that the introduction of high technology "Green Revolution" strains favored large "agribusiness" enterprises over smaller family farms. The results were frequently socially undesirable.

Fabricating New Plants

Plant breeding

Our current high crop yields are made possible largely as a result of the plant breeder's art, which has frequently given us productive and disease resistant varieties. One of the outstanding success stories has been with corn. Modern corn is *hybrid*, containing genes from several inbred lines. Geneticists have long known that successive inbreeding diminishes the vigor and productivity of most agricultural species, both plant and animal. However, if certain highly inbred lines are crossed, *hybrid vigor* frequently results. The offspring are much larger, more vigorous, and more productive than their parents, presumably because of the presence of particular combinations of genes within a single cell. Despite newer knowledge in the field of molecular biology, we do not have an accurate biochemical explanation for the improved growth of hybrids.

In modern hybrid corn, four distinct strains are involved: these strains are hybridized two by two; the final step, involving a hybridization of the two hybrids, produces the "double cross" corn product (Fig. 16-1). Clearly, seed produced by self-pollination of such stock will not "breed true," but will be highly variable. Thus, new seed, produced on specialized seed farms, must be purchased each year by the farmer. Despite the added expense, the American farmer has learned to follow this practice, since it pays off well in increased yield when sufficient fertilizer is added. The annual higher yields produced by the introduction of hybrid corn *alone* would pay for the total cost of all plant research ever done in the United States. Early increases in yield per acre of corn since the 1930's are correlated with the introduction of these genetic hybrids, while the later yield increases are related to the increasing use of chemical fertilizers and finally of pesticides. Many other plants now grown are hybrids, and almost universally higher yields and incomes result from the use of such seed, despite the high cost of hybrid seed production.

Selection and breeding also have the potential to increase yield by developing varieties adapted to particular local situations. Many old native varieties of traditional agriculture, while low yielding, produce better at low fertilizer levels than high yielding varieties developed for intensive agriculture, and are more resistant to diseases. Such varieties contain a valuable gene pool waiting to be tapped for marginal agricultural conditions.

A currently vigorous area of selection and breeding involves the search for salt-tolerant plants to grow in areas where salinity has built up in the soil

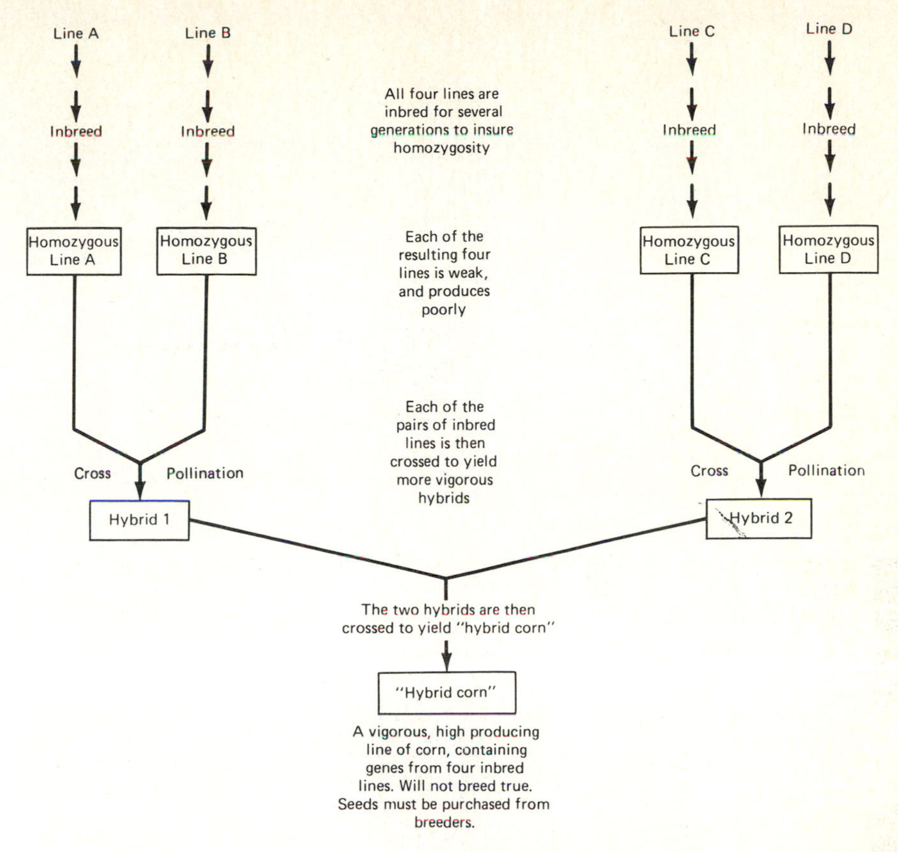

Figure 16-1 The production of modern "hybrid corn" involves the progressive merging of genetic material from four different inbred lines. The nature of the final product will of course depend on the genetic nature of the starting inbred lines.

(see also Chapter 14). Some wild species of tomato, for example, grow in maritime areas close to the salt spray of the ocean. Genes from these varieties might improve salt tolerance of commercial tomato varieties. The contents of cereal grain seed banks are also being screened for salt tolerance and considerable differences have already turned up which promise to be the foundation of salt-tolerant varieties. This illustrates the value of maintaining a seed bank containing even "outdated" varieties. Similar programs involve screening for cold or drought tolerance.

Modifying old plants by new techniques

Clearly the breeding of existing plants has not exhausted the potential of new combinations of genomes. Conventional plant breeding must proceed, not only to increase yield and quality of crop, but also to protect against new strains of microbial and insect predators. In addition, the new techniques suggested by the use of tissue and cell culture and by somatic hybridization through fusion of protoplasts, while only laboratory curiosities at the

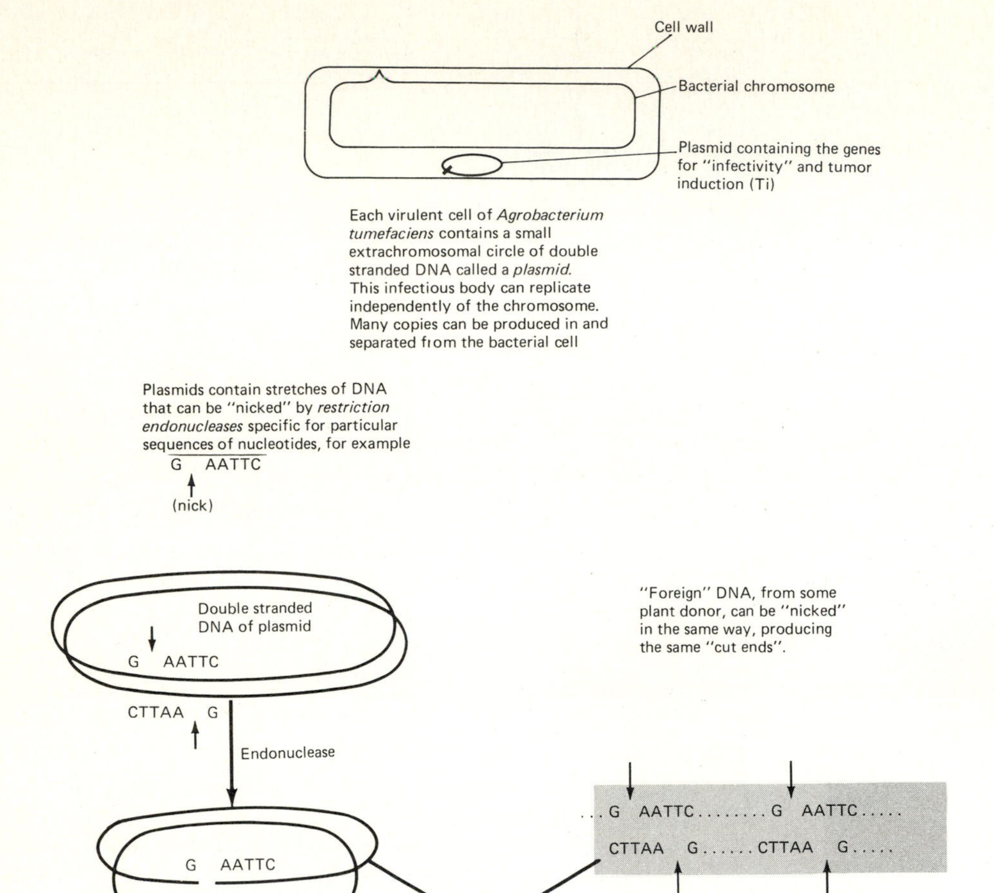

Opened plasmids

The dotted regions contain desired genes one wishes to transfer to a receptor plant

Through other enzymes called *ligases*, the complementary pieces can be joined to make "hybrid" plasmids into which the "foreign" DNA is incorporated.

G AATTC......G AATTC

C TTAAG......C TTAAG

These plasmids, containing desired genes in the dotted regions, are replicated in bacteria. They are then used to infect the host plant. If they succeed in entering a host cell, the plasmid continues its replication, and the "desired genes" may be expressed in the host cell.

Ultimately, through "crossover" between the plasmid and the host chromosome, the "desired genes" may be incorporated into the host chromosome, producing a stable new genotype.

Figure 16-2 Through the use of restriction endonucleases and bacterial plasmids, higher plant genes can be cloned and possibly reintroduced into new host plants.

moment, should be further exploited. Recently, a true somatic hybrid between potato and tomato has been produced by the fusion of protoplasts. The resulting hybrid cells have been cultivated to form calli and eventually entire plants with some characteristics of both parents. Hybridization was proved by isolation and analysis of RuBP carboxylase, a complex enzyme consisting of large protein subunits specified by chloroplast genes and smaller protein subunits specified by nuclear genes. In the hybrid, the two protein subunits were shown to come from different parents. This is the first example of a plant produced by somatic hybridization that could not be produced, or at least has not yet been produced, by conventional sexual techniques. This technique will certainly yield more surprises in the near future.

Protoplasts will almost certainly be the objects of choice in experiments designed to accomplish plant cell transformation. Several vectors are already available to transfer genes from one cell to another, the most promising of which is the Ti (tumor-inducing) *plasmid* of *Agrobacterium tumefaciens*. This small circle of DNA, which causes the transformation of normal cells to crown gall cells, is capable of being "nicked" and opened up by the use of *restriction endonucleases* derived from various microorganisms (Fig. 16-2). Other pieces of DNA, obtained in the same way from cells whose genes one wishes to transfer, can then be mixed with the opened-up plasmid DNA. In the presence of appropriate energy sources and enzymes called *ligases*, the cut ends of the DNA can be made to rejoin. In some instances, hybrid plasmids result, in which desired eukaryotic genes are included in the closed ring. These plasmids can then be used to infect protoplasts of a host plant into which gene transfer is planned. If the experiment is successful, the plasmid will enter the cell and replicate along with chromosomal DNA. Parts of the plasmid DNA may then escape from the ring and be incorporated into the receptor genome, becoming then a regular part of the genetic make-up of the now altered host plant. Many schemes have been proposed for the transfer of genes for nitrogen fixation, disease resistance, more abundant amino acid production, and sturdier habit. Although at present fanciful, these schemes must be taken seriously for the future. Plant genetic engineering will almost certainly play a role in the agriculture of the twenty-first century.

Protoplasts derived from leaves of potato have already been used to select variants more resistant to stress and disease (Fig. 16-3). It appears that any leaf is a mosaic of cells of differing genetic makeup; these differences could result from mutations, chromosomal changes such as deletions, or inequitable plastid segregations at cell division. Whatever the cause of the genetic change, the cells can be isolated as protoplasts, caused to regenerate into cells, calli and intact plants, and the resultant individuals screened for differences. This powerful technique seems certain to be much employed in the future.

A related technique is the culture of cells in the presence of metabolite antagonists (Fig. 16-4). Any cell that grows in the presence of the antagonist

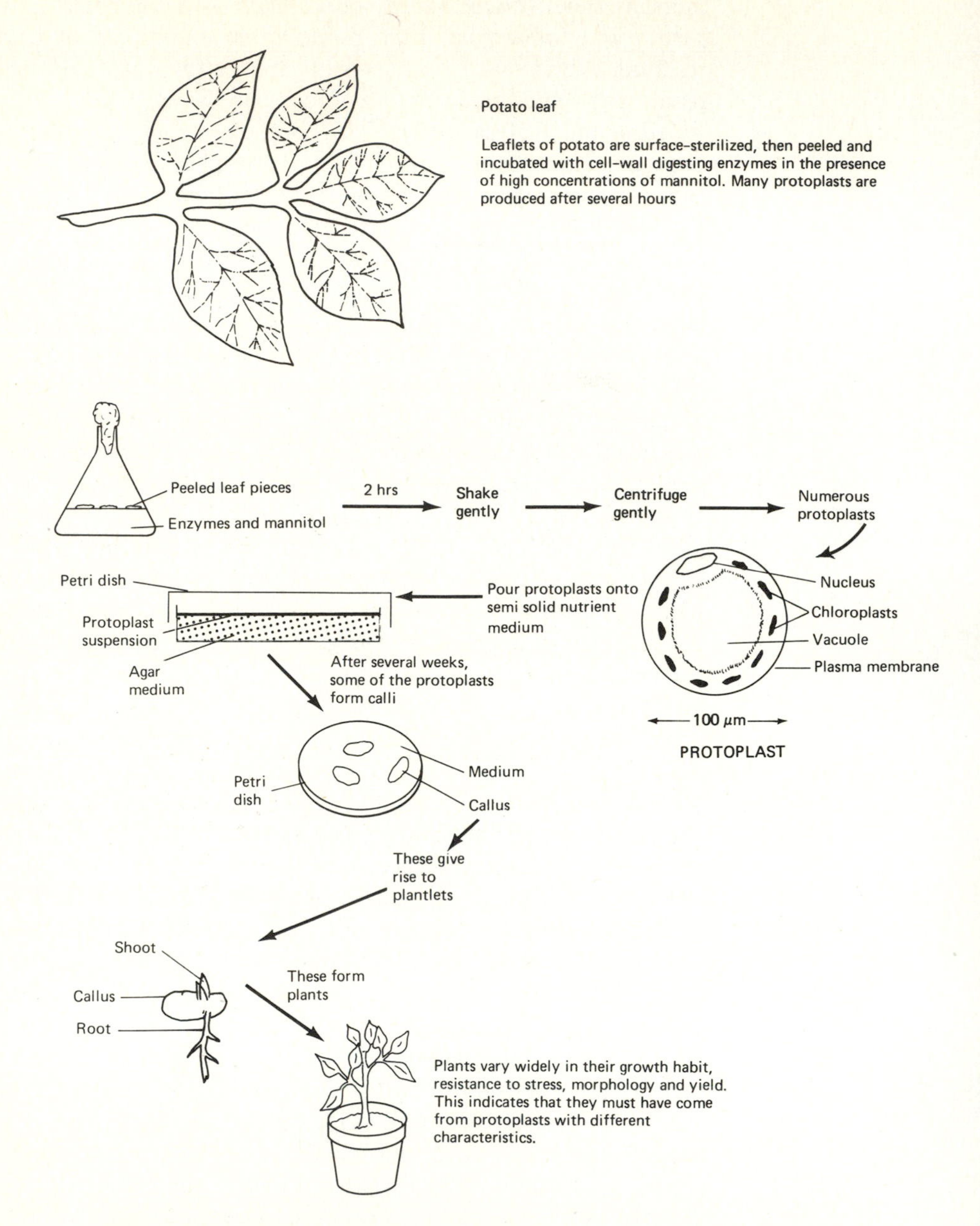

Figure 16-3 Protoplasts isolated from potato leaves can be used to regenerate entire plants. Such plants show great variability, showing that the cells (protoplasts) from which they came were variable.

Figure 16-3 continued. Sensitive (RB) and 'resistant' (RES) plants derived from single protoplasts isolated from a leaf of potato. They have been inoculated with the spores of the fungus *Phytophthora infestans* (potato blight). The resistant plant survives while the sensitive plant dies. Note the extensive fungal growth and blackening of the leaf of the sensitive plant (Courtesy of J. F. Shepard, Kansas State University).

is likely to be overproducing the normal metabolite. If it can be selected and cultured, it is likely to give rise to a plant with a higher production of the desired metabolite. This experiment has been successfully carried out with carrot cells. Inclusion of 5-methyl tryptophan (MT) in the medium causes most of the cells to die, since MT is an effective tryptophan antagonist. But some cells do not die; these turn out to have high endogenous tryptophan levels, because their anthranilate synthetase enzyme, involved in the produc-

Figure 16-4 The normal variability of cells can be put to good use in selecting desired cells and regenerating improved plants from them. Metabolite antagonists for amino acids or vitamins are very useful in the process.

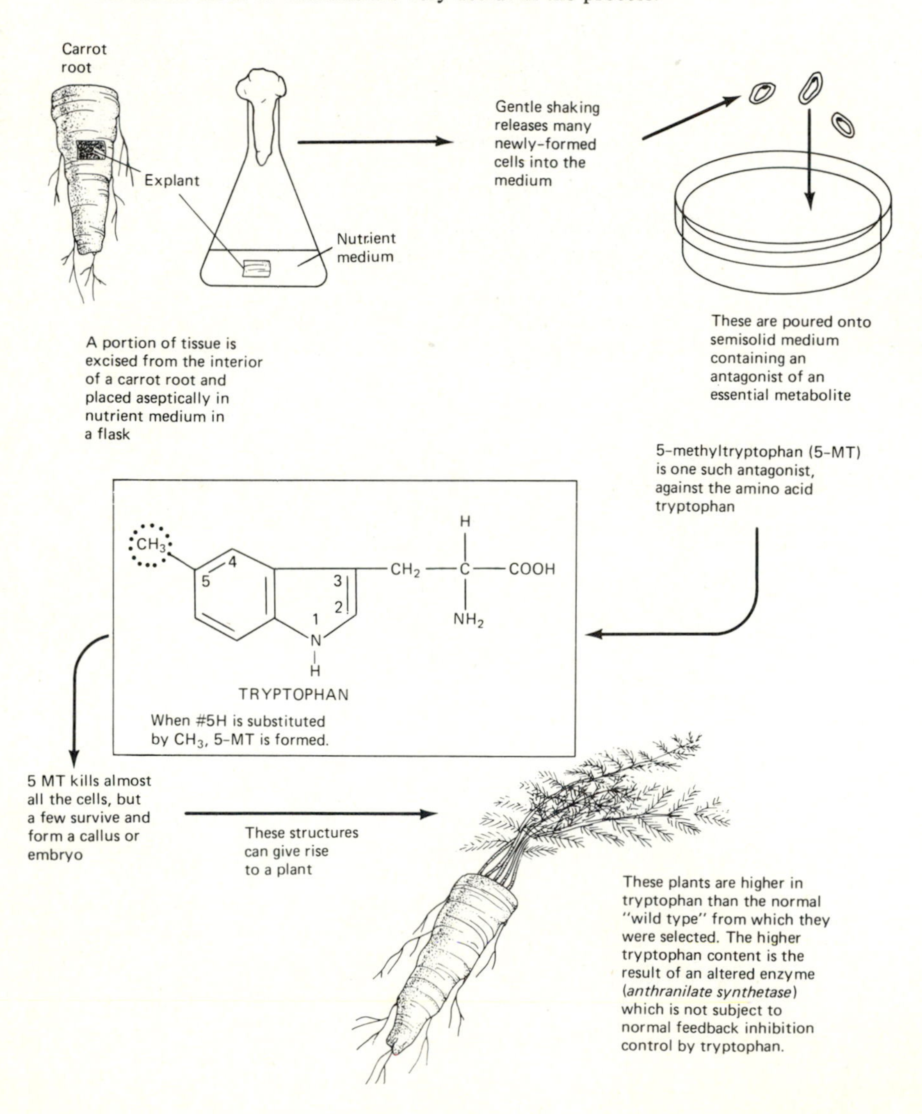

tion of the indole ring of tryptophan, does not show normal feedback inhibition by the product of the reaction. Thus, the enzyme "overproduces" tryptophan, and the plants regenerated from these selected cells are richer in tryptophan than the strains from which they originated. This technique will certainly be used in the future to develop strains of important crop plants richer in vitamins, amino acids, steroids, and other desirable products of plant metabolism.

Introduction of New Wild Plants to Agriculture

One area of agricultural advancement is the introduction of new plants from the wild into cultivation. Of the approximately quarter of a million angiosperm species in existence today, only a few hundred are employed as economic crops, and the major crops are numbered only in the dozens. There must be, among the incredible variety of yet unexploited seed plants, others that could be developed into important plants for man. A great many of these occur in tropical and subtropical areas and have thus been overlooked by our temperate agricultural system. With the expansion of agriculture into new, often tropical, regions, new species might well be added to our crop repertory, to provide both food and new products.

Guayule

One suggestive example is guayule, a shrubby relative of the daisy native to the Sonoran desert of Mexico and certain parts of the arid southwestern United States (Fig. 16-5). This hardy plant produces abundant rubber, stored as particles in its cortical cells. Guayule was known to the pre-Columbian Indians of Mexico and Central America, who chewed the plant to extract the rubber from the fibrous material and used it to make small rubber balls used in games resembling soccer and basketball.

During World War II, when Japanese conquest of the Malaysian peninsula cut off almost all of the *Hevea* tree rubber from the United States, an Emergency Rubber Project was set up to develop guayule as an agricultural plant. With the application of abundant funds and available scientific and horticultural labor, the plant was rapidly converted from a wild desert shrub to a respectable rubber producer under plantation agriculture conditions. The project was suddenly and unceremoniously ended when synthetic rubber produced from petroleum products became a reality in the midst of the war. Now, with petroleum products in short supply and increasingly expensive, and with new emphasis on the use of non-polluting renewable resources, the cultivation of guayule seems once again attractive. Old seed stocks are being exhumed from storage, new plantations are being set up, tissue culture methods of studying rubber production are being employed, and an entirely

Figure 16-5 A plant of guayule (*Parthenium argentatum*). (Courtesy of N. Vietmeyer, National Academy of Sciences, Washington, D. C.)

new approach to a potentially useful crop is emerging. Since guayule grows well in land that is not of much other use in agriculture, the prospect is good that it may once again be an important economic plant.

Echinochloa

Another promising plant is *Echinochloa*, a little known wild Australian grass (Fig. 16-6). This plant has never been cultivated by man, but those who know its habits predict it could be easily adapted as an important forage and grain crop for arid and semiarid regions. *Echinochloa*'s most significant characteristic is that a single watering suffices to cause the plant to develop from germination to harvest. The seeds will not germinate after light rains (possibly because of an inhibitor in the seed coat), but require deep flooding. Such flooding causes prompt seed germination and rapid growth, permitting the plant to complete its entire life cycle before the ground dries out. The

Figure 16-6 A plant of *Echinochloa* (*E. turnerana*). (Courtesy of N. Vietmeyer, National Academy of Sciences, Washington, D. C.)

grain it produces is palatable, nutritious, readily eaten by cattle, sheep and horses, and the vegetative parts of the plant are savored by livestock and can be made into hay. Currently limited to a region of inland Australia, there is no reason why *Echinochloa* could not be further developed.

Jojoba

Considerable attention has recently been focused on an obscure plant which, by providing a vegetable substitute for whale oil, could save the remainder of the world's dwindling whale population. The jojoba (pronounced ho-hó-ba) (*Simmondsia chinensis*) is a hardy shrub that grows in arid regions of northern Mexico (Fig. 16-7). Its main attraction is a liquid wax composed of long-chain fatty acids esterified to long-chain alcohols, as in whale oil. The current crisis in whaling, occasioned by the diminishing supply of whales in the oceans of the world, has caused scientists to look

Figure 16-7 A plant of jojoba (*Simmondsia chinensis*). (Courtesy of N. Vietmeyer, National Academy of Sciences, Washington, D. C.)

elsewhere for such products, and since the jojoba, like guayule, can be cultivated under conditions of semiarid agriculture in lands not readily adaptable for other uses, it seems an attractive alternative. The jojoba can tolerate extreme desert temperatures of up to 35–45°C in the shade, and is a drought-resistant plant that can thrive under soil and moisture conditions not suitable for most other agricultural crops.

The seeds contain about 50% of the liquid wax, which can be obtained by pressure or by solvent extraction. It is so pure that it requires little refining before use as a lubricant, has very good keeping properties, and can withstand repeated heating to high temperatures without changing its viscosity. It so closely resembles whale oil as to make it easily useable wherever that product is now employed. Not only is the oil potentially very valuable, but the residual seed meal left after the oil is expressed has considerable protein, i.e., about one-third of its total weight. There are some problems in that chemicals present in jojoba seed meal suppress the appetite of test laboratory rats, but systematic research can probably tell us how to get rid of this noxious material, thus making the plant palatable for feeding to livestock. Conversely, the appetite depressant might find use in combatting human obesity.

Winged bean

Another tropical plant of high promise is the winged bean (*Psophocarpus tetragonolobus*) (Fig. 16-8). This tropical legume is relatively free of pests and disease, and is able to thrive on relatively poor soil, as long as it receives abundant water. It has large nitrogen-fixing nodules on its roots, produces edible seeds, pods, tuberous roots and leaves with high protein levels, and can be processed to yield an edible seed oil. Since it is a fast growing perennial, it could be useful to small farmers in tropical regions where protein deficiency in the human diet is difficult to remedy. Most people who know the plant say that the winged bean could easily become for the tropics what the soybean has become for the temperate zones. It is currently grown in quantity only in southeast Asia and in Papua-New Guinea. Nutritionally, the

Figure 16-8 A plant of winged bean (*Psophocarpus tetragonolobus*). (Courtesy of N. Vietmeyer, National Academy of Sciences, Washington, D. C.)

seeds of winged bean are like those of soybean, containing a high-lysine protein, considerable unsaturated oil, and abundant vitamins. Over a ton of seed per acre yields are apparently not uncommon. The tuber-like roots, which can be eaten much as we eat potatoes, have the added advantage of containing about 20% protein. Even the vegetative parts of the plant can furnish good nutrition to animals.

With all such new plants, considerable research in both plant physiology and agriculture is needed to determine how best to grow the crops. Little is known of the biochemistry of rubber production in guayule or oil production in jojoba, the dormancy and growth in *Echinochloa*, or the nitrogen metabolism and protein synthesis of winged bean. Such knowledge would direct selection and breeding programs and suggest optimal cultural conditions. Further trials on growing conditions would be needed to provide optimum yields of products for mankind.

Forests of the Future

Until recently forests were not cultivated but simply cut. With the increasing demand for lumber and paper, the cultivation of rapidly growing trees has increased in importance. Since forests are usually of very mixed genetic stock, intense selection and breeding is needed to develop the fastest growing trees. The main problem is that many trees are slow to mature and are limited in their propagation potential. If fast-growing trees can be found or developed, how can they be bred in a program of reasonable duration? And how can the desired product be propagated to provide rapidly the number of trees to cover hundreds of thousands of acres? Here plant physiology has provided some important answers, in the form of rapidly growing, rapidly maturing, disease resistant trees, particularly conifers. These new types of trees grow to a useable size in a fraction of the time previously required. Although the rapid growth habit usually decreases the strength of the lumber for construction purposes, the new tree lines are ideal for the manufacture of paper, which depends only on the bulk of wood produced. One problem is that such conifers are extremely hard to root from cuttings, and their seeds will not produce uniform populations with desired characteristics. The increasing use of tissue culture has in some instances provided large numbers of new clonal seedling trees, genetically identical to the original desired tree. Although efforts to bring all the desirable species into tissue culture has met with mixed success, it is virtually certain that tissue culture propagation of many tree lines will be a standard commercial operation in the near future.

To obtain further improvements in genotype, standard tree breeding programs must continue. This is hampered by the fact that conifers, like many other trees, have an extended period of juvenility before they become "ripe-to-flower." This juvenile period may last up to twenty years and is prolonged by conditions, such as high rainfall, favoring vegetative growth. Recently the control of conifer juvenility has been shown to reside in changes in the nature of their endogenous gibberellins, and sexual maturity can thus be hastened by application of certain of these hormones. Gibberellic acid (GA_3), the common commercially available gibberellin, is effective only on certain species, not including the *Pinaceae*, the family to which most of the valuable timber species belong. In fact, in Douglas fir, high GA_3 content is correlated with juvenility, not sexual maturity. It has recently been found, however, that the nonpolar gibberellins increase in such trees at maturity; juvenility can be broken and reproductive activity induced by applying these particular gibberellins to trees only four years old. Though such gibberellins are currently difficult and therefore expensive to obtain, their demonstrated commercial utility should lead to increased production and lowered costs. A mixture of the effective nonpolar gibberellins GA_4 and GA_7 is now available; optimal results are obtained by continuous feeding over a long time through an inserted tube.

Once reproductive activity has been initiated, it is even possible to control the number and sex of the cones. Long days promote maleness, short days promote femaleness, and a cold period frequently increases total cone production. Finally, seed production can be speeded by application of an auxin such as NAA together with nonpolar gibberellins.

Plants and Pollution

Air and water pollution are products of our modern industrial age, resulting from the assumption that the environment has an infinite capacity to absorb waste chemicals. In the last several decades, we have learned that some of our wastes remain with us, injuring many life forms, including plants. Relevant examples are too numerous for comfort.

Atmospheric pollutants

In mining areas, where ores containing sulfur are smelted, vast regions, extending many miles downwind of the smelter smokestacks, are often completely denuded of vegetation because of the emission of sulfur dioxide. Life forms are extremely sensitive to both the SO_2 and the strong acids formed when the sulfur dioxide contacts the moisture in the atmosphere, forming sulfurous and eventually sulfuric acid. Even in sublethal concentrations, SO_2 is noxious to trees, in which it causes chlorosis and reduced growth. The effects of SO_2 are manifest not only in plants growing near smelters, but also in regions and even countries far removed from the combustion operation. SO_2 can travel vast distances in the atmosphere, being eventually returned to earth as "acid-rain," with a pH about 3.5 or below. Acid rain, derived from industrial emissions many hundreds, even thousands of miles distant, now falls over all of the northeast United States, England, and Scandinavia. This acid rain damages and stunts the vegetation, leaching nutrients from the soil so that the growth of future plants, largely forests, is made less vigorous; some acid sensitive plants may disappear completely. Fluoride (F^-) is another phytotoxic atmospheric component released by combustion of fossil fuels.

Some air pollutants with severe effects on plants, like ozone (O_3) and peroxyacetyl nitrate (PAN), are common components of "smog" that arise through interaction of intense ultraviolet irradiation and gasoline vapors. The components that actually produce plant damage are unstable peroxides (products of the reaction between ultraviolet-generated ozone and unsaturated hydrocarbons) and PAN (formed through the reaction of unsaturated hydrocarbons with nitric oxide, nitrogen dioxide and oxygen in the atmosphere, also under the influence of ultraviolet light in sunlight). These pollutants are particularly prevalent in areas of dense automobile traffic such as the Los Angeles area in California. As a result of these high pollutant concentrations in and around cities, many familiar plants can no longer survive. First to be affected are many street trees, and ornamental plants in gardens. The symptoms of damage are the development of a bronze color on the leaves together with chlorotic and necrotic spots. Some abnormalities assumed to be infectious "plant diseases" have in fact been shown to be due to air pollution. Certain trees such as the London plane tree are more tolerant of air pollution than others, and therefore are preferentially planted in cities. Agricultural and horticultural plants in areas outside the city may also be affected, particularly, as in Los Angeles or Mexico City, if the area is ringed by mountains which prevent the air turbulence that dissipates the pollutants. Whereas the Los Angeles basin used to be covered with orange groves, air pollution is so toxic to orange trees that it will eliminate those groves that have not already fallen victim to real estate subdivisions.

A related problem in industrial regions is the increasing production of CO_2 from the combustion of fossil fuels. When combined with massive deforestation, which prevents the normal large-scale removal of CO_2 from the air in photosynthesis, this has caused an increase in the CO_2 content of the atmosphere. Will this result in a "greenhouse effect" with a concomitant increase in the earth's temperature, a shifting climate and an increase in deserts in the world? Or will the increase in CO_2 and temperature be compensated by more luxuriant growth of vegetation, leading to a lowered CO_2 and ultimate temperature stabilization? There are many and varied opinions but we really do not know the answer.

Water pollutants

Most of man's industrial wastes find their way into the waters of the world, rather than into the atmosphere. Such pollutants do not affect plant growth directly unless the water is used for irrigation. However, water pollution can be extremely dangerous to most forms of life, including man; carcinogenic, mutagenic, teratogenic, and directly toxic compounds abound in many waterways used as sources of drinking water. One water-soluble pollutant distributed exclusively by man that has disastrous affects on vegetation is the ordinary salt used to melt snow on highways in colder parts of the country. When the snow melts, the salt in the runoff water damages sensitive plants within a considerable distance of the highway. In fact, salt marsh conditions may be set up adjacent to the road, and may even cause formation of a local desert. So much salt is poured onto the roads in some regions that much of the surrounding waters become saline. For example, in regions of Lake Ontario next to the city of Rochester, New York, the salinity of the water in winter approaches that of sea water. Except on crucial highways and steep hills salt should be used more conservatively and replaced by the more ecologically acceptable alternatives of snow ploughing and sanding.

Some chemical pollution of waterways results directly from agricultural practice. For example, the heavy use of soluble nitrogenous fertilizers may cause nitrogen enrichment in surrounding streams and lakes after a heavy rain. This could cause eutrophication of previously clear waterways. Excessive nitrate in drinking water is also a source of danger, especially to young children, since reduction of nitrate to nitrite in the body may be coupled to oxidation of hemoglobin from the Fe^{2+} to Fe^{3+} condition, leading to *methemoglobinemia* and deficient oxygen transport in the blood. A partial solution to this problem lies in the use of more slowly soluble nitrogenous fertilizers.

Sewage

Mineral nutrient deficiencies frequently limit crop production in numerous areas of the world. Although increased use of fertilizers could undoubtedly raise production, fertilizers are becoming increasingly expensive and scarce. Since supplies are limited and often too costly for "third world" agriculture, one cheap and ecologically beneficial way to improve yields lies in the utilization of resources currently wasted.

For example, many of the heavily populated centers of the world are currently facing difficult and expensive problems in the disposition of the abundant waste produced by humans. In the old days, when countries were thinly populated, raw sewage was flushed into streams and lakes, but as population pressures increased, such a practice became unacceptable. Sewage was then converted by microbial action to a sludge, which was disposed of by dumping, frequently into the ocean. Once permissible, this practice now leads to many unacceptable adverse effects, including the fouling of beaches and the formation of large anaerobic undersea deserts with an altered and diminished fish catch. Thus we must now look for new solutions. One obvious possibility is a practice followed by Orientals for millenia, i.e., adding the properly fermented organic matter back to the soil for use in growing crops. We balk at the practice of using "night soil," partly for aesthetic reasons and partly because human waste could carry numerous disease-producing dangers. Although infectious viruses, bacteria and cysts of disease-producing organisms do exist in human manure, these problems have been successfully solved in the Orient. Large anaerobic fermenters, in which temperature rises markedly, convert the human manure into a pasteurized, unobjectionable odorless amber liquid, which can be applied directly to rice fields and other agricultural plantations without untoward effects. In the United States today, experiments are in progress in the application of such sludge materials to forests and other nonfood plantations. One problem that has arisen in the use of sludge from municipal sewage systems is that it frequently contains high levels of toxic industrial materials such as heavy metals. Such materials inhibit the growth of plants and would be unacceptable in food. The answer does not lie in discarding the valuable nutrients in sewage but rather in ensuring that industrial wastes are not deposited into our sewage systems. As population and energy pressures increase, the use of human sewage on food crops will have to be contemplated, since the organic matter and mineral nutrients contained in this resource may become too precious to waste.

Not only can crops benefit from sewage, but in certain situations vegetation can form a natural sewage treatment plant. Two basic methods have met with success. Sewage has been distributed by sprinklers over fields of grass or other vegetation, which remove many nutrients from the sewage and facilitates the bacterial degradation of other compounds so that virtually pure water percolates into the ground water system. Marshes are also effective purifiers if sewage in dilute solution is kept flowing through the dense vegetation. This process is particularly useful in areas such as campsites, that are too small for sewage treatment plants. The resulting plant matter in a field can be ploughed under as green manure or used for livestock feed.

Aquatic plants, especially algae, can help to bridge the gap between the present cultural resistance to the use of human sewage for fertilizer and the increasing fertilizer needs of the world. Waste waters, heavily contaminated with organic matter, have been led into lagoons seeded with algae; this facilitates the production of tremendous quantities of green matter that can be harvested and safely fed to livestock, circumventing the use of other plants like soybeans that are suitable both for human nutrition and livestock feeding. One blue-green algae, *Spirulina*, can even grow in alkaline saline waters. Natives of the Lake Chad area of Africa eat *Spirulina*, using it as their main source of protein. Experiments in Mexico and in the United States are now probing its possible use as a human foodstuff. It apparently can be added to cereals and other food products without changing their flavor or creating objectionable side effects.

Plants in Enclosed Environments and Outer Space

In Chapter 14 we discussed the culture of plants under artificial lights, with minerals supplied by the nutrient film technique. This method of growing plants, although expensive and complex, is destined to increase, particularly in areas such as Alaska, where the favorable growth season is too short or too cold, and where the cost of transportation makes the cost of vegetables prohibitively high. The advantages of such a system are an accurate control of conditions for maximal growth rate and high yield combined with a totally-enclosed, pest-and disease-free environment. Such developments have led naturally to enquiries about the possibility of growing plants in space stations outside the gravitational field of the earth.

Space travel and life in space have become possibilities in our time. With the moon landings already accomplished, it is inevitable that man will eventually try to venture further afield. As other planets become the goal, space journeys will last not just a few days or weeks, but months and years. Whereas on short journeys all necessary food and water can be carried along, long distance journeys or permanent space stations will require a more self-contained life-support system. Plants would form a valuable and probably essential component in such a system, because not only could they provide a continuous supply of food, they could also recycle human wastes. Human space travelers would consume oxygen and exhale carbon dioxide during respiration; green plants could reverse the process through photosynthesis. Human wastes could provide part of the plant nutrient requirements, and the transpired water of plants, suitably condensed, could be used as the drinking water supply for humans. Experiments have already been carried out in enclosed capsules on earth to show that such closed systems can operate successfully.

Plants in space could be grown like those in controlled earthbound environments, with some modifications. Light would be provided by electric lamps powered by solar cells outside the space craft. Although the plants would be in a gravity-free environment, they could be oriented by low centrifugal forces; phototropism could also aid in plant orientation. The problems of keeping water around the roots could be solved by the artificial centrifugal field, or in a gravity-free environment, by keeping the roots in sealed containers, with only the shoots emerging from the containers. An additional precaution that might be needed would be the trapping of the many volatile compounds emitted by plants on activated charcoal filters, which would probably be needed for air purification in any case. Single-celled algae such as *Chlorella* could also be grown in giant sealed illuminated containers for food, air and waste purification. It is probable that both algae and whole plants would be used to provide both bulk and variety in the diet and to optimize the purification properties of the two systems.

In an unanticipated return on our scientific investment, space technology is now providing new information that directly affects man's utilization of plants on earth. Vegetation surveys from satellites orbiting the earth, employing photography at different wavelengths, can define the nature of the vegetation, the extent of disease, and large-scale ecological trends. From such surveys important agricultural and foresting strategies can be planned. Satellite surveying is still in its infancy and we can anticipate increased use in the future.

Plants as Nonfood Renewable Energy Sources

With the energy crisis growing in urgency, the use of plants as solar energy converters becomes increasingly important. Plants grow over large areas of the earth and continually convert sunlight into fixed carbon products that can be oxidized as fuels. The great advantage of plants as an energy source is that they are renewable, because they either reproduce, keep growing after partial harvest, or can be reseeded. An increased interest in wood as a fuel has sprung up, particularly in the forested northeast United States, where wood is readily available. Not only can wood be used for home heating in new efficient wood burning stoves, but some electric utilities are even converting to the burning of wood chips. Wood can thus be used to provide a portion of our heating needs in the future. Care must be taken that in converting to this seemingly endless supply of fuel, we do not destroy its source by overexploitation. It must be recalled that the whole of the European continent used to be forested, but Europe now boats only small groves of trees. The rest were cut down for fuel in the Middle Ages, and have not been replaced. Unless the situation is carefully managed, that could happen elsewhere again. It is already occurring in many underdeveloped countries, where wood forms an almost exclusive fuel. With an increasing human population, tree removal is outpacing tree growth. Unless increased tree planting, improved management, and some population stabilization occur, a third-world wood energy shortage will arise, whose impact will be far more difficult to overcome than that of an oil shortage in the western world.

Another way of using plants as a nonfood energy source is to convert their products to alcohol. Virtually any plant material can be enzymatically or chemically degraded to form sugars, and these sugars can then be fermented by yeast to produce ethanol. Surplus grain, sugar cane, corn stalks, forest wastes, algae, or even garbage containing much plant material can be converted into alcohol in this manner. A by-product is frequently a protein-rich mass of yeast which, together with unconverted plant material, can be used as animal feed. The alcohol can be mixed with gasoline in a 1: 9 or even 1: 4 ratio to give a blend ("gasohol") capable of powering current internal combustion engines with little design modification. If otherwise wasted or useless plant materials are used, this could provide a significant contribution to alleviating the oil shortage. It is unlikely, however, that much would be gained from raising crops solely for fuel alcohol production, as the amount of fuel needed for machines and fertilizer to raise the crop often consumes as much energy as can be obtained from the final crop.

Epilogue: Plant Research and the Future

Research in the plant sciences, in the form of both basic scientific discoveries and cultural practice, has led to many of the advances through which modern agriculture has become capable of feeding the growing population of the world. What happens to our food supply in the future depends on continued research into how plants grow and function, and into management practices that guarantee their growth in an optimal manner. The intensity and level of such research in turn depends on the importance that the people, through their governments, place on agricultural and plant research. During times of either plentiful food or of financial depression, plant research is often given a relatively low priority. This is short-sighted, for only with continued endeavors in this field can mankind be adequately fed and maintained in an ecologically balanced world.

SUMMARY

The domestication of plants permitted human society to change from the nomadic habit to fixed population centers, to decrease the acreage required to support each person, and to increase its population. Subsequent improvements in plant productivity have come from selection, breeding, fertilizers, irrigation, pesticides and growth regulatory chemicals. The "Green Revolution" has not realized its potential because of unforeseen difficulties, especially the increased cost of energy sources and the intractability of tropical soils and arid regions.

In addition to conventional plant breeding, the modern botanist can avail himself of molecular genetic techniques for crop improvement. This could involve introduction of new genetic information by *plasmids*, selection of new types from isolated protoplasts, and *somatic hybridization* of protoplasts. Man uses few plants in agriculture, and many unexplored plants await extensive exploitation. Prominent possibilities include *guayule* for rubber, *jojoba* for wax and oils, *Echinochloa* for grain, and *winged bean* for leguminous protein. New techniques of plant propagation involving tissue culture and controlled reproduction can also help forestry.

Plants are adversely affected by atmospheric, soil and water pollutants, but can also help to rid the environment of some noxious materials. Human sewage, currently dumped or wasted in the Western world, is fermented and applied to crop growth in the Orient. Because of the mineral and organic richness of this material, its greater use in Western agriculture is probable. Plants are also useful for food and environmental purification in enclosed environments such as space capsules, and as possible sources of renewable nonfood energy. Future progress will depend greatly on continued research.

SELECTED READINGS

A. Early agriculture

CARRIER, L. *The Beginnings of Agriculture in America.* (New York: McGraw-Hill, 1923).

LEONARD, J. N. *The Emergence of Man: The First Farmers.* (New York: Time-Life Books, 1973.

B. The Green Revolution, plant breeding, plant research

ABELSON, P. H. (Ed.). *Food: Politics, Economics, Nutrition & Research.* (Washington, D. C: American Association for the Advancement of Science, 1975).

The Green Revolution: Symposium on Science & Foreign Policy. Proceedings of the Subcommittee on National Security Policy and Scientific Developments of the Committee on Foreign Affairs, House of Representatives, Ninety-First Congress, First Session, Dec. 5, 1969. U. S. Gov't. Printing Office, Washington, D. C. (1970).

WORTMAN, S., and others. "Food and Agriculture." *Scientific American.* (September 1976).

C. New plants for agriculture

The following pamphlets have been published by the National Academy of Sciences, Washington, D. C.

Underexploited Tropical Plants with Promising Economic Value. 1975.

Agricultural Production Efficiency. 1975.

The Winged bean: A high protein crop for the tropics. 1975.

Guayule: An alternative source of natural rubber. 1977.

D. Pollution

EHRLICH, P. R., A. H. EHRLICH and J. P. HOLDREN. *Ecoscience: Population, Resources, Environment* (San Francisco: W. H. Freeman and Co., 1977).

MANSFIELD, T. A. *Effects of Air Pollutants on Plants.* (Cambridge: Cambridge University Press.) 1976.

NAEGELE, J. A. *Air Pollution Damage to Vegetation.* (Washington, D. C.: American Chemical Society.) 1973.

ORMROD, D. P. *Pollution in Horticulture.* (New York: Elsevier, 1978).

E. Forests of the Future

PHARIS, R. D. and S. D. ROSS. "Gibberellins: Their Potential Uses in Forestry. *Outlook on Agriculture* **9,** 82–87 (1976).

SPURR, S. H. "Silviculture." *Scientific American* (February 1979).

QUESTIONS

16-1. Several sequential steps are involved in the conversion of a wild plant to a a highly productive crop. Discuss some of these steps, describing the probable order in which they should be applied to maximize yield.

16-2. Describe the procedure used in the production of modern hybrid corn. What is the theoretical basis, if any, for these procedures?

16-3. How might molecular genetic techniques developed for microorganisms be used to aid in genetic modification of higher plants?

16-4. Protoplasts isolated from a single potato leaf regenerate plants with vastly different characteristics; they may resemble neither each other nor the original plant. Since all somatic cells of a multicellular organism are supposed to contain identical genetic information, present a possible explanation for the different phenotypes expressed by presumably identical protoplasts.

16-5. What is meant by each of the following?
a. Plasmid
b. Restriction endonuclease
c. Crown gall
d. *Agrobacterium*

16-6. What techniques can be used to recover desired variants from a large population of cells or protoplasts?

16-7. How can biochemical techniques be used to establish hybridization *in vitro* of cultures of protoplasts?

16-8. Many of the promises of "The Green Revolution" have failed to materialize. Discuss.

16-9. Describe several promising unexploited plants that might be introduced into large scale agriculture in the near future. Why have they not been massively utilized until now?

16-10. Plants react to, and modify the environmental effects of, several major classes of pollutants. Discuss.

16-11. What major role may plants play in the energy economy of the future that they do not now play?

16-12. What are some of the problems and possible benefits associated with plant growth during prolonged space-flight?

INDEX